NASA SP-4007

ASTRONAUTICS AND AERONAUTICS, 1966

Chronology on Science, Technology, and Policy

Text by
Science and Technology Division
Library of Congress

Sponsored by
NASA Historical Staff
Office of Policy

Scientific and Technical Information Division   1967
NATIONAL AERONAUTICS AND SPACE ADMINISTRATION
Washington, D.C.

At the opening of the tenth year in the era of man's mobility in outer space, we can look back on 1966 as offering convincing evidence that the United States had gained great competence. This evidence included: five orbital space flights by ten Gemini astronauts; four lunar missions undertaking the orbiting of and softlanding on the moon; numerous contributions to scientific knowledge by unmanned spacecraft and sounding rockets; and further demonstrations of the practical utility of operational space systems, including weather and communications satellites.

During 1966, a record 100 American spacecraft were placed into earth orbit or on escape trajectories. Thousands of revealing and useful pictures of the earth were taken from space and of the moon from lunar orbit and on its surface. The Gemini program ended with rendezvous and docking experiments and extravehicular activity by the Gemini test pilots as the Apollo R&D test flights leading to the manned lunar mission came into the schedule. Thirty-five major scientific, technological, and operational milestones were cited for 1966 by the President in his *Report to the Congress* on aeronautical and space activities of the United States. This was one measure of the American commitment to share in the peaceful exploration of space for all mankind. Another was our support of the final steps toward a United Nations space treaty, undertaken to ensure that the peaceful exploitation of space had juridical basis in international law.

Spectacular as some events in the space venture were in 1966, they nonetheless came to have diminished novelty in the eyes of many laymen. The multitude of both important and unspectacular space activities attained almost an accepted and routine place. This volume, as well as its predecessor chronicles, offers a ready reference on the major as well as the less-well-known events. Beyond the welter of documented details on the complex nature of aeronautical and space related events and their impact, this volume helps to provide a better perspective upon today as we contemplate tomorrow. Such intent also serves future historians and analysts who cannot be unmindful of what is herein presented.

When the Congress created the National Aeronautics and Space Administration in 1958, it charged NASA with the responsibility to "contribute materially to . . . the expansion of human knowledge of phenomena in the atmosphere and space" and to "provide for the widest practical and appropriate dissemination of information concerning its activities and results thereof." NASA has attempted to do this and to include documentation for the historical record. The relating of NASA history begins with this day-by-day chronicle and leads to more specialized studies and histories of the unprecedented task of extending man's mobility and understanding beyond his planet. While a chronology cannot in itself serve as a full-fledged history, the size of this annual volume alone is illustrative of the scope and complexity of the historical task yet to be completed as memories fade and the records disappear.

EUGENE M. EMME
NASA Historian

Contents

	PAGE
FOREWORD	III
PREFACE	VII
JANUARY	1
FEBRUARY	39
MARCH	75
APRIL	125
MAY	161
JUNE	202
JULY	231
AUGUST	255
SEPTEMBER	283
OCTOBER	310
NOVEMBER	336
DECEMBER	364
APPENDIX A: SATELLITES, SPACE PROBES, AND MANNED SPACE FLIGHTS, 1966	387
APPENDIX B: MAJOR NASA LAUNCHES, 1966	417
APPENDIX C: SUMMARY CHRONOLOGY OF MANNED SPACE FLIGHTS, 1966	423
APPENDIX D: ABBREVIATIONS OF REFERENCES	427
INDEX	431

Preface

Before the commentator or the administrator or the scholar can analyze and interpret the record in his search for understanding and lessons, there must be a factual account. This chronology, like its predecessors, is the beginning of such a record. Drawn from open public sources, it has the serious limitations of any work based largely on quickly available sources and further of being done virtually on the heels of the events it chronicles. But these limitations are offset by the assets of the historical method; by the very catholicity of material and interests included it helps open new vistas of relationship, of cause and effect. Because a chronology is limited for the user who does not have a date for the particular item he seeks, an extensive crossreference system is presented in the detailed index. And to provide the broadest possible context, we have included worldwide space-related material, to the extent it was available to us.

The entire NASA Historical Staff in Headquarters participated in source collection, review, and publication. The Science and Technology Division of the Library of Congress was responsible for drafting of the text proper, in the persons of Miss Lynn Catoe, Mrs. Anne Horton, and Miss Shirley Medley. Miss Jane Adams and Mrs. Carmen Brock-Smith translated comment edition into final manuscript. The index was prepared by Arthur G. Renstrom, also of the Library of Congress. General editor of the volume was Frank W. Anderson, Jr.; Mrs. Helen T. Wells was technical editor. Creston Whiting (USS) provided timely translations of Russian materials. Historians and historical monitors throughout NASA contributed useful materials. Validation was the constant concern of many busy persons throughout NASA.

Appendix A, "Satellites, Space Probes, and Manned Space Flights, 1966," and Appendix B, "Major NASA Launches, 1966," were prepared by Frank W. Anderson, Jr. Appendix C, "Summary Chronology of Manned Space Flights, 1966," was prepared by William D. Putnam, Assistant NASA Historian for Manned Space Flight. Mrs. Wells prepared Appendix D, "Abbreviations of References."

This preliminary chronicle is but a first step in the historical process of documenting the dynamic and complex events of space exploration and exploitation. Comments, additions, and criticism are welcomed.

FRANK W. ANDERSON, JR.
Deputy NASA Historian
Office of Policy

January 1966

January 1: MSFC Director Dr. Wernher von Braun named Dr. Jerry C. McCall to organize and manage the Experiments and Applications Office—MSFC's portion of Apollo Applications; Stan Reinartz would assist him. (MSFC Release 65-307)

- The strongest nation will be the one that applies chemistry most effectively, wrote *Science* editor Philip Abelson in *Saturday Review*: "The long-range interests of this nation require a strong chemical profession and basic to it are strong chemistry departments in the universities. Yet, while NASA was receiving an annual $5 billion, a sum that represented almost all the money asked by the agency, chemistry received only a small fraction of its requests. . . . In my estimation, chemistry, for the long haul, is ten to 100 times as important as space, yet it receives only about one percent as much money." (*SR*, 1/1/66, 102)

January 2: Changes in organizational structure of NASA Hq. were effective this date. Most important change was establishment of an organizational unit called the Office of the Administrator, which included the Deputy Administrator and his staff, the Associate Administrator and his staff, and the Executive Secretariat. Dr. Robert C. Seamans, Jr., newly appointed Deputy Administrator, was continuing to serve as Associate Administrator. Willis H. Shapley, Associate Deputy Administrator, serving as the principal assistant to Administrator James E. Webb and to Dr. Seamans, was responsible for policy planning and for general supervision of the agency's work processes. The Executive Secretariat, headed by Col. Lawrence W. Vogel (USA), was responsible for channeling and scheduling work within the Office of the Administrator. Earl D. Hilburn continued as Deputy Associate Administrator.

Director of the Office of Tracking and Data Acquisition, Edmond C. Buckley, became Associate Administrator for Tracking and Data Acquisition. (Other Associate Administrators were Dr. George E. Mueller, Manned Space Flight; Dr. Homer E. Newell, Space Science and Applications; and Dr. Mac C. Adams, Advanced Research and Technology.)

Four Deputy Associate Administrators became Assistant Administrators: William B. Rieke, Industry Affairs; D. D. Wyatt, Programming; Adm. W. Fred Boone, Defense Affairs; and John D. Young, Administration. These were in addition to the four existing Assistant Administrators: Richard L. Callaghan, Legislative Affairs; Arnold W. Frutkin,

International Affairs; Julian W. Scheer, Public Affairs; and Breene M. Kerr, Technology Utilization. Kerr also was serving as Assistant Administrator for Policy Analysis (formerly Policy Planning). (NASA Release 66-3; Organ. Chart, 1/2/66; NN 1132)

January 2: Dr. Clark Blanchard Millikan, director of Cal Tech Graduate Aeronautical Laboratories and pioneer in the development of multi-engine, high-altitude airplanes, jet propulsion, and guided missiles, died of a heart ailment. Millikan, chairman of the board of JPL in its early days, had been awarded the Presidential Medal of Merit for his work in wind tunnel research during World War II and had been given the British King's Medal. (*NYT*, 1/3/66, 27; AP, Balt. *Sun*, 1/3/66; *M&R*, 1/10/66, 13)

January 3: NASA awarded two nine-month study contracts to determine the impact of a proposed improved J-2 rocket engine on the S-II and S-IVB stages of the Saturn V launch vehicle: (1) $148,000 contract to North American Aviation, Inc., developer of S-II stage; and (2) $225,000 modification to an existing contract with Douglas Aircraft Co. MSFC Propulsion and Vehicle Engineering Laboratory was seeking to simplify the J-2 and give it, and the stages it powered, more flexibility. (MSFC Release 66-1)

- COSMOS LXXX, LXXXI, LXXXII, LXXXIII, and LXXXIV, launched Sept. 3, 1965, by U.S.S.R., would orbit the earth for more than 1,000 yrs., reported *Pravda*. COSMOS C, launched Dec. 17, 1965, would orbit for 10 yrs. and ELECTRON I, launched Jan. 30, 1964, and ELECTRON III, launched July 11, 1964, would orbit for at least 200 yrs. Report said Soviet space tracking stations were receiving data from more than 20 sputniks. (AP, Wash. *Eve. Star*, 1/3/66)

- GSFC Nimbus Project Manager Harry Press told *Aviation Week* that Nimbus B would move NASA's weather satellite program into a second generation research and development effort by starting the measurement of fundamental quantities in the atmosphere necessary in new analytical weather-forecasting formulas. Nimbus B, third and largest in NASA's meteorological satellite series, would carry seven major experiments: infrared interferometer spectrometer; satellite infrared spectrometer; image dissector camera system; high-resolution infrared radiometer; monitor for ultraviolet solar energy; medium-resolution infrared radiometer; and interrogation, recording, locating system. In addition, Nimbus B would carry Snap-19 radioisotope thermal generators capable of producing 60 w of power to supplement solar panels and determine feasibility of radioisotope generators for ultimate operational weather satellite system. Scheduled for launch in 1967 with Extended Thrust Augmented Thor-Agena D launch vehicle, Nimbus B would use basic Nimbus spacecraft configuration, but would require strengthened structural elements because of its additional weight.

    Press said the two major control difficulties which had shortened the August 1964 NIMBUS I mission had been analyzed and corrected on Nimbus C, scheduled for launch in 1966. A larger solar array drive motor, equipped with heat-conducting strap that would lower operating temperature of motor almost 100°C and with improved grease, would be substituted on Nimbus C for smaller motor whose failure had left NIMBUS I's solar panels locked in near-vertical position when it apparently overheated and its grease failed to lubricate. Also, attempts would be made during Nimbus C mission to rely on the flywheels to dampen spacecraft oscillations and to use the gas thruster system as little as possible, Press said. NIMBUS I had spun about its roll axis when its depreciating power supply had caused its batteries to feed lower voltages to the attitude control electronics system and the thrusters had fired excessive amounts of cold freon gas. (*Av. Wk.*, 1/3/66, 52, 53)

January 3: Report of the White House Committee on Meteorology concluded that global weather experiment using meteorological satellite and network of sensors on constant-altitude balloons, remote land stations, and ocean buoys could be deployed in four years for $50 million. Assignment of overall management responsibility given to single agency to begin design study was discussed in *Aviation Week.* Designated Ghost (global horizontal sounding technique), project would attempt to extend atmospheric measurements to that 90% of the earth's surface not yet adequately covered, using meteorological satellite in 600-mi.-altitude polar orbit to obtain pair of readings from all global points every 12 hrs. In addition, 1,000 to 3,000 super-pressure balloons would be floated at six different atmospheric levels to measure wind, pressure, temperature, humidity, and overpressure.

NASA Nimbus B meteorological satellite, scheduled for launch in 1967, would carry interrogation, recording, and locating system experiment (Iris) designed to perform ranging and data collection function on which Ghost was based. However, NASA would attempt only limited experiments with Iris involving a check of the system's ability to locate fixed sensor packages, and track them on slow-moving and fast-moving platforms, Dr. Morris Tepper, Director, Meteorological Programs Div., NASA Office of Space Science and Applications, told *Aviation Week.* Tepper said any agency providing its own transponders and funding its own equipment could participate in the Iris tests. (*Av. Wk.*, 1/3/66, 55)

- *Missiles and Rockets* reported prediction of Japanese rocket engineer Dr. Hideo Itokawa that Japan might launch 60-lb. probes to Mars and Venus after orbiting its first satellite in 1968. Itokawa said the Mu series launch vehicle, under development by Univ. of Tokyo Institute of Space and Aeronautical Science, would produce enough thrust—400,000 lbs.—to launch interplanetary probes. (*M&R*, 1/3/66, 7)
- T-38 aircraft of M/G Irving H. Branch (USAF), Commander of Flight Test Center, Edwards AFB, crashed over Puget Sound when attempting an instrumented approach to Boeing Field, Seattle, during rainy weather. Branch was enroute from Edwards AFB; his body was later recovered from Puget Sound by divers. (*Seattle Times*, 1/6/66, 31)

During week of January 3: Representatives of U.S. Dept. of Agriculture's Plant Quarantine Div. met with NASA's Planetary Quarantine Committee to discuss handling and disposal of soil and rock samples brought to earth from the moon or other planets. A Dept. of Agriculture spokesman warned that without precautions, insects and microorganisms from outer space could conceivably be brought to earth on returning spacecraft and multiply unchecked in an environment that would not control them. (*NYT*, 1/9/66, 58)

January 4: NASA MARINER IV, launched from ETR Nov. 28, 1964, for Mars flyby July 14, 1965, was 216 million miles from earth—greatest distance it would reach before gradually returning to within 30 million miles in 1967—and still operating. Goldstone Tracking Station of NASA's Deep Space Network (DSN) tracked a signal from the spacecraft, but no attempt was made to receive data because of the extreme distance. DSN would continue attempting to contact MARINER IV once a month to see if it was still operating. Daily engineering and scientific data from the spacecraft had been suspended Oct. 1, 1965, by ground command that set new space communications record of 191,059,922 mi.—more than twice the distance from earth to the sun. (NASA Release 66-7)

January 4: Apollo spacecraft—command module, service module, and adapter section—is raised up the gantry for mating to Saturn IB launch vehicle at Launch Complex 34, Eastern Test Range.

January 4: NASA awarded five contracts to Saturn contractors to continue studies to improve Saturn concepts of an intermediate vehicle with a payload capability between that of Saturn IB and the Saturn V. On Saturn V, Boeing Co. received $370,500; North American Aviation, Inc., $329,624; and Douglas Aircraft Co., $96,730. Eight variations of the Saturn V vehicle would be examined. Boeing would be study systems integration contractor for all prospective configurations involving the S-IC stage, NAA for all others. On Saturn IB, Chrysler Corp. received $450,000; Douglas, $116,345. Six Saturn IB configurations would be studied. All five contracts would be directed by MSFC. (MSFC Release 66-3)

- Albert Sehlstedt, Jr., writing in the Baltimore *Sun*, observed that the "next twelve months in space should give earthbound observers one very positive impression: men are really going to land on the moon in a few years. . . ." He discussed NASA missions for 1966: (1) four or five Gemini missions—two including Eva and none longer than two days—would attempt to rendezvous and dock with unmanned Agena target vehicles; (2) first of seven Project Surveyor attempts to softland on the moon a spacecraft to relay to earth high-resolution television pictures of the lurain and texture; (3) first of five Lunar Orbiter attempts to orbit the moon at low altitude taking pictures to determine the height and slope of moon's mountains and craters by photometric techniques; (4) four Saturn IB launches—two to launch Apollo command and service module models into nearspace, one to test the booster's upper stage—S-IVB, and one to orbit the Apollo command and service module around the earth for two weeks; (5) first unmanned test flight of Saturn V launch vehicle; (6) launch of four Tiros Operational Satellites into polar orbit for Weather Bureau; (7) launch of Pioneer probes into solar orbit; and (8) launch of Orbiting Geophysical Observatories, Orbiting Astronomical Observatories, and Orbiting Solar Observatories. (Sehlstedt, Balt. *Sun*, 1/5/66)

- USAF XB-70 No. 2, flying at 2,000 mph, reached 70,000-ft. altitude for three minutes in 112-min. test flight from Edwards AFB to determine how aircraft would handle during increasingly longer periods at top speed. Flight broke record set by XB-70 No. 1 Oct. 15, 1965. (*N.Y. Her. Trib.*, 1/4/66)

- NASA awarded two contracts for continuation of engineering studies relating to a manned reusable space transport system: (1) nine-month, $237,000 contract to Lockheed Aircraft Corp. to study possibilities of developing a reusable transport system based on presently approved launch and space vehicles; (2) six-month, $51,000 contract to Martin Marietta Corp. for comparison study of launch modes for reusable launch vehicles. Both contracts would be under direction of MSFC. (MSFC Release 66-2)

January 5: A ten-year lunar exploration program emphasizing manned missions was recommended in reports released by six working groups of scientists who had attended NASA 1965 Summer Conference on Lunar Exploration and Science in Falmouth, Mass., July 19-31, 1965. Program, which would utilize capabilities already under development by NASA, would gradually include scientific experiments in geology, geochemistry, geophysics, and bioscience. One or two manned lunar missions and one manned lunar orbital mission would be conducted annually through 1979, with astronauts remaining on the moon's surface for longer periods after 1974.

Reports recommended four requirements for period through 1974: (1) assurance of overlapping operation of instruments placed on moon by astronauts; (2) sufficient frequency of missions to provide continuity in experiments; (3) scheduling of adequate time between missions to permit necessary modification of experiments; and (4) gradually increasing staytimes of astronauts on moon. Top priority for early Apollo landing missions would be return of lunar material samples; second priority, emplacing Apollo Lunar Surface Experiment Package (Alsep); third, traversing lunar surface to describe topographic and geological features, supplemented by stereoscopic photography. Report urged immediate testing to determine amounts and effects of atmosphere escaping from Lunar Excursion Module (Lem) and outgassing of astronaut suits. One Apollo Applications-Manned Lunar Orbiter mission (Aa-mlo) was recommended prior to first Apollo Applications-Manned Lunar Surface mission (Aa-mls), with minimum of three manned spacecraft missions in polar orbit to make complete lunar survey; five or six Aa-mls missions through 1974 were suggested, with larger scientific payloads and stays for the astronauts of up to 14 days.

Reports listed basic types of equipment necessary for exploration program: automatic position-recording systems to relay to earth or Lem movements of astronauts and roving vehicles; roving vehicles to carry one or two astronauts and scientific payload of at least 600 lbs.; manned lunar flying vehicle able to carry 300-lb. scientific payload about 10 mi. to study features not accessible with roving vehicle; one-inch drill capable of penetrating at least 10 ft. in rubble or solid rock to assist heat-flow studies and provide samples of possible biological interest. Scientific training for astronauts in specific rather than general areas was recommended, with emphasis on geology.

For post-1974 period, the scientists recommended surface travel by astronauts up to 500 mi. from landing point and fixed-site investigations lasting from two months to a year for measuring geophysical phenomena that could vary with time; studying lunar surface processes; deep-drilling studies to obtain information on early moon history and crustal composition; construction and manning of large radio and optical telescopes. Roving vehicles for the post-Apollo Applications phase should have 500-mi. range, shelter for three-man crew, two-month mission capability, and no requirement to return to starting point. (NASA Release 66-4; *Conference Rpt.*, NASA SP-88)

January 5: Twenty-one Long Tank Thors were purchased from Douglas Missile and Space Systems Div. by USAF for use in its programs at Vandenberg AFB. Expected to succeed the present Thrust Augmented Thor (Tat), the Long Tank Thor would provide increased payload capability by enlarging the volume of Thor's liquid propellant tanks; result would be a 20% increase in payload capability over Tat. Although its total thrust of 330,000 lbs. would be essentially the same as that for Tat, the Long Tank Thor would have a burn time of 216 sec. as compared to 146 sec. for Tat. (AFSC Release 205.65)

• NASA announced appointment of Robert E. King, formerly director of labor relations for General Dynamics/Convair, as NASA Director of Labor Relations. King would provide overall direction for NASA's external labor relations program; personally coordinate major policy issues and decisions affecting labor relations with DOD, AEC, and Dept. of Labor; and administer a positive labor relations program to prevent or resolve labor disputes between NASA contractors and their labor unions. He would report to William Rieke, Assistant Administrator for Industry Affairs. (NASA Ann.)

January 6: 170 usable terrain photographs from GEMINI V mission flown by Astronauts L. Gordon Cooper and Charles Conrad, Jr., Aug. 21-29, 1965, included pictures of South Africa which would provide valuable information about continental drift, GSFC geologist Dr. Paul D. Lowman said in interim report at NASA Hq. on 13 scientific and technological experiments carried aboard GEMINI V. Color photos of South Africa's Namib Desert revealed 7-million to 8-million-year-old rock formations. Presence of similar formations along east coast of South America could indicate two continents had once been closer together, Lowman explained. Photos also pinpointed "a major diamond area" between rock formations and the sand dunes. Kenneth Nagler, U.S. Weather Bureau, told the meeting that GEMINI V pictures, some in color, had provided meteorologists with more detail of cloud systems than had orbiting Tiros satellites. (Text)

- NASA officials accepted Gemini 8 spacecraft from McDonnell Aircraft Corp., St. Louis, after studying data supplied by nearly 6,000 subcontractors involving close inspection of over 6,000 test points. The spacecraft, scheduled for launch in March, would be encased in a plastic shroud and flown to KSC in USAF C-124 cargo plane Jan. 7. (Wilford, *NYT*, 1/7/66)

- Of 351 astronaut applicants, 192, including all 6 female candidates, failed to meet minimum requirements, an MSC spokesman reported. The successful 15 would be announced in May. (MSC *Roundup*, 1/21/66, 3; AP, *New Haven Register*, 1/7/66)

- DOD and NASA signed agreement specifying funding arrangements, responsibilities, and general procedures to be followed in acquisition, modification, test, and associated administration of the Apollo/Range Instrumentation Aircraft (A/Ria), which would provide telemetry and communication relay support for Apollo and other national missile and space programs. National Range Division (NRD) of AFSC would exercise overall management responsibility for A/Ria project. (Text; NASA NMI 1052.67)

- Dr. Seibert Dudley of the Univ. of California's Visibility Lab., San Diego, said test equipment on the ground and in the GEMINI V spacecraft had definitely demonstrated that Astronauts L. Gordon Cooper and Charles Conrad, Jr., had seen what they claimed they had seen below on earth during their Aug. 21-29, 1965, mission. On a one-man Project Mercury spaceflight May 15-16, 1963 (MA-9), Cooper had drawn skepticism for reporting he could see dust clouds, roads, and other small features. (AP, *Houston Post*, 1/6/66)

- NASA's decision to launch Mariner E—MARINER IV's backup probe—to Venus rather than Mars was criticized by William Hines in the Washington *Evening Star*: "As a Venus probe, just about everything conceivable is wrong with Mariner-E. Its solar panels are too big, but without them its center of gravity would be all wrong; its earth-seeking radio antenna—essential for long-distance communication—is oriented the wrong way; it is protected against cold rather than heat; its instruments are designed to explore a small planet with a thin atmosphere rather than an earth-sized object with a dense one.

    "Another Mariner flight to Mars promises far greater return . . . than a jury-rigged mission to Venus." (Hines, Wash. *Eve. Star*, 1/6/66)

January 7: U.S.S.R. launched COSMOS CIV with scientific instruments aboard for investigation of outer space, Tass announced. Orbital data: apogee, 401 km. (248.4 mi.); perigee, 204 km. (126.1 mi.); period, 90.2 min.; inclination, 65°. All systems were reported to be functioning normally. (*Krasnaya Zvezda*, 1/9/66, USS-T Trans.)

- It was one minute after Command Pilot Walter M. Schirra, Jr., decided not to eject in the Dec. 12, 1965, Gemini VI launch failure at ETR, that his pulse rate increased, reported Gemini medical director Dr. Charles A. Berry in the American Medical Assn. *News*. Delayed reaction was viewed as supporting theory that fear follows action in an emergency—particularly in the case of a highly trained subject. (MSC *Morning News Flyer*, 1/7/66)

- NASA and USN completed an agreement calling for studies by U.S. Naval Oceanographic Office to define a series of experiments aimed at enhancing oceanographic and marine-technology research from manned orbiting spacecraft. Experiments would be considered for use in NASA Apollo Applications earth-orbiting missions and would employ existing or modified Apollo hardware. (NASA Release 66-6)

- Augmented Target Docking Adapter (Atda)—backup target vehicle for the Gatv—was in final stages of assembly at McDonnell Aircraft Co., and would be shipped to KSC in early February for possible use in Gemini VIII rendezvous mission scheduled for March, the *New York Times* reported. Atda would cost less than $1 million and could duplicate Gatv's performance except for maneuvers in space. (Wilford, *NYT*, 1/8/66, 5)

- Weather modification is a reality, concluded the National Academy of Sciences' Panel on Weather and Climate Modification after two-year review of scientific aspects of weather control. In a report submitted to NAS President Dr. Frederick Seitz, recommendations included: (1) immediate study by one agency of research and development in weather modification; (2) raising agency support from 1965 level of $5 million to at least $30 million in 1970; (3) early establishment of several carefully designed, randomized seeding experiments to assess seedability of different storm types; (4) careful monitoring and regulation of operational programs; and (5) full U.S. support and leadership in establishing advanced global observation system. Recognizing that research in the atmospheric sciences would contribute to the goals of weather modification, the committee urged that highest priority be given to studies of: atmospheric water budgets, initially on vapor transport over U.S. areas where potential of cloud seeding is important; boundary-layer energy-exchange processes; theoretical models of condensation and precipitation machines; and meteorological effects of atmospheric pollution, including carbon dioxide. Report strongly urged that appropriate recognition be given to international implications of weather control. (Text; NAS Release)

- High-altitude test chamber at USAF Arnold Engineering Development Center was being expanded for test of "battleship" model S-IVB stage of NASA's Saturn IB and Saturn V launch vehicles to verify stage's reignition capability under simulated high-altitude conditions. Stage would arrive at Arnold from Sacramento in late January; series of 15 tests expected to last six to nine months would begin in mid-1966. (MSFC Release 66-4)

January 7: NASA awarded RCA two-year $7,837,000 contract for logistic support of computers used to check automatically Saturn IB and V booster stages during manufacture and at launch site. Major equipment for the Saturn checkout systems would be the RCA 110A computer produced under separate contract with NASA MSFC. (MSFC Release 66-8; *Marshall Star*, 1/12/66, 1)
- Unidentified USAF satellite launched by Thor-Altair booster from Vandenberg AFB did not achieve orbit. (*U.N. Public Registry*, 5/31/66)
- Thorough knowledge of cosmic dust, particularly interplanetary dust particles near the asteroid belt, was necessary for space engineers to successfully design manned and unmanned probes, wrote M. J. S. Belton of Kitt Peak National Observatory in *Science*. He stressed the importance of cosmic dust in theoretical and observational astrophysics: "It is apparently an important factor in the process of star formation. The polarization of starlight which is induced by aligned particles of interstellar dust gives information about the magnetic field in galactic systems, including our own. . . . In the solar system, the presence of dust in planetary atmosphere may lead to misinterpretations of photometric and polarimetric data that would otherwise give important information about the total mass of the atmosphere. . . . And finally, dust is responsible for such interplanetary phenomena as the F corona, zodiacal light, gegenschein, and type-II comet tails. . . ." (Belton, *Science*, 1/7/66, 35-43)
- Murray Klein, Head of Drafting Section, GSFC, died of cancer. Klein had supervised the design and construction of buildings and facilities at NASA installations throughout the Nation. (*Wash. Post*, 1/8/66)
- USAF/Lockheed SR-71B trainer version aircraft was delivered to SAC's 4200th Strategic Reconnaissance Wing at Beale AFB. (O'Lone, *Av. Wk.*, 1/17/66, 33)

January 8: GEMINI IV Astronaut Edward H. White II and GEMINI V Astronaut Charles Conrad, Jr., were among the ten outstanding young men of 1965 announced by U.S. Junior Chamber of Commerce in Tulsa. (UPI, *Wash. Post*, 1/9/66)
- Radio signal from Soviet deep space tracking station in the Crimea was bounced off Venus and received minutes later by Jodrell Bank Experimental Station, U.K. Experiment, part of a program arranged in 1963 by Sir Bernard Lovell, director of Jodrell Bank, and Vladimir Kotelnikov, director of Soviet space tracking station in the Crimea, was repeated Jan. 9. (AP, *NYT*, 1/14/66, 12)

January 9: Project managers for the Mariner series deep-space exploration missions to Venus and Mars were announced by Alvin R. Luedecke (M/G, USAF-Ret.), Deputy Director of JPL: (1) Dan Schneiderman would manage the Venus project involving a single spacecraft to be launched in 1967, in addition to continuing his duties as MARINER IV project manager through 1967; (2) H. M. Schurmeier, former Ranger project manager, would manage the Mars project involving two new Mariner spacecraft scheduled for launch in 1969. Effective Feb. 1, Geoffrey Robillard would replace Schurmeier as Voyager capsule system manager in addition to continuing his present duties as deputy Voyager project manager. (JPL Release 372)
- U.S. Bureau of Standards chemists had developed an advanced rocket fuel which might be used in spaceflights to the moon and other planets. A mixture of liquid and solid hydrogen called "slush hydrogen," the fuel

involved a freeze-thaw process which could be used to produce large quantities for storage. (NANA, *Boston Sunday Globe*, 1/9/66)

January 9: National Science Foundation must abandon its passive role and be more dynamic if scientific progress is to keep pace with society's growing technological needs, concluded a report released by the Subcommittee on Science, Research, and Development of the House Committee on Science and Astronautics. Subcommittee viewed the revitalization of NSF as "an area of the government-science relationship in which Congress should provide a large share of the leadership essential to significant improvement in the Nation's resources and development." Some recommendations of the report: (1) NSF supplement its traditional philosophy with positive plans and programs; (2) character and functions of the National Science Board be strengthened; (3) authority of NSF's director be widened; and (4) important executive and policy problems of NSF be acted upon administratively or legislatively in the near future. (Text, Committee Print)

- President Johnson was asked to invite a Russian cosmonaut to join an American astronaut in a space mission in an open letter in *Parade* magazine written by editor Jess Gorkin. Gorkin admitted that, officially, there had been little cooperation in space science between the two countries, but said that "Russian and American scientists, guided by a tradition that scientific research belongs to all men, had been eager to cooperate. On a person-to-person basis, they had already exchanged more space data than their governments have officially sanctioned." Gorkin envisioned joint space mission as proof that U.S. and U.S.S.R. could work together for "the good of all men on earth" despite disagreement on "how they want to live." (*Parade*, 1/9/66)

- Since AEC's plans to build a 200-billion electric volt (bev) atom smasher were being complicated by site and cost considerations, the Commission was considering three alternate proposals: (1) Cornell Univ. physicist R. R. Wilson's suggestion for a "poor man's" 200-bev machine costing about $150 million; (2) Columbia Univ. physicist S. Devons' suggestion to increase the energy of existing 33-bev machine at Brookhaven National Laboratory to 150-bev, costing about $150 million; and (3) incorporating changes in the 200-bev machine being designed by Univ. of California to lower costs. (Sullivan, *NYT*, 1/9/66, E9)

- Soviet rocket pioneer Vladimir Vasilyevich Razumov, presumed to have died in Stalinist purges of Leningrad during World War II, had been located and interviewed by *Leningradskaya Pravda* reporters. Razumov had headed the Leningrad Jet Study Group of space enthusiasts until Stalin abolished it in 1935. His residence was not disclosed, possibly because of his continued association with missile manufacture. (*NYT*, 1/10/66, 46C)

January 10: Nike-Apache sounding rocket launched from NASA Wallops Station carried 51-lb. payload to 115-mi. (185-km.) altitude and impact 88 mi. downrange in the Atlantic. Payload instrumentation failed to function properly, however, and desired results were not obtained. Flight was last in a series of seven similar experiments conducted by NASA for Univ. of Illinois to measure seasonal changes and other effects in the D- and E-regions of the ionosphere. Previous six launches had been conducted successfully from Wallops Station in April, July, and November 1964; and June, September, and December 1965. (Wallops Release 66-1; NASA Rpt. SRL)

January 10: An explanation of white dwarf stars—based on study of spectra of four stars in constellation Orion gathered during October 1965 NASA Aerobee sounding rocket flight—appeared in report released by Dr. Donald C. Morton, head of Princeton Univ. rocket program. Results indicated that the four stars are hurling matter toward the earth at about 4 million mph. This loss of mass, extended over a long period of time, would eventually remove up to 95 per cent of the stars' mass, according to Dr. Morton. They would then be white dwarfs. (*NYT*, 1/11/66, 6)

- Reported that the European Space Research Organization (ESRO), which met in Paris Nov. 24-26, had agreed that the TD-2 satellite, scheduled for launch in 1969 by Douglas Thor-Delta booster purchased from NASA, would carry nine solar-ionospheric experiments: (1) topside counter to measure radio waves; (2) solar Lyman-alpha radiation experiment; (3) a far ultraviolet spectroheliograph experiment; (4) x-ray crystal spectrometer; (5) solar particle flux experiments; measurements of (6) low energy electron fluxes; (7) energy and particle distribution of charged ionospheric particles; (8) oxygen and nitrogen light emissions; and (9) neutral ionospheric constituents. The 880-lb. satellite would be launched into near-polar orbit with 620-mi. apogee and 218-mi. perigee. ESRO planned to launch TD-1 stellar astronomy satellite in 1968. (*Av. Wk.*, 1/10/66, 33)

- U.S. domestic and international scheduled airlines would carry nearly 160 million passengers and fly nearly 113 billion revenue passenger miles by FY 1971, according to FAA's five-year aviation forecasts. Predicted airline activity would be almost twice that of FY 1965 when U.S. airlines flew 95-million passengers and 63 billion revenue passenger miles. (FAA Release 66-3)

- NASA was considering landing unmanned Lunar Excursion Module (Lem) equipped with an Apollo-type lunar television camera on the moon before the first manned mission, *Aviation Week* reported. Photographs from the camera, which would be mounted externally and coupled with a scanning device to permit broad coverage, would provide mission planners with information on lunar surface characteristics. (*Av. Wk.*, 1/10/66, 36)

- Negotiations by U.S. Army Corps of Engineers to acquire the 14,981-acre Sudden Ranch properties adjacent to Vandenberg AFB marked beginning of massive expansion operation for Manned Orbiting Laboratory (Mol) program, *Missiles and Rockets* reported. B/G Joseph S. Bleymaier, Deputy Commander for Manned Systems, AFSC Space Systems Div., predicted that within five years, Vandenberg would have "continuous manned operations" involving 40 or more launches annually. (*M&R*, 1/10/66, 35)

- Soviet life-support systems used in Vostok and Voskhod spacecraft appeared to use a sodium superoxide compound as a source of oxygen, Dr. A. W. Petrocelli, General Dynamics Corp., Electric Boat Div., told *Missiles and Rockets*. Petrocelli estimated the Russians had published three times more basic research papers than U.S. scientists on these materials and were continuing efforts to improve life-support systems by studying compounds such as new superoxides, peroxides, and ozonides. He said they were also searching for new and better carbon dioxide absorbers. (*M&R*, 1/10/66, 33)

January 10: Reported that the Westinghouse Apollo TV camera, designed to provide standard earth receivers near-continuous, live, black-and-white coverage of manned lunar missions, was also adaptable for: (1) use in Eva to aid in rendezvous and docking or close inspection of satellites; (2) full-color transmission to show true colors of lunar surface and space objects; and (3) instant transmission showing effects of continuing spaceflight in a biosatellite on monkeys and other subjects. (*Av. Wk.*, 1/10/66, 35)

- Synchronous-orbit television relay satellite capable of providing direct service to home receivers for an area equivalent to one sixth of continental U.S. could be launched before 1970, according to preliminary company-funded design studies conducted by RCA and GE and presented to NASA Hq. officials, *Aviation Week* reported. Both companies conceded the existence of technical problem areas, including the design of solar cell arrays 15 to 30 times larger than any previously attempted and the lack of a suitable launch vehicle for orbiting 6,000-lb. to 8,000-lb. satellite, but believed solutions could be devised within the existing state-of-the-art. (*Av. Wk.*, 1/10/66, 115)

- Sen. Harry F. Byrd, Jr. (D-Va.), replaced Sen. Joseph D. Tydings (D-Md.) on the Senate Committee on Aeronautical and Space Sciences. House Science and Astronautics Committee membership remained the same. (NASA LAR V/1)

- Rep. Donald Rumsfeld (R-Ill.) inserted into the *Congressional Record* a tribute to the late Dr. Hugh L. Dryden, NASA Deputy Administrator: "Dr. Dryden was highly respected by his colleagues both as an administrator and as a scientist-engineer. He will hold an eminent place when the history of man's accomplishments in space is written. Although the country mourns his passing he will be long remembered for his many contributions to the good of his country." (*CR*, 1/10/66, 29)

- NASA appointed George J. Howick Director of Technology Utilization Div. He would succeed Dr. Richard L. Lesher, Deputy Assistant Administrator for Technology Utilization who, in addition to his other duties, had been serving as Acting Director of the Technology Utilization Div. Howick was formerly manager of industrial technology services for Midwest Research Institute. (NASA Release 66-5)

- GEMINI VII Astronaut Frank Borman and his family were welcomed by a parade in their hometown of Tucson, Ariz. Borman, who was given the keys to the city and told that a $300,000 planetarium would be built in his honor, was also praised in a special resolution passed unanimously by the Arizona legislature. He told a press conference he had joined the space program to go to the moon and "that's what I want to do." (AP, *Chic. Trib.*, 1/9/66; AP, Balt. *Sun*, 1/11/66)

- No new communication-intelligibility problems were created in a reduced pressure environment when helium was added to same oxygen partial pressure, reported USAF School of Aerospace Medicine, according to *Missiles and Rockets*. After conducting a series of 13 simulated flights involving 25 subjects and four chamber operators using 8,300 numbers and words and 2,200 words in sentences, researchers discovered that problems were no greater in 50-50 oxygen-helium atmosphere at 7 psi than in pure oxygen at 3.5 or 5 psi, and were often less. (*M&R*, 1/10/66, 33)

- Prof. Victor Bazykin, member of U.S.S.R. Academy of Sciences, said that Soviet VOSKHOD II Cosmonaut Pavel Belyayev had received slightly more radiation (75 millirads) than had his copilot Alexey Leonov, who had received about 65 millirads, during the 26-hr. spaceflight Mar. 18, 1965. Bazykin said the walls of the spacecraft did not absorb radiation evenly and Belyayev was in a more exposed area. (*M&R*, 1/10/66)

January 10: USN "Southern Cross" program, attempt to convert fleet communications to secure, long-range over-the-horizon communications using techniques such as satellite relay or troposcatter, was undergoing reevaluation because of competitive pressure for funds for Vietnam war, *Aviation Week* reported. "Southern Cross" was sponsored by USN Bureau of Ships and administered by Naval Electronics Laboratory. (*Av. Wk.*, 1/10/66, 36)

January 11: Prime goal of space exploration was the development of scientific capabilities, but it was important to recognize the many other benefits accrued from the space program, NASA Administrator James E. Webb told YMCA meeting in Oklahoma City: "We are developing, in government, in industry, in our universities, the sinews and nerves our Nation needs for healthy growth. Scientific discovery is no longer the sole province of the lone inventor. It requires an all-systems endeavor that calls on all the resources of a nation. . . .

"NASA reaches across all industry. About 90 per cent of NASA's budget goes out under contract to American industry—to pay for work by almost 400,000 men and women in the factories and laboratories of 20,000 prime and subcontractors. . . ." (Text)

- AEC deliberately destroyed a Snap-10A reactor at National Reactor Testing Station, Ida. Safety test, designated SNAPTRAN-2, was conducted to verify ability to "predict the consequences of an accidental nuclear excursion of a Snap reactor." (*Atomic Energy Programs, 1966*, 200)

- Potential of manned space flight in Apollo Applications program was examined by Peter Chew in *National Observer*. He suggested ". . . earth-orbiting astronauts . . . may be able to locate and track icebergs . . . spot fish-feeding grounds . . . support oceanography . . . produce agricultural, mineral, and fresh water surveys." Manned satellites could provide "worldwide television coverage" and serve as "relay points for deep-space communications." (*Nat'l. Obs.*, 1/11/66; *CR*, 2/8/66, A609-10)

- FAA announced that Washington (D.C.) National Airport would be open to short-haul jet traffic April 24. (FAA Release 66-5)

January 11-12: USN was formulating plans for a Sealab III mission in an effort to expand and intensify its manned undersea activities, Under Secretary of the Navy Robert H. B. Baldwin reported at the "Symposium of Man's Extension into the Sea" in Washington, D.C.

James W. Miller, Office of Naval Research, reported that results of the Sealab II experiment had indicated that at ocean depths 600 ft. and below man could perform useful tasks such as salvaging ships and aircraft if he were equipped with electrically heated diving suits, diver-to-diver communications, and underwater swimming aids. Outside of Sealab II, the aquanauts had been cold, frustrated, unable to talk to one another, limited in their visibility, and constantly in danger. Inside the capsule, they had experienced difficulty in sleeping and had had speech problems because of the helium atmosphere. There were no serious medical problems, and "an unprecedented amount of useful work was accomplished." (Text)

January 12: President Johnson, in his State of the Union address to the joint session of Congress, proposed a "program to construct and to flight-test a new supersonic transport airplane that will fly three times the speed of sound—in excess of 2,000 miles per hour." No mention of the space program was made in the address. This fact received comment in *Aviation Week*: "Mentioned or not, the National Aeronautics and Space Administration has decided to stake out a top spot in domestic prestige after Apollo. . . . The agency now will take the initiative." (*Pres. Doc.*, 1/17/66, 31; *Av. Wk.*, 1/17/66, 25)

- Successful launching of Boosted-Dart sounding rockets carrying chaff payloads from Natal, Brazil, and NASA Wallops Station marked first meteorological sounding rocket launchings conducted from North and South America the same day under coordinated NASA-Comissao Nacional de Atividades Espaciais (CNAE) experimental program. Natal launch was first of a series of 32 launches planned by CNAE for 1966 as part of Inter-American Experimental Meteorological Sounding Rocket Network (EXAMETNET). Brazil, Argentina, and U.S., cooperating in EXAMETNET, were planning a north-south chain of stations through the Western Hemisphere to conduct coordinated sounding rocket launchings to obtain experimental data on hemispheric weather patterns. (NASA Release 66-12)

- Ground-test version of S-II, North American Aviation, Inc.-built 2nd stage of NASA's Saturn V booster, was static-fired at Santa Susana Field Laboratory for 354 sec. in successful test of its engine-gimballing and liquid-oxygen-cutoff systems. S-II, which developed 1-million-lbs. thrust from five Rocketdyne J-2 engines, had been previously ground-tested for 18 sec. on Dec. 29, 1965. Firing was monitored by MSFC engineers. (MSFC Release 66-17)

- NASA Administrator James E. Webb, discussing applications of space technology at a meeting of Dallas County Medical Society in Dallas, said that "by adding to our knowledge of space, and of man in space, . . . we are building a very large bank of knowledge that will serve generations of man in all major fields of activity." In a news conference in Dallas, he said that the Soviet space program was ahead of the U.S.'s "and will be for the next year or two." He said he felt the U.S. would beat the U.S.S.R. to a lunar landing, but was not as confident as 12 mos. ago: "A year ago I would have said the Saturn V would be far bigger than anything they could do. I'm not so sure now. They have a very vigorous program. . . . This year they have flown PROTON I and II—both four times the size of Gemini—and they have announced a big new rocket." Asked if the Proton series promised ability to maneuver like U.S.'s Gemini spacecraft, Webb replied: "Anytime you can put up a spacecraft that weighs 27,000 pounds, you've got a capability to do a lot of things you cannot do with a smaller craft. You can develop the maneuverability, but it's hard to develop the large capacity of large spacecraft—and this is where we have been behind and are still behind." (Text; AP, *Denver Post*, 1/13/66)

- Three NASA Gemini astronauts—Charles A. Bassett II, Elliot M. See, Jr., and Eugene A. Cernan—scheduled for a spring mission, completed a training program at Univ. of North Carolina's Morehead Planetarium. They used a device that projected simulated Gatv against a stellar background. (UPI, *NYT*, 1/13/66, 44)

January 12: FRC awarded Martin Co. a six-month $200,000 contract to design, fabricate, and test a thermal protection system for X-15 experimental aircraft that would fly at speeds above 5,000 mph. The 300-lb. system would be based on Martin's MA-25S silicon ablator and would maintain air-friction heating at well below the 1,200°F safety limit. Contract also called for establishment of refurbishment procedures. (NASA Release 66-11; Martin Release)

- Royal Aeronautical Society of Great Britain marked 100th anniversary. (*J/RAS*, 1/66)

January 12-13: "Technology gap" between U.S. and Western Europe was discussed at Paris meeting of industry's scientific officials, sponsored by Organisation for Economic Co-operation and Development (OECD), which comprised most non-Communist industrial countries. Recent OECD report had stated that U.S. research effort was at least twice as great as Western Europe's and that Europe could catch up in civilian research only by extraordinary effort, but it could not catch up in military and space work. Specific proposals for international research centers, international fellowships, and general increase in sharing of scientific effort and results were discussed. U.S. was represented by Office of Science and Technology Director Dr. Donald F. Hornig. (Mooney, *NYT*, 1/13/66, 63)

January 13: Experimental Hibex (High-E Boost Experiment) missile booster was successfully fired on a vertical trajectory and responded in flight to preplanned maneuvers in last of series of 10 tests. ARPA said "all test objectives of the program were met": to demonstrate feasibility of high-acceleration booster and ability to guide and control a high performance vehicle while undergoing trajectory maneuvers; to prove feasibility of igniting, burning, and launching a booster from an underground hardened silo; to institute turning maneuvers to demonstrate structural integrity of both vehicle and propellant. (*M&R*, 1/24/66, 9)

- National Park Service team inspected 18-sq.-mi. tract of proposed parkland for Cape Canaveral National Seashore, adjacent to KSC. Team later met with NASA and USAF officials, Florida state personnel, and representatives of the East Central Florida Planning Council to prepare development plans. (Titusville *Star Advocate*, 1/14/66; *Miami News*, 1/25/66)

January 14: Ground-test stage S-IC-F, Boeing-built first stage of NASA's Saturn V booster, was placed aboard the barge *Poseidon* at NASA Michoud Assembly Facility for shipment to KSC, where it would be assembled at Launch Complex 39's Vertical Assembly Building with other ground-test vehicle elements: a 2nd stage (S-II-F); a 3rd stage (S-IVB-F); an instrument unit (IU-F); and a dummy Apollo spacecraft. Stage was expected to arrive at KSC Jan. 21. (NASA Release 66-10; MSFC Release 66-18)

- Memorandum of understanding for cooperative meteorological sounding rocket experiments was signed by the Spanish Comisión Nacional de Investigación del Espacio (CONIE) and NASA. Purpose of project would be to obtain synoptic information on wind, temperature, and pressure at altitudes between 18 and 36 mi. Sixteen Boosted-Dart and Arcas sounding rockets carrying chaff (tuned dipoles) or instrumented payloads would be launched from a range in Spain to be operated by the Instituto Nacional de Tecnica Aeroespacial (INTA).

CONIE would procure an Arcas launcher, sounding rockets, payloads, and other equipment. NASA would provide, on a loan basis, an MPS-19 radar, van, Boosted-Dart launcher and additional equipment. NASA also would train Spanish personnel in equipment operations and maintenance. Scientific results of project would be made available to world scientific community. (NASA Release 66-20)

January 14: NASA invited aerospace industries to propose definition studies of integrating experiment equipment to be used in spacecraft for manned Apollo Applications missions. Studies would define experiments integration work in the Apollo Lunar Excursion Module, Saturn launch vehicle instrument units, and top stages of Saturn IB and V vehicles; they would include consideration of mission analysis, experiment equipment, installation and integration of equipment, specialized crew requirements, launch facility requirements, tracking, and other support requirements. Two or more firms would be selected for negotiation of parallel nine-month study contracts. (NASA Release 66-14)

- NASA Inventions and Contributions Board granted United Aircraft Corp. a waiver of title rights in all inventions made, or to be made, under NASA contract, which involved the RL-10 rocket engine. (KB)
- Two major changes effected by reorganization of activities of NASA LRC: appointment of Dr. Seymour C. Himmel, former head of the Agena Project office, as Assistant Director for Launch Vehicles and establishment of an Air-Breathing Engine Div., to be headed by J. Howard Childs. (LRC Release 66-2)
- Astronaut Frank Borman returned to Gary, Ind., to help the city celebrate Frank Borman Day and start a year-long observance of its 60th anniversary. He said at a news conference that "it is inevitable" that the U.S. would lose a crew in its space program and that he hoped the public was "mature enough to accept the sacrifice." (UPI, *N.Y. Her. Trib.*, 1/15/66; AP, *NYT*, 1/15/66, 22; AP, Wash. *Eve. Star*, 1/14/66, A2)

January 15: Sir Bernard Lovell, director of Jodrell Bank Experimental Station, U.K., announced that Soviet space tracking station in the Crimea would send signals to Jodrell Bank via Venus for an entire 225-day Venusian year. Sir Bernard said that by following Venus during one complete solar revolution, scientists hoped to gain further information on exact distances of planets from each other and the sun; nature of the hidden surface of Venus; changes in the very slow rotation of Venus; and conditions in space between earth and Venus. (*NYT*, 1/16/66, 24)

- Dr. Robert C. Seamans, Jr., NASA Deputy Administrator, predicted in response to questions in an interview with the *Chicago Tribune* that there would be at least four Gemini flights during the year: "We may even get off a manned Apollo flight before the end of the year." Duration of the flight would depend on success of test flights of unmanned Apollos. (*Chic. Trib.*, 1/16/66)
- Second-stage flight prototype of Lockheed Propulsion Co.'s 156-in. solid motor was successfully static-fired at the company's Potrero, Calif., facility. Burn-time was 72 sec. for total 1-million-lb. thrust. Pressure was 655 psi and thrust vector control system made three complete cycles as programed. This was Lockheed's second and final firing under a $5.5-million USAF contract. (*M&R*, 1/24/66, 10)
- USAF XB-70 research bomber flew at 45,000-ft. to 47,000-ft. altitudes in test flight from Edwards AFB, to collect data for development of supersonic transports. (UPI, *NYT*, 1/16/66, 60)
- Sergey Pavlovich Korolev, "chief constructor of space systems for the Soviet Union," died in Moscow during surgery, Tass reported. Korolev had been active in rocketry since the early 1930's and had directed

gathering of information from captured German rocket scientists after World War II. As head of space system design for the Soviet Union, Korolev's responsibilities had ranged from SPUTNIK I in 1957 to VOSKHOD II manned spacecraft. He also designed the Luna series of unmanned spacecraft and possibly the Zond series. (Tass, 1/15/66; UPI, *Wash. Post*, 1/16/66, A6; *NYT*, 1/16/66, 82)

January 15: Boeing Co. announced plans to build and equip an $11-million laboratory at its Vertol Div. facility, Morton, Pa., for R&D of vertical takeoff and landing aircraft. (Boeing Co. Release)

January 16: Astronaut David R. Scott would attempt—during about 90 min. outside earth-orbiting Gemini 8 spacecraft—to crawl about using hand holds, move through space using a maneuvering gun with nine times the capability of that used by Astronaut Edward H. White II, and unscrew bolts at small mechanic's station on the spacecraft, NASA announced at MSC news conference. Gemini VIII mission, designed to test man's ability to maneuver in space and the first rendezvous and docking mission, would take place before the end of the first quarter of 1966. Charles W. Mathews, Gemini program manager, said "the main significance of the mechanical work is that he [Scott] will be evaluating a tool that is specifically designed for space." The special tool contained a motor to eliminate need for twisting motion by the astronaut that would swing his body around when he was weightless. Target for the rendezvous and docking—Gatv or Augmented Target Docking Adapter (Atda)—had not been chosen. (AP, *NYT*, 1/17/66, 12)

Dr. Joseph F. Shea, manager of MSC's Apollo Spacecraft Office, said in an interview in Houston that the U.S. would achieve its goal of landing men on the moon and returning them safely to earth before the end of this decade: "No new inventions are needed. No breakthrough is needed. This certainly doesn't mean that it is going to be easy. It is not going to be, but the technology needed to do the job is known now." He stressed that the pace that had been set would have to be maintained until the goal was achieved. (Maloney, *Houston Post*, 1/16/66)

- CBS newsman Walter Cronkite received the annual TWA award for outstanding reporting of aerospace subjects at a ceremony in Phoenix, Ariz., sponsored by the airline. (AP, *NYT*, 1/18/66, 62)
- NAS Space Science Board issued the first in series of three reports entitled *Space Research: Directions for the Future*. Report, prepared at NASA's request, outlined research objectives for planetary and lunar exploration in 1970-1985. Compiled by working group led by Dr. Gordon J. F. MacDonald of the Institute of Geophysics and Planetary Physics at UCLA, Part I repeated NAS' 1964 recommendation that top priority in post-Apollo program be assigned to unmanned exploration of Mars. Report suggested secondary importance be divided equally between detailed investigation of lurain and unmanned exploration of Venus, and completed priority list with other major planets, comets and asteroids, Mercury, Pluto, and interplanetary dust. Major recommendations were: launching an orbiting planetary observatory, delivering a drop sonde to resolve question of high Venusian surface temperature, developing fine-pointing control, accurate to 10 sec. of arc, for rocket guidance, and considering enlargement of NASA deep-space communication capabilities beyond present plans. To support outlined scientific program, report recommended: (1) increasing manpower and facilities available to lunar and planetary science programs; (2) development by NASA of substantial

program for observation of planets through ground-based telescopes; (3) early opportunities for experienced observational scientists to explore the lurain; (4) joint study by NASA OSSA and OMSF of early use of Saturn V for planetary exploration, with special emphasis on a Martian capsule landing in the early 1970's; (5) increased attention by NASA to development of research and educational capabilities in theoretical sciences related to study of the solar system. Regarding funding of space research, the report said: ". . . since we believe that the exploration of the solar system bears so directly on the major central scientific questions of our time, we anticipate that unmanned experiments will probably provide the most significant contribution to the program of planetary exploration . . . we recommend that the percentage of support of lunar and planetary exploration be maintained over the 1965-1985 time period and be devoted predominantly to scientific objectives and that programs whose objectives are other than scientific be started only as additional resources become available." It said there should be a gradual shift in emphasis from the moon to the planets, progressing toward nearly equivalent expenditures for lunar and planetary exploration in the 1970-1985 period. (Text; NAS-NRC Release)

January 17: NASA Aerobee 150A sounding rocket launched from NASA Wallops Station carried scientific payload to estimated 83-mi. (134-km.) altitude in GSFC experiment designed to measure spectral irradiance of the stars Sirius and Rigel. Control system, star tracker, and ultraviolet stellar spectrometer performed satisfactorily. Brief data periods, caused by failure of fine jet pressure regulator on control system gas supply resulting in high thrust, had possibly aided establishment of flux values at discrete wavelength intervals. (Wallops Release 66-4; NASA Rpt. SRL)

- Urgency of NASA decision on post-Apollo program was discussed by Karl Abraham in Philadelphia *Evening Bulletin.* He noted Saturn V's potential for solar system exploration and many scientists' fear of overcommitment to investigation of the moon. Among other factors involved, Abraham said, NASA "must consider how particular programs will affect its farflung field centers and their specialized work . . . some are so highly specialized that a sharp change of direction in rocketry or spacecraft and satellite objectives would tear them apart unless sufficient advance time were allowed for gradual changeovers to the new missions." (Abraham, Phil. *Eve. Bull.*, 1/17/66; CR, 2/10/66, A704)

- Air traffic control radar equipment which could automatically display the third dimension of aircraft position—altitude—would go into service in New York City metropolitan area by late 1967, according to FAA. Present radar equipment employed by FAA air traffic controllers provided only two-dimensional—distance and direction—picture of air traffic. (FAA Release 66-6)

- AEC Chairman Dr. Glenn T. Seaborg said in a speech at Georgetown Univ., Washington, D.C., that man might be able to create life "before the end of this decade." Science, he said, has amplified man's senses so that he could now "see" stars beyond what was once thought to be the outer edge of the universe, "hear" radio signals from the farthest galaxies (and one day might hear communications from intelligent beings far out in space), and observe magnetic fields and radiations to which his original senses were blind. Human beings are learning how to replace diseased organs and understand the nature of life itself, Dr. Seaborg noted. He said that even before man managed to create life, science might discover how to control mental illness, live and work under the sea and in the depths of space, and use the power of the atom to cut passes through mountains or fight cancer. (Text; AEC Release S-3-66)

January 17: Of President Johnson's State of the Union message, Robert Hotz said in *Aviation Week*: "The President's statement that he will 'propose a program to construct and to flight test a new supersonic transport plane that will fly three times the speed of sound, in excess of 2,000 mph.' came as somewhat of a surprise in view of the stretched out program he had been sponsoring. Apparently, this statement means that some funds for actual prototype supersonic transport construction will be requested in Fiscal 1967, although the size of this request may dismay the management entrusted with the task of building the prototype in time to remain competitive with the Anglo-French Concorde." (Hotz, *Av. Wk.*, 1/17/66, 21)

January 17-18: NASA launched five Nike-Apache sounding rockets from NASA Wallops Station between 5:39 p.m. and 6:31 a.m. Each rocket was programed to eject a vapor cloud of reddish or bluish color as its altitude increased from 30 to 130 mi. Experiments were to measure wind velocities and directions at various altitudes; motion of the trails was photographed from five camera sites within a 100-mi. radius of Wallops Island. Launchings were conducted for GCA Corp., under contract to NASA Goddard Space Flight Center. (NASA Release 66-13; Wallops Release 66-4; NASA Rpt. SRL)

January 18: Scheduled test of Apollo Launch Escape System (LES) at WSMR was postponed indefinitely due to heavy cloud cover. Countdown for the Little Joe II had proceeded to minus 30 min. [See Jan. 20.] (UPI, *NYT*, 1/19/66, 38)

- NASA Langley Research Center awarded a $5-million contract to Virginia Associated Research Center to operate NASA Space Radiation Effects Laboratory—a $15-million facility established by LaRC for research in support of national programs in space technology—through mid-1970. (LaRC Release)

- HL-10 lifting body vehicle, designed and developed by NASA Langley Research Center to help solve control problems of future manned spacecraft entering earth's atmosphere, was formally delivered to NASA by its builder, Northrop Corp., in ceremonies at company's Hawthorne, Calif., plant. The 22-ft.-long, 4,600-lb., wingless, tri-finned research vehicle would be dropped from B-52 aircraft for flight tests at 45,000-ft. altitude and mach .8 (530 mph). (LaRC Release)

- Rep. Joseph E. Karth (D-Minn.), emphasizing the need for expanded efforts in the field of oceanography, said before the National Space Club in Washington, D. C., that oceanography was becoming increasingly related to space. He noted the recent NASA-Naval Oceanographic Office agreement whereby NOO would coordinate all investigations about possible applications of manned earth orbital operations in the field of oceanography. He suggested that this program "be expanded formally to include unmanned operations as well." He urged a national program with national goals, "coordinated by one agency like . . . [NASA]" and given adequate funds—"an all-wet NASA." Karth said he did not think that post-Apollo programs, such as manned expeditions to Mars or to the moons of Jupiter would gain "general acceptance" until "we have solved the continually worsening home planet problems of hunger and poverty." (Text)

January 18: U.N. Committee on Peaceful Uses of Outer Space had proposed a world conference on outer space to be held in New York. The conference should examine the impact of space data on education and communications—issues of major importance even to poor and economically retarded states. (*NYT*, 1/20/66, 12; *SBD*, 1/31/66, 185)

- Funeral of Soviet space designer Sergey Korolev was held in Red Square. After eulogies by Leonid Smirnov, deputy chairman of the U.S.S.R. Council of Ministers; Mstislav Keldysh, President of the U.S.S.R. Academy of Sciences; Nikolay Yegorychev, Moscow Gorkom Secretary; and Cosmonaut Yuri Gagarin, the funeral procession moved to the Kremlin wall where the urn containing Korolev's ashes was placed in a niche.

  Commentary in the *New York Times*: "Death has finally declassified the role and identity of Academician Sergei P. Korolev, the man who provided the scientific and technical leadership of the vast Soviet rocket program. The extraordinary pomp of his state funeral in Red Square contrasted sharply with the almost complete anonymity imposed on him while he was alive. In the most fruitful years of his career, the man who led the Soviet Union's march into the cosmos from the first sputnik to Leonov's 'walk in space' was normally referred to in the Soviet press only as the mysterious 'chief designer.' Korolev's rockets were powerful enough to send men into orbit and to put cameras into position to photograph the back side of the moon. But they were too weak to break the chains of secrecy that denied him, while he lived, the world applause he deserved." (Tass, 1/18/66; *NYT*, 1/20/66, 31)

January 19: On the floor of the Senate, Sen. Daniel B. Brewster (D-Md.) paid tribute to the late Hugh L. Dryden and proposed "that Wallops Island be renamed Dryden Island, as an appropriate memorial to his immense contribution to space exploration." Dr. Dryden was born on the Maryland Eastern Shore in Pocomoke City. (*CR*, 1/19/66, 583)

- ComSatCorp awarded a $60,000 study contract to France's Centre National d'Etudes des Télécommunications for engineering analysis of a system of phased communications satellites operating in 12-hr. 30°-inclined orbits. It would include analysis of a 12-satellite or 15-satellite system, amount of coverage, orbit stability, launch conditions, types of communications, and other factors required to maintain a given quality of service over a five-year period. Work would be performed within 180 days. (ComSatCorp Release)
- USAF launched two unidentified satellites with Atlas-Agena D booster from WTR. (*U.S. Aeron. & Space Act., 1966*, 147)
- NASA's contributions to multidisciplinary research in support of development were discussed by NASA Administrator James E. Webb at a meeting of the West Point Society of the District of Columbia. Webb described how NASA programs fulfilled need pointed out by the late NASA Deputy Administrator Dr. Hugh L. Dryden in 1949 for cooperative research channeled "into directions permitting early application." (Text)
- Fred J. Drinkwater III, aerospace research engineer, scientist, and test pilot at NASA Ames Research Center, received the Flight Safety Foundation's Richard Hansford Burroughs Test Pilot Award for his "significant contributions to the safety and efficiency of flight testing." Award was established by United Aircraft Corp. (*Av. Wk.*, 1/3/66, 13)
- Radiation, Inc., announced receipt of a $2-million contract to design and build a "data acquisition center" for the British-French 1,450-mph

Concorde supersonic aircraft scheduled to make its first flight in 1971. During early performance tests, data tape recorded in flight would serve as design checks for the aircraft's builders. (*NYT*, 1/20/66, 58)

January 19: Electronic multispectral sensors—high-altitude scanners so sensitive they could tell when apples were ripe enough to be picked—had been developed for agricultural and forestry use, *Boston Globe* reported. Sensors were intended for use on orbiting spacecraft to give reports on an entire forest or farmland on one pass; they were being programed for a wide variety of applications, such as measurements of an entire district of orchards and grain fields. Signals would be converted into photographic form for quick analysis. (Barton, *Boston Globe*, 1/19/66, 6)

January 20: Little Joe II launch vehicle boosted a five-ton unmanned Apollo spacecraft to 15-mi. altitude from WSMR in test of ability of Launch Escape System (LES) to rescue the spacecraft from an abort at medium altitude. Planned pitch-up maneuver was executed and satisfactory launch escape vehicle performance demonstrated; successful canard operation caused command module to stabilize in proper attitude. After jettison of the LES, the hard part of the boost protective cover, and the forward heat shield, command module made safe descent on its three main parachutes. Test, conducted at maximum-Q conditions, was slightly marred by telemetry failure. (NASA Release 65-348; UPI, *NYT*, 1/21/66, 10; AP, *Balt. Sun*, 1/21/66)

- NSF's Special Commission on Weather Modification issued a report recommending greatly increased research on international weather modification and the behavior of the weather and climate in general. Report also recommended thorough studies to gauge the biological, social, economic, political, and legal consequences of all kinds of weather modification—both "deliberate attempts to modify local weather conditions for the good of the total population" and changes that "man's increasingly large and complex society bring about inadvertently." Foundation's report was a companion to one published Jan. 7 by NAS. (Text)

- H.R. 8210, a bill to provide U.S. tax immunity to the European Space Research Organization, was reported to the Senate by the Finance Committee. Floor action had not yet been scheduled. (*CR*, 1/20/66, 719)

- Dr. Pyotr Kapitsa, director of the Institute of Physical Problems of the Soviet Academy of Sciences, deplored the productivity gap between U.S. and Soviet science in an article in *Komsomolskaya Pravda*. He urged U.S.S.R. to weed out unproductive research workers periodically: "It would . . . be possible to transfer 15 to 20 percent of our staffs from science to industry every year and to take into research well-prepared and qualified youth. In this manner we would improve the quality of the scientific staffs and also not close the door of science in the face of the young." He noted that the number of scientific research workers was almost the same in the two countries—800,000 Americans compared with 700,000 in the Soviet Union—and said that U.S. scientists were responsible for one third of the world's scientific output (in terms of research papers published) compared with only one sixth for Soviets. (*Komsomolskaya Pravda*, 1/20/66, 2; *Science*, 1/28/66, 432)

- NASA Assistant Administrator Breene M. Kerr presented a charter to the recently formed Northern Alabama Chapter of the American Society for Public Administration. Kerr announced NASA Administrator James E. Webb had been elected to succeed Harlan Cleveland as president-elect of ASPA. Cleveland had resigned when appointed ambassador to NATO. (*Marshall Star*, 1/26/66, 1)

January 20: Blanche W. Noyes, FAA air marking specialist and veteran pilot, was awarded the Brazilian Medal of Merit of Santos Dumont at the Brazilian chancery in Washington, D.C., for outstanding and meritorious services to Brazilian aviation. Mrs. Noyes was the first American aviatrix to receive the award. (FAA Release 66-8)

January 21: Full 2¼-min. duration test of fourth Saturn IB booster (S-IB-4) was conducted on MSFC's East Test Area facility. S-IB-4 would leave MSFC in late January for Michoud Assembly Facility for post-firing checkout. (MSFC Release 66-25)

- Astronaut Donald K. Slayton received an honorary Doctor of Engineering degree from Michigan Technological Univ. in ceremony at MSC. Degree was conferred "in recognition of high attainments in engineering." As MSC Assistant Director for Flight Crew Operations, Slayton directed the Astronaut Office, Aircraft Operations Office, and the Flight Crew Support Div. (MSC *Roundup*, 1/21/66, 2)

- Discussing Federal subsidization of scientific research in *Science*, Don K. Price, dean of Harvard Univ. Graduate School of Public Administration and new AAAS President, said: "We have to learn how to fit the research interests of free scientists into a pattern of public policy, and to take account of the need for balanced national development while building up our existing centers of high scientific quality. And we need, equally obviously, to devote our knowledge to the service of human welfare, as effectively as it has been enlisted in the service of national defense. We obviously have not yet learned how to do all these things. But we can at least begin, if we are not afraid to make some changes in some of our most stubborn political and administrative habits." (*Science*, 1/21/66, 290)

- France was planning to build rockets bigger than the missiles now being developed for her nuclear strike force, reported an article in *Air et Cosmos*. One purpose was to enable France to orbit a European comsat without the help of an American-built launch vehicle. A second goal would be to provide France with advanced, solid-fuel boosters that could carry one-ton thermonuclear warheads over intercontinental distances or orbit military surveillance satellites. (*Air et Cosmos*, 1/21/66)

- K. T. Keller, retired president and board chairman of Chrysler Corp., died in London. He had been a special adviser to DOD, in 1950 had been appointed director of guided missiles by Gen. George C. Marshall, then Secretary of Defense. (AP, *Wash. Post*, 1/22/66, B4)

January 22: COSMOS CV containing "scientific equipment . . . designed to continue the exploration of outer space" was launched by the Soviet Union, Tass announced. Orbital data: apogee, 324 km. (201 mi.); perigee, 204 km. (126.7 mi.); period, 89.7 min.; inclination, 65°. Instruments were functioning normally. (Tass, 1/22/66)

- Dr. George Veis of the Smithsonian Astrophysical Observatory revealed that the earth's mean radius, measured from satellite observations, was 3,963.203 mi. Radius determination was based on analysis of more than 46,500 observations of 13 different satellites. (Sci. Serv., *NYT*, 1/22/66)

- Advanced Minuteman ICBM was launched from a modernized Minuteman I silo at Vandenberg AFB as part of a series of tests to verify modernization procedures for upgrading of Minuteman I silos and control centers. (UPI, *NYT*, 1/23/66, 57)

January 23: Environmental Science Services Administration (ESSA) of the U.S. Department of Commerce announced creation of an interagency committee to coordinate Government research efforts in clear air turbulence (Cat). (ESSA Release 66-3)

- NASA Lewis Research Center, first Government laboratory devoted to flight propulsion in all its aspects, celebrated its 25th anniversary. Authorized by Congress in 1940, the NACA facility was originally called the Aircraft Engine Research Lab.; it was renamed the Lewis Flight Propulsion Laboratory in 1948 in honor of Dr. George W. Lewis, former NACA Director of Aeronautical Research (1924-47). In 1958, when NACA facilities and personnel were absorbed by the newly created NASA, the facility became the Lewis Research Center. From an initial Congressional appropriation of $2 million in 1941, LRC had emerged as a major research and development laboratory valued at $200 million.

  Marking the anniversary, LRC Director Dr. Abe Silverstein said: "Man isn't going to be held to a single conquest or success. The past decade has seen aviation and rocketry undergo phenomenal change, but we've just started to realize its potential. The present generation will live to fly better than 2,000 mph in commercial aircraft and to observe at close range the interplanetary exploits of U.S. astronauts. The past 25 years have witnessed truly remarkable strides in aeronautics and space activities. In 1941, man had the capability to fly 300 mph and to attain altitudes of perhaps 30,000 ft. Today, he orbits the earth at 17,500 mph and in a few years will set foot on the moon and possibly later Mars." (LRC Release 66-1; *Lewis News*, 1/23/66, 1-12; UPI, *Miami Herald*, 1/24/66)

- NASA Nike-Cajun meteorological sounding rocket launched from Churchill Research Range carried 19-grenade payload to 78-mi. (125-km.) altitude. GSFC experiment obtained temperature, wind, density, and pressure data during auroral condition at altitudes to 56 mi. (90 km.) by recording sound arrivals of explosions of all 19 grenades. (NASA Rpt. SRL)

January 24: President Johnson's message to Congress on the FY 1967 budget said in part: "Just over 60 years ago man entered the age of controlled flight. Today, men orbit the earth at speeds measured in thousands of miles an hour. In 1967, less than 6 years after this Nation set the goal of a manned landing on the moon within the present decade, we will begin unmanned test flights of the giant Saturn V rocket and the Apollo spacecraft—the complete space vehicle required for achieving that goal. Later on in the 1960's, we will undertake the manned lunar mission itself.

  "Our many space achievements—both manned and unmanned—have dramatically advanced our scientific understanding and technological capabilities. They have also demonstrated our remarkable progress in the peaceful exploration of space. In 1967, our large space projects will be progressing from the more expensive development phase into operational status, and new projects of equivalent cost will not be started. Accordingly, expenditures of the National Aeronautics and Space Administration are estimated to decline by $300 million in 1967 [as opposed to the new authorization requested in the FY 1967 budget]. This level will sustain our progress in space exploration and continue the advancement of science and technology." (Text, CR, 1/24/66, 861; *Pres. Doc.*, 1/31/66, 88-9)

- President Johnson sent FY 1967 budget request to Congress including total space budget of $6.8 billion. Of this sum, NASA would receive $5.012 billion (NASA had originally requested $5.58 billion); DOD, $1.65 billion; AEC, $173.7 million; Environmental Science Services Administration, $27 million; and National Science Foundation, $2.8 million.

NASA R&D program was a major area of reduced expenditure with reduction of approximately $186 million (compared with FY 1966 appropriation) mostly absorbed by the sharp drop in requirements for Project Gemini. With only three flights remaining to be made in FY 1967, the total budget dropped from $226.6 million in FY 1966 to $40.6 million requested for FY 1967. Sum included $19.1 million for the spacecraft and $8.5 million for mission support. There was also a reduction of some $54 million in lunar and planetary exploration and $12 million in physics and astronomy. Funding for Project Apollo increased slightly—from $2.967 billion in FY 1966 to $2.974 billion, including $1.2 billion for spacecraft (as compared with FY 1966's $1.17-billion level). $100 million requested for Apollo Applications (Aa) program—once considered the single program with any potential for new major funding—was not approved.

Major portions of the DOD space budget would be spent on development of the Manned Orbiting Laboratory (Mol)—which would be slowed down with no increase in funds over the $150 million allowed in FY 1966; development and launch operations costs of the Titan III launch vehicle; continuation of the reconnaissance and Vela nuclear detection satellite programs; navigational satellite launches; and initial launch of a satellite by the Defense Communications Satellite System. DOD budget also provided for initiation of procurement of the FB-111B aircraft and development of a new short-range air-to-surface missile (Sram).

AEC's budget allotment for space programs included $78 million for Project Rover; $39.5 million for testing of the Nerva engine system; and $8.2 million for testing of the Snap-8 nuclear reactor.

ESSA's Tiros weather satellite program would involve expenditures in FY 1967 of $17 million for spacecraft and launchings and approximately $10 million for data acquisition, processing, and archiving.

FAA's FY 1967 budget request for $758 million contained $80 million for six more months of supersonic transport research. (Text; *Wash. Post*, 1/25/66, A10; Schmeck, *NYT*, 1/25/66, 1; *Av. Wk.*, 1/31/66, 23, 29, 30; *M&R*, 1/31/66, 25)

January 24: NASA Deputy Administrator Dr. Robert C. Seamans, Jr., said at NASA FY 1967 budget briefing that the "extremely stringent budget" ($5.012 billion) approved by President Johnson and the BOB was the result of pressing needs of the Vietnam war and the Great Society. Major loser in the NASA budget reduction was the Apollo Applications (Aa) program: "The option to go ahead or not [on the Aa program] will be made in 1968 [budget]," he said. Although NASA had wanted $5.58 billion to carry out its FY 1967 program, Seamans said, the $5.012 billion would make it possible to continue "a good space program" with most of the projects intact. Yet, NASA did "not have the funds we would like to see" for Apollo. He said the approved FY 1967 Apollo budget of $2.974 billion (compared to $2.967 billion in FY 1966) would give the U.S. "a chance of lunar landing in this decade," but there were no funds "for major difficulties that might occur." (Transcript)

- NASA had begun contract negotiations with TRW Systems to convert the prototype Orbiting Geophysical Observatory (Ogo) into a flight observatory. Negotiations might also include option to fabricate an additional flight observatory. Contract was expected to be for $9 million. (NASA Release 66-16)

January 24: Major decisions on the future of space exploration facing the Administration received comment by William J. Coughlin in *Missiles and Rockets:* "History teaches us . . . that such exploration is more likely to come in waves, one overlapping the other, than in single units, each with a distinct beginning and a distinct end.

"Man strives for the most distant objective within reach of his technology. And as the first explorers try and try again on that far reach, others—at first other explorers and scientists, then those scouts of commerce, miners and trappers, then businessmen, settlers and shopkeepers—follow in their wake to prove and develop what courage has proven attainable." (Coughlin, *M&R*, 1/24/66, 46)

- NASA Nike-Cajun sounding rocket launched from NASA Wallops Station carried instrumented payload to estimated 104-mi. (167-km.) altitude. Experiment conducted for Univ. of Michigan measured ambient air density from 19–75-mi. (30–120-km.) altitudes by radar track of two falling Mylar spheres ejected at plus 72 sec. Flight also measured solar heating effects by comparing diurnal variations of measurements. (NASA Rpt. SRL)

- NASA Associate Administrator for Manned Space Flight Dr. George E. Mueller said in a speech at the 51st annual meeting of the National Dairy Council in Washington, D.C.: "We in NASA do not feel that the planning of our space programs is the prerogative of NASA alone. . . . The participants should include men from all walks of life—national, political, and intellectual leaders, the business and agricultural communities, and all informed citizens—not just scientists and engineers nor those who are presently engaged in the space program." Describing the launch vehicles and spacecraft in the Apollo program, Dr. Mueller said: "The uprated Saturn I launch vehicle [formerly Saturn IB] . . . is capable of placing 40,000 pounds in orbit. . . . The Saturn V launch vehicle . . . provides the power to place in orbit loads weighing 280,000 pounds. . . . The Apollo spacecraft itself consists of three major systems or modules. The Command Module . . . houses the astronauts during most of their journey. The Propulsion Module [formerly Service Module] has rocket engines and abundant fuel for mid-course maneuvering, for braking into lunar orbit, and for the thrust to return to earth. The Lunar Module [formerly Lunar Excursion Module] is for descent to and return from the lunar surface. . . . We have the ability to fly this equipment at an annual rate of six Saturn I launch vehicles, six Saturn V launch vehicles and eight Apollo spacecraft." (Text)

- Astronaut M. Scott Carpenter said surgeons would have to remove a callus, formed on his left forearm after he broke it in a July 1964 motorcycle accident in Bermuda, before he could be considered for another spaceflight. Ability to rotate the arm properly was hampered. (AP, Wash. *Eve. Star*, 1/25/66, A2; AP, *Chic. Trib.*, 1/25/66)

- Prof. Philip Handler of Duke Univ., member of the President's Science Advisory Committee, said at the Asia-Pacific Seminar that the U.S. was the only nation in the scientific revolution and that the U.S.S.R., U.K., and West Germany had not yet entered it. He described the scientific revolution as dependent upon man's brains and not upon water, coal, or iron, as was the industrial revolution. The computer, he said, was the chief symbol of the scientific revolution. (AP, *NYT*, 1/27/66, 3)

January 24-26: At AIAA Third Aerospace Sciences Meeting in New York City, Dr. Raymond L. Bisplinghoff was elected 1966 AIAA president. Dr. C. Stark Draper, chairman of MIT's Dept. of Aeronautics and Astronautics, gave 29th annual Wright Brothers Lecture on the role of "informetics" in modern flight systems.

NASA Associate Administrator for Space Science and Applications Dr. Homer E. Newell chaired a special "Space Sciences Report" plenary session which considered priorities in planetary exploration and the interaction of solar wind with the planets. Donald L. Hunter of Kitt Peak National Observatory said Venus exploration was likely to be as interesting as exploration of Mars; high Venusian surface temperature did not rule out possibility of rudimentary life forms, and Venus' environment might be less hostile than that of Mars. Manned round trips to Mars could be shortened to 450-560 days by a Venus "swingby," according to Rollin W. Gillespie and Stanley Ross of NASA Office of Manned Space Flight; direct flights would take 800 to 900 days.

Dr. Robert C. Seamans, Jr., NASA Deputy Administrator, was principal speaker at the Honors Convocation. He said: "Yesterday, President Johnson presented his budget for the coming fiscal year to the Congress. . . . This year, the process of sorting out priorities and allocating resources has been especially difficult. Abroad, there is the growing commitment to Vietnamese freedom. At home, there is the growing commitment to freedom from poverty, ignorance, ugliness, and unemployment. In addition to these and other commitments, the President's budget once again reiterates the commitment made in 1958 and repeated year after year: to undertake the exploration of space for the benefit of mankind. Even in a year of extraordinary budget pressures, slightly more than $5 billion for NASA's part in this enterprise have been requested. . . .

"It is within this framework that the NASA program has been hammered out—not representing the sum of our program demands. However, as we indicated in our budget briefing for the Press this past weekend, we believe we have a balanced program of aeronautics and space research, of science and technology, of large scale manned exploration and unmanned probes and satellites.

"We recognize there is an enormous responsibility attached to the planning, justification, and direction of a sum so large during a period so rich in other needs. . . .

"We have reached a point where the exploration of space is no longer a promise but a hard fact, where scientific data and tested capabilities are in hand instead of in the future.

". . . The program that we have laid out for NASA for the coming year will meet the highest priorities of scientific and technological needs. It will permit flexible evaluation of the next major steps open to us in space. It will maintain the forward momentum that has been built up with energy and care over the past years. It will permit strong competition for those who wish to outstrip us and meaningful cooperation for those who can find communities of interest with us. It will provide an honest and practical balance between the many elements of scientific investigation, technological application and operational experience. It provides a challenge that will require hard work, hard judgment, and careful appraisal at every step of translation from plan into execution. It does not close to future exploitation the many options that have been built for the nation in the past. Most importantly, it recognizes that true national security is best measured by the useful knowledge in men's minds and that the great arena of space remains a battleground between the intellect of man and the unknown."

Tribute to late Dr. Hugh L. Dryden was made at the Honors Convocation by L/G James Doolittle (USAFR): "On the second of December 1965 the world became poorer; our country lost one of its foresighted leaders; aeronautics lost a man who had devoted a major part of his life to its advancement; astronautics lost its great ambassador at home and at large and those fortunate of us—who had been privileged to know and work with him—lost a beloved, revered and respected friend and associate.

"Dr. Hugh Latimer Dryden dedicated his whole life to the service of his fellow man. He was a doer, but a gentle person—a man who strove prodigiously, but instinctively did the kindly thing. He was eager and courageous. He had been critically ill for several years, but he worked hard and effectively up to a few days before his death. . . .

"Hugh Latimer Dryden: engineer, scientist, administrator, public servant, patriot, theologian, realist. With his immediate family and all the nation we, his close friends and associates, mourn his passing. We will long remember him with affection, admiration, respect, and gratitude, for excellence endures."

Goddard Award, AIAA's highest, was presented jointly to Hans J. P. von Ohain, Chief Scientist, Aerospace Research Lab., Wright-Patterson AFB, for "his contributions to the achievement in 1939 of the first successful application of turbojet propulsion to aircraft"; and to A. W. Blackman, Chief of Propulsion, United Aircraft Corp., and George D. Lewis, Project Engineer, UAC Pratt & Whitney Aircraft Div., for "their contributions to the understanding of the phenomenon of combustion instability and for their recognition of acoustical liners as a method of suppressing such instabilities in turbojet afterburners and rocket engines." Sylvanus Albert Reed Award went to Clarence L. Johnson, vice president of Lockheed Aircraft Corp., and designer of the U-2 reconnaissance aircraft and the YF-12A interceptor, for his "production of two triple-sonic military aircraft, for continuing personal design innovation, imaginative engineering, and practical manufacturing techniques that over the years have aided immeasurably in maintaining U.S. ascendency in defensive aerial weaponry." 1966 Research Award was given to Prof. Shao-Chi Lin, Univ. of California (San Diego), for "basic research in the electric and electromagnetic properties of ionized gases, and for significant contributions to reentry physics." (Texts; AIAA Release; AIAA *Bulletin*, 12/65, 693, 709, 710)

January 25: U.S.S.R. launched COSMOS CVI with instrumentation to continue studies of outer space. Initial orbital data: apogee, 564 km. (350 mi.); perigee, 290 km. (180 mi.); period, 92.8 min.; inclination, 48.4°. Onboard equipment was functioning normally. (Tass, 1/25/66)

- NASA MSC announced that Astronaut John Young would be command pilot of Gemini 10; pilot would be Astronaut Michael Collins. No date had been set for the Gemini X mission. (UPI, Wash. *Daily News*, 1/25/66)
- FS-3 nuclear reactor electric power system designed for use in space completed a full year of generating electricity continuously and without external control. Test far exceeded known continuous power operation of a nuclear reactor power plant of any type. FS-3 was a prototype nu-

clear reactor space power system derived from the 500-watt Snap-10A unit that became the world's first nuclear reactor system to operate in space in April 1965. (AEC Release J-19)

January 25: Dr. Philip Morrison, MIT physics professor, called for $1 million from NSF for five telescopes identical to the one on Mt. Palomar, presently the world's largest. He said that "properly situated in various parts of the world, these instruments could peer almost to the edge of creation ... and perhaps unlock the secret of the origin of the universe." Dr. Morrison spoke in a panel discussion on cosmology at MIT. (Hines, Wash. *Eve. Star*, 1/26/66, A20)

- A 2,000-mph SR-71 (YF-12A) jet reconnaissance aircraft crashed in northeastern New Mexico after an explosion during a routine flight from Edwards AFB. Lockheed test pilot James T. Zwayer was killed. Aircraft had been built for USAF by Lockheed as successor to the U-2 and had been undergoing tests since December 1964. (UPI, *Wash. Post*, 1/26/66, 12; AP, Wash. *Eve. Star*, 1/26/66, A5)

- *Wall Street Journal* editorial on the FY 1967 budget: "For the first time since ... [NASA] was formed, space spending will decline. The $300 million drop, to $5.3 billion [in expenditures], is attributed to lower development costs for some major programs and a decision not to initiate big new efforts." (*WSJ*, 1/25/66, 6)

- International conference on outer space to mark 10th anniversary of first artificial satellite (1967) was proposed by U.S.S.R. Academician A. A. Blagonravov to working group of U.N. Committee on the Peaceful Uses of Outer Space. European site was suggested for conference, which "would provide information for peoples in all countries and training for the developing countries." (Rossi, *CSM*, 1/26/66)

- Dr. Donald F. Hornig, Director of Office of Science and Technology, said in an interview with Robert Cahn of *Christian Science Monitor:* "There are two kinds of breakthroughs. There are those which come through completely new insight and understanding, which is what we usually mean by a scientific breakthrough. You might consider as scientific breakthroughs the discovery of a new kind of nuclear force or the discovery of a new fundamental particle. Then there are the more common breakthroughs in technology—the harnessing of principles in new and unusual ways to achieve the ends you want. Sending a man to the moon and back is one of these." (Cahn, *CSM*, 1/25/66)

January 25-27: House Committee on Science and Astronautics and its panel of advisers on science and technology held 7th annual meeting on "Government, Science, and Public Policy." In keynote address, Vice President Hubert Humphrey urged closing of "gap between public policy and advancing science and technology ... when NASA comes to a great university with a grant, it is like planting a field of gold. There is a harvest soon to be reaped, a whole new partnership that creates new wealth, the wealth of brain power, of scientific knowledge, and the end product."

Guest panelist Lord C. P. Snow, joint parliamentary secretary of U.K.'s Ministry of Technology, told Committee that U.K. had no intention of becoming totally dependent on U.S. technology: "By and large, our technological position is uncomfortable, and we have got to change it ... before we get our economy sturdy again. ... It will take ten years to do all that has to be done. But it will be done." Lord Snow also made these points: (1) if U.K. could not maintain a high competence in certain areas of advanced technology, she would lose her talented young men, "slide into somnolence and be no good" to herself, America, or the world; (2) the computer or cybernetic revolution would be the biggest revolution mankind had known; (3) mathematics would "take on a new relevance in all advanced societies." Noting danger that allocational decisions for research in pure and applied science may involve "too much concentration of power," Lord Snow added: "The scientific results—not the technological results—... of observational and radioastronomy have been greater by an order of magnitude than the scientific results of space exploration."

In discussion period, Dr. Lee A. DuBridge, Cal Tech president, commented: "... though I think I would agree with you generally, I hope you do not minimize the scientific results that have come from ... recent space exploration flights—the understanding of the earth's magnetosphere, the nature of the solar wind, our closeup views of Venus, the Moon, and Mars have added a very substantial amount to our understanding ... of these objects and of interplanetary space ..." When Lord Snow replied that "space things are interesting, but not at the moment to the same degree" as radioastronomy, Dr. DuBridge said: "Give us another 20 years."

Dr. Charles C. Price, Univ. of Pennsylvania chemist, pleaded for support of more basic research in the areas of living systems and suggested that "the more applied agencies such as NASA, NIH and AEC put increased emphasis on the support of fundamental research and education in chemistry ..." (*Proceedings*, House Comm. on Science and Astronautics, 1/25-27/66)

January 25-27: John S. Foster, Jr., DOD Director of Defense Research and Engineering, and Dr. Robert C. Seamans, Jr., NASA Deputy Administrator, told Military Operations Subcommittee of the House Committee on Government Operations, which had questioned DOD/NASA cooperative agreements, that cooperation was necessary to effect coordination and to plan for best possible use of national ranges. Cooperation had been established through May 1965 DOD/NASA Agreement on Tracking, Data Acquisition, and Communications Facilities which stated that the two agencies were to "achieve a maximum of mutual assistance, to avoid unwarranted duplication, and to realize economies wherever practical and consistent with mission requirements, by means of coordination and planning and efficient division of responsibilities." (House, *Hearings, Missile and Space Ground Support Operations*)

January 26: Seven geophysical experiments were selected by NASA to be included in three 150-lb. Apollo Lunar Surface Experiments Packages (Alsep) and one backup on initial Apollo landing missions: (1) passive lunar seismic experiment; (2) lunar triaxis magnetometer; (3) medium-energy solar wind experiment; (4) suprathermal ion detector; (5) lunar heat-flow measurements to provide information on distribution of radioactive elements and thermal history of the moon; (6) low-energy solar wind experiment; (7) active lunar seismic experiment to study tremors resulting from firing of mortar projectiles, designed to gather data on depths to 500 ft. (NASA Release 66-17)

- Dr. Alexander G. Smith, Univ. of Florida physicist and astronomer, discussed applications of orbiting platforms and space probes in radioastronomy at MSFC Space Science Seminar. Of current theories explaining powerful outbursts of radio noise from Jupiter, the "most plausible" was that radiation was due to a cyclotron process, Smith said. "Recent thermal work seems to indicate that Jupiter is probably a hot planet, . . . emitting more heat than it is receiving from the sun—2 to 4 times as much." Smith, asked if his emphasis on orbital experiments meant it was unnecessary to go to the moon, replied: "If I had to choose my favorite site for low frequency radio astronomy, it would be the back side of the Moon, where one could put up large arrays and be screened by the bulk of the Moon from terrestrial interference." (Transcript)

January 26: NASA Associate Administrator for Manned Space Flight Dr. George E. Mueller spoke of reliability and quality assurance in the manned space flight program at the Annual Symposium on Reliability in San Francisco: "We have . . . had to organize our mission reliability around the concept of a very exhaustive ground test program, with a detailed, recorded tracking of the history and experience of every part, from the original developer-fabricator's plant, through the entire chain of subcontractor and contractor activities, to the final, completed article on the launch pad. By this thorough and repeated testing under every environmental condition that it is possible to simulate, and by understanding and correcting each test failure or performance discrepancy, we believe we can achieve the required high reliability whereby we will be justified in conducting major missions on the second, third or fourth vehicle, rather than on the thirtieth or the hundredth as in Gemini and Mercury." (Text)

- USN had issued to General Dynamics Corp. a $20-million supplemental agreement to an existing contract to install satellite communications systems aboard each of three former Navy tankers that would serve as ocean-going tracking stations for NASA's Project Apollo. (Gen. Dynamics Release 1267)

- Astronaut Walter M. Schirra, Jr., GEMINI VI command pilot, received a tiny gold harmonica from the Italian National Union of Mouth Organists and Harmonica Musicians for being "the first mouth organist in outer space." (AP, Phil. *Eve. Bull.*, 1/26/66)

- Commentary in *New York Times* on difficulty of allocating limited funds for scientific research: "The eminent Soviet physicist P. L. Kapitsa has published an article [see Jan. 20] acknowledging that American science is still contributing far more to mankind's knowledge than its Soviet counterpart. He attributes the United States lead largely to the wiser American technique for Government subsidy of research. . . . Ironically, several months ago President Johnson . . . directed that Federal funds be more equally distributed among the nation's universities and geographic areas and that a larger proportion of this money be allocated to universities for distribution as their administrators see fit, rather than directly to research scientists.

  "If the Kapitsa analysis is correct, the President's order was a mistake and may lead to reduced productivity in American science." (*NYT*, 1/26/66, 32)

- Detailed report on "Scientific and Engineering Manpower in Communist China, 1949-1963," by Dr. Chu-yuan Cheng, of the Univ. of Michigan, was made public by NSF. Communist China had been strengthening its scientific and engineering fields by putting thousands of students through colleges and universities. Yet, these masses of young scientists were getting less comprehensive training than older generations of scientists and also lacked sufficient experience to engage in top level research. In

1963, many of China's leading scientists were 60 or 70 yrs. old and engaged in basic and creative research. The success or failure in bridging the gap between the mass of young scientists and the older scientists would determine whether Communist China would become a scientifically advanced power in the next decade or two, Dr. Cheng said. (Text)

January 26: U.K. had made a successful underground test of a major component of its own nuclear warhead for American-supplied Polaris missiles, it was disclosed. Prime Minister Harold Wilson told the House of Commons the test would lead to a considerable savings in costs. An earlier test had been unsuccessful. (*NYT*, 1/26/66, 39)

January 26-29: American Physical Society met in New York. Dr. Charles H. Townes, Nobel prize-winner and MIT provost, said in a dinner speech that the manned lunar landing planned for the end of this decade was both economically and scientifically feasible. "Furthermore," he continued, "the task probably will be accomplished within time and money limitations set four years ago. These envisage a moon landing by late 1969 at a total cost of about $20 billion." Townes announced the appointment of "an important ad hoc Science Advisory Committee" aimed at obtaining "the most fruitful relationship between NASA and the scientific community." Headed by Prof. Norman Ramsey, the new group would "consider many . . . problems, concentrating particularly on how to carry out the most efficient and rich scientific program and what ways will allow the scientific community, including universities, professors, and students to participate most easily and effectively in the interesting possibilities which are now emerging."

In a panel on the origin and possible death of the universe, P. J. E. Peebles and David Wilkinson of Princeton Univ. and Robert W. Wilson of Bell Telephone Labs. told about evidence indicating that space was permeated by black body radiation. If this were so, it would support the "big bang" theory of the origin of the universe which held that at first, all was a mass of dense energy; then came expansion powered by much of the energy. The original energy would have the properties of the energy emitted by a black body—light, heat, and radio waves proportionate to its temperature. Because no other explanation existed for the energy the Princeton and Bell scientists had detected uniformly distributed over space, and because black body characteristics were so clearcut, they believed confirmation could be achieved within a year.

Yuval Ne'eman of Tel Aviv Univ. proposed a new theory to explain the quasar (quasi-stellar radio source): the force represented by the tendency of the universe to expand had been suppressed in some parts of the universe; thus, these areas were highly dense. But because this tendency is not suppressed under all conditions, it sometimes bursts forth and matter tears free of the intense gravitational attractions. When this happens, enormous amounts of matter and energy are emitted and observers on earth detect the event as a quasar.

Scientists in a panel discussion on scientific data said the task of coping with the scientific information explosion should not become exclusively a Government responsibility. Discussion was based on December 1965 report of Federal Council for Science and Technology recommending that the Government—with participation of private groups—establish a national document handling system to help keep scientists abreast of new knowledge.

In a discussion of planetary exploration, Cal Tech Prof. Robert B.

Leighton, scientist in charge of the televising by MARINER IV of Mars (July 14, 1965), said there was still hope the TV pictures would yield information about Mars' canals: "When we know more nearly where the camera was pointed—because it is not known very clearly now . . . we may be able to say that this picture was taken at a place very near where a canal goes across." Leighton said he had agreed to collaborate with Prof. Gordon J. F. MacDonald of UCLA on a television experiment for the Mariner spacecraft scheduled to fly by Venus in 1967. (Texts; Wash. *Eve. Star*, 1/27/66, A3; Hines, Wash. *Eve. Star*, 1/28/66, A4; Weil, *Wash. Post*, 1/27/66, A2, 1/29/66, A6, 1/30/66, A7)

January 27: NASA's PIONEER VI interplanetary space probe had reported its first major solar events and the highest solar wind velocity it had measured since launch Dec. 16, 1965: a stream of solar particles was rushing past the spacecraft and the earth at about 1,440,000 mph. This compared with the highest solar wind speed ever measured by a NASA spacecraft—EXPLORER XVIII—of about 1,675,000 mph. (NASA Release 66-21)

- Arnold Frutkin, NASA Assistant Administrator for International Affairs and U.S. representative to the Working Group of the U.N. Committee on the Peaceful Uses of Outer Space, supported U.S.S.R. proposal for an international conference in 1967 to commemorate the first decade of space exploration, further international cooperation in space, and promote the practical benefits of man's exploration of space. He suggested four topics for the conference: (1) appraisal of contributions of scientific research in space to practical applications and international cooperation; (2) opportunities for participation in space exploration through space research and cooperative projects; (3) applications and economic implications of space research; and (4) impact of space exploration on education. (*SBD*, 1/27/66, 170)

- Sen. J. W. Fulbright (D-Ark.) spoke satirically on the floor of the Senate on priorities of space, as opposed to social problems, in allocation of natural resources. He labeled as "small minded, selfish, and sentimental" the Americans who "tell us there is no real hurry about going to the moon when tens of millions of Americans . . . spend their lives in dirty, dangerous, and unhealthy cities, when hundreds of millions more are condemned to degraded lives of poverty and early deaths of preventable diseases." He cited a recent article in Britain's *Punch* that reproached those Englishmen who would complain about giving up an iron lung and artificial limbs to provide the aluminum for space vehicles "as if a nation's prestige lay in its free artificial legs, and not up there"—in the sky. (*CR*, 1/27/66, 1209)

- North American Air Defense Command (NORAD)—nerve center of North American air and space defense—took formal possession of new headquarters built 1,200-1,400 ft. into the solid granite of Cheyenne Mountain, Colo., to insure survival of a retaliatory capability in the event of nuclear attack. Eleven interlinked, steel-framed, steel-sheathed, three-story buildings constituting the command post were mounted on coil springs that would protect equipment from ground shock of nearby thermonuclear blast. (*Eng. Opp.*, 2/66, 20-28)

January 28: Senate Committee on Aeronautical and Space Sciences unanimously approved nomination of Dr. Robert C. Seamans, Jr., as NASA Deputy Administrator. (Transcript)

- Glenn Reiff, Manager of Mariner/Pioneer program in NASA Hq. OSSA, summarized in an article in *Science* the "steps taken to combine the scientific instruments with the other vital elements of the [MARINER IV] spacecraft to create a spaceworthy craft capable of a reliable 10-month mission." He said over "1600 type-approval and flight acceptance tests were performed on the assemblies, and some 113 failures were reported. As a result, more than 75 design changes were incorporated." (Reiff, *Science*, 1/28/66, 413-18)

January 28: Saturn V crawler completed first successful load-carrying run at KSC since track support roller bearings had been crushed during 7.9-million-lb.-load test July 25, 1965. Crawler lifted 447-ft., 10.6-million-lb. launcher umbilical tower No. 1 (Lut) and moved it ¾ mi. in about nine hrs. KSC Procurement Director Michael Haworth revealed costs for two crawlers ordered would total $14.2 million, $5.1 million more than original $9.1-million cost-plus-incentive-fee contract with Marion Power Shovel Co. (*M&R*, 2/7/66, 34)

- NASA Administrator James E. Webb said in a letter to MSFC Director Dr. Wernher von Braun that after "careful review" he had approved an agreement between MSFC and Lodge 1858, American Federation of Government Employees (AFL-CIO). Mr. Webb noted that agreement, negotiated after MSFC employee vote May 19, 1965, for AFGE representation, was "a forward step in the NASA Employee-Management Cooperation Program." (*Marshall Star*, 2/2/66, 1-2)
- USAF launched unidentified satellite from WTR with Scout booster. (*U.S. Aeron. & Space Act.*, 1966, 147)
- Commenting on the Federal budget for FY 1967, *Science* said it was the first budget since the middle 1950's which had not carried a request for an increase in total funds for research and development: "This downturn in what for about a decade has been a steadily ascending curve reflects the major rationale of the new budget: to provide funds to support a military buildup in Southeast Asia without underfinancing new education and welfare programs enacted during the past two sessions of Congress. . . ." (*Science*, 1/28/66, 425)
- Possibilities of studying the ocean, its storms, currents, shoals, and marine life from orbiting spacecraft were being actively explored by the Naval Oceanographic Office in cooperation with the NASA Natural Resources Program Office, reported Harold M. Schmeck, Jr., in the *New York Times*. Early studies had included observations that might have direct practical usefulness and others that could give man a more comprehensive picture of the forces determining climate: photographs taken from manned spacecraft at altitudes over 100 mi. could be used to correct maps and give accurate indications of water depths over shoals and near shore; a global iceberg patrol might be possible from manned or unmanned spacecraft; monitoring of sea roughness by radar would be useful for shipping; measurements to detect major upwellings of deep ocean currents might be valuable to the ocean fishery industry; studies of river outflow could be valuable in water pollution control. (Schmeck, *NYT*, 1/29/66, 7)
- Close tie between President Johnson and Dr. Donald F. Hornig, Director of the Office of Science and Technology (OST) and science adviser to the President, was discussed in *Science*. During the first six or so months of Hornig's term "the President and the White House inner circle had an unclear notion of how they might employ OST" with the result that "some of the science staff members complained that they were bored,

and that OST was underemployed or involved with trivia." But a transition had taken place, and as Hornig began his third year of service, "OST and its surrounding bodies—the President's Science Advisory Committee and the Federal Council for Science and Technology—are heavily involved in policy formulation, coordination of the technical aspects of the Great Society programs, and troubleshooting in a variety of areas." (*Science*, 1/28/66, 431)

January 28: A "commercial missile" that could transport 170 passengers from Bombay to New York in 39 min. was described at the Pacific Area Travel Association Conference in New Delhi by B. F. Coggan, corporate vice president of Douglas Aircraft Co. He said the rocket—which would reach a speed of 17,000 mph—would be a bell-shaped vehicle 114 ft. high with a diameter of from 33 to 49 ft. It would weigh more than three million pounds at liftoff. Passenger compartment would have 45 individual passenger couches arranged on each of four decks; during the period of weightless flight, passengers would be held to the couches by belts. On a typical flight plan, Coggan said, the rocket would reach 125-mi. peak altitude. Then, it could hover and maneuver horizontally before landing on four extendable legs. (*NYT*, 1/29/66, 54)

- FAA and the Agency for International Development (AID) had signed a new agreement under which FAA would continue to provide aviation assistance to foreign countries through AID. Civil Aviation Assistance Groups (CAAG) manned by 84 FAA aviation specialists would work in 17 countries under AID sponsorship. They would advise and assist these countries in planning and organizing management and development of all phases of their national civil aviation. (FAA Release 66-9)

January 29: Senate confirmed appointment of Dr. Robert C. Seamans, Jr., as NASA Deputy Administrator. (*CR*, 1/29/66, 1469)

- Fourth flight Saturn IB booster (S-IB-4) left NASA MSFC for Michoud Assembly Facility aboard the NASA barge *Palaemon* for post-captive-firing checkout. (MSFC Release 66-27)
- Many first-rate scientists in the U.S. did not receive enough financial support to allow them to develop their fullest effectiveness in research work, NSF Director Dr. Leland J. Haworth said in NSF's annual report. "This lack of support is a detriment to the national scientific capacity," he added.

    Discussing progress in the biological and medical sciences, the report said that science was in the midst of a fourth biological revolution. It termed this revolution comparable in importance and ultimate impact on mankind to the three earlier revolutions in mankind's knowledge that led to development of agriculture, to understanding of infectious disease and its control, and to the understanding of biological evolution. Report referrred to current biological research clarifying mechanisms of heredity and development of all earth's creatures. (Text)

January 30: *Minneapolis Tribune* polled 600 Minnesotans on whether U.S. or U.S.S.R. led in space exploration. 79% believed U.S. was ahead; 11% gave lead to U.S.S.R. June 1965 poll had indicated that 59% thought U.S. led, while 24% considered U.S.S.R. in front. Increased confidence in U.S. position was attributed to success of GEMINI VII-VI rendezvous Dec. 15, 1965. 68% thought men could reach the moon and return safely, 56% believed U.S. would be first to do it, and 25% considered trip impossible. (*Minn. Trib.*, 1/30/66)

January 31: LUNA IX was launched by U.S.S.R. on a trajectory toward the moon, Tass announced. All systems were functioning normally. Launch was apparent attempt to softland on the lurain. (*Pravda*, 2/3/66, USS-T Trans.)

- Eighth anniversary of first U.S. satellite, EXPLORER I. Despite predictions of a lifespan that would not exceed five years, satellite continued to orbit the earth every 103.9 min. Other orbital parameters, as of Jan. 15, were: apogee, 956.9 mi. (15,406 km.); perigee, 211.9 mi. (341 km.). (MSFC Release 66-26)
- President Johnson transmitted his annual *Report to the Congress on United States Aeronautics and Space Activities*, calling 1965 "the most successful year in our history." He said in the transmittal letter: "As our space program continues, the impact of its developments on everyday life becomes daily more evident. It continues to stimulate our education, improve our material well being, and broaden the horizons of knowledge. It is also a powerful force for peace." He cited U.S. accomplishments in the Gemini, Ranger, Mariner, and Early Bird programs. "Research and development in our space program continued to speed progress in medicine, in weather prediction, in electronics—and, indeed, in virtually every aspect of American science and technology." Among chief aeronautical developments report cited "design refinement and wind-tunnel testing" of Sst and decision to build C-5A jet transport. Among successful U.S. launches during 1965, DOD had 67 satellites, 34 of which were unidentified. Military space programs were discussed, including Project West Ford; Manned Orbiting Laboratory (Mol); Initial Defense Communications Satellite Project; and the geodetic Secor satellites. Research was under way on the best means of determining physical characteristics of "uncooperating objects in earth orbit" through observation by ground-based radar. The section on AEC said that a third pair of Vela radiation-detection satellites launched in July 1965 was watching for nuclear explosions in space and measuring x-ray emissions from the sun. In a separate chapter of the report, the NASC said: "This Nation's space achievements during the year were so distinctive in quality and quantity as to cause people, at home and abroad, to conclude the United States had attained the space leadership of the world." The Council said that in many respects this conclusion was correct "but it would be incorrect and unwise to minimize the vitality and size of the space program of the U.S.S.R." (Text; *CR*, 1/31/66, A419; *Pres. Doc.*, 2/7/66, 149)
- Highlights of AEC's annual report to Congress for 1965: studies were made on large nuclear-powered water desalting facilities, including study of requirements of New York-New Jersey metropolitan area; Snap-10A, first nuclear power reactor unit to be placed in orbit, functioned in space for 43 days before electronic component failures shut it down; two AEC-instrumented Vela satellites for nuclear test detection were launched, raising the total now in orbit to six; 27 underground nuclear tests were conducted. (Carey, *Wash. Post*, 2/1/66, A2)
- NAA's Rocketdyne Div. at Edwards AFB was struck by AFL-CIO United Auto Workers in a wage dispute. An estimated half of the 227 workers in UAW bargaining unit stayed off their jobs. (L.A. Times, *Wash. Post*, 2/1/66, A9)

During January: Series of tests was being completed by NASA MSC's Landing and Recovery Div. to check out direction and finding equipment for homing an Apollo spacecraft for post-landing location and recovery. Tests were performed off Galveston Island in Gulf of Mexico using Apollo boilerplate spacecraft with the Apollo Recovery Beacon and Apollo Survival Radio transmitting signals. Homing runs were made from altitudes of 28,000 ft. to 500 ft. to determine range of direction finder receiver onboard a USAF HC-130H air rescue aircraft with the Apollo Recovery Beacon and Survival Radio on the spacecraft in the Gulf. Equipment had performed as expected with line-of-sight acquisition of boilerplate Apollo on all homing runs by the aircraft. (MSC *Roundup*, 1/21/66, 3)

During January: Eulogies for the late Dr. Hugh L. Dryden, NASA Deputy Administrator, and the late Dr. W. Randolph Lovelace II, NASA Director of Space Medicine, appeared in *Air Force and Space Digest*:

Walter T. Bonney said of Dr. Dryden: "Hugh Dryden was a deeply religious man. He held that man's life at its fullest is a trinity of activity, physical, mental, and spiritual. Throughout his life he preached the compatibility of religion and science, and he lived what he preached. No less, he was a realist; and so, as in 1954, he would often remind his listeners that: 'I am not one of those few who believe that we can abolish the use of force in the world. As a nation we find it necessary to build great military power. I am confident that such strength is a greater contribution to the peace of the world . . . than military weakness.'

"In 1947, he was called to direct the work of the National Advisory Committee for Aeronautics. His tasks were enormously difficult—radically new and expensive research facilities had to be provided, but even more important, learning the nature of the intricate problems of flight in the new speed ranges and methods for their solution demanded new concepts of team activity and functional coordination. His leadership, disarmingly light and unobtrusive, was amazingly perceptive and effective. Facilities were obtained and changes were made; the period 1947-58 became what might be truly called NACA's golden age.

"With establishment of the National Aeronautics and Space Administration (of which NACA was the nucleus), he undertook the role of scientific leader, reporting to the Administrator, and made of it a position of paramount influence. He rejoiced in the US policy that activities in space should be devoted to peaceful purposes and worked to promote international space cooperation; but no less, he saw in US exploration of space an activity essential to the strength and security of the nation. As perhaps never before, his great wisdom and his unexcelled mastery of the art of the possible were in those later years valued and used to their fullest."

Editor John Loosbrock said of Dr. Lovelace: "In widening circles, the full import of Randy's untimely death becomes apparent. The Air Force Association and the Aerospace Education Foundation have lost a distinguished leader. The medical profession is poorer for the passing of a brilliant surgeon, an imaginative researcher. The United States Air Force will miss his sound advice and pioneering efforts in the special kind of medical knowledge that is required for those who fly—in air and in space. The National Aeronautics and Space Administration must find new leadership for its space medical effort. The entire nation and the world itself will find that Randy Lovelace's death leaves a large and well-nigh unfillable niche." (*AF/SD*, 1/66)

During January: Dr. Raymond L. Bisplinghoff, Special Assistant to NASA Administrator, paid tribute to the late Dr. Hugh L. Dryden, in *Applied Mechanics Reviews* on Dr. Dryden's scientific and engineering accomplishments. Dr. Bisplinghoff noted that the NACA, under Dr. Dryden's leadership, "produced a vast body of new knowledge which made possible routine supersonic flight and laid much of the ground for space flight that was to come. We discern here, perhaps as much as any other place, the impact of Dryden's leadership on the nation's destiny. The development of high-speed wind-tunnel facilities and a companion competence for theoretical research within the NACA contributed, in the opinion of the author, more than any other single factor to the leadership of the United States in supersonic flight. . . ." (*AMR*, 1/66, 1-5)

- An air-transport vehicle that had attracted the special interest of NASA aeronautics research was the personal airplane, Charles Harper, Director of Aeronautics Div., NASA OART, wrote in *Astronautics & Aeronautics*. Attempts were being made to devise aircraft whose control commands would provide a response "wholly natural to the operator" and which could be operated with no more special training than required for the automobile. Recent scientific developments spurred by spaceflight activity were being assessed with regard to the problem. (Harper, *A&A*, 1/66, 24)

- In *Reader's Digest* article, William Hines, science editor of the Washington *Evening Star*, charged and documented what he described as a banding "together in a unique alliance, [of] NASA and the Urban Renewal Administration" to build NASA Electronics Research Center in Cambridge, Mass., at the cost of several citizens' rights. He said: "It was a splendid arrangement. The glamor of a Federal space center would grease the way for urban renewal which, realistically, didn't otherwise stand a chance of tearing up this thriving industrial area. NASA had captured a prize piece of real estate at a bargain-basement price— roughly $4 million—thanks to the fact that urban renewal would pay most of the 'real' cost of the land and would take on the expense and dirty work of removing the occupants. And the Cambridge-Boston area was going to have all that money spent there.

    "There was only one catch: nobody had bothered to ask the opinion of the 94 businessmen and their 3,500 employees who were on the Kendall Square chopping block. Most didn't realize what had happened until they read the newspaper headlines on August 21: $60-Million NASA Plant To Be Built in Kendall Square.' " (*Reader's Digest*, 1/66, 99-104)

- European Space Research Institute (ESRIN) had begun operations in a temporary location in the Park Hotel, Frascati, Italy. A center of the European Space Research Organization (ESRO), it would work initially on plasma physics and its application to spacecraft control as well as plans for proposed ESRO astronomical observatory. Permanent facilities were planned near the Italian Atomic Energy Authority's research center. (*M&R*, 1/24/66, 7)

- Report of the first U.S. Space Industry Trade Mission, which visited Europe under sponsorship of the National Space Club, indicated that European space industry offered "a potentially attractive market," and concluded that U.S. aerospace industry could best develop its European market potential by "the sale of materials, components and subsystems;

licensing or technical support; outright investment in European firms; and joint ventures with European firms." (Text)

During January: Dr. Donald F. Hornig, Director of Office of Science and Technology, said in *International Science and Technology*: ". . . a technological society has to have first-rate research in progress—in some cases for immediate application and in other cases simply because we realize more and more that we have to depend on how much we know and understand in order to build a new society . . . So research, as research, is an important goal, and in accomplishing it we end up supporting two-thirds of all the research going on in the graduate schools of the country. . . . We're supporting some 26,000 students a year, indirectly, through research grants. And now through direct federal grants and fellowships we're supporting a little more than that many again. So that even if we had no plan whatever, we already have a mighty effect on the higher educational system." (*Int. Science and Tech.*, 1/66, 66-70)

- NSF *Reviews of Data on Science Resources* indicated, in a preliminary report on research and development in American industry for 1964, that the aircraft and missile industry was responsible for 44% of total net increase in R&D performance funds from $6.6 billion in 1956 to $13.4 billion in 1964. Report attributed 17% of increase to the electrical equipment and communication industry, noting that performance of R&D in those two fields was "strongly influenced by the requirements of the Federal Government in the areas of defense and space exploration." Only those two industries received more Federal than company financing for R&D; in 1964, 90% of aircraft and missile industry's R&D funds came from Federal Government, and 62% of those of the electrical equipment and communication industry. Aircraft and missile industry employed 30% of total number of R&D scientists and engineers; January 1965 figure of 101,200 for these personnel showed decline of 7,700 from January 1964. Electrical equipment and communication industry accounted for 22% of total—74,800 scientists and engineers as of January 1965, 600 less than previous January. (NSF, *Reviews of Data on Science Resources*, 1/66)

- Dr. George P. Woollard, president of the American Geophysical Union and director of the Hawaii Institute of Geophysics, said in AGU *Transactions* that there was a "crisis in science" requiring bold Government action. Stating that less and less of NSF's budget was going into basic research, Dr. Woollard called on Congress to assess the overall scientific needs of the country, including "ambitious" programs in various fields proposed by NSF and NAS panels. "Research proposals that were good enough five years ago to obtain NSF support now would not stand a chance," he said. "They would be turned down not because the projects lacked significance but because funds are short and only the most established and influential researchers win out." (AGU *Transactions*, 1/66)

February 1966

February 1: Two large telescopes orbiting 20,000 mi. above earth and occasionally requiring man to service them in space should be the ultimate goal for the national program in space astronomy, according to *Space Research: Directions for the Future* (Part II), second report of NAS Space Science Board's Summer Study (Woods Hole, Mass., 1965). Report, compiled by a working group of astronomers headed by Dr. George P. Woollard, Univ. of Hawaii, said 120-in.-dia. optical telescope and radiotelescope with extended aperture of 12 mi. would aid study of many basic scientific problems in astronomy. It recommended that NASA support, in addition to Orbiting Astronomical Observatories (Oao) and Orbiting Solar Observatories (Oso): more rapid growth of ground-based solar astronomy in accordance with 10-yr. projection of needs for large optical and radiotelescopes contained in a 1964 report by the Academy; specialized instrumentation for ground-based solar astronomy; doubling of sounding rocket launches for research in optical, solar, and x-ray and gamma-ray astronomy and for improvements in their guidance and flight stability; laboratory studies for more efficient detection and recording of radiation in all readings of the electromagnetic spectrum; substantial increases in balloon flights for radar and for x-ray and gamma-ray astronomy; and development of systems to reach altitudes higher than present 100,000- to 150,000-ft. limit for balloon-borne payloads. (Text; NAS Release)

- Surveyor soft-landing retrorocket system was successfully tested by Hughes Aircraft Co. and JPL at Holloman AFB. After test vehicle was released from balloon at 1,450-ft. altitude, three throttlable liquid engines operated by terminal descent guidance system's radar and autopilot control slowed vehicle to simulated landing 500 ft. above ground, partially demonstrating performance required to softland on the moon. Three engines were then shut down and parachute system deployed to lower vehicle to ground. In actual mission, Surveyor would cut off its engines 13 ft. above moon's surface and free-fall remainder of distance to avoid unsettling lunar dust with rocket engines' exhaust. Test was second successful operation of the three liquid-fueled engines under radar control. (NASA Release 66-24; *Marshall Star*, 2/9/66, 9)
- NASA launched three Nike-Cajun sounding rockets, each carrying payload of 19 acoustic grenades, from three launch sites—Point Barrow, Alaska, Churchill Research Range; and NASA Wallops Station—to gather data on atmosphere between 20- and 60-mi. altitudes over widely spaced

geographic areas. Grenades were ejected and detonated at programed altitudes, yielding information on wind directions and speeds and atmospheric densities, pressures, and temperatures. Rocket launchings were first in series of nine planned for February. Similar experiments would be conducted throughout the year to observe atmospheric conditions during different seasons. (Wallops Release 66-9)

February 2: NASA Aerobee 150 sounding rocket with 300-lb. Princeton Univ. payload was launched from WSMR to 108-mi. (175-km.) altitude to measure ultraviolet radiation from the star Alpha Virginis, popularly known as Spica. Although rocket performed satisfactorily and payload was recovered, no stellar spectra were obtained because of arcs in the guidance voltage system which may have prevented acquisition of the star. (NASA Rpt. SRL)

- USAF launched unidentified satellite by Thor-Agena D booster from Vandenberg AFB. (*M&R*, 2/7/66, 11; *U.S. Aeron. & Space Act.*, 1966, 147)
- Secretary of Defense Robert S. McNamara and U.K. Minister of Defence Denis Healey signed memorandum providing for cooperation in R&D for U.S. Initial Defense Communications Satellite Project—system of up to 22 comsats in near-synchronous, equatorial orbit. U.K. would provide, operate, and fund several ground terminals for communications tests and experiments using DOD comsats free of charge. (DOD Release 87–66)
- Prototype unfolding tool box containing 16 geological tools for astronauts to examine and obtain samples of lunar rock was delivered to MSC by Martin Co. Box would also provide storage area for rock samples. Tools included battery-powered drill; dust scoop; sample-weighing device; hand-held magnifier designed for use through spacesuit visor; surveying instruments and rangefinder; and retriever to pick up small specimens of rock which pressure-suited astronaut could not reach. (MSC Release 66-9)
- MSC awarded $70,000 contract to Rodana Research Corp. to develop emergency medical kits that would "satisfy all inflight and training requirements for the Apollo Command Module and the Lunar Excursion Module." Two training units would be delivered for each flight, in addition to one mock-up and six prototype models; they would contain loaded injectors, tablets, capsules, ointment, inhalers, adhesives, and compressed dressings. (MSC Release 66-8)
- Proposal for an "international research program for the establishment of contact with extraterrestrial civilizations" was sent to International Astronomical Union by unidentified Soviet astronomers, Tass reported. The scientists believed search for extraterrestrial life should concentrate first on radiation from stars within 1,000-light-yr. radius from the three nearest galaxies. *Kommunist*, party's major theoretical journal, said that question of the existence of other civilizations, and the establishment of contact with them was one of three "fundamental problems" confronting astrophysics. (Zorza, *Wash. Post*, 2/5/66, A7; AP, *NYT*, 2/3/66, 9)
- Parachutist Nick Piantanida, launched in a balloon from Sioux Falls, S.D., reached 123,000-ft. altitude—unofficial manned-balloon-altitude record —but failed to achieve world's free-fall record when his spacesuit oxygen hose froze and he was ordered to descend in the balloon's gondola. Piantanida, who had planned to jump from 120,000-ft. altitude, free-

fall to 7,000 ft., and then parachute to earth, said he had wanted to prove that a trained parachutist could free-fall from 100,000-ft. altitude without stabilizing devices. Experiment was second failure by Spaco, Inc., private firm developing space survival and bail-out equipment. (Balt. Sun, 2/3/66)

February 2: AEC announced plans to develop a radioisotope-powered cardiac pacemaker for treatment of heart block. Proposed instrument, fueled with plutonium-238, would be inserted in the body by surgery and would have a minimum operating life of 10 yrs. Currently available battery-powered cardiac pacemakers had operating lifetimes of three to five years. (AEC Release J-26)

- European Launcher Development Organization (ELDO) Secretary-General Renzo di Carrobio, addressing members of "Britain in Europe" committee in London, contrasted "present hesitancy" on space programs in Europe with positive U.S. position. Americans, Carrobio said, were aware of economic benefits from technological applications of space research and were "making great efforts to ensure that the spin-off from their heavy investment in space research is put to good use in the industrial sector." He noted that U.S. $5-billion space budget meant $26 per capita annual expenditure, while Europe spent $200 million annually or $1.15 per person. (*Science*, 3/18/66, 1372-4)

February 3: ESSA I (Tiros OT-3), first meteorological satellite in Tiros Operational Satellite (TOS) series, was launched from ETR for Environmental Science Services Administration (ESSA) by NASA with three-stage Thor-Delta booster into nearly polar, sun-synchronous orbit to provide cloud coverage of entire sunlit portion of the earth at least once daily. Before achieving orbit, spacecraft performed three precise "dogleg" maneuvers initiated by airborne autopilot system, injecting it into orbit: apogee, 523 mi. (842 km.); perigee, 433 mi. (697 km.); period, 100 min.; inclination, 97.9°. During first orbit, ESSA I was spin-stabilized at 10 rpm with spin axis in plane of orbit. Entering second orbit, spacecraft responded to ground control command to begin turning on its side into cartwheel position. Maneuver was expected to be completed 24 hrs. after launch; television cameras would be turned on 36 hrs. after launch. Cartwheel shaped, ESSA I contained two ½-in. vidicon cameras mounted 180° apart on each side so they could view earth every six seconds. Camera system could send pictures directly to command and data acquisition stations at NASA Wallops Station and Gilmore Creek, Alaska, or store photos on its tape recorder for readout when satellite passed within 1,500-mi. radius of a ground station.

ESSA I (Environmental Survey Satellite No. 1) was 11th successful Tiros (Television Infrared Observation Satellite) to be launched by NASA in 11 attempts. It was first of two TOS operational meteorological satellites planned and financed by ESSA to provide daily cloud-cover pictures to ground stations on a global basis. ESSA (at that time U.S. Weather Bureau) had also funded TIROS X (OT-1), launched by NASA July 2, 1965. (NASA Proj. Off.; ESSA Release 66-7; UPI, *NYT*, 2/4/66, 37)

- First soft-landing on moon was made by U.S.S.R. when LUNA IX spacecraft softlanded an instrument package near Ocean of Storms west of the Reiner and Marius craters. Instrument package began relaying telemetric signals to earth immediately and Tass said "information received from the station is being processed and studied."

Landing sequence, as later described by Tass, began 8,300 km. (5,155 mi.) above moon approximately one hour before touchdown when LUNA IX was oriented vertically to the moon. At 77-km. (47-mi.) altitude, 48 sec. before touchdown, radar system aimed at lunar surface turned on the retrorocket and shock-absorbing system was prepared. Less than one second before impact, after spacecraft had been slowed to less than 10 mph, instrument capsule was ejected. Rocket crashed and 4 min. 10 sec. after instrument capsule had landed, its petal-like covering and antennas opened and radio transmissions began.

Jodrell Bank Experimental Station announced LUNA IX had transmitted facsimile pictures of the moon from the time it landed at 1:45 p.m. EST until 2:05 p.m. EST, gone off the air, and then resumed signals shortly after 9:00 p.m. EST. Soviet spokesman declined to confirm report. British scientists were unable to translate signals into pictures but recorded them on magnetic tapes.

U.S.S.R.'s four previous attempts to softland spacecraft on the moon had failed: LUNA V, launched May 9, 1965; LUNA VI, launched June 8, 1965; LUNA VII, launched Oct. 4, 1965; and LUNA VIII, launched Dec. 3, 1965. (Tass, *Pravda*, 2/4/66, 1, USS-T Trans.; Simons, *Wash. Post*, 2/4/66, A1, A13; Grose, *NYT*, 2/4/66, 1, 37; AP, Wash. *Eve. Star*. 2/4/66, A1, A6; 2/6/66, A14)

February 3: President Johnson sent a telegram to Nikolay Podgorny, Chairman, Presidium of the Supreme Soviet, U.S.S.R., on the success of LUNA IX: "You and the people of the Union of Soviet Socialist Republics are to be congratulated for the great success of Luna 9. Your accomplishment is one that can benefit all mankind. And all mankind applauds it. Your scientists have made a major contribution to man's knowledge of the moon and of space." (*Pres. Doc.*, 2/7/66, 166)

- Breadboard version of Nerva (Nuclear Engine for Rocket Vehicle Application) was successfully tested for the first time by NASA and AEC at Nuclear Rocket Development Station, Jackass Flats, Nev. Four hours after initial ignition, engine restarted itself under test conditions simulating lower temperatures of space, thereby demonstrating its capability to start using only self-contained energy. Engine system was tested below peak power and temperature levels: exhaust temperature was 2,000°F compared with 3,500°F design temperature; reactor power was 440 mw compared with 1,100-mw design power. Test was first of a series scheduled by NASA-AEC Space Nuclear Propulsion Office as part of the Rover program to develop nuclear propulsion for space exploration. (AEC Release J-35; *Atomic Energy Programs, 1966*)

- First static test firing of production version of Apollo spacecraft launch-escape solid-propellant rocket motor, key component of emergency Launch Escape System (Les), was conducted at Redlands, Calif., by Lockheed Aircraft Corp. All test objectives were met, with motor producing scheduled 155,000 lbs. thrust for programed eight seconds to become first major Apollo spacecraft subsystem to complete NASA qualification tests for future manned flights. (Lockheed Release)

- Two NASA Nike-Cajun sounding rockets launched 7 hrs. 23 min. apart from NASA Wallops Station reached estimated 81-mi. (131-km.) and 95-mi. (152-km.) altitudes. Experiment, conducted for Univ. of Michigan, measured ambient air density from 19-75 mi. (30-120 km.) by radar track of two falling Mylar spheres ejected from each rocket at +72 sec. Spheres ejected as programed in first launch, but radar was unable to track first sphere until +14 min. Preliminary indications were that it did not inflate properly. Radar tracked second sphere for 18 min. During second launch, both spheres ejected and inflated properly, and radar tracking was continuous for 20 min. Diurnal variations of the measurements were compared to measure solar heating effects. (NASA Rpt. SRL)

February 3: Statement reaffirming U.S. demand that liability for individual air passenger fatalities be raised from $16,600 to $100,000 was delivered by chief U.S. delegate A. F. Lowenfeld at a special meeting of International Civil Aviation Organization in Montreal. Lowenfeld said U.S. might withdraw from Warsaw Convention if higher liability limit were rejected. (*NYT*, 2/4/66, 53)

- "Informed British and French sources" predicted costs of developing supersonic airliner Concorde would reach $1,112,000,000 compared with August 1964 estimate of $924,000,000. Development costs were being divided equally between the two countries and sources said that "despite rising costs . . . the two Governments are going ahead." (*WSJ*, 2/4/66, 19; Farnsworth, *NYT*, 2/4/66, 53)

- According to count by North American Air Defense Command's (NORAD) Space Defense Center, NASA's launch of ESSA I (Tiros OT-3) meteorological satellite brought total of objects currently in orbit to 1,000. Of 209 earth-orbiting payloads, U.S. had 162; U.S.S.R., 41; France, two; U.K., two; and Canada, two. NORAD counted nine U.S. deep-space probes and debris of eight others; U.S.S.R. had ten, with debris of two more. Remaining objects were "space debris" or "junk." Space Defense Center had catalogued 1,982 objects, but nearly half had decayed or had been intentionally deorbited. (NORAD Release 66-2-2)

February 4: Vice President Hubert Humphrey attended ceremony at NASA Hq. celebrating successful launch of ESSA I, ESSA's first operational meteorological satellite. During ceremony, first pictures taken by ESSA I, launched by NASA from ETR Feb. 3, were released; NASA Administrator James E. Webb administered the oath of office to NASA Deputy Administrator Dr. Robert C. Seamans, Jr.; and the Vice President presented NASA Exceptional Service medals to Dr. Morris Tepper, NASA Director of Meteorological Programs; Herbert I. Butler, NASA Chief of Operational Satellites, GSFC; and David S. Johnson, ESSA Environmental Satellite Center Director. Vice President Humphrey praised the spirit of cooperation between NASA and Dept. of Commerce, noting that "science leaves no room for jurisdictional disputes." He said of NASA's Administrator and Deputy Administrator: "I know of no two men who are more dedicated to our space program than Jim Webb and Bob Seamans. Their devotion is almost sacrificial." (NASA Release 66-29)

- NASA Administrator James E. Webb and Secretary of Defense Robert S. McNamara signed a Memorandum of Understanding establishing Manned Space Flight Policy Committee, joint NASA-DOD committee which would meet regularly to determine top-level policy, ensure coordinated planning, and resolve matters concerning mutual participation in and support of manned spaceflight programs which could not be resolved at lower levels. Co-chairmen of the committee would be Dr. Robert C. Seamans, Jr., NASA Deputy Administrator, and Dr. John S. Foster, Jr., DOD Director of Defense Research and Engineering. Memorandum superseded Jan. 21, 1963, agreement between the two agencies on Gemini program. (NASA Release 66-26)

- Reorganization of NASA Office of Space Science and Applications (OSSA) "to involve its operating offices more deeply in planning the use of manned spaceflight capabilities for scientific exploration of space" was announced by Associate Administrator for Space Science and Applications Dr. Homer E. Newell. Several individual programs were reassigned: Pioneer interplanetary probe program from Lunar and Planetary Programs to Physics and Astronomy Programs; geodetic satellite program, including Geos and Pageos, from Physics and Astronomy Programs to Space Applications; management of programs leading to acquisition of scientific data through remote sensing from Manned Space Science Programs to Space Applications; Manned Space Science Programs office was renamed Manned Flight Experiments Office/OSSA, to supervise all OSSA manned flight support. Dr. John R. Clark, who had been Director of OSSA (Sciences), was named Deputy Associate Administrator for SSA (Sciences). Robert F. Garbarini, who had been Director of OSSA (Engineering), was named Deputy Associate Administrator for SSA (Engineering). All OSSA advanced study efforts under Advanced Missions Staff would report to Dr. Clark. Acting director of that staff would be Dr. John E. Naugle. (NASA Release 66-27)

February 4: Soviet officials acknowledged that LUNA IX photos were being received in U.S.S.R. and said first LUNA IX photo would appear in Feb. 5 edition of *Izvestia*. (*Wash. Post*, 2/5/66, A6)

- Two photographs of the moon produced at Jodrell Bank Experimental Station by feeding radio transmissions from LUNA IX into newspaper telephoto machine were shown on British television. Each picture took eight minutes to receive and was recorded at 100 lines to one inch for finished 10- by 11-in. photo. Sir Bernard Lovell, station director, said in a television interview that photos "seem to destroy the theory that the moon's surface is covered with dust several feet thick. There may be a few inches of dust, but the pictures tend to confirm the view of the moon's surface as hard, spongelike pumice stone substance.

    "It would be perfectly satisfactory for landing not only heavy vehicles but also men."

    Lovell pointed out rock-like pinnacles estimated to be 10- to 20-ft. tall; small stones, some less than one inch in diameter; the curved horizon of the moon; and sharp shadows which indicated the sun was casting the shadow. He emphasized, however, necessity of knowing the proper scale before photos could be meaningfully interpreted. Lovell said LUNA IX had landed in a location which would receive sunlight for 14 days, thereby permitting solar power to replenish its batteries. (*Wash. Post*, 2/5/66, A1, A6; Sullivan, *NYT*, 2/5/66, 1, 10)

- Dr. Gerard P. Kuiper, director of Univ. of Arizona Lunar and Planetary Laboratory, told the *New York Times* in a telephone interview that he thought from photos released by Jodrell Bank Experimental Station LUNA IX was resting in a shallow crater and the 10- to 20-ft.-high rocks described by Sir Bernard Lovell were the result of distorted foreshortening of the most distant rocks and features. (Sullivan, *NYT*, 2/5/66, 1, 10)

- U.S. space officials did not feel that LUNA IX's successful soft-landing on the moon indicated U.S. was behind U.S.S.R. in the "moon race," reported John Wilford in the *New York Times*. Vice President Humphrey, NASC Chairman, told Wilford in an interview that "so many systems and programs are involved that it is very difficult to say who's ahead and who's behind." Dr. Robert C. Seamans, Jr., NASA Deputy Administrator, admired the Soviets' "solid, aggressive and expanding" space program and conceded that although U.S. program was broader-based "the Russians are awfully good at picking goals and going after them." Wilford said U.S. space officials were unwilling to speculate on who would make first manned lunar landing, but foresaw no significant changes in U.S. program because of LUNA IX's success. Dr. Seamans said Surveyor project would not be curtailed and could not be speeded up: "We've got to gather our own data because we can't count on the Russians to release their own data. Besides, we want to land in other places on the moon. Hopefully, we can exchange data with the Russians." (Wilford, *NYT*, 2/5/66, 11)

February 4: International scientific community commented on LUNA IX's soft-landing on the moon. Dr. William Pickering, Director of JPL, offered congratulations and said: "The difficulty of the task is attested to by the number of previous attempts to accomplish a soft landing on the moon. The success of this mission has already added to our knowledge of the lunar surface. We now await with interest the scientific data which will be received in the next few days." Heinz Kaminski, chief of Bochum Observatory, West Germany: "American space flight technology is on a high level of perfection which the Russians just can't match." Dr. L. R. Shepherd, president of British Interplanetary Society: "[Soft-landing on the moon is] much more difficult than the orbit rendezvous achieved by the Americans." Dr. Shotaro Miyamoto, chief of Kyoto Univ. astronomical observatory, Japan: "Soviet landing means men and materials will now be sent to the moon and its development accelerated." (*NYT*, 2/5/66, 11; 2/6/66, E11)

- International press commented on LUNA IX's soft-landing on the moon. *New York Times*: "Two facts indicate the magnitude of the advance that has now been made. One is the known failure of four earlier Soviet rockets to make the soft landing finally achieved by Luna 9. The second is the unhappy history of efforts in this country to build the Surveyor vehicle, which is intended to perform moon reconnaissance of the type in which the Soviet ship is pioneering. The Surveyor project is currently far behind schedule. The first flight is not planned until later this year, and it is impossible even to forecast when a Surveyor will duplicate the soft landing of Luna 9.

  "The whole process demonstrates again the wastefulness of the Soviet-American moon race. Were the United States and Russia cooperating fully, rather than competing, the hundreds of millions of dollars spent on Surveyor would have been applied to progress in fields still inadequately explored by either country."

  Yomiuri Shimbun, Tokyo: "It is possible to say that the Soviet Union is considerably ahead of the U.S. in the moon development program. It is regrettable that the two nations have failed to exchange information and technology for the development of outer space."

  La Nazione, Florence: "If the Russians already know how to land on the moon, they still don't know how to effect a 'rendezvous' in orbit. The inverse holds, naturally, for the Americans. This means that both are still faced by formidable problems and that the great contest is still open."

  Times of India, Bombay: "At this rate it is not unreasonable to presume that a Russian-manned probe will land on the moon and return in about a year from now, beating the Americans to it by about three years." (*NYT*, 2/6/66, E11)

February 4: Speech by the late H. J. Bhabha, director of Tata Institute, Bombay, and chairman of the Indian Atomic Energy Commission, at January 1965 meeting of the International Council of Scientific Unions in Bombay, appeared in *Science.* Bhabha explained the Indian Radio-astronomy Group and described new 32-parabolic-dish facility near Bombay for solar radioastronomy work and project to build large cylindrical radiotelescope at Ooty. Bhabha was killed in an airplane crash at the top of Mt. Blanc, Jan. 21. (*Science*, 2/4/66, 544-5)

February 5: U.S.S.R. released LUNA IX's photographs of the moon for first time and criticized distortion of photos publicized by Jodrell Bank Experimental Station. Photos, shown on an eight-minute television program, were analyzed by scientists Vladimir Yurovsky, Yuri Mbelouson, and Ivan Koval. Moon's surface appeared very uneven with many small depressions and mounds. Surface near the station had not sunk into the soil to any substantial degree. No visible traces of dust were evident. At bottom of first picture, part of spacecraft, including antenna and two-faced mirror which reflected the lunar surface, could be seen. A small stone believed about 6½ ft. from the craft could have been a meteorite or could have fallen from a hill, in which case "we could presume there are some processes of movement on the moon's surface," scientists concluded.

Tass announced details of LUNA IX's structure and flight: 3,482.6-lb. spacecraft was an airtight package carrying radio system, program-timer, thermal-regulation system, power-supply sources, and television system ensuring 360° view of the landscape. Antennas mounted on outside of craft were set to open automatically after it had settled on lunar surface. Complete with shock-absorbing system and metal screens shaped like petals to protect it during touchdown, 220-lb. instrumented payload detached itself from engine immediately before landing, and two units landed at some distance from each other.

Anatoli A. Blagonravov, chairman of the Soviet Commission for Exploration and Use of Outer Space, was critical of Jodrell Bank's publication of LUNA IX photos and said that Jodrell Bank lacked the correct vertical and horizontal scales: "The horizontal scale of the British photographs had been shrunk by approximately 2.5 times."

Photos released by Moscow, vivid in detail and contrast, showed a landscape less pinnacled and strange than that in British versions. (Sullivan, *NYT*, 2/6/66, 76; Grose, *NYT*, 2/6/66, 1, 76; AP, Wash. *Eve. Star*, 2/6/66, A14)

- U.S. monitors had received photographs transmitted from the moon by LUNA IX, Dr. Edward C. Welsh, NASC Executive Secretary, told the press: "We have obtained some information regarding those Luna 9 pictures from our own sources. Such data are currently under study and evaluation. I believe it is important for the public to know that we do not rely solely on Soviet releases or Jodrell Bank pronouncements for our information." (NASC)
- ComSatCorp requested FCC permission to launch satellite which would relay network television programs from their point of broadcast to affiliated stations for transmission to home sets. Request also asked that similar proposal submitted to FCC by American Broadcasting Co. (ABC) May 13, 1965, be dismissed. (Denniston, Wash. *Eve. Star*, 2/5/66, A5)
- Mystery of how the universe began could be solved within 15 to 20 yrs. if U.S. would use the $100 million spent on one manned Gemini mission

for five new 200-in. telescopes, MIT professor Dr. Philip Morrison suggested at New York meeting on cosmology. Morrison said three or four telescopes equal to the one at Mt. Palomar could be built for the $60- to $80-million cost of one Orbiting Astronomical Observatory (Oao) and would provide more information on the origin and nature of the universe than the satellite. Meeting was sponsored by American Institute of Physics and National Assn. of Science Writers, Inc., with NSF support. (*NYT*, 2/5/66)

February 6: "The lunar research program by means of Luna 9 has been fulfilled and successfully concluded," Tass reported after LUNA IX had completed 48 hrs. of active life transmitting a total of 10 panoramic photographs of lunar surface. Several hours after the announcement, Jodrell Bank Experimental Station reported it had received 2½ more photos of poor quality.

Third picture transmitted by LUNA IX's instrument package Feb. 4 was shown on Moscow television. During transmission, camera was inclined toward horizon and sun was to the east. Part of spacecraft was visible in the lower left-hand corner of the picture and another part in the lower right, but out of focus because of short distance from camera lens. Soviet astronomer Nikolay Barabashov told Tass photographs confirmed his theory that moon's surface was "highly pitted, porous and probably covered with numerous rocks and fragments." LUNA IX "has proved beyond doubt that the upper layer of lunar soil is a spongelike rough-textured mass scattered with individual sharp-edged fragments of various sizes." (*NYT*, 2/7/66, 1, 15; AP, Wash. *Eve. Star*, 2/7/66, A1, A5)

February 7: In summary of NASA-USAF X-15 flight operations, FRC revealed that the three rocket-powered aircraft had completed 253 flight operations since first glide flight June 8, 1959: 156 were actual flights, eight were planned captive flights, and 89 were canceled because of weather or mechanical problems. The three aircraft had flown over 24 hrs. of free flight; 6 hrs. 29 min. faster than mach 3; 4 hrs. 13 min. above mach 4; 56 min. above mach 5; and 11 sec. above mach 6. On nine separate occasions, five different pilots had flown over the 50-mi. astronaut qualifying altitude: L/Col. Robert M. White, L/Col. Robert A. Rushworth, and Capt. Joseph H. Engle, all USAF; and Joseph A. Walker and John B. McKay, both NASA civilians. Four flights terminated in emergency landings, two of which seriously damaged the aircraft, but all three aircraft were still being flown. One flight had been made with modified X-15 No. 2 using external propellant tanks. Six current X-15 pilots included: L/Col. Robert A. Rushworth, Maj. William J. Knight, and Capt. Joseph H. Engle, all USAF; and William H. Dana, John B. McKay, and Milton O. Thompson, all NASA civilians. (FRC Release 2-66)

- Jodrell Bank Experimental Station received signals transmitted from Soviet space tracking station in the Crimea via Venus. Experiment, similar to one conducted Jan. 8, was part of a program between Jodrell Bank and Soviet station to send signals via Venus for an entire 225-day Venusian year. (Reuters, *Wash. Post*, 2/8/66, A8)

- NASA Lunar Orbiter, scheduled to orbit moon in spring 1966, might photograph LUNA IX to gain perspective dimensions of lunar surface and determine spacecraft's size and extent of damage caused by landing, the *New York Times* speculated. LUNA IX's landing area, the Ocean of Storms, was not included in original flight plan of 10 regions to be photographed. (*NYT*, 2/7/66, 4)

February 7: Research pilots of the X-15, in front of X-15 No. 2 at Edwards, Calif.: Capt. Joseph H. Engle (USAF), Maj. Robert A. Rushworth (USAF), John B. McKay (NASA), Capt. William J. Knight (USAF), Milton O. Thompson (NASA), and William H. Dana (NASA).

February 7: Contact Feb. 6 with LUNA IX after its scheduled research had been completed was explained by Tass: "Since the craft still had a certain amount of power left in excess of rated level an additional 2-hr. commutation session was held. . . .

"During the transmission period the station exhausted practically all its remaining supply of electric energy and further radio communications with Luna 9 will be discontinued."

Moscow television showed photos of a circular panorama of the lunar landscape obtained from LUNA IX's television system during communications period Feb. 6. Panorama showed LUNA IX to be resting on comparatively even area of the lunar surface close to the equator on the eastern extremity of the Oceanus Procellarum. Surface of moon around station was very uneven with many hollows and hillocks. Some stone-type formations were scattered at different places. Small cavities could be seen in different parts of the panorama, specifically at its right end. Slightly hilly area could be seen along the horizon. Parts of the spacecraft, including antennas of receivers, transmitters, and dihedral mirrors reflecting sections of the lunar surface, were visible in forefront of panorama. At the bottom, lying on the ground, was section of rocket which had been thrown aside during landing. (Grose, *NYT*, 2/8/66, 1, 20; AP, Wash. *Eve. Star*, 2/7/66, A1, A5)

February 7: House passed H.R. 6125 amending Air Museum Act to authorize construction of a National Air and Space Museum in Washington, D.C., to replace the National Air Museum. Museum, to be completed in five years at an estimated $40 million, would house permanent exhibits of space exploration and be part of the Smithsonian Institution. The NASA Administrator would serve on its advisory board. (NASA LAR V/20)

- Dr. Gerald W. Johnson, Lawrence Radiation Laboratories, was named director of Naval Laboratories. (DOD Release 114-66)

- AEC had developed standby capability to test nuclear weapon research devices at high altitudes and in space if the 1963 international test ban treaty were abrogated, *Aviation Week* reported. Standby capability, based on large launch vehicles developed as nuclear device carriers and small instrumentation rockets for diagnosis, would concentrate on atmospheric testing of ICBM and anti-ICBM systems and effects of nuclear detonations on these systems. (*Av. Wk.*, 2/7/66, 31)

- LUNA IX's revelations of the moon's surface had deprived Iranian poets of a source of inspiration thousands of years old, Ebrahim Sahba, speaking for the Iranian Poets Assn., told AP: "Now, with the ugly and coarse surface, Iranian poets must search for something else to describe beauty." (AP, *Wash. Post*, 2/8/66, A8)

February 8: Boeing Co. delivered first of eight Lunar Orbiter spacecraft—an 85-lb. test model—to KSC for engineering ground tests. (Boeing Co.; *Bus. Wk.*, 2/12/66)

- Organisation for Economic Co-operation and Development (OECD), meeting in Paris, was urged by an advisory group on fundamental research to establish from 15 to 20 new centers for advanced study and research which would serve scientists from any member country. Each center would concentrate on one field and would be established at the university in a member country that had shown the highest achievement in that field. In addition, a program of international fellowships could provide funds to outstanding researchers. (Sci. Serv., *NYT*, 2/8/66, 14)

February 9: NASA launched 210-lb. Reentry E payload by Scout booster from NASA Wallops Station to evaluate char integrity of low-density phenolic nylon heat-shield material and test its effectiveness in withstanding actual reentry conditions. In addition to four stages of Scout launch vehicle, 17-in. spherical rocket was attached to instrumented payload as velocity package. First two Scout stages lofted payload to 54-mi. (87-km.) altitude. Remainder of vehicle coasted upward to 110-mi. (177-km.) altitude and at peak of trajectory, during 3rd-stage burn, guidance system began positioning vehicle for desired reentry angle. Fourth stage fired during descent, and then velocity package motor ignited, driving reentry payload to a velocity of more than 18,000 mph through earth's atmosphere. Payload impacted 1,150 mi. downrange in the Atlantic; recovery was not attempted.

Throughout flight, continuous telemetry channel furnished data before and after the one-minute communications blackout caused by reentry, and aircraft photographed visible portion of reentry. Delayed telemetry system, used successfully on previous flights to transmit during blackout period, failed to operate because of a transmitter malfunction, but data furnished by continuous channel on post-blackout conditions would permit evaluation of total performance and correlation with previous ground tests. Flight, fifth in series of experiments in Scout Reentry Heating Project designed to investigate heating environment of a body reentering earth's atmosphere at 18,400 mph, was directed by LaRC. (NASA Proj. Off.; NASA Release 66-18; Wallops Release 66-8)

February 9: NASA Nike-Apache sounding rocket launched from Churchill Research Range reached 111-mi. (178-km.) altitude. GSFC experiment, first in series of three, successfully measured number and energy distribution of electrons in energy range of 1-300 kev, which produce visual aurora. (NASA Rpt. SRL)

- USAF launched unidentified satellite with Thor-Agena D booster from WTR. (*U.S. Aeron. & Space Act., 1966*, 147)
- Soviet aircraft designer Oleg K. Antonov, interviewed in Kiev by five Western European journalists, said U.S.S.R. was developing An-154, a double-decked, 724-passenger aircraft. It would be larger version— nearly 50 ft. longer—of An-22 and would use four turboprop motors. Two preproduction models of An-22 were undergoing flight tests, Antonov said, and Aeroflot expected it to be in cargo service by 1968. (*Flying Review International*, 5/66, 544)

February 10: In a White House ceremony, President Johnson awarded the National Medal of Science to 11 scientists, including the late Dr. Hugh L. Dryden, former NASA Deputy Administrator. Dr. Dryden was cited for his "contributions as an engineer, administrator, and civil servant for one-half century to aeronautics and astronautics which have immeasurably supported the Nation's preeminence in space." Mrs. Dryden accepted the medal—first ever awarded posthumously—for her husband.

President Johnson told recipients that "in a nation of millions and a world of billions, the individual is still the first and basic agent of change. Without the unfettered curiosity of individual men probing and reaching for new truth our planet would be a dry and dreary place." (*Pres. Doc.*, 2/14/66; *NYT*, 2/11/66, 5; *Wash. Eve. Star*, 2/11/66, A2; *Balt. Sun*, 2/11/66)

- Mstislav Keldysh, president of Soviet Academy of Sciences; Alexander Lebedinsky, professor; and Alexander Vinogradov, geochemist, held 2½-hr. televised press conference in Moscow on the LUNA IX mission. Keldysh said U.S.S.R. had mastered the soft-landing technique and was planning more soft-landing missions in 1966 to obtain information on physical conditions on the moon, composition of lunar rock, and variations in lunar temperature. He said the next major challenge would be designing a spacecraft capable of returning men to earth from the moon and called for U.S. "cooperation not competition" in a program for manned lunar landing. LUNA IX's mission was "only to photograph the surface of the moon and measure cosmic radiation," Keldysh said. The full panorama of the lunar surface included 6,000 lines and was transmitted to earth in 100 min. Spacecraft contained no solar batteries.

Lebedinsky revealed that LUNA IX had "shifted between the second and third transmissions of the panorama, and as a result, the inclination of the phototelevision camera changed several degrees, while the camera itself shifted its position several centimeters." Keldysh added: "It may be that the station landed on an unstable small stone, or perhaps the ground settled slightly. Such a very small deformation could occur because of temperature changes on the moon or a certain mechanical impact by the station itself." He expressed confidence that the lunar surface could support manned spacecraft.

Lebedinsky said LUNA IX had measured rate of radiation in outer space at 30 millirads a day and that the spacecraft had detected additional radiation on lunar surface—apparently produced by nuclear reactions from cosmic rays hitting moon's upper layers—but he did not disclose amount. Vinogradov reported that lunar surface was hard, porous, volcanic cracked rock, but acknowledged that other areas might be different. Keldysh announced that VENUS II and VENUS III spacecraft, launched Nov. 12 and Nov. 16, 1965, respectively, were scheduled to approach Venus March 1. (*Wash. Post*, 2/11/66; *Balt. Sun*, 2/11/66; Sullivan, *NYT*, 2/11/66, 17)

February 10: Apollo/Saturn AS-201 successfully completed countdown rehearsal at ETR in preparation for scheduled Feb. 22 suborbital flight test. (AP, *Wash. Post*, 2/10/66)

- NASA launched two sets of three Nike-Cajun sounding rockets, each carrying payload of 19 acoustic grenades, about 12 hrs. apart from three launch sites—Point Barrow, Alaska, Churchill Research Range, and NASA Wallops Station—to gather information on atmosphere between 20- and 60-mi. altitudes over widely spaced geographic areas under day and night conditions during winter. Grenades were ejected and detonated at programed altitudes, yielding information on wind directions and speeds, atmospheric densities, pressures, and temperatures. (Wallops Release 66-9)

- U.S.S.R. launched COSMOS CVII containing scientific equipment for investigation of outer space. Orbital parameters: apogee, 322 km. (200 mi.); perigee, 204 km. (126.7 mi.); period, 89.7 min.; inclination, 65°. Equipment was functioning normally. (Tass, 2/10/66)

- U.K. would build $84-million prototype nuclear-power reactor of the breeder type—designed to produce large amounts of new fissionable atoms—at Dounreay in northeast Scotland, Technology Minister Frank Cousins told the House of Commons. Sir William Penney, head of British Atomic Energy Authority, told news conference he hoped to have commercial models of the breeder power station in operation by 1980. (*NYT*, 2/10/66; Wash. *Eve. Star*, 2/11/66, A1)

February 11: U.S.S.R. launched COSMOS CVIII carrying scientific instruments for continued space research into orbit with 865-km. (537.2-mi.) apogee; 227-km. (141-mi.) perigee; 95.3-min. period; and 48.9° inclination. Equipment was functioning normally. (Tass, 2/12/66)

- NASA and Swedish Space Research Committee (SSRC) signed Memorandum of Understanding to conduct pulsed laser radar experiments to determine relationship of noctilucent clouds to height, distribution, and scattering properties of cosmic dust and aerosol particles. SSRC would provide optical equipment and design and construct light transmitter optical system; NASA would provide laser head, power supply, and other equipment. Each agency would bear the cost of the equipment it would supply and the responsibilities it would undertake. Results of experiments would be made available to scientific world. (NASA Release 66-35)

- Device to alter wavelength of a laser beam to tune it for specific tasks such as communications, surgery, analysis of materials, or industrial applications was patented by Bell Telephone Labs. physicists Drs. Joseph A. Giordmaine and David A. Kleinman. In patented method, laser beam would be passed at a calculated angle through a special crystal so that frequency of the emerging light would be a multiple of original beam's. (Jones, *NYT*, 2/12/66, 33)

February 11: Breadboard of Nerva (Nuclear Engine for Rocket Vehicle Application) was tested by NASA and AEC at Nuclear Rocket Development Station, Jackass Flats, Nev. Initial fixed-drum start tested control drum response; maximum power reached was 230 mw, compared with reactor's 1,100-mw design power. Test duration was nine minutes, and maximum exhaust temperature was 1,540°F. After a second low-pressure start was aborted, system was given normal start; in 22-min. test it reached 350-mw maximum power and 2,130°F maximum temperature. Nerva engine had been successfully started twice on Feb. 3. (*Atomic Energy Programs, 1966,* 186)

- Theoretical possibility of self-supporting tapered cable extending 22,000 mi. into space from earth and from extraterrestrial rotating or revolving bodies was examined in *Science* report by John Isaacs, Hugh Bradner, and George E. Bachus, Scripps Institute of Oceanography, and Allyn C. Vine, Woods Hole Oceanographic Institution. Cable, referred to as a "skyhook," would be extended downward from a satellite in synchronous orbit, become an extension of the satellite, and be held in place by the balance between centrifugal and gravitational forces that keeps any satellite in orbit. Authors of the report conceded in a telephone interview with Harold Schmeck of the *New York Times* that "present-day engineering is not yet up to the task of making a cable strong enough and light enough for the sky hook project from earth." (*Science,* 2/11/66, 682-3; Schmeck, *NYT*, 2/12/66, 12)

- DOD would use Titan III booster to launch 23 comsats in three launch groups to form a global communications system, M/G Ben I. Funk, Cmdr., AFSC's Space Systems Div., told National Assn. of Manufacturers in New York. Orbiting about 18,200 mi. above earth, satellites would form system "of high reliability and reasonably long life, secure against interference, and with sufficient capacity to handle a high volume of unique and vital military traffic," Funk said. He also revealed that USAF was seeking solutions to problems of reusable boosters, maneuvering reentry, lighter-weight materials, and higher performance fuels. (Text, *Av. Daily,* 2/15/66)

February 12: France failed to launch research satellite from Hammaguir Range because of malfunction in ignition system of three-stage Diamant booster. At T−0 in countdown, the Diamant booster remained on launching pad. (Reuters, *NYT*, 2/13/66, 3)

- NASA Associate Administrator for Manned Space Flight Dr. George E. Mueller, outlining the NASA manned spaceflight program at American Assn. of School Administrators meeting in Atlantic City, discussed the benefits to man on earth of the Apollo Applications program: ". . . man would have the capability of establishing world-wide, real-time television by placing large antennas and powerful transmitters into synchronous orbit; of establishing control towers in space both for aircraft and ocean-going ships; of using space observations to identify and exploit earth's natural resources; and of controlling the weather." (Text)

- Three U.S. scientists were elected to Soviet Academy of Sciences: Dr. Severo Ochao, Nobel Prize-winning biochemist, New York Univ. College of Medicine; Dr. Richard Courant, mathematician, New York Univ.; and Dr. Herman F. Mark, chemist, Polytechnic Institute of Brooklyn.

Total U.S. membership was now six; five Soviet scientists were members of U.S. National Academy of Sciences. (Schwartz, *NYT*, 2/12/66, 2)

February 12: U.S. embassy confirmed that West Germany was negotiating with several unidentified American aircraft companies for technical assistance in removgin flaws from the F-104G Starfighter jet aircraft. Since beginning of 1965, 29 Starfighters had been destroyed in accidents and 16 Luftwaffe pilots killed. (*NYT*, 2/20/66, 9)

February 13: Agena rocket stage identical to one at KSC which NASA planned to use as target vehicle in the March Gemini VIII mission suffered an explosive start in ground firing at USAF Arnold Engineering Development Center. Failure, which followed five successful test firings, was similar to explosion which canceled Gemini VI mission Oct. 25, 1965. (UPI, *Wash. Post*, 2/14/66)

February 14: Benefits of U.S. space program were emphasized by NASC Executive Secretary Dr. Edward C. Welsh at AIAA meeting in St. Paul. He cited in particular space program's contributions to education, saying, "it has added immeasurably to total available knowledge of man, of the earth, and of the solar system. . . . Funds have flowed from NASA . . . to our educational institutions in the form of scholarships and fellowships, in the form of new laboratories, and research opportunities. It is estimated that from this one source about 1,000 new Ph.D.'s are added annually to this resource of the Nation . . . and . . . it has been a catalyst and a stimulus to education throughout the whole country. . . ." (Text)

• Threat of contamination of Mars made search for that planet's indigenous life, at any stage of evolution from prebiotic to extinct, an urgent undertaking, according to Biology Working Group's section on exobiology in third and final report of NAS Space Science Board Summer Study (June-July 1965) at Woods Hole, Mass. Report, *Space Research: Directions for the Future, Part Three*, said: ". . . the exploration of Mars, with initial emphasis on the detection and characterization of possible Martian life, should constitute the major scientific goal of the United States space program in the period following the manned lunar landing." High priority in field of environmental biology was given to study effects of weightlessness, with particular attention to "synergistic or antagonistic effects from simultaneously imposed stresses of radiation and weightlessness."

Orbiting research facilities with six- to eight-man crews to study effects of prolonged space flight preceding extended manned missions were recommended by Medicine and Physiology Working Group. Report urged series of manned flights, ground-based studies, and biosatellite experiments to yield data for planetary exploration.

Group examining NASA-university relations urged increases in training budget and facilities and research grants, plus a program providing graduate students with flight opportunities. (Text)

• D. Brainerd Holmes, former NASA Deputy Associate Administrator for Manned Space Flight (1961-1963) and now with Raytheon Co., reflected upon his NASA experience at National Space Club luncheon in Washington, D.C. "NASA has done well" to date but today decisions must be shaping up for continuing space exploration beyond the lunar landing by an American by 1970. "Whither now NASA?" was the question, Holmes said, and in the democratic process of decision an informed public must play a vital part. He asserted that it was not only NASA's job to inform but the duty of everyone aware of the significance and challenge of space exploration to the future of the U.S. (EH)

February 14: NASA Apollo Lunar Excursion Module (Lem) ascent engine test completed at Arnold Engineering Development Center. (AEDC)
- Col. Jack Bollerud (USAF) was named Acting Director, Space Medicine, NASA Office of Manned Space Flight, and would assume all duties assigned to the late Dr. W. Randolph Lovelace. He would also continue to discharge the functions of Deputy Director, Space Medicine, a position he had held since May 1965. (NASA Hq. *Bull.*, 2/14/66, 2)
- Vice President Hubert Humphrey designated Col. Donald W. Paffell (USAF) his Assistant for Aeronautics and Space Matters. (USAF Hq. *Bull.*, 2/14/66, 2)
- Development of sealed cities enclosed by domes "utilizing the sealed [space] cabin concept on a more grandiose scale, such as the astrodome environmental module" to assure mankind an environment free from air and water pollution, was urged by Dr. Eugene Konecci, of NASC staff, at the National Cybernetics Foundation symposium on "Ecological Technology—Space, Earth, Sea," at the Smithsonian Institution. Konecci envisioned self-sufficient community "modules" of several thousand persons consisting of "a series of astrodomes enclosing the atmosphere," which would be cleansed and regenerated continuously. Each community would have "its own self-contained, recirculating water supply" and facilities for converting sewage back into useful products. Food would be largely synthetic, but there would be "meat factories" in which animals would be raised "in modern air-conditioned surroundings and fed the high nutrient feeds provided through biosynthesis." (UPI, *NYT*, 2/15/66, 23)
- DOD awarded General Dynamics Corp. Convair Div. $4-million contract to refurbish and update 23 retired Atlas ICBM's to launch upper-stage vehicles for Abres (Advanced Ballistic Reentry Systems), Nike target, and scientific satellite programs. (General Dynamics Release)

February 15: President Johnson announced he would send Astronauts Walter M. Schirra, Jr., and Frank Borman, command pilots of the Dec. 4-18 GEMINI VII/VI mission, on a three-week, eight-country tour of the Far East to demonstrate scientific, technological, and educational values of U.S. space program and to visit Far Eastern countries which had cooperated with U.S. in space programs. Astronauts would leave Feb. 21, and visit Japan, Korea, Taiwan, Malaysia, Thailand, the Philippines, Australia, and New Zealand.

The President had sent GEMINI IV Astronauts James A. McDivitt and Edward H. White II, to the Paris Air Show in June 1965; GEMINI V Astronauts L. Gordon Cooper and Charles Conrad, Jr., on an African tour in September 1965; and John Glenn, first American to achieve orbit, on a European tour in October 1965. (*Pres. Doc.*, 2/21/66, 221; *NYT*, 2/16/66, 54)
- NASA converted its contract with Grumman Aircraft Engineering Corp. for development of Lunar Excursion Module (Lem) from cost-plus-fixed-fee to cost-plus-incentive-fee. Under new four-year contract, which would bring total cost to $1.42 billion, Grumman would deliver 15 flight articles—four more than in previous contract—10 test articles, and two mission simulators. (NASA Release 66-37)
- USAF launched three unidentified satellites with Atlas-Agena D booster from WTR. (*U.S. Aeron. & Space Act., 1966*, 148)
- Rep. Albert Thomas (D-Tex.), Chairman of House Committee on Appropriations' Independent Offices Subcommittee, died of cancer. A member

of the House for almost 30 yrs., Thomas had been deeply interested in Congressional funding for NACA, NSF, and NASA. (Wash. *Eve. Star*, 2/15/66, B8)

February 15: In speech at Univ. of Missouri on impact of Federal research and development expenditures on American colleges and universities, Leo S. Tonkin, Executive Director of D.C. Commissioners' Council on Higher Education, cited statement by two NASA employees on problems involved in awarding research grants. Dr. Thomas P. Murphy, Deputy Assistant Administrator for Legal Affairs, and Dr. Thomas W. Adams, Socio-Political Specialist for Office of Policy Planning, said in a paper presented at 1965 annual APSA meeting: "Can a mission-oriented agency use geography as a criterion in awarding grants, recognizing full well that it concomitantly will sacrifice time, quality, and money in the process? When grants are given to one institution on grounds other than merit, there is little basis for denying similar grants to other universities which would not normally merit them." (Text; *CR*, 2/23/66)

- Two gallons of fuel exploded inside drum near Saturn IB launch pad at ETR, but there were no injuries and no damage to the booster or launch complex, AP reported. Saturn IB was scheduled to launch unmanned Apollo spacecraft on Feb. 22. (AP, *NYT*, 2/16/66, 52)

February 16: NASA Nike-Apache sounding rocket launched from Churchill Research Range reached 130-mi. (209-km.) altitude in second of three GSFC experiments to measure number and energy distribution of auroral-producing electrons in energy range of 1-300 kev. Satisfactory data were recovered despite malfunction of one detector. (NASA Rpt. SRL)

- Multi-foil high-performance insulation system to reduce supercold propellant "boil off" in space had been successfully tested for 96 hrs. at MSFC in a vacuum chamber at pressures approaching the vacuum of space. Since liquid hydrogen propellant, already used in Saturn IB and Saturn V upper stages, was being considered as a propellant on longer-duration missions, cryogenic technology was being extended at MSFC to meet the needs of future space exploration programs. (MSFC Release 66-41)

- Cosmonaut Valery Bykovsky, who orbited earth in U.S.S.R.'s VOSTOK V June 16-19, 1963, discussed problems of flight to the moon in *Aviatsiya i Kosmonautika* and suggested three possible methods for flight: (1) a launch directly from earth with a very powerful carrier rocket to provide escape speed for a huge payload including the spacecraft and facilities for return flight; (2) a launch from earth orbit with separately fueling low-power rockets orbiting the lunar spacecraft; (3) a landing from an orbit near the moon (such as that planned for U.S. Apollo project) which would "insure more dependable contact between landing capsule and the main ship." Bykovsky said work "is in full swing" to design maneuvering instruments, but U.S.S.R. would send a manned spacecraft to the moon only when the cosmonauts' return to earth was guaranteed. (Tass, 2/16/66)

- Joseph C. Satterthwaite, former State Dept. foreign service officer, was sworn in as a consultant to NASA Administrator James E. Webb for international affairs "in the general area of Western Europe." (NASA Release 66-39)

February 17: French-built instrumented scientific satellite, DIAPASON I (D-IA), was successfully launched from Hammaguir Range with three-stage Diamant booster. Initial orbital data: apogee, 2,753.5 km. (1,710 mi.); perigee, 503.1 km. (312 mi.); period, 118.6 min.; inclination,

34.04°. Designed with a three-month lifetime, the 20-in.-dia., 49-lb. satellite would transmit information on earth's magnetic field and test France's tracking stations. Bretigny Tracking Center near Paris was receiving satellite's radio transmissions normally.

This was second launch of a French-made satellite by France's Diamant booster. First was Nov. 26, 1965, when A-I satellite carrying radio and radar transmitter, but no scientific equipment, was orbited. (French Embassy Release; AP, Wash. *Eve. Star,* 2/17/66, A1; *Wash. Post,* 2/18/66, A13)

February 17: NASA launched 49-lb. payload with ionosphere experiment for Rice Univ. to 116-mi. (186-km.) altitude from NASA Wallops Station with Nike-Apache sounding rocket to test theory that intense electrical currents flow at high altitudes over the earth and are strongest at about 120-mi. altitude. Data from magnetometer designed to detect magnetic field caused by the currents were telemetered from the payload to ground stations at Wallops during flight. (Wallops Release 66-11; NASA Rpt. SRL)

- NASA Deputy Associate Administrator for Space Science and Applications Edgar M. Cortright, testifying before the House Committee on Science and Astronautics' Subcommittee on Space Science and Applications, stressed importance of selectivity in planning space science program: "We have been looking at Apollo applications for some time to identify those areas of scientific activity where the man can be a real asset to the experiment, and the areas that interest us most are astronomy; natural resources, which is looking down at the Earth with various types of detectors; biology, which is concerned with long-duration weightless flight, from both a fundamental biological point of view and in preparation for longer flights; and, of course, continued lunar exploration.

    ". . . man is not an asset in space science experiments in all cases. You have to be extremely selective to find those cases where he will be an asset, and they do exist. We think that astronomy in the seventies is one such area; also biology, Earth resources, and continued lunar exploration. But in the particles and fields work it is unlikely that man will be an asset, and it is unlikely that he will play the major role in applications where what we really want is long lifetime." (*NASA Auth. Hearings,* 57-8)

- NASA Associate Administrator for Manned Space Flight Dr. George E. Mueller told the Business Council in Washington, D.C., that "so far as overall manned space flight manpower is concerned, we have passed our peak" in engineering, we're "reaching our peak in manufacturing," and we're "at our peak in total manpower costs." He added: "One of the anomalies of the space program is that we must begin to 'go out of business' before we fly our first operational vehicle. . . . By the time manned flights begin, the employment level will be quite low in comparison with the peak, which occurred in 1965. . . ."

    Commenting on post-Apollo plans, he said: "The budget for the coming fiscal year permits NASA to hold open the option for a program to procure additional flight vehicles beyond those now programmed, so as to employ the Apollo hardware and capabilities at least through 1971. If we do not exercise this option, in the decision for the [FY] 1968 budget we will have to begin a phase down of the manned space flight activities and the 'mothballing' of some of our facilities." (Text)

- AEC Chairman Dr. Glenn T. Seaborg predicted at Univ. of Texas lecture that if man decided to colonize the moon or planets, he would need nuclear energy to make the environments habitable for human beings and large nuclear reactors to provide "planet conditioning"—an outer space version of air conditioning. In addition, reactors could furnish power "to extract water from the rock, to help produce synthetic food and to extract minerals and other materials from the surface." Dr. Seaborg said he foresaw in the more immediate future the use of nuclear power plants on manned orbiting laboratories: ". . . it is desirable to operate these stations at the relatively low altitudes of between 100 and 200 nautical miles. For these low orbits the smaller exposed area of nuclear power systems results in greater reductions in the atmospheric drag—and in the pounds of propellant required to keep the station in its assigned orbit. An unmanned scientific or defense system which has to operate close to the Earth will similarly have the advantage of low atmospheric drag if nuclear power systems are used." (Text; AEC Release S-6-66)

February 17: Program to seek basic principles of sonic booms—including flights over U.S. communities—was being prepared by a top-level, four-man Government committee, Evert Clark reported in the *New York Times.* Program hoped to obtain at least preliminary answers before Sept. 1 so that limitations on boom making could be written into Sst development contracts. Committee members: Dr. Donald F. Hornig, Director of Office of Science and Technology (OST) and Presidential science adviser; Dr. Nicholas Golovin, OST; Secretary of Defense Robert S. McNamara; and Dr. John R. Dunning, dean of Columbia Univ.'s School of Engineering and Applied Science. (Clark, *NYT,* 2/18/66, 8)

- Use of personal radar units by astronauts to counteract optical illusions which they might encounter while assembling a space station or exploring the moon was recommended by Cambridge Univ. Prof. Richard L. Gregory, author of *Eye and Brain: The Psychology of Seeing,* in a New York press interview. Gregory believed astronauts might experience spatial distortion when seeing spacecraft's structural elements gleaming luminously in sunlight or sharply etched against a jet black background. In addition, he thought they might experience reversed depth perception which would cause distorted images. Gregory recommended aiding astronauts with tones, heard through earphones, of a variable pitch which would be determined by a beam from radar unit each man would wear. Variations in pitch would indicate distance to different objects and astronaut's rate of approach or retreat from them. Gregory was conducting his research at New York Univ. on a senior foreign scientist fellowship financed in part by NASA. (Osmundsen, *NYT,* 2/17/66, 31; *Houston Chron.,* 2/16/66)

- Dr. Wilmot N. Hess, Chief of Laboratory for Theoretical Studies at GSFC, received one of the Arthur S. Flemming Awards presented annually to 10 outstanding young men in Government service. (NASA Notice)

- Certificate promoting the late Dr. William Randolph Lovelace II, to rank of major general in U.S. Air Force Reserve was presented to his eldest daughter, Mrs. John Sellman, at USAF Satellite Test Annex, Sunnyvale, Calif., by M/G Ben I. Funk, AFSC's Space Systems Div. Cmdr. Dr. Lovelace, former NASA Director of Space Medicine, died with his wife and their pilot in an aircraft accident near Aspen, Colo., December 1965. (AFSC Release 43.66)

- Question whether "superrockets" for the Apollo manned space program should have liquid or solid stages was no longer debatable, said William

Hines in the Washington *Evening Star*; the solids had been "forced through studied neglect to default." He continued: "This is not to suggest, however, that the whole liquid-solid debate is moot.

"Between Saturn I-B's 35,000-pound orbital capability and Saturn V's 250,000 pounds is a huge gap; between Atlas-Centaur's 2,300-pound escape capability and Saturn V's 95,000 pounds is a gap that is even greater, proportionately.

"A wide variety of scientific and manned payloads can be envisioned as fitting within these gaps. A flexible booster system based on large solid rockets of varying lengths and flight characteristics could fill the gaps quickly and at reasonable cost; liquid systems could not." (Hines, Wash. *Eve. Star*, 2/17/66, A6)

February 17: T. R. May, vice president of Lockheed-Georgia Co., disclosed plans to build commercial version of the C-5A military fan jet which could carry up to 900 passengers and thus reduce fares drastically. May, speaking at an Air Force Assn. seminar in Salt Lake City, said Lockheed "fully expects" to have the aircraft, designated the L-500, "flying commercial routes" in the early 1970's.

Lockheed-Georgia Co. had received a $1.4-billion DOD contract Sept. 30, 1965, to develop and build the C-5A—a military carrier that would be world's largest aircraft. Delivery was scheduled for mid-1968. (DJNS, Wash. *Eve. Star*, 2/18/66, D1; *NYT*, 2/18/66, 53M)

- FAA proposed a series of changes in cockpit procedures and crew training on the Boeing 727 jet aircraft that had had four fatal crashes since August 1965—all as the aircraft were preparing to land—"to assure ourselves that an aircraft coming in visually at night will not suffer the consequences of the ... others." Changes, designed to ensure crew awareness of their descent rate, were discussed at an all-day meeting of experts from FAA and 17 airlines operating the Boeing 727.

  FAA Deputy Director of Flight Standards Cliff W. Walker said: "We do not know what caused these accidents. We have searched, searched, searched . . . and we have not found any cause to be suspicious of the 727's airworthiness." (Hudson, *NYT*, 2/18/66; AP, Wash. *Eve. Star*, 2/18/66, A2; Simons, *Wash. Post*, 2/18/66, A1)

- Soviet Tu-114—world's largest aircraft—crashed on takeoff from Moscow's Sheremetyevo Airport for Brazzaville, Congo Republic, when its wing hit a snowbank. Unofficial Soviet sources said 48 of 70 persons aboard were killed. Tass reported the "catastrophe" was "caused by sharp worsening of weather conditions." It was the first known crash of a Tu-114, which could carry 220 passengers, had longest range of any commercial aircraft, and had been in regular service about eight years. (Tass, 2/18/66; AP, Wash. *Eve. Star*, 2/18/66, A5; AP, *NYT*, 2/18/66, 1)

- U.S.S.R. charged at 17-nation disarmament conference in Geneva that the crash of an American nuclear bomber in Spain in January was "flagrant violation of international law and of the 1963 Moscow Treaty" which banned all but underground tests. U.S.S.R. claimed flight violated the treaty because "it is common knowledge that the most important purpose of this treaty was to prevent radioactive contamination of the atmosphere, outer space, and water. . . . The southern coast of Spain and adjacent sea areas have been radioactively contaminated by American nuclear weapons." U.S. chief delegate William C. Foster rejected the allegations as "false" and countercharged that U.S.S.R. was guilty of a "propaganda" maneuver. (*Wash. Post*, 2/18/66, A16)

February 18: President Johnson submitted NSF's report on weather modification research to Congress and said "highly encouraging steps are being taken toward establishing safe and effective programs for modifying the weather." The President, a supporter of weather-control research, said: "We can now begin to see the day when such programs may become operationally feasible." (*Pres. Doc.*, 2/21/66, 233; *NYT*, 2/19/66, 29)

- NASA Deputy Administrator Robert C. Seamans, Jr., testifying on NASA FY 1967 budget authorization before the House Committee on Science and Astronautics' Subcommittee on Manned Space Flight, discussed use of Apollo/Saturn systems in proposed Apollo Applications (Aa) program: "In the last half of FY 1966 and during FY 1967 we will define those new and useful missions that will be able to take advantage of exciting opportunities provided by the capabilities of the developed and available flight hardware. Together with Project Voyager, a major new unmanned system for the exploration of the planets, the Apollo Applications effort represents that next family of major flight missions which we will be recommending for approval and authorization in the coming years."

  He cited three basic elements of Aa effort foreseen by NASA: "We believe we can improve the basic Apollo space vehicle capabilities with minor modifications to extend the manned time in orbit from 2 weeks to 45 days and longer; we can procure additional spacecraft and launch vehicles for new, or follow-on, missions beyond the time frame of the current Apollo schedule; if the program can be carried out along the lines of our most optimistic scheduling, we will find that, within the approved and programmed Apollo schedule, there are up to nine vehicles that can be used for alternate missions in the period 1968 to 1970." (*NASA Auth. Hearings*, 5-6)

- 25 scientific experiments for sixth Orbiting Geophysical Observatory (OGO-F), scheduled for launch into polar orbit in 1968, were selected by Dr. Homer E. Newell, NASA Associate Administrator for Space Science and Applications, on recommendation of OSSA's Space Science Steering Committee. Fifty-five scientists at U.S. and foreign universities, industrial firms, and Government laboratories would make correlative experiments of near-earth space phenomena based on data from the OGO-F experiments. (NASA Release 66-36)

February 19: U.S.S.R. launched COSMOS CIX with "scientific instruments to continue the study of outer space." Orbital parameters: apogee, 309 km. (192 mi.); perigee, 209 km. (130 mi.); period, 89.5 min.; inclination, 65°. Instrumentation was functioning normally. (Tass, 2/19/66)

- Television system on U.S.S.R.'s LUNA IX, which transmitted first pictures from moon Feb. 3-6, weighed less than 1.5 kgs. (3.3 lbs.), disclosed the Soviet designers of this equipment in an *Izvestia* interview. Since the TV camera had to "distinguish minute details of the surface and transmit to earth simultaneously the landscape around the spot where it landed," a miniature Vega photocamera with lens that could "photograph at distances from two meters to infinity without focusing" had been used. First "radio communication session with the moon" began within 4 min. 10 sec. after the soft-landing and first television reportage began within seven hours. Complete panoramic scanning took 100 min. Future probes of moon's surface would use camera coupled with microscope for minute study of the lurain. (*Izvestia*, 2/19/66, 1)

February 20: MSFC engineers Michael J. Vaccaro and Haydon Y. Grubbs, Jr., entered lunar vehicle simulator to begin 14- to 21-day final laboratory validation tests on interior chamber design and investigation of human factors of an enclosed cabin environment. Cabin, 10-ft. cylinder with 7-ft. diameter and 5-ft. ceiling, had stations for performing various tasks—eating, sleeping, taking sample measurements, and completing geophysical assignments. Study, conducted by Honeywell, Inc., under NASA Office of Advanced Research and Technology (OART) sponsorship, would determine: (1) functional relationships between physiological responses and behavior and the time required to perform assigned tasks; (2) how to design lunar roving vehicles to provide crew mobility and utilize cabin space; and (3) how to increase proficiency in performing required tasks during extended lunar exploration. Engineers would be isolated as if on moon, wear long-johns or spacesuits, and eat dehydrated food, but would have radio communications for receiving instructions and equipment to record physical condition continuously during test. (MSFC Release 66-44)

- NASA Aerobee 150 sounding rocket launched from Churchill Research Range carried instrumented payload to 103-mi. (166-km.) altitude. Conducted for Johns Hopkins Univ., experiment measured auroral spectral emission lines in the upper atmosphere as a function of altitude. Rocket and instrumentation performed satisfactorily. (NASA Rpt. SRL)
- GEMINI V spacecraft in which Astronauts L. Gordon Cooper and Charles Conrad, Jr., orbited the earth Aug. 21-29, 1965, went on display at Buenos Aires (Argentina) Municipal Airport as part of a U.S.-sponsored exhibit on U.S. advances in space. (USIA; Reuters, *Wash. Post*, 2/21/66, A6)
- Soviet cosmonauts had had their "psychic state" monitored by an instrument that also registered "the slightest nervousness or lack of confidence," *Ogonyek* reported. Instrument, first tried on Cosmonauts Andrian Nikolayev and Pavel R. Popovich during Aug. 12, 1962, VOSTOK III-VOSTOK IV tandem spaceflight, showed them nervous before launch, but with nervousness disappearing during flight. (AP, *Phil. Eve. Bull.*, 2/21/66)

February 21: Former Astronaut John H. Glenn, Jr., consultant to NASA Administrator James E. Webb, attended dinner in Rangoon, Burma, given in his honor by Burma's Foreign Minister U Thi Han. In Burma on a good-will tour, Glenn earlier had given an illustrated lecture on U.S. space program. (AP, *NYT*, 2/22/66, 4)

- Russian-born Igor I. Sikorsky, "father of the helicopter," was honored as Air Force Man of the Year by the New York chapter of the Air Force Assn. He was cited especially for the outstanding work helicopters had done in the Vietnam war. (*N.Y. News*, 2/21/66)

February 22: U.S.S.R. launched COSMOS CX containing two dogs—Veterok and Ugolyek—on whom "biological investigations" would be conducted, Tass announced. Initial orbital data: apogee, 905 km. (562 mi.); perigee, 187 km. (116 mi.); period, 96.3 min.; inclination, 51.54°. Equipment was functioning normally.

Tass noted COSMOS CX launch marked first time since period prior to Cosmonaut Yuri Gagarin's April 12, 1961, VOSTOK I spaceflight that dogs had been launched and added that "further study of outer space, the moon and the inner planets of the solar system" was impossible without further development of "space biology and medicine." Announcement was first indication that satellites in Cosmos series were equipped to carry anything but instruments. (Tass, 2/23/66; *Pravda*, 2/24/66, 1, USS-T Trans.)

February 22: Launch from ETR of first unmanned Apollo spacecraft, scheduled for Feb. 23, was postponed because of cloudy, uncertain weather that would hamper visual and long-range photographic coverage of liftoff. (*NYT*, 2/23/66, 20; Stern, *Wash. Post*, 2/23/66, A1; Hines, *Wash. Eve. Star*, 2/23/66, A6)

- Astronauts Frank Borman and Walter M. Schirra, Jr., began eight-nation good-will tour of Asia with four-day visit to Japan. (*NYT*, 2/23/66)
- White Paper released by U.K. Defence Minister Denis Healey said RAF would buy 50 F-111 variable wing bombers from U.S. for 2.1 million pounds ($5,880,000) each and that plans for new British aircraft carrier had been dropped. F-111 aircraft operating from land bases would take over the strike-reconnaissance and air-defense functions of carriers. First Sea Lord, Adm. Sir David Luce, resigned in protest; Navy Minister Christopher Mayhew had resigned Feb. 19. (AP, *Wash. Eve. Star*, 2/22/66, A1; Lewis, *NYT*, 2/23/66, 1, 5; Meyer, *Wash. Post*, 2/23/66, A1)

February 22-23: American Astronautical Society met in San Diego.

Reconnaissance satellites, adapted for peaceful observations of earth's surface, would bring revolutionary scientific and economic advances, Willis B. Foster, Director of Manned Space Science in NASA's Office of Space Science and Applications, predicted: "Remote sensing from space has unique capabilities, particularly in the opportunity it provides for repetitive synoptic coverage of the earth's surface. . . . Spaceborne sensors will add to the knowledge of the figure of the earth, its mass distribution, and its magnetic and gravity fields. Movement of glaciers, growth of deltas, growth of crops, and even growth of populations can all be followed with this type of survey." (Text)

Austin M. Stanton, president and board chairman of Varo, Inc., said features of space such as airlessness, weightlessness, extreme temperature, and radiation could be "exploited as a completely new realm" for manufacturing. He said the earth might ultimately be encircled by a "vast doughnut-like ring 150,000 miles in circumference," of articulated orbiting manufacturing facilities: "As a businessman, I propose that the time is ripe for a profitable industrial venture in space. . . . It will not be difficult to persuade investors to venture into buying stock in the new companies for three reasons: the predictions of the early advocates of astronautics have been fulfilled; the efforts of space agencies have been spectacularly successful; and . . . the communications satellite venture seems sound." (Hill, *NYT*, 2/27/66, 60)

The massive investment in technology and manpower created by space programs could be used to handle pressing problems associated with anticipated population growth on earth, suggested Dr. Peter A. Castruccio, director of IBM's advanced space programs. Satellites could survey earth from space, pinpoint tillable land and examine staple crops using infrared techniques to detect diseases invisible to the naked eye. Seas could be mapped, isolating concentrations of fish and edible seaweed, and growing consumption of water offset by using satellites to map snow fields and glaciers. Underground rivers could be detected from space by measuring the tiny difference in soil temperatures above them; such streams were estimated to have 3,000 times the water present in all known rivers. Other fields, such as weather control and geological survey searches for fuel and minerals, could also be handled from space. (*M&R*, 2/28/66, 9, 12)

"The ultimate goal of weather modification and control may be realized only if man himself is placed in a Space Meteorological Lab," suggested Sidney Sternberg, vice president and general manager of Electro-Optical Systems, Inc. He said the "ability and bandwidth of transfer of information between experiments and experimenter could not be duplicated by unmanned observations." (*M&R*, 2/28/66, 12)

February 23: Second 260-in.-dia. solid-propellant rocket motor (SL-2), Nation's largest, was successfully test-fired for NASA in Dade County, Fla., by Aerojet-General Corp. Fueled with polybutadiene, aluminum and ammonium-perchlorate and fired nose down in a 150-ft.-deep pit, motor developed within 1% of the predicted 3.5-million lbs. thrust and burned for almost precisely the planned 126 sec. Test showed repeatability of the firing, proved that manufacturing techniques were reliable, and demonstrated that the motors could be produced at a predictable cost. First 260-in. motor (SL-1) was successfully fired for two minutes Sept. 25, 1965, producing 3.6-million lbs. thrust. Strength of the maraging steel motor case, structural integrity of the cast propellant, insulation, and the ablative nozzle were tested.

Large solid motor project was designed to demonstrate feasibility of building and operating solid motors of greater size than those in current use for multistage launch vehicle systems carrying heavy payloads into space. (NASA Release 66-31; Stern, *Wash. Post*, 2/24/66, A5; Hines, Wash. *Eve. Star*, 2/24/66)

- Testifying before House Committee on Science and Astronautics' Subcommittee on Advanced Research and Technology, NASA Associate Administrator for Advanced Research and Technology Dr. Mac C. Adams said: "During 1965 we flew the three X-15 airplanes 32 times, setting a new activity record. These aircraft comprise a unique research facility capable of providing data during actual hypersonic flight. . . .

"At the end of 1965, the three X-15 airplanes had been flown 156 times. . . ."

Referring to rocket reactor tests, Adams said: "The KIWI-NERVA-Phoebus operations during 1965 demonstrated an altitude equivalent specific impulse of over 750 lb-sec./lb., equivalent thrust of over 50,000 lbs., and total duration of 4200 seconds, or over an hour. The experiments have shown that rapid starts and restarts can be made and that operation is stable over a wide range of conditions.

"On February 3, 1966, we made the first power run with a complete nuclear rocket engine system. Although the major components were spread out and connected in a 'breadboard' fashion, the system started and operated as a unit with no external power. On February 11, the system was operated again. We feel that these runs are very significant milestones in the nuclear rocket programs." (*NASA Auth. Hearings*, 50, 57)

- NASA Associate Administrator for Space Science and Applications Dr. Homer E. Newell summarized achievements and status of NASA's physics and astronomy programs at hearings before the House Com-

Feburary 23: Test firing of NASA's 260-in. solid-propellant rocket motor at Aerojet-General Corporation's Dade County, Fla., facility.

mittee on Science and Astronautics' Subcommittee on Space Science and Applications. Dr. Newell said that from a scientific point of view, Explorer program had been "one of the most cost effective means of accomplishing the scientific objectives in space physics and astronomy." During 1965, "seven spacecraft were successfully launched including Alouette II, our second international project with Canada, and FR-1, the first cooperative satellite project with France. . . ."

Of the OGO program he said: ". . . as of early January 1966, data acquisition from OGO-I included 3,522 hours of real-time wideband telemetry, approximately 800 hours of special-purpose telemetry, and 2,800 hours of onboard data storage. . . . Good quality magnetic field data" were contributed by the OGO II mission to the World Magnetic Survey of the International Quiet Sun Year (IQSY). Results of OGO I and OGO II missions had proven the operational capability of the spacecraft and underscored the validity of the concept of a laboratory for correlative geophysical studies. (*NASA Auth. Hearings*, 114, 147, 159)

February 23: NASA had awarded a $38,000 contract to U.S. Public Health Service on best approach to control and early diagnosis of viral infection in astronauts during space flight. Dr. Walter Kemmerer, MSC's monitor on the virus control study, said in an interview: "We want to have techniques to enable us to prevent this from happening, and to detect that it is happening. Also, we want to know how to prevent one astronaut from transmitting it to another inside the spacecraft." (AP, *Wash. Eve. Star*, 2/23/66, A1)

- Charles H. Ruby, president of the Air Line Pilots Assn., charged FAA had implied pilots were to blame for four recent Boeing 727 jet aircraft crashes although investigations were incomplete. He referred to "premature and seriously questionable public assessment" of the 727 during a Feb. 17 FAA meeting at which FAA said the aircraft was airworthy and had no serious defects: "We don't believe this [cause determination] can be expected to be achieved with total objectivity by closed door meetings of parties with a substantial stake in the outcome." Ruby added pilots felt the 727 was basically a "good and safe aircraft." (UPI, *Wash. Post*, 2/24/66, A2)

February 23-24: Gemini Mid-Program Conference was held at MSC to summarize Gemini program to date with emphasis on results obtained from first seven Gemini missions. Conference, attended by more than 600 persons, was conducted in the middle of the Gemini program rather than at the end so that scientists and the aerospace industry would have earlier access to Gemini experiences for application to other space programs.

Gemini program director Charles W. Mathews announced that NASA had canceled plans for land landings of Gemini spacecraft because of "development time constraint"; U.S. spacecraft would continue to make ocean landings "for the foreseeable future." (Text; *Houston Post*, 2/24/66)

February 24: First Saturn V launch vehicle (S-IC-1) underwent second successful captive firing at MSFC for 85 sec. (MSFC Release 66-48)

- NASA's second space-tracking station near Canberra, Australia, was officially opened by Minister of Supply Denham Henty. Rep. George P. Miller (D-Calif.), chairman of House Committee on Science and Astronautics, attended ceremony and praised Australia's contribution to U.S. space program saying: "We will explore space together for the benefit of mankind." (*NYT*, 2/25/66, 3)

- Dr. George E. Mueller, NASA Associate Administrator for Manned Space Flight, testified in support of NASA's FY 1967 budget authorization before the House Science and Astronautics Committee's Manned Space Flight Subcommittee. Questioned by Congressman Emilio Q. Daddario (D-Conn.) about the Apollo program schedule, Dr. Mueller said current budget request left no room for a major problem; any test failure would be directly reflected in a schedule slip. In connection with possible further use of Apollo/Saturn V hardware beyond the lunar landing mission, Dr. Mueller stressed necessity for early decisions on the FY 1968 budget. Industrial research, development, and manufacturing employment had already reached its peak and, ironically, would begin to decline sharply before the first manned flights with Apollo/Saturn hardware. Facilities now becoming operational would be scheduled for "mothballing" within two years if no further use were planned for capability beyond the lunar landing, Dr. Mueller said. (*NASA Auth. Hearings*, 279-384; MSF Historian)

February 24: NASA's lunar and planetary programs were discussed by NASA Deputy Associate Administrator for Space Science and Applications Edgar M. Cortright, testifying before the House Committee on Science and Astronautics' Subcommittee on Space Science and Applications: "It is important to note that the Mariner IV photographs do not determine whether life exists on Mars. Even before the flight of Mariner IV, it was known that pictures taken at this resolution could not give any direct evidence of life forms. . . .

"We propose to continue the exploration of Venus and Mars begun so well with Mariners II and IV, respectively, by utilizing for the 1967 Venus opportunity an appropriate modification of the spare spacecraft from the Mariner IV program. This will be launched on an Atlas-Agena vehicle."

Cortright presented prepared statement by Associate Administrator for Space Science and Applications Dr. Homer E. Newell summarizing comsat and sounding rocket launches: "The first launching by NASA of a satellite for the Corporation [ComSatCorp] took place on April 6, 1965, when the Early Bird satellite was placed in orbit using a thrust-augmented Delta launch vehicle. The launching was carried out under an agreement between Comsat and NASA, signed in December 1964, which detailed the services to be performed by NASA, financial arrangements, etc. . . .

"Although the original agreement gave Comsat options for as many as seven additional launchings through March 1967, none of these options has been exercised. In the meantime, Comsat has inaugurated the HS-303A program, and has advised NASA that it will require two or three HS-303A launchings in 1966, beginning in August. The HS-303A is larger and heavier than Early Bird, and will need the capabilities of the thrust-augmented improved Delta launch vehicle. Otherwise, launch services are expected to be similar to those provided in support of the Early Bird.

"The Corporation is now negotiating for development and procurement of spacecraft for the eventual global system. These spacecraft are expected to be deployed beginning in 1968."

Referring to the NASA sounding rocket program, Dr. Newell said: "Since 1959 to the present NASA has launched 668 sounding rockets for scientific investigations in space and 516 of these were for investigations in the areas of physics and astronomy." He mentioned as significant the International Quiet Sun Year (IQSY) Mobile Launch Expedition carried out in spring 1965 using an aircraft carrier as a launching platform: rockets carrying instrumented payloads were launched off west coast of South America from 5° north latitude to 60° south latitude with the greatest concentration of flights at the geomagnetic equator near Lima, Peru. 45 Nike-Cajun/Nike-Apache rockets and 32 single-stage Arcas rockets were flown carrying experiments furnished by NASA, USN, USAF, universities, and industry. (*NASA Auth. Hearings*, 233-4, 359, 375-6)

February 24: Charles W. Harper, Director, Aeronautics Div., NASA Office of Advanced Research and Technology (OART), testified before the House Committee on Science and Astronautics' Subcommittee on Advanced Research and Technology on the NASA aeronautics program. Citing the need for close coordination between scientific groups within Government, he said: "Recognizing the Air Force interest in hypersonic flight, a proposal was made to the Air Force Systems Command Headquarters that the two agencies prepare a single coordinate program representing a national effort in hypersonic technology with each agency supporting that portion most appropriate to its capabilities. The proposal was accepted. . . . Technical approval of general plans is through the NASA OART Office and the AFSC RTD Office and generation and coordination of program details are developed through appointed technical groups from the NASA and Air Force research centers. To date this coordination activity has been very effective. . . ."

As major elements of NASA's 1967 aeronautics program, Harper listed: (1) a continuing effort in advanced research; (2) major effort in support of Sst with emphasis on flight operations and advanced propulsion; (3) a substantial increase in research in propulsion systems appropriate to the whole speed range, from V/Stol to hypersonic speeds with special attention directed to the noise problem and supported by two new facilities; (4) a substantially increased effort in V/Stol research supported by construction of the first research facility to be built expressly for conducting research on problems unique to V/Stol aircraft. (*NASA Auth. Hearings*, 170-1, 179)

- Astronauts Frank Borman and Walter M. Schirra, Jr., met with Japanese space scientists in Tokyo and presented Japanese Premier Eisaku Sato with a letter from President Johnson and a color photograph of their GEMINI VII/VI rendezvous in space. (Wash. *Eve. Star*, 2/25/66, A1; *Wash. Post*, 2/25/66)
- Two solid-fueled Minuteman ICBMs were launched simultaneously from Vandenberg AFB, to evaluate multiple-firing techniques which could be used "under combat conditions at operational bases." This was first salvo launch down WTR. (SAC Release 6-17)
- AFSC announced that all-weather landing systems (Awls) developed by USAF and FAA would be installed on all C-141 Starlifters within two or three years to enable the huge fan jet cargo-troop carriers to land safely under adverse weather conditions with a minimum of ground-landing aids. Awls would program aircraft's flight path, speed, angle of approach, and attitude; generate and provide data to pilot or autopilot; and be nearly self-sufficient. Lockheed Aircraft Corp. had been contracted by AFSC in May 1965 to design, install, test, and obtain FAA certification for Awls and produce a flying prototype aircraft by spring 1966. (AFSC Release 6.66)
- Lunar surface photograph from U.S.S.R.'s LUNA IX mission "fully coincided with the photograph of a model of the moon designed by scientists at Kharkov University's laboratory several years ago," Tass announced. Nikolay Barabashov, chief of university's observatory, hypothesized that "the relief at the landing site of the Soviet station is typical of the entire lunar surface," which he claimed was "porous, spongelike, and sufficiently firm to hold a rather heavy body." (Tass, 2/24/66)

February 25: NASA postponed third attempt to launch unmanned Apollo spacecraft with Saturn IB booster 44 sec. before scheduled liftoff because of continuing bad weather. (*Wash. Post*, 2/25/66, A7; *WSJ*, 2/25/66, 1)

- Testing of Agena stage—halted following Feb. 13 hard start of Agena prototype—was resumed at Arnold Engineering Development Center in preparation for Gemini VIII/Agena docking mission. (*SBD*, 2/28/66, 1)
- ComSatCorp asked FCC for authority to build six advanced synchronous satellites—four to be launched and two as backups for development of global commercial communications system. Proposed satellites would weigh 250 lbs., and have 1,200 voice-quality telephone channels and five-year operational life compared with 85-lb. EARLY BIRD I's 240 voice-quality channels and 18-mo. planned life. (ComSatCorp Release; *WSJ*, 3/1/66, 32)
- NASA EXPLORER XXIX (Geos-A) geodetic satellite, orbited Nov. 6, 1965, had developed problems which affected planned investigations: accurate measurement of spacecraft's attitude had become difficult; had precise attitude information could not be included in Mutual Visibility Program for observing stations. Failure of one of four flashing lights had reduced utility of some observing cameras, and interference had occurred between SECOR transponder and Doppler beacons. (NASA Proj. Off.)
- Estimated 7.5 billion microscopic meteorites hit earth daily, but only about 90 large meteorites weighing up to 10 lbs. struck earth each year, National Geographic Society said. (NGS Release, 2/25/66)
- Pan American Grace Airways (Panagra) asked FAA to refund $200,000 which company had deposited to reserve delivery on Sst being developed under FAA contract. This was first Sst cancellation request. (AP, *NYT*, 3/28/66, 57)

February 26: NASA's Apollo/Saturn AS-201 mission was successfully launched from ETR at 11:12 a.m. EST. Unmanned Apollo spacecraft (009) was boosted into suborbital flight by two-stage Saturn IB launch vehicle (SA-201) to qualify command module (Cm) heatshielding and service module (Sm) of spacecraft-launch vehicle combination and major spacecraft systems. Liftoff and powered flight events were as programed. After ignition of liquid-hydrogen-powered 2nd stage (S-IVB), launch escape tower was jettisoned and recoverable cameras ejected. After seven-minute burn, 2nd stage and instrument unit separated from spacecraft; spacecraft reached 310-mi. (499-km.) altitude. Descending, service module performed ullage maneuver, firing reaction-control system rockets for 30 sec. to increase spacecraft's reentry speed; Sm's main propulsion engine was fired twice—once for 100 sec. and once for 10 sec. Sm was jettisoned and Cm reentered at 27,500 fps—2,000 fps faster than Apollo reentry from earth orbit—testing ability of spacecraft's ablative heat shield to withstand reentry heat of about 4,000° F. Three main parachutes deployed at 12,000-ft. altitude, lowering Cm to splashdown in Atlantic at 11:51 a.m. EST within 35 mi. of target. Recovery was by helicopter from carrier U.S.S. *Boxer*. Spacecraft was in "good" condition.

Mission was first launch of Saturn IB and Apollo spacecraft. Saturn IB, uprated version of Saturn I, was being developed under direction of MSFC and Apollo spacecraft, under direction of MSC. (NASA Release 66-32; NASA Proj. Off.; MSC *Roundup*, 3/4/66, 1-2; Stern, *Wash. Post*, 2/27/66, A1, A8; Hines, Wash. *Sun. Star*, 2/27/66, A1, A6; Wilford, *NYT*, 2/27/66, 1, 60)

February 26: NASA's first Saturn IB booster launches unmanned Apollo spacecraft down the Eastern Test Range.

February 26: At KSC post-launch conference, NASA Associate Administrator for Manned Space Flight Dr. George E. Mueller commented on AS-201: "This test was of extreme importance to the eventual manned lunar landing in that it provides us with the first test of the ablative characteristics of the heat shield protection for the Apollo spacecraft. It is the one test that we were unable to perform on the ground—the one test, therefore, that was a qualification test and which will be repeated again on 202, which is the next flight in the Apollo system." (MCS *Roundup*, 3/4/66, 1-2)

- NASA Nike-Apache sounding rocket launched from Sonmiani, Pakistan, carried sodium payload to 124-mi. (200-km.) altitude in experiment conducted for Pakistan Space and Upper Atmosphere Research Committee to measure wind direction and speed at 50- to 124-mi. (80- to 200-km.) altitudes. No photography was possible since sodium flash, observed at +60 sec., was not followed by sodium cloud release. (NASA Rpt. SRL)
- Prerecorded pictures of dogs Veterok and Ugolyek, transmitted from Soviet satellite COSMOS CX launched Feb. 22, were shown on Moscow television. Announcement described the dogs as subjects of biological studies and said such studies "will precede every new and important step of man in space." Tass said "the condition of the space dogs is quite excellent." (*Krasnaya Zvezda*, 2/27/66, USS-T Trans.; *Pravda*, 2/27/66, 6, USS-T Trans.)
- Astronauts Frank Borman and Walter M. Schirra, Jr., received rousing welcome on arrival at Seoul, South Korea. Many flag-waving schoolchildren had to be restrained by policemen during welcoming ceremonies at City Hall. (*NYT*, 2/27/66, 14; *N.Y. J/Amer.*, 2/27/66, 28)
- Astronaut James A. Lovell, Jr., and Astronaut Donald K. Slayton (MSC Assistant Director for Flight Crew Operations) received gold medallions from Gov. Warren P. Knowles in Madison, Wis., after two-day homecoming celebration. Lovell, who had attended high school in Milwaukee, in turn presented state with Wisconsin flag he had carried for 14 days in space during GEMINI VII flight Dec. 4-18, 1965. Slayton was born in Sparta, Wis. (AP, *Chic. Trib.*, 2/25/66)

February 27: Two Nike-Apache sounding rockets were launched from Ascension Island, 24 and 36 hrs. after Feb. 26 Apollo reentry test. Experiment, conducted by GSFC and Univ. of Michigan's Space Physics Research Laboratory, was designed to measure pressure, temperature, density, and winds at altitudes of 19-75 mi. (30-120 km.) using pitot-static probe. First flight obtained necessary data for Apollo support program, attaining 82-mi. (131-km.) altitude. No useful data were obtained by second flight, because of breakup during 2nd-stage burn. (NASA Rpt. SRL)

- Scientists were considering "capturing an asteroid" and, through nuclear propulsion, placing it in earth orbit where it could be analyzed, reported John Wilford in the *New York Times*. He said "experts" believed "capture of an asteroid will be possible as soon as it is possible to send manned expeditions to the surface of Mars or Jupiter." Astronauts would take with them a separate rocket engine which they would anchor to asteroid's surface. Rocket would have to "produce long-duration burn but because of asteroid's low gravity would not have to be very powerful to push the asteroid toward a new orbit." (Wilford, *NYT*, 2/27/66, E7)

February 28: ESSA II (Tos OT-2), second meteorological satellite in Tiros Operational Satellite (Tos) system, was launched by NASA from ETR for ESSA with three-stage Thor-Delta booster. Launch vehicle performed three "dogleg" maneuvers before injection into near-polar, near-circular, sun-synchronous orbit. Initial orbital parameters: apogee, 885 mi. (1,425 km.); perigee, 843 mi. (1,357 km.); period, 113.5 min.; inclination, 101.16°.

A cylindrical, 18-sided polygon, 290-lb. ESSA II was equipped with two Automatic Picture Transmission (Apt) camera systems for photographing local cloud cover. During first orbit, spacecraft was spin-stabilized at 10.9 rpm with bottom of spacecraft toward earth; during second orbit, orientation maneuvers initiated by ground command began to turn satellite on its side; after 18 orbits—approximately 33 hrs. after launch—ESSA II would be in cartwheel position. Combination of cartwheel configuration and near-polar, sun-synchronous orbit would permit 100% photographic coverage of earth's cloud cover during daylight hours with a given area photographed at the same local time each day. Primary objective of ESSA II (Environmental Survey Satellite No. 2) flight was to provide direct readout to worldwide network of Apt receiving stations.

ESSA financed, managed, and operated the Tos systems. GSFC managed NASA participation in the project. Tracking operations were responsibility of STADAN. First satellite in Tos system, ESSA I, was launched Feb. 3. (ESSA Release 66-14; NASA Proj. Off.; AP, Wash. *Eve. Star,* 2/28/66, A1; UPI, *Wash. Post,* 3/1/66, A4)

- NASA Administrator James E. Webb, testifying before the Senate Committee on Aeronautical and Space Sciences as it began hearings on NASA's FY 1967 budget authorization, said U.S.S.R. would become unchallenged leader in space exploration unless U.S. soon decided to support post-Apollo missions. Webb reviewed U.S.S.R.'s very active year in space and said: "More important to us than any other indication from Soviet space activity is the clear commitment to a long-term effort. The fact that they are making such a commitment shows the importance they attach to advancing their capabilities in space. Unless we soon decide to follow through with a strong program in the years after we have achieved a manned lunar landing, we cannot avoid a gap in our flight schedule while they forge ahead as the unchallenged contestant in the field."

Webb noted FY 1967 budget for NASA reflected a Presidential decision "to hold open for another year decisions which cannot be delayed beyond the period of the FY 1968 budget." He added: "The programs we are now carrying out so successfully, the new knowledge we are acquiring so rapidly, the ending of the period of uncertainty as to what both Russian Cosmonauts and American Astronauts can achieve in space, and the growing utilization of the long term values that our investments have created in reliable launch vehicles, spacecraft, and assembly test and launch complexes, all point toward decisions to use rather than lose the values we have worked so hard and spent so much to create. And so I believe that our presentation of this 1967 Budget marks not just assertion to you that we have built well as we have created a foundation for space power, but also marks the beginning of a momentous era in which we must decide how we will use this power in future years."

He assessed the national space program: ". . . we have demonstrated that we can do the tasks we set out to do. An important value of the space program is this demonstration of national capability in science and technology. That we have this capability is no longer a matter of opinion but a demonstrated fact that is clear to all who concern themselves with the power nations can and do develop for many purposes from the mastery of a new environment or a new technology. We have put it on view for all the world to see. The world knows today that the United States can digest the space-related problems of long lead-times, can hold its position in a demanding competition, and can forge ahead simultaneously in both aeronautics and space and in those advanced technologies of which space is the leading symbol." (*NASA Auth. Hearings*, 15-17)

February 28: Testifying to Senate Committee on Aeronautical and Space Sciences on NASA's FY 1967 budget, NASA Deputy Administrator Dr. Robert C. Seamans, Jr., said: "We are presenting the budget that reflects our best balanced provision for four essential priorities, priorities which must rank together as the core of the NASA program: The need to press forward the development of the Nation's capability for major Manned Space Flight operations, to be demonstrated by a manned lunar landing within this decade.

"The need to continue those important projects in space sciences and space applications that are already underway and that have already made such valuable contributions to our fundamental understanding of the universe and to the use of space systems for human benefit.

"The need to maintain the flow of advanced research and technology effort that is at the heart of the Nation's ability to undertake future projects in aeronautics and space explorations.

"The need to take certain steps now to avoid an otherwise certain gap in future space capabilities and achievements." (*NASA Auth. Hearings*, 64-65)

- Summarizing NASA's launch vehicle and propulsion program in testimony before the House Committee on Science and Astronautics' Subcommittee on Space Science and Applications, NASA Associate Administrator for Space Science and Applications Dr. Homer E. Newell reported that in 1965 "there were 18 successful orbits out of 20 attempts for a 90% success record." Looking to the future, Newell said: "in order to assure that the launch vehicle space transportation evolves a logical and economical fashion for the more demanding. . .missions which are beyond the capability of present vehicles, it is essential that some effort continuously be put into advanced studies of future vehicles and into some of the critical launch vehicle technology areas."

He described research in behavioral biology, outlining prospective experiment to determine whether for pocket mice in earth orbit "circadian rhythm of either temperature or locomotion changes when all geophysical variables other than light and temperature are either removed or sensed by the animals with a period of about 90 minutes (orbital period) instead of 24 hours."

By the end of FY 1966, NASA's sustaining university program would be supporting 64 projects at 54 institutions throughout U.S., Dr. Newell testified. In FY 1967 it would support approximately 70 projects at 60 universities at a cost of $12 million. It was expected that 3,681 students would be in the predoctoral training element of the program by September 1966 and the number of participating universities would be 152. (*NASA Auth. Hearings*, 487, 568, 581, 589)

February 28: Gemini Astronauts Charles A. Bassett II and Elliot M. See, Jr., were killed when their T-38 aircraft overran St. Louis Municipal Airport in rain and fog and crashed into the McDonnell Aircraft Corp. building which housed Gemini 9 and 10 spacecraft. The jet, cleared for an instrument landing, was left of center in its approach to the runway when it turned toward the McDonnell complex 1,000 ft. from instrumented landing strip. It struck the roof of Building 101, bounced into an adjacent courtyard, and exploded. Several McDonnell employees in Building 101 were treated for minor injuries. See and Bassett, enroute to McDonnell for two weeks training in space simulator, were followed later by Gemini IX backup crew Thomas P. Stafford and Eugene A. Cernan, who landed safely. NASA Hq. officials announced that the backup crew would fly the Gemini IX mission on schedule. (Wagner, *Wash. Post*, 3/1/66, A1, A4)

- Discussing the Apollo/Saturn IB program's future, Dr. Joseph Shea, director of Apollo spacecraft development at MSC, told *Missiles and Rockets* that the hoped-for early end of the program (after a total of seven flights, including four manned missions) would depend on availability of the Saturn V booster, as well as success of Saturn IB missions: "We plan to end the series at Apollo-207 if the Saturn V is available on schedule. If not, we will probably continue with Saturn IB flights for the Apollo-208 and 209 missions." (Taylor, *M&R*, 2/28/66, 17)

- Joint Chiefs of Staff believed U.S.S.R. was accomplishing much more in advanced weaponry than Secretary of Defense Robert S. McNamara was giving them credit for, reported *Aviation Week*. Gen. John P. McConnell, USAF Chief of Staff, was said to be especially concerned about Soviet developments in antimissile missiles and their radars, maneuverable warheads for ICBMs, military space systems, and advanced aircraft. During hearings on the FY 1966 budget, General McConnell had said the decision on whether to proceed with full-scale development of the advanced manned strategic aircraft (Amsa) should be made in FY 1967. *Aviation Week* pointed out McNamara did not make that decision in the the FY 1967 budget and dropped Amsa one category—from engineering development to advanced development. Also, in a posture statement before the Senate Armed Services Committee's Defense Appropriations Subcommittee, McNamara appeared to contradict Secretary of the Air Force Dr. Harold Brown's view that the FB-111 was only a stop-gap bomber until Amsa was approved. Comparison of Boeing B-52C through F models, McNamara reportedly said, showed "the FB-111 is not an interim aircraft but is, indeed, a truly effective strategic bomber." (Wilson, *Av. Wk.*, 2/28/66, 16)

- Astronauts Walter M. Schirra, Jr., and Frank Borman and their wives met with Chinese Nationalist President Chiang Kai-shek in Taipei, and gave him a letter from President Johnson and a color photograph taken in space. Chiang gave each astronaut wings of the Chinese National Air Force. Astronauts then flew to Bangkok, Thailand, to begin a three-day visit during which they would give a series of lectures and be received by the King and Queen of Thailand. (AP, Wash. *Eve. Star*, 2/28/66, A2; *NYT*, 3/1/66, 19)

- AFSC awarded General Electric Co., Pratt & Whitney Div., United Aircraft Corp., and Curtiss-Wright Corp. contracts totaling $41,750,000 for development of vectored-thrust cruise propulsion system (lift-cruise engine) for aircraft systems including V/Stol. (AFSC Release 13.66)

February 28: William J. McWilliams, president of Republic Supply Co., was sworn in as consultant to NASA Administrator James E. Webb and NASA Secretariat. (NASA Release 66-46)
- ComSatCorp requested FCC authority to build six 250-lb. advanced synchronous satellites, for global commercial communications system. Launches would begin in 1968. (ComSatCorp Release)
- Comparing U.S. and U.S.S.R. Mars exploration efforts in *Science*, Bruce C. Murray, Cal Tech Associate Professor of Planetary Science, and Merton E. Davies, RAND Corp., said of NASA's MARINER IV July 14, 1965, flyby: "The American scientific community should, and evidently does, feel proud and gratified that the Mariner mission to Mars, through successful utilization of national resources, not only accomplished a remarkable technological feat but also discovered extraordinary new facts... beyond the reach of earth-based instruments." Authors termed "disappointing" U.S. decisions to terminate Mariner program, fly extra Mariner C spacecraft to Venus in 1967, drop plans for 1971 Voyager launch, and "fill the gap" with an abbreviated 1969 Atlas-Centaur program. U.S.S.R., they noted, had spent "between five and ten times as much on flight programs for planetary exploration" as U.S. and, in spite of "six consecutive failures in their efforts to explore Mars," may have developed a reliable 2,000-lb. spacecraft for Mars flyby. Authors predicted U.S.S.R. would launch two Zond spacecraft in December 1966: "It will be surprising to us if at least one... does not reach the planet and return new facts about it." They suggested 1971 as earliest likely date for Soviet landing of survivable payload on Martian surface and concluded that by 1967 "the U.S. will be ready to make an enduring national commitment to the exploration of the planets, and be ready to demonstrate intellectual appreciation of discovery as well as enthusiasm for technological achievement." (*Science*, 2/25/66, 945-54)

Week of February 28: NASA delegation extended President Johnson's invitation for European participation in U.S. manned and unmanned space programs in discussions with officials of West Germany, France, U.K., and Italy. Delegation had been instructed not to make any offers or commitments but to express potential opportunities of participation in future U.S. space flights on equal basis with U.S. scientists. (Normyle, *Av. Wk.*, 2/28/66, 23)

During February: Dr. Ernst Stuhlinger, Director of MSFC Research Projects Laboratory, summarized progress in development of electric propulsion in *Astronautics & Aeronautics:* "The first laboratory experiments with electrostatic thrustors began in 1958. In the early thrustor models, power efficiencies and propellant-utilization efficiencies were 20-30%; lifetimes did not exceed a few hours; ion-current densities were less than 1 milliamp/cm^2; the problem of beam neutralization was unsolved—was not even well understood. Mission applications seemed far in the future. Today, only seven years later, electrostatic thrustors are very close to flight readiness. Two systems, the ion-bombardment system and the surface-ionization system, have reached power efficiencies and propellant efficiencies above 90%; they produce ion currents of about 20 milliamp/cm^2. Beam neutralization has proven in the laboratory as well as in actual space tests, and engine lifetimes of several-thousand hours have been demonstrated. With a few more years of refinement and quality engineering, these ion thrustors will be ready for extended missions into interplanetary space...." (*A&A*, 2/66, 71-2)

During February: Arnold Frutkin, NASA Assistant Administrator for International Affairs, noted two viewpoints on achieving Soviet cooperation in space research in *Astronautics & Aeronautics*: ". . . the one holds that the answer . . . lies in finding a suitable device, either bilateral or international—a new idea, an imaginative approach, a twist . . . for example, a proposal for an international space laboratory. . . .

"The other view is that the device is substantially irrelevant and that it is the basic relationship between the two countries, the basic political and military realities, which will determine at any given time whether cooperation is feasible or not." (*A&A*, 2/66, 20)

- Dr. Raymond L. Bisplinghoff, former NASA Associate Administrator for Advanced Research and Technology and new AIAA president, wrote his first "President's Message" in *Astronautics & Aeronautics*: "One of the dilemmas of modern times is that science and technology can do a multitude of things—many of which the community may not necessarily desire and so will not support. As professionals we have tended to preoccupy ourselves with the potential of science and technology rather than the expressed needs of the community. A responsible professional society of the future must find . . . the wisdom among its members to relate its technologies to human needs. A wise professional society . . . will convey the legitimate aspirations of its thinkers and men of action to the community at large. . . ." (*A&A*, 2/66, 29)

- Japan scheduled launch of her first earth satellite for March 26 with four-stage, 8.6-ton Lambda booster. Planned orbit would have 723-km. (480-mi.) apogee and 290-km. (180-mi.) perigee. If successful, Japan would become fourth nation to orbit spacecraft. (*NYT*, 2/27/66, 33; UPI, *Wash. Post*, 2/27/66)

- Commenting on national space goals in the post-Apollo period, Henry Simmons said in *Astronautics & Aeronautics*: "NASA over the past year quietly shifted its emphasis from an AAP exploiting already financed Apollo hardware for alternate missions to a much vaster AAP involving outlays on the scale of the Apollo lunar effort itself to support an intensive follow-on effort extending perhaps to 1975. Some NASA Centers engaged in the Apollo program, and particularly the Marshall Space Flight Center, are already confronted with the prospect of unemployment among their most skilled and valuable design personnel. 'We are right now in the position where over the next six months the engineers who did the original design on the Saturn and probably a year from now those who did it on the Apollo will be out of work,' NASA Administrator James Webb told the Senate Committee on Aeronautical and Space Sciences during its hearings last August on post-Apollo goals. . . ." (*A&A*, 2/66, 5)

- ARC scientists who had produced organic material in simulated Jupiter atmosphere said presence of some form of life on Jupiter and other planets was possible, but unlikely. They would continue experiments to find out more about chemical processes that occur at very low temperatures. (UPI, *NYT*, 2/19/66, 44)

- MSFC was using seven vessels to transport Saturn IB and Saturn V rocket stages and components too large for conventional road, rail, or air transport between manufacturing, test, and launch sites. MSFC's marine transportation had grown from one craft in 1960 to present size of six river and ocean-going barges—*Promise, Palaemon, Little Lake, Pearl River, Orion,* and *Poseidon*—and one ship—USNS *Pt. Barrow*. (MSFC Release 66-36; *Marshall Star*, 2/16/66)

March 1966

March 1: U.S.S.R. VENUS III spacecraft "reached" the surface of Venus at 1:56 a.m. EST to become man's first space probe to land on another planet, Tass announced. Observers believed the 2,116-lb. spacecraft, launched Nov. 16, 1965, had crash-landed. "Regular radio communications" had been maintained with VENUS III throughout the flight but were lost as the probe approached the planet; no telemetric data were received in final moments before impact. Announcement said companion spacecraft VENUS II, launched Nov. 12, 1965, had passed within 14,900 mi. of Venus Feb. 27, and was continuing in solar orbit. (Later, on March 5, U.S.S.R. revealed that communications with VENUS II were lost as the spacecraft approached Venus.) 2,123-lb. VENUS II had flown its entire course without a corrective maneuver "insured only by its correct launching on an interplanetary trajectory," but the course of VENUS III had been adjusted in Dec. 27, 1965, midcourse maneuver to make possible precise rendezvous with planet. "The experiment conducted through the work of the automatic stations Venus 2 and 3 allowed the solution of a number of absolutely new tasks of interplanetary flights and the obtaining of new scientific data," Tass said. Man's closest previous contact with Venus had occurred Dec. 14, 1962, when NASA's MARINER II (launched Aug. 27, 1962) had passed 21,648 mi. from its surface and had relayed back scientific data. The brief Tass statement said VENUS III carried pennant bearing Soviet coat of arms to the Venusian surface. Further details of the spacecraft's instrumentation were to be revealed later [see March 5].

Sir Bernard Lovell, director of Jodrell Bank Experimental Station, called the landing "a vivid technical feat" but regretted that "the Russians should have endangered the future biological assessment of Venus by contaminating the planet." (*Pravda*, 3/2/66, 1, 4; Grose, *NYT*, 3/2/66, 1; *Wash. Eve. Star*, 3/1/66, A1; Simons, *Wash. Post*, 3/2/66, A1)

- U.S.S.R. launched COSMOS CXI unmanned satellite with scientific equipment to continue space investigations, Tass announced. Orbital data: apogee, 226 km. (140 mi.); perigee, 191 km. (119 mi.); period, 88.6 min.; inclination, 51°. (Tass, 3/1/66)
- Edgar M. Cortright, NASA Deputy Associate Administrator for Space Science and Applications, testifying before the House Committee on Science and Astronautics' Subcommittee on Space Sciences and Applications in support of NASA's FY 1967 authorization, discussed the merits of the Apollo and Surveyor programs: "The scientific advantages of having Surveyors in the program along with Apollo is the ability to

sample more sites and to determine what areas of the moon are worth visiting by man, for example, in addition to going to sites that man is not likely to for the next decade or more—up near the poles and certain of the mountain regions.

"We would like to help develop this total strategy of exploration of the Moon in a sensible way. We think automated equipment continues to have a role to play. . . .

"I think there is nothing we can fly on Surveyor that would improve or even come close to equalling what can be done by the Apollo once it gets there. In other words, bringing a sample back is so much better than doing analyses on the lunar surface." (Transcript)

March 1: Astronaut Alan B. Shepard, Jr., headed six-man panel of investigators that began attempt to reconstruct St. Louis crash of T-38 jet plane in which Astronauts Elliot M. See, Jr., and Charles A. Bassett II died Feb. 28. MSC Public Affairs Officer Paul Haney told UPI that full report would not be completed for "at least six weeks."

The tragedy prompted Rep. Thomas M. Pelly (R-Wash.) to urge the House Committee on Science and Astronautics to investigate flight training of astronauts, particularly the number of hours of flying time required and the justification for such requirements. (UPI, *NYT*, 3/2/66, 22; NASA LAR V/36)

- The Feb. 28 deaths of Astronauts Elliot M. See, Jr., and Charles A. Bassett II received editorial comment.

Washington *Evening Star*: "Yesterday's tragedy is all the more shocking in that it needn't have happened at all. . . .

"It is assumed that Bassett was 'logging flying time' . . . See, a civilian, was not eligible for flight pay but as an astronaut was encouraged to keep his hand as an airplane jockey.

"It is difficult to see why it is necessary for military men on astronaut duty to fly airplanes to qualify for 'hazardous duty' pay . . .

"Let's save our spacemen for space, and let's make it a bit easier for them by amending the military pay act to give uniformed astronauts those few hundred extra dollars a month just for being astronauts." (Wash. *Eve. Star*, 3/1/66, A6)

Houston Post: "There is . . . an element of irony in the fact that two men preparing for a highly hazardous space mission should die in this prosaic manner, but actually, under the procedures that are followed in space flights, there are just as great risks on the surface of the earth and in its atmosphere as in space, and they probably are much more numerous. If air travel is safer than land travel today, space travel is probably even safer . . ." (*Houston Post*, 3/2/66)

Missiles and Rockets: "By emphasizing the everyday risk taken by all the astronauts in the course of training, the deaths of Capt. Freeman, Mr. See and Maj. Bassett have almost certainly lessened the outcry against the space program in the event of similar losses during space operations.

"Each time an astronaut steps into a military jet aircraft, he is running a risk perceptibly greater than that he takes in space. . . . [NASA] takes the greatest of care to assure that manned spacecraft launches are as safe as the best engineering talent in the nation can make them. A launch atop a Titan II, hazardous as it may be, cannot compare with the danger inherent in groping through low fog for an elusive runway in a jet aircraft." (Coughlin, *M&R*, 3/14/66, 54)

March 1: NASA had selected Fairchild Hiller, General Electric Co., and Lockheed Aircraft Corp. for negotiation of contracts totaling $450,000 for parallel six-month feasibility studies of missions for five possible second-generation Applications Technology Satellites (Ats). Spacecraft would contribute information useful for orbiting data-relay satellites, deep-space tracking, navigation satellites, data collection, galactic probes, and communication with aircraft, other satellites, ships and ground vehicles, and broadcast satellites. (NASA Release 66-45)

- Establishment of National Natural Disaster Warning System to provide warnings of hurricanes, tornadoes, tidal waves, floods, and other natural hazards was announced by the White House in response to proposal by Secretary of Commerce John T. Connor. Warning System would "make effective use of present technology and existing facilities for the distribution of all warnings over a single authentic channel between the warning agency and the public"; it would eventually provide around-the-clock weather intelligence reaching hundreds of communities which had not previously had adequate warning service. (*Pres. Doc.*, 3/7/66, 303)

- The advancement of military technology contributed greatly to the civilian economy with "higher reliability, new management techniques, new materials, improved manufacturing methods, better instrumentation, and the expansion of technical fields such as cryogenics and microelectronics," AFSC Commander Gen. B. A. Schriever told a Rotary Club meeting in Honolulu. He cited in particular the development of long-range air transportation "which clearly has both military and commercial value" and discussed the progress that had been made in some key technical areas: ". . . current experiments with oxide-dispersed metals for strength at high temperatures show a 40 percent improvement over the findings of two years ago. Boron fibers for composite materials have already been produced with a 20 percent greater average strength than they were predicted to attain by 1970. Their monthly production is more than 20 times the total supply that existed three years ago . . . .

   "When we turn to the propulsion area, we find that a new aircraft engine is presently under test, which surpasses the gains predicted for 1970 in bypass ratio, high temperature capability, and reduced specific fuel consumption. The supersonic combustion ramjet, or SCRAMJET, is showing great promise for airbreathing vehicles at speeds of mach 6 or greater. Supersonic combustion has already been repeatedly demonstrated at simulated speeds of mach 6 in test facilities. . . ." (Text)

- Dr. Jocelyn R. Gill, Chief of In-Flight Sciences for NASA OSSA's Manned Space Science Div., was one of six Government career women who received the Federal Woman's Award at a banquet sponsored by Woodward and Lothrop, Inc., department store. (*Wash. Post*, 2/7/66, C2)

- British Royal Society had been quietly conducting exchange of scientists with Chinese Communist Academy of Sciences for last three years, Royal Society spokesman told the press in London. About 15 British scientists had visited China during three-year period and varying numbers of Chinese had visited U.K. for periods ranging from several months to two years. At present 25 Chinese were studying wide range of scientific subjects at British institutions. (Schmidt, *NYT*, 3/2/66, 13)

March 1-3: "Space Maintenance and Extravehicular Activities" meeting sponsored by USAF Aero Propulsion Laboratory and Martin Co. was held in Orlando.

Goodyear Aerospace Corp. engineer Charles Brownell, together with three USAF space engineers, told the meeting that construction in space would cost approximately $100,000 per hour, but added: "These costs will be reduced drastically in the next few years." Brownell said that because space labor would be so expensive, efforts were being made to automate as much repair work as possible. (Hines, *Wash. Eve. Star,* 3/1/66, A4)

Man must learn how to rescue orbiting astronauts stranded in a disabled spacecraft because "it is a problem we all know will one day confront us," Col. Emil G. Beaudry, vice commander of USAF Air Rescue Service, told the conference. He said one expert had concluded there was a 62 per cent probability of at least seven emergency situations in space involving 22 men in the next 20 yrs., and a 58 per cent probability of two or three emergencies within 10 yrs. Beaudry continued: "There is no reason today—or certainly in the near future—why an operational rescue spacecraft could not be ready and standing by in the event of distress or tragedy during manned flights.

"For example, a launch vehicle such as the Titan 3 with its great weight-lifting capability and mission versatility could easily carry a rescue crew, medical facilities, and technicians aloft to assist in any way required...." (AP, *New Haven Register,* 3/7/66)

NASA engineer David F. Thomas reported on his "jet shoes" to aid astronauts during extravehicular activity: "This device may be thought of as a pair of shoes with jets attached to the shoe soles in such a manner as to produce a thrust vector along the subject's leg when the jets are activated." Projecting from the front of each shoe would be hinged toe-plate which astronaut could depress or release at will to operate an electrical switch which controlled gas jets. Thomas said shoes, which would be slipped on over space boots, had been tested successfully on counterbalanced harness device at LaRC. (*Wash. Eve. Star,* 3/2/66, A18)

Douglas Aircraft Co. engineers Thorne L. Runyan and John M. Dick reported that intensity of sunlight and blackness of shadows would cause astronauts working in space to "confuse spheres with points of light, cones with wedges, and pipes with strange shapes that appear and disappear." To counteract these distortions, they recommended a bracket of strong lights "looking over the shoulder" of a working astronaut to fill in shadows and lessen contrast with sunlight. (*Wash. Eve. Star,* 3/3/66, A5)

Bell Aerosystems Co. test pilot Robert F. Courter reported that Bell had developed a rocket propellant "flying pogo stick" for transporting men and equipment on earth or over the surface of the moon. Pogo stick would be powered by hydrogen peroxide jets and its two rocket nozzles would be controlled directly by pilot's hand, arm, and shoulder movements. (UPI, *Miami Her.,* 3/3/66)

The space tool kit to be used by Astronaut David R. Scott during Gemini VIII mission was exhibited for the first time. Developed by Martin Co. and Black & Decker Tool Co. at a cost of $124,000, kit's power tool used impact instead of torque as its motive principle because man in space, with neither friction nor gravity to give him a firm footing, would be sent spinning counterclockwise by the action of a hand-held tool that rotated clockwise. (Hines, Wash. *Eve. Star*, 3/3/66, A1)

In a luncheon address, Dr. Eugene B. Konecci, NASC staff, warned that U.S.S.R. might soon recapture the lead in manned spaceflight: "They don't lead now in manned spaceflight. They may again in the near future, but they don't now." Konecci said space experts expected Russia to launch a "major manned program in the near future," possibly a manned laboratory. (UPI, *Miami Her.*, 3/3/66)

March 2: NASA's PIONEER VI interplanetary probe, launched Dec. 16, 1965, passed near earth after 76 days of near-perfect operation in solar orbit. Probe was 84,190,800 mi. from the sun, passing 7,775,600 mi. closer than earth does, and in earth's orbital plane on a course which would ultimately take it within 75.6 million mi. of the sun. Devised to study interplanetary space, PIONEER VI had telemetered 250 million readings from its six scientific instruments covering 3,300 separate measurements and taken 2.5 million readings of 100 engineering measurements. It had also registered solar wind speed of 1,636,000 mph on Feb. 23, close to highest speed previously recorded of 1,675,000 mph by EXPLORER XVIII in earth orbit. Preliminary scientific findings would be presented to American Geophysical Union meeting in Washington, D.C., April 19-22. Pioneer project was managed by ARC. (NASA Release 66-43.)

- Vice President Hubert Humphrey, accompanied by NASA Administrator James E. Webb and NASC Executive Secretary Edward C. Welsh, flew by helicopter to GSFC for three-hour tour, during which they watched GSFC technicians receive first photos from ESSA II meteorological satellite. Successful operation of ESSA II officially created world's first fully operational weather-forecasting satellite network. (AP, Wash. *Eve. Star*, 3/2/66; AP, Balt. *Sun*, 3/3/66)

- NASA successfully launched two companion sounding rockets five minutes apart from NASA Wallops Station for simultaneous measurements of properties of the neutral atmosphere, ionosphere, and solar radiation intensities at 120- to 380-mi. altitudes. Aerobee 150A sounding rocket with 180-lb. AFCRL-instrumented payload to measure solar ultraviolet radiation and electron temperatures as function of altitude reached 142-mi. (229-km.) altitude. While Aerobee 150A was still in flight, four-stage solid-propellant Javelin boosted instrumented payload to 380-mi. (611-km.) altitude to obtain simultaneous measurements of neutral gas densities, temperatures, and composition; ion densities and composition, and electron temperatures and densities. Data from both flights were telemetered to the ground and would be correlated to study effects of solar radiation on ionosphere and upper atmosphere. (Wallops Release 66-12; NASA Rpt. SRL)

- Discovery of two powerful x-ray galaxies which might indicate existence of a whole family of celestial objects was reported by Naval Research Laboratory physicist Dr. Herbert Friedman at a news conference in Washington, D.C. The two galaxies—Cygnus A and M-87—were among the strongest of the 10,000 galaxies that emit radio signals and had been detected previously only by optical telescopes and by their radio emissions. "The startling thing about these extra-galactic sources is that their x-ray output is 10 to 100 times greater than their combined output of light and radio waves," Dr. Friedman said, and he suggested that an x-ray satellite or a "fence-like array of 1,000 Geiger counters erected on the moon" could identify the sources. The discovery supported theory of an "exploding universe" whose evolution over billions of years, and in the future, would be characterized by violent cosmic explosions far beyond the relative calm of the Milky Way. Dr. Friedman, E. T. Byram, and T. A. Chubb obtained data from an April 1965 Aerobee sounding rocket flight. (Clark, *NYT*, 3/3/66, 2; Simons, *Wash. Post*, 3/3/66, A3)

March 2: In a special message to Congress, President Johnson urged the establishment of a Cabinet-level Dept. of Transportation consolidating FAA and several other Federal agencies and functions "to serve the growing demands of this great Nation, to satisfy the needs of our expanding industry and to fulfill the right of our taxpayers to maximum efficiency and frugality in Government operations." The President repeated the proposal made in his State of the Union message Jan. 12 for U.S. program to construct and flight-test 2,000-mph supersonic aircraft. He requested $200 million in FY 1967 to initiate prototype phase, including completion of design competition, expanded economic and sonic boom studies, and start of prototype construction. U.S. planned to conduct first flight tests by 1970, he said, and to introduce the aircraft into commercial service by 1974. The President also directed his Science Adviser to work with NASA, FAA, Dept. of Commerce, and Dept. of Housing and Urban Development to devise an "action program" to attack the problem of aircraft noise—"a growing source of annoyance and concern to the thousands of citizens who live near many of our large jetports." (*Pres. Doc.*, 3/7/66, 304-12)

- In statement presented to the Senate Committee on Aeronautical and Space Sciences, Dr. Homer E. Newell, NASA Associate Administrator for Space Science and Applications, summarized significant 1965 mission results: "The 1965 successful Space Science and Applications missions totalled 16. Of these, eight were scientific satellites; three, applications satellites; three, deep-space probes; and two were vehicle developments. Science and applications experiments have also been carried on five manned missions.

  "The list of U.S. space firsts was extended during 1965. Ranger IX took the first close-up pictures of the floor of a lunar crater. For the first time, as television pictures of the lunar surface were being obtained from the Moon, they were sent on a real-time basis to home receivers. Mariner IV achieved the first successful mission to Mars. It obtained the first close-up pictures of the Martian surface, revealing the cratered character of the planet. By October 1, 1965, Mariner IV had extended the long-range communication record to 191 million miles; it has since been tracked to its maximum distance of 216 million miles. The various measurements of the Mars magnetic field, atmosphere, and ionosphere constitute significant firsts. Early Bird became the first operational commercial communication satellite. The first full global picture of the Earth's cloud cover in a single day as it will be observed by TOS was provided by TIROS IX early in the year. In the Gemini program excellent pictures were obtained of the terrain and the gegenschein, a faint reflection of sunlight from materials dispersed throughout interplanetary space. It was clearly established that astronauts can perform scientific experiments in space, producing excellent usable results.

  "During 1965, 138 sounding rockets were launched to continue the investigation of our atmosphere, and to carry out exploratory space experiments. In support of the International Quiet Sun Year, the ship

USNS *Croatan* launched some 77 rockets while sailing down the west coast of South America. Over 90 per cent of the sounding rocket launches and about 80 per cent of the payloads were successful, yielding data on cosmic rays, and the Earth's atmosphere, ionosphere, and magnetic field. A Goddard Space Flight Center sounding rocket observed the planet Mars in the ultraviolet, indicating an atmosphere pressure of about five millibars at the planet's surface, which agrees well with the Mariner IV data. Other rocket flights made ultraviolet studies of the [Ikeya-Seki] Comet of 1965, and of the Sun in ultraviolet and X-ray wavelengths." (Testimony; *NASA Auth. Hearings*, 341-5)

March 2: NASA Deputy Associate Administrator for Tracking and Data Acquisition, Gerald M. Truszynski, testified before the Senate Committee on Aeronautical and Space Sciences: "To provide effective support of flight projects, NASA has established a worldwide network of tracking and data acquisition stations.... This network, at the present time, consists of ground stations which are operational or under construction at 28 locations. These stations are located in the United States and 16 foreign countries and territories. They represent a total capital investment of over $600 million and are operated by more than 2,500 highly trained NASA, contractor, and foreign national personnel....

"Many of the stations being used for Gemini support will also be used for Apollo. Capabilities are being added at these stations while other stations are being established. The network for support of Apollo missions . . . will consist of 10 fixed land stations with 30-foot antennas, three stations with 85-foot antennas, 1 transportable station, 5 instrumentation ships, and 8 instrumentation aircraft. In addition, three 85-foot antenna stations normally used for deep space missions will be used for support of Apollo. . . .

"Our efforts are directed toward having the network fully qualified for lunar mission support in 1967." (Testimony; *NASA Auth. Hearings*, 383, 387)

- Dr. Hermann H. Kurzweg, NASA Director of Research, OART, in testimony before the House Committee on Science and Astronautics' Subcommittee on Advanced Research and Technology, discussed the development at NASA Marshall Space Flight Center of a new automatic programmer called AMTRAN (Automatic Mathematical Translator): "In this system, keyboards and cathode ray oscilloscopes, by which mathematical commands and replies to and from a central digital computer may be transmitted and received, can be located near the desk of the mathematician, and operated directly by him. Experiments have indicated that this new AMTRAN computer system may reduce the time required for solutions of certain problems by a factor of 100. An additional economical advantage of AMTRAN is that sets of keyboards and display screens, all working through the same central computer, may be located in a large number of locations for direct and immediate use." (Testimony; *NASA Auth. Hearings*, 600-02)
- FCC adopted a Notice of Inquiry inviting comments by interested parties to be submitted by Aug. 1, 1966, on five specific questions concerning legality, compatibility with the Communications Satellite Act of 1962, and economic and technical feasibility of proposed FCC authorization of nongovernmental communications satellite facilities. (Text)
- U.S.S.R.'s ZOND III interplanetary probe, launched July 18, 1965, was 153,520,000 km. (95,397,328 mi.) away from earth in heliocentric

orbit gradually receding from the sun, Tass announced. All equipment was said to be functioning normally. (Tass, 3/2/66)

March 2: Harold B. Finger, Director of NASA Nuclear Systems and Space Power Div., speaking at the Franklin Institute, Philadelphia, summarized the history and development of electricity in space exploration and predicted it would play an increasingly important role in the future of the space program: "It is important to remember that the program of space exploration is still young and there is still a vast sea and many space objectives to explore and to learn about. The highest power levels used in space so far have been only a few kilowatts...." Orbital research laboratories, lunar bases, broadcast satellites, and interplanetary spacecraft, he said, would require increasing levels of electric power. (Text)

- U.S. scientists were questioning whether VENUS III's landing on the planet had been a mistake, reported Walter Sullivan in the *New York Times*. He noted that the Nov. 28, 1965, issue of *Izvestia* had reported the two probes would pass on different sides of Venus and that Mstislav Keldysh, president of Soviet Academy of Sciences, had told a Moscow press conference Feb. 10 that both probes would "approach" the planet about March 1, but that no "soft" landing would be attempted. Marvin Miles of the *Los Angeles Times* reported that U.S. scientists, emphasizing the difficulty of knowing the exact track of distant planets, said that a small trajectory error could have caused a crash-landing even though a flyby might have been programed. (Sullivan, *NYT*, 3/2/66, 22; *Wash. Post*, 3/3/66, A21)

- Report of Feb. 15 firing of French ballistic missile 1st stage from Les Landes launching site south of Bordeaux, France, published in *Air et Cosmos* and confirmed by French Ministry of Armed Forces, indicated base was operational, *Washington Post* reported. Development of the base, intended to bridge the gap between the scheduled 1967 closing of Hammaguir Range and the 1968 opening of the French Guiana base, had been kept secret and "it had not been realized that the base was anywhere near ready for use," *Post* suggested. (*Wash. Post*, 3/3/66, A16)

- Rupert Wildt, professor of astrophysics at Yale Univ., had been named to receive the Eddington Gold Medal—highest award of Royal Astronomical Society of England. (*NYT*, 3/3/66, 58M)

- Home-built photo-receiver set of RCA electronics engineer Wendell Anderson in Morristown, N.J., received photos of North Atlantic storm taken by ESSA I meteorological satellite—first known usable pictures ever received from an orbiting satellite on home-made set. Equipment, which included 25-yr.-old ham radio, 20 ft. of copper tubing, television antenna, mast, sheet of chicken wire, wooden laundry hangers, sawed-off rolling pin, and $25 microscope, cost $600—$31,000 less than an ESSA ground station. Anderson used antenna to pick up signals on ham radio, taped signals on recorder, reducing them to series of electrical impulses that could be changed into light signals, then beamed light through reverse end of a microscope onto spinning sheet of film to record photos. (O'Toole, *NYT*, 3/3/66, 31)

March 3: Breadboard version of Nerva (Nuclear Engine for Rocket Vehicle Application) successfully ground-tested for first time at full power—1,100 mw, corresponding to 55,000-lb. thrust in space—by NASA and AEC at Nuclear Rocket Development Station, Jackass Flats, Nev. Two and one-half hours after initial ignition, engine was restarted and run for 15 min. at varying power levels, reaching a maximum at slightly less than full power. Nerva engine had operated successfully up to partial power—40 mw—Feb. 3. (AEC Release J-54; *Atomic Energy Programs, 1966,* 186)

March 3: Seven scientific experiments for the 1967 Mariner mission had been selected by Dr. Homer E. Newell, NASA Associate Administrator for Space Science and Applications, on recommendation of NASA's Space Science Steering Committee: (1) ultraviolet photometer, (2) S-band occultation, (3) dual-frequency radio occultation, (4) trapped radiation detector, (5) helium vapor magnetometer, (6) plasma probe, and (7) celestial mechanics. (NASA Release 66-42)

- President Johnson submitted ComSatCorp's third annual report to Congress. In transmittal message he said that many satellite communications facilities would be used by the Government only "to meet its unique and vital national security needs which cannot be met by commercial facilities." He said the capacity of Government facilities "shall at all times be limited to that essential to meet such unique needs."

  Report cited ComSatCorp's 1965 accomplishments: operation of EARLY BIRD I comsat in synchronous orbit above Atlantic; initiation of program to furnish communications service for Apollo manned space program and to expand commercial service; beginning of construction of terminal stations in Washington state and Hawaii; pursuit of expanded research and development programs, including program to establish global commercial satellite system by launching synchronous satellites in 1968; and issuance of requests for proposals to study multipurpose satellites. (Text; ComSatCorp Release; *Pres. Doc.,* 3/7/66, 315)

- Harold B. Finger, Director, NASA Nuclear Systems and Space Power Div. and Manager, AEC-NASA Space Nuclear Propulsion Office, discussed NASA and AEC programs to provide nuclear energy systems for space applications before the Senate Committee on Aeronautical and Space Sciences: "A major new program initiated by the AEC at NASA's request during 1965 is the development of the SNAP-27 isotope-thermoelectric generator to power experiments that will be left on the Moon by the Apollo astronauts. This Apollo Lunar Surface Experiment Package (ALSEP) is intended to transmit data . . . for at least a year. The SNAP-27 isotope power supply will provide 50 watts to the experiments and will weigh about 40 pounds. It will use a single plutonium loaded fuel capsule which will be inserted into the generator on the lunar surface by the astronauts." (Testimony; *NASA Auth. Hearings,* 541)

- Edmond C. Buckley, NASA Associate Administrator for Tracking and Data Acquisition, in testimony before the House Committee on Science and Astronautics' Subcommittee on Advanced Research and Technology, reviewed NASA's tracking and data acquisition network program: "During fiscal year 1965, the network, under the management of Goddard Space Flight Center, successfully supported four Gemini missions, two of which were manned. So far in fiscal year 1966, three manned Gemini missions have been supported, including the highly successful Gemini VII/VI rendezvous in December. . . .

  "During 1965, an average of 32 satellites was supported at any given time. This included satellites launched in prior years which were still transmitting useful data plus new satellites which were launched during the year. In 1966, the average number of satellites to be supported is expected to increase to more than 35 including flight of the Orbiting Astronomical Observatory, the Biosatellite, and Applications Technology Satellite.

"In 1967, the average number of satellites to be supported is estimated to exceed 40 with the spacecraft, collectively, contacting the network stations more than 650 times daily. Contributing to the extensive workload in 1967 will be the additional flights of the Orbiting Astronomical Observatory, Biosatellite, Applications Technology Satellite, and Orbiting Geophysical Observatory. . . .

"During fiscal year 1965, the [Deep Space] network successfully supported the Ranger VII, VIII, and IX missions and the cruise phase of the Mariner IV mission. The precise operation of the facilities at Goldstone and the control center at Pasadena in support of the Ranger missions was essential to the accuracy achieved in impacting on the surface of the moon, and to the activation of the cameras and transmission of live television pictures of the moon's surface. In fiscal year 1965, 85-foot antenna facilities at both Canberra, Australia, and Madrid, Spain, became operational and participated in Mariner IV support. . . .

"The excellent support provided the Mariner IV mission permitted the spacecraft, after a flight of 325 million miles and an elapsed time of 228 days, to pass Mars with a deviation of only 1,000 miles from the projected fly-by distance of 5,000 miles and to take the first close-up pictures of the planet. The Deep Space Network continued to support the Mariner IV spacecraft after encounter and established a new two-way communications record of 191 million miles between network stations and the spacecraft. . . . On January 4, 1966, the Goldstone facility received a signal from the Mariner IV spacecraft over a distance of 216 million miles." (Testimony; *NASA Auth. Hearings*, 608, 727-9)

March 3: Canine alternates of Soviet space dogs Veterok and Ugolyek orbiting earth aboard COSMOS CX appeared on Moscow TV with physician-cosmonaut Dr. Boris Yegorov. Wearing space jackets to which pickups and feeding devices were attached, dogs showed no signs of anxiety when Yegorov placed them in a cabin similar to COSMOS CX and demonstrated a number of instruments and apparatus used onboard COSMOS CX such as pharmacological preparations painlessly introduced into dog's aorta through tube and cuffs with built-in sensors. (Tass, 3/3/66)

- NASA Aerobee 150A launched from NASA Wallops Station carried AFCRL-instrumented payload to 136-mi. (219-km.) altitude in experiment to measure incident solar radiation. Rocket vehicle coning motion made accurate solar pointing and consequent data acquisition possible only 60 per cent of the time. (NASA Rpt. SRL)

- Legislation (H.R. 13266) to provide for building prototype of a commercial supersonic transport (Sst) without using Federal funds was introduced in the House by Rep. Frank T. Bow (R-Ohio). Bow proposed that responsibility for development of aircraft be shifted from FAA to a new Supersonic Transport Authority empowered to sell public $1.5 billion in development bonds. Bonds would not be obligations of the Government, but Government would guarantee payment in event of default. Bow said his plan would avoid large appropriations of public funds and provide sound basis for determining industry's share of development costs when commercial product was realized. (*CR*, 3/3/66, 4640)

- U.S. plan for halting spread of nuclear weapons was unacceptable to

U.S.S.R. because it did not bar non-nuclear nations from sharing in an international atomic force or ban transfer of atomic arms to army units of non-nuclear nations, Soviet negotiator Semyon K. Tsarapkin told 17-nation disarmament conference in Geneva. Tsarapkin joined U.S. delegate Adrian S. Fisher in rejection of demand by nonaligned nations that nonproliferation treaty contain pledge by nuclear powers to liquidate their atomic arms. (AP, *Wash. Post*, 3/4/66, A17)

March 4: NASA had signed supplemental agreement with Boeing Co. converting Saturn V 1st stage (S-IC) contract from fixed-price to incentive-fee. Contract, currently valued at $850,114,303, was first Saturn stage contract to be converted to incentive type. (MSFC Release 66-49; NASA Release 66-48)

- Five NAA Rocketdyne J-2 engines designed to power Saturn V's S-II stage successfully underwent second consecutive full-duration test firing aboard "battleship" test vehicle at Santa Susana, Calif. Engines were fired for 360 sec. and developed 650,000-lbs. thrust—comparable to 1,000,000-lbs. thrust in flight operation. (NAA *Skywriter*, 3/11/66, 1)

- Prof. Emilio Segre of Univ. of California, Berkeley, reviewed postwar physics in *Science*: ". . . many illustrious men of science, physicists in particular, have made the mistake of thinking that the end of physics was in sight. They have consistently been proved wrong by the opening up of completely new fields. . . . Space exploration and the study of the interior of the earth are new departures. Here we do not yet see any new phenomena, but we are penetrating in unexplored regions. It is possible that these regions will not yield anything extraordinary, such as extraterrestrial life. However, they present phenomena on scales impossible to reproduce in the laboratory, and a change in orders of magnitude is a well-known source of surprises. Furthermore, we must not forget that particle physics originated with the study of cosmic rays." (*Science*, 3/4/66, 1052-5)

- U.S. and other scientifically advanced nations have a solemn and urgent duty to help less-developed countries on their way by providing both knowledge and materials, wrote Frederick Seitz, president of National Academy of Sciences, in *Science*, but the "success of this partnership . . . will ultimately depend on the extent to which those in the less-developed lands are willing to sacrifice themselves and their way of life to the pursuit of technological development. One of Homi Bhabha's [late director of Tata Institute, Bombay] great contributions to India lies in the willingness he had to mold old traditions to new patterns of development." (Seitz, *Science*, 3/4/66)

- Astronauts Elliot M. See, Jr. (Cdr., USNR), and Charles A. Bassett II (Maj., USAF), killed Feb. 28 when their T-38 jet training plane crashed in St. Louis, were buried in Arlington National Cemetery with full military honors in two separate ceremonies. Fifteen astronauts attended, 12 of whom served as pallbearers. See and Bassett had been selected as prime crew for the Gemini IX mission scheduled for spring 1966. (Hoagland, *Wash. Post*, 3/5/66, B1; AP, *NYT*, 3/5/66, 9)

- FAA would advance from Oct. 15 to Sept. 1, 1966, date for submission of final design proposals by two airframe and two engine manufacturers participating in supersonic transport (Sst) development program, announced B/G J. C. Maxwell, director of FAA Sst Devleopment. FAA expected to complete its evaluation of the competing designs by Dec. 31 as scheduled. Boeing Co. and Lockheed Aircraft Corp. were competing for airframe contract; General Electric Co. and Pratt & Whitney Div. of United Aircraft Corp., for engine contract. (FAA Release 66-23)

March 4: Launch Umbilical Tower No. 3 at KSC was lifted by crawler transporter and moved to Pad A of Launch Complex 39—three miles away—for placement on pedestals in checkout of procedures for Saturn V launch operations. (KSC Historian)

- Dr. Norman F. Ness of GSFC Laboratory for Space Sciences described in *Science* the new view of earth's magnetic field afforded by satellite mapping. Among developments in space physics, he pointed out studies in planetary magnetism leading to comparison of earth with a "magnetic comet" and results obtained by MARINERS II and IV indicating the unique strength of earth's magnetic field, compared with those of Venus and Mars. He noted the importance of these data and other satellite measurements for paleomagnetic research. (*Science*, 3/4/66, 1041-52)

- FCC authorized ITT World Communications, Inc., to provide live television coverage of recovery operations for March 15 Gemini VIII mission. A portable station aboard U.S.S. *Boxer* would use EARLY BIRD I comsat as it did for GEMINI VII and VI recoveries. (UPI, *NYT*, 3/5/66, 13)

March 5: Agena target vehicle similar to one to be used on March 15 Gemini VIII rendezvous and docking mission had apparently passed series of test firings necessary to qualify for flight, Charles W. Mathews, NASA Gemini program manager, told AP. Data from tests conducted at USAF Arnold Engineering Development Center, Tenn., were being evaluated. (AP, Balt. *Sun*, 3/6/66)

- After trip from Seal Beach, Calif., aboard USNS *Pt. Barrow*, S-II-F facilities stage—nonflight version of Saturn V/Apollo launch vehicle's 2nd stage—arrived at KSC for use in testing launch facilities. Stage would be mated with two other facilities stages and an Apollo spacecraft mockup to form the 365-ft.-tall Saturn V launch vehicle. This would mark first time all major components would have been joined in the Saturn V configuration. (MSFC Release 66-47)

- Failures of VENUS II and III space probes were admitted by U.S.S.R. Tass's original announcement had reported that a "softlanding" had been planned for VENUS III, but a corrected announcement which appeared two hours later eliminated the word "soft" and failed to confirm that the spacecraft had ejected its sterilized capsule. Tass said the descending apparatus had consisted of parachute system and 23.6-in. sphere containing instruments to measure Venusian surface temperature and pressure, to which "heat-resistant substance had been applied ensuring protection from high temperatures when braking in the dense layer of the atmosphere." VENUS II had failed to transmit photos or televise the planet's sunlit side as programed because of a breakdown in its communications system: "The last radio communications with stations Venus 2 and Venus 3 as they approached the planet were not held for reasons that had not been established."

Both VENUS II and VENUS III carried equipment to measure interplanetary magnetic fields, cosmic rays, and streams of low-energy charged particles, in addition to special "traps" to determine magnitude of solar plasma and micrometeorites; however, slightly different instrumentation was used to obtain cross-reading of the data. Both spacecraft were similarly constructed with one section for flight equipment and a second for auxiliary equipment. Second section in VENUS II carried television photographic device for use during flyby. Tass said VENUS II was to have flown "close to Venus on the side lighted by the sun at a distance not exceeding 40,000 km. (25,000 mi.); VENUS III was to have landed on the surface of Venus in the center of the visible disc of the planet." 89 successful radio exchanges had been held with the spacecraft; data had been recorded to "obtain a high accurate forecast of the movements of both stations," Tass said. (*Pravda*, 3/5/66, USS-T Trans.; *NYT*, 3/6/66, 78; *Wash. Post*, 3/6/66, A14)

March 5: Secretary of Defense Robert S. McNamara's statement to House Armed Services Committee that the "Manned Orbiting Laboratory development program should proceed on a deliberate and orderly schedule" received comment by James J. Haggerty, Jr., in *Journal of the Armed Forces*: " 'Deliberate' is an excellent word in this context; according to one definition . . . it means 'leisurely in movement or action. . . .' In these terms MOL . . . shows promise of becoming one of the most deliberate projects of all time.

" 'Orderly', though, is not such a good word. Last summer DOD approved hardware development and set a manned flight target of 1968, then in a matter of months, reneged on the schedule by whacking almost a quarter of a billion dollars out of the planned funding program. McNamara made no mention of a new schedule, but other sources indicate that late 1969 is now an optimistic target . . .

"The SecDef reported that the MOL funding figure for fiscal 1967 is $159 million. This, together with remaining funds from the $150 million allocated in the current fiscal year, will provide for 'design, definition, system integration, development of specifications and determination of firm cost proposals . . . scheduled for completion during this coming spring and summer, after which contract will be awarded for full-scale development of hardware.' . . ." (Haggerty, *J/Armed Forces*, 3/5/66, 8)

- Over 100 scientists, including U.S. space officials, would visit Greece to observe annular eclipse May 20, during which lunar shadow on earth would make 99 per cent of the sun invisible in that area. Joint NASA-Greek National Committee for Space Research project would launch seven GSFC-instrumented Boosted Arcas rockets from USNS *Range Recoverer* stationed several miles off-shore from Koroni in southern Peloponnesos. Experiments would investigate ionization below 56 mi. due to eclipse-caused "changes in solar ultraviolet and X-ray flux." Data would be coordinated with those obtained by nine Arcas and Centaure rockets which European Space Research Organization (ESRO) would launch from Euboea, Greek island in Aegean Sea. (*NYT*, 3/6/66, 10; NASA Release 66-108; Wallops Release 66-22)

- Astronaut James Lovell and nuclear submarine U.S.S. *Benjamin Franklin*'s crew exchanged mementos of a joint experiment conducted during Dec. 4-18 GEMINI VII mission. *Benjamin Franklin*, with two seconds to fire a Polaris missile so it could be seen by GEMINI VII Astronauts Lovell and Borman as they passed over Cape Kennedy, fired perfectly, enabling the astronauts to track the missile visually. In a ceremony aboard the submarine docked in Newport News, Va., Lovell was presented gas generator plate from the Polaris missile and crew received the NASA patch from Lovell's GEMINI VII spacesuit. (AP, *Wash. Sun. Star*, 3/6/66)

March 6: USAF announced it was designing an expandable sidewall folding aircraft tire which would deflate after aircraft had taken off, and inflate before landing. New tire concept, being developed by B. F. Goodrich Co. for AFSC's Flight Dynamics Laboratory, would give large aircraft better ground flotation capability through increased contact between tire and ground, making rough field landings easier, and permitting control of tire pressure. (AFSC Release 197.65)

March 6: USAF awarded Sylvania Electric Products, Inc., $46,000 contract to continue studying methods of eliminating six-minute communications blackout caused when friction of the atmosphere envelops reentering spacecraft in a ball of hot, electrically charged gases. Dr. Ronald Row, Sylvania Electronic Systems Applied Research Laboratory, told the *New York Times* in a telephone interview of the development of a device to simulate the flow of charged gases around typical spacecraft antenna. He said Sylvania was considering shortening pulses of signals so they could penetrate gases or covering antennas with plastic or ceramic shrouds that would shield them from gases and heat but still permit radio signals to penetrate. (Wilford, *NYT*, 3/7/66, 19)

During week of March 6: Possibility that residents of communities afflicted by jet noise might develop psychotic symptoms because their dreams were interrupted at night was suggested by Dr. Julius Buchwald, psychiatrist, and Dr. Howard M. Bogard, psychologist, at a hearing held by New York State Assembly's Mental Hygiene Committee. Dr. Buchwald said that everybody dreams at least five times a night. If a person is awakened and prevented from having his dreams, psychotic symptoms from mild to "more severe" could occur, including paranoidal delusions, psychoses, hallucinations, and suicidal and homicidal impulses. Dr. Bogard said persons prevented from dreaming would "tend toward true psychoses." He urged study of whether residents near airports "lose out on dream time," whether such communities become "disoriented" by people leaving, and whether family life is disrupted because of interruptions of ordinary communications. (Hudson, *NYT*, 3/13/66, 66)

March 7: All primary mission objectives of first unmanned Apollo Saturn IB mission Feb. 26 were achieved and flight was very satisfactory according to initial evaluation of test results, NASA announced. Guidance and control systems performed well; both S-IB and S-IVB trajectories and velocities were normal and no structural problems were found in either stage or instrument unit. Quality of data received at ground stations was excellent and few losses occurred in expected 1,300 measurements telemetered.

All spacecraft systems and subsystems performed as planned with two exceptions: (1) one of the main parachute harness legs failed to sever after landing, thus retaining chute attached to module; and (2) Service Module (SM) propulsion system exhibited below nominal performance during both firing periods resulting in reentry speed about 500 mph less than 18,500 mph expected, but sufficient to evaluate performance of heatshield for orbital missions. Earth landing system functioned properly to safely recover CM, and unsevered parachute was recovered for post-mission examination. Spacecraft would be flown to Downey, Calif., for detailed inspection by NASA and NAA Apollo officials and tests for improvements. (NASA Proj. Off.; MSFC Release 66-51)

• NASA Assistant Administrator for Technology Utilization Breene M. Kerr, reviewing NASA technology utilization program for 1965 in testimony before the House Committee on Science and Astronautics' Subcommittee on Advanced Research and Technology, stated that the total number of NASA Tech Briefs now published was more than 600. He continued:

"We anticipate publication of approximately 700 Tech Briefs during calendar year 1966. . . . we estimate the average Tech Brief is seen by more than 10,000 persons. . . . Industry interest in our program is . . . indicated by the fact that approximately 1,500 persons visited our Field Center Technology Utilization Offices during calendar 1965. . . . Two additional Regional Dissemination Programs have been started in the last year, one at the University of New Mexico in Albuquerque and the other at Southeastern State College in Durant, Okla. One hundred and thirty-one companies are now paying annual membership fees at our established computer-based centers. . . . Of real significance . . . is the fact that approximately 3,000 companies have received some degree of service from these centers. . . . We have also recently concluded an interagency agreement with the Small Business Administration whereby several of that organization's field offices, under an experimental program, will be providing NASA technology to selected small business concerns." (Testimony; *NASA Auth. Hearings*, 634-9)

March 7: Series of small balloons was launched from Christchurch, New Zealand, in joint U.S.-New Zealand pilot project to examine feasibility of Global Horizontal Sounding Technique (Ghost)—plan for using satellite to locate and read out information from large number of balloons floating around earth at constant altitudes. Balloons, circling hemisphere at 20,000-ft., 40,000-ft., and 80,000-ft. altitudes, would be tracked by six stations equipped with high-frequency receivers, stop watches, and sun tables. Program would determine whether small superpressure balloons could fly for extended periods; test temperature, humidity, pressure, and radiation sensors which might be used in operational Ghost system; learn whether balloons would cluster in certain areas; and acquire data on planetary circulation in Southern Hemisphere at various altitudes. Called Southern Hemisphere balloon experiment, project was sponsored by ESSA and NSF and had the endorsement of U.N. World Meteorological Organization. Data would be analyzed by National Center for Atmospheric Research and Information and shared with all participants and interested nations. (ESSA Release 66-17)

- Sen. Fred R. Harris (D-Okla.), speaking on Senate floor in support of Sen. Carl T. Curtis's (R-Neb.) resolution calling for National Science Foundation to set guidelines for wider and more equitable distribution of Federal R&D funds, said: ". . . expenditure of this money can and should be an investment in the future of each of the several states, without regard to geographic location or the accident of history." (*CR*, 3/7/66, 4917)

March 7-11: Third Annual Space Congress of the Canaveral Council of Technical Societies held at Cocoa Beach, Fla., was attended by more than 1,000 scientists and engineers. Astronaut-aquanaut M. Scott Carpenter was principal opening day speaker. (KSC Historian)

March 8: NASA Aerobee 150 sounding rocket launched from WSMR carried high angular resolution x-ray collimators to estimated 103-mi. (165-km.) altitude in NASA-American Science and Engineering, Inc.-MIT experiment to determine size and location of strong x-ray sources in Scorpius. Rocket and instrumentation performed successfully [see May 23]. (NASA Rpt. SRL)

- Solar electric propulsion of unmanned spacecraft to Mars appeared entirely possible by the mid-1970's, reported JPL researchers John Stearns and Donald W. Ritchie at AIAA meeting in San Diego. Four 55-ft.-long folding rectangular solar-cell panels would furnish power that could be harnessed to a set of small ion-chamber engines to propel spacecraft with approximate 7,000-lb. payload toward Mars. Trip would take about 300 days and culminate in orbit around Mars. Report also envisioned 850-day mission by solar-electric-propelled spacecraft to and beyond Jupiter in late 1970's. (JPL Release; NASA Release 66-53)

March 8: NASA Langley Research Center awarded General Electric Co. $3,000,000 incentive-type contract to build the sixth spacecraft in NASA's Scout-launched reentry heating flight series. The 13-ft. conical spacecraft would be launched to measure heating rates associated with turbulent air flow close to vehicular surface, which could not be simulated with ground equipment. (LaRC Release; NASA Release 66-54)

- ComSatCorp petitioned FCC to deny ITT's request to construct a fourth satellite earth station in Puerto Rico. ComSatCorp said "public interest would best be served" if it, rather than ITT, owned the station, and revealed plans to seek permission to construct and begin operating the facility by mid-1968. (Wash. *Sun. Star,* 3/1/66, E11; *NYT,* 3/12/66, 22)

- Reps. William Green (D-Pa.) and Earle Cabell (D-Tex.) were appointed to House Committee on Science and Astronautics, replacing Reps. Bob Casey (D-Tex.) and Brock Adams (D-Wash.), who resigned. (NASA LAR V/41)

- Lt. Valentina Tereshkova, who made 48-orbit VOSTOK VI spaceflight in June 1963, was promoted to major in the Soviet Air Force, reported *Krasnaya Zvezda.* March 8 was U.S.S.R. International Woman's Day. (Reuters, *NYT,* 3/9/66, 6; AP, Wash. *Eve. Star,* 3/8/66, A2)

March 9: Seven U.S. scientists had consented to serve on joint U.S.-U.S.S.R. Editorial Board provided for in agreement on space biology and medicine reached by NASA and Soviet Academy of Sciences representatives Oct. 8, 1965, in New York City: Melvin Calvin, Univ. of California, Berkeley— U.S. co-chairman; Loren D. Carlson, Univ. of Kentucky; Dr. Robert W. Krauss, Univ. of Maryland; Robert B. Leighton, Cal Tech; Dr. John P. Marbarger, Univ. of Illinois; Dr. Orr E. Reynolds, NASA; and Wolf Vishniac, Univ. of Rochester. Agreement provided that Editorial Board, co-chaired by a Soviet and a U.S. scientist, would supervise collection, preparation, and publication of materials on space biology and medicine. It would select compilers to assemble and prepare materials and authors —approximately equal number of Soviet and U.S.—to write the material. Work was expected to be published in several volumes, in both English and Russian, between 1967 and 1968. Cost of work performed by each country would be borne by that country. (NASA Release 66-49)

- Successful mock launching of Gemini 8 spacecraft at ETR in preparation for March 15 three-day rendezvous and docking mission. Astronauts Neil Armstrong and David Scott spent two and one-half hours in Gemini capsule during simulated spaceflight. (UPI, *NYT,* 3/10/66, 10)

- NASA had awarded Lockheed Propulsion Co. $96,000 contract to evaluate solid-propellant motor which could stop and restart spacecraft propulsion units on command. Current solid-propellant rocket motors did not have this capability. On future space missions concept could be applied to midcourse correction, probe separation, and orbit and space-probe injection. Lockheed, under contract to JPL, would conduct 18 test firings. (NASA Release 66-56)

- USAF had established and would implement in FY 1967 procedures for the management of DOD's secondary payload space activities, Dr. John S.

Foster, Jr., DOD's Director of Defense Research and Engineering, told closed hearing of Senate Committee on Aeronautical and Space Sciences. Program would provide for operation of a central business office which would accept requests for DOD launch support of space experiments from NASA, other armed services, and other DOD nonservice R&D activities. Arrangements for secondary payloads on USAF vehicles had previously been handled on an informal basis. (*M&S Daily*, 3/9/66)

March 9: ComSatCorp proposed to FAA an aeronautics satellite to handle complex communications for airlines and aid in FAA's air traffic control. Two-channel, 210-lb. satellite would be launched in late 1967—two or three years earlier than previous proposals—into 22,300-mi.-altitude synchronous orbit to provide more reliable very-high-frequency radio communications over North Atlantic area currently served by high-frequency radio system affected by bad weather and sunspot activity. (UPI, *Wash. Post*, 3/11/66, D6; Cooke, *WSJ*, 3/10/66, 32; Clark, *NYT*, 3/10/66, 19; Wash. *Eve. Star*, 3/11/66, A8)

- USAF launched unidentified satellite from WTR with Thor-Agena D booster. (*U.S. Aeron. & Space Act., 1966*, 148)
- Flight operations through national airspace system in 1965 surpassed all previous years, according to annual FAA Air Traffic Activity Report. The 292 FAA airport traffic control towers reported 37.9 million takeoffs and landings during the year—an 11% increase over 1964. Largest increase —nearly 34%—was in general aviation (nonairline) flying; airline activity increased by only 14%. (FAA Release 66-25)

March 10: NASA Administrator James E. Webb, testifying on the NASA FY 1967 budget request before the House Committee on Science and Astronautics, said that the gap between the U.S. and Soviet space programs still remained and "this budget will not close it. We are as much as 2 years behind the Soviet Union in certain important aspects of space power. In . . . 1965, they launched 52 Cosmos satellites; successfully orbited a 3-man spacecraft; demonstrated a communications satellite capability with 2 Molniya spacecraft; and orbited the heaviest payload by anyone in the world to date, indicating they have developed a new launch vehicle with some 2½ to 3 million pounds of thrust. Since the beginning of 1966 they have achieved a successful soft landing on the moon and they have reached Venus with two probes.

"The Soviet program shows every evidence of a continuing major commitment to long-term, large-scale operations in space. . . . The massive Soviet commitment to a rapid buildup and a long-term program underlines the importance they attached to advancing their space capabilities. It will require a strong and increasing effort initiated no later than fiscal year 1968 and vigorously pushed in the years after the United States has achieved a manned lunar landing to prevent them from forging ahead as the unchallenged leader in space. . . ." (Testimony; *NASA Auth. Hearings*, 5)

- Soviet space dogs Veterok and Ugolyek, approaching the end of their 16th day in space aboard COSMOS CX, were in "quite satisfactory condition," Tass reported. "This is indicated by telemetered data such as pulse, respiration, electrocardiograms, and others—as well as observation of the dogs during television transmission session." (Tass, 3/10/66; Reuters, *Chic. Trib.*, 3/11/66)
- MSFC engineers Michael J. Vaccaro and Haydon Y. Grubbs, Jr., ended 18-day stay inside lunar vehicle simulator to test interior chamber design

and investigate human factors of an enclosed cabin environment. During experiment they performed assigned tasks inside chamber, went outside in spacesuits to simulate going onto lunar surface to gather rock samples, and performed emergency tasks when the compartment walls were "punctured by meteoroids." Results of experiment, conducted by Honeywell, Inc., might be used in developing hardware needed in future exploration of moon. (MSFC Release 66-56)

March 10: NASA awarded Honeywell, Inc., a $230,000 contract to determine alternate approaches and concepts for measuring horizon radiance characteristics over earth's surface—knowledge essential for design of space vehicle guidance and control systems. LaRC-directed study, including investigation of factors affecting horizon radiance profiles, analysis of measurements and evaluation of possible flight techniques, would support Project Scanner rocket flight tests. (LaRC Release; NASA Release 66-57)

- USAF Agena D rocket launched from Vandenberg AFB with Atlas booster Feb. 15 exploded into at least 40 pieces which fell from orbit during seven-day period, UPI reported. There was no indication that the payload itself was damaged: "It is no longer in orbit, but it may have been a reconnaissance satellite which was brought down and recovered," UPI explained. Explosive gas in the otherwise empty Agena apparently was detonated accidentally after booster had fired. (UPI, *Chic. Trib.*, 3/11/66)

- GEMINI IV Astronaut Edward H. White II, first American to walk in space, told AP that the most important word of advice he had given Astronaut David Scott, scheduled for extravehicular activity during the March 15 Gemini VIII mission, was: "Don't get in a hurry.

  "Take things slow and try not to hurry and rush, because you can't do things in a real rapid manner up there. It's better to go on the slow, deliberate approach." (AP, *Chic.Trib.*, 3/11/66)

- U.S.S.R. would soon launch single spacecraft with six cosmonauts farther into space than man had ever ventured, possibly to the inner Van Allen radiation belts about 500 mi. (805 km.) above earth, *Veda a Technika Mladezi*, Czechoslovakian technical journal reported. Last manned spaceflight was that of VOSKHOD II with cosmonauts Alexey Leonov and Pavel Belyayev on March 18, 1965. (*Av. Wk.*, 3/14/66, 34; Simons, *Wash. Post*, 3/10/66, A4; AP, Wash. *Eve. Star*, 3/10/66, A1)

March 11: Dr. Mac C. Adams, NASA Associate Administrator for Advanced Research and Technology, outlined the highlights of NASA's 1965 advanced research and technology program before the House Committee on Science and Astronautics, citing significant contributions to such national programs as the supersonic transport (Sst) and the USAF C-5A transport, and listing major accomplishments: landing studies of reentry lifting bodies, second Fire experiment to measure heat transfer at high reentry velocities, two successful firings of 260-in. solid-propellant rocket motor, series of successful nuclear rocket firings, 2,600-hr. endurance test of an ion engine, and endurance tests of the Snap-8 nuclear engine which "demonstrated altitude specific impulse of over 750 lb.-sec./lb.—almost twice chemical rocket values—and a thrust of over 50,000 pounds. Accumulated running time for the reactor tests was over 4200 seconds. . . ." (Testimony; *NASA Auth. Hearings*, 169-78)

- Quasars are relatively small objects 10 million to 1 billion yrs. old and 30 million to 300 million light years from earth that have been shot at super speeds out of another galaxy by a titanic explosion, astronomer Dr.

Halton C. Arp, Mt. Palomar Observatory, contended in *Science*. Arp's theory contradicted prevailing view that quasars could enable man to see the ends of time and space and possibly glimpse the secrets of creation because they were on the rim of the universe—10 or more billion light yrs. away—and were almost as old as the universe—8 billion yrs. Astronomers believed that the quasars' "red shifts" in light—changes toward red end of the light spectrum as objects speed away—were caused by extremely rapid motion away from our own galaxy. Arp suggested that quasars might be very dense and compact, exerting a strong gravitational attraction for their size which could shift wavelengths to produce "red shifts." (*Science*, 3/11/66; Weil, *Wash. Post*, 3/12/66, 2; Sullivan, *NYT*, 3/11/66, 28; Ubell, *N.Y. Her. Trib.*, 3/11/66)

March 11: NASA Associate Administrator for Manned Space Flight Dr. George E. Mueller, testifying on FY 1967 budget authorization before the House Committee on Science and Astronautics, stated that NASA had reached a manpower peak of 300,000 people in the manned spaceflight program: "We have passed our peak in engineering, we are reaching our peak in manufacturing, and we are at our peak in total Manned Space Flight manpower resources. . . .

"A striking anomaly of the space program is that we begin to 'go out of business' before we fly our first operational vehicle. . . . The program to develop the uprated Saturn I launch vehicle illustrates this situation. The decline in manpower on this program has been underway for several months, but the first flight took place only 2 weeks ago—and the first manned flight is almost a year away. By then, the employment level will be almost half the peak level of 1965."

Reviewing NASA plans for a manned project after Apollo, Dr. Mueller cited five major program alternatives under study. The alternatives "placed emphasis on (1) direct economic benefits to mankind, (2) lunar exploration and science, (3) planetary exploration, (4) maximum effort in all program areas, and (5) a program balanced over the whole spectrum of alternatives, which could be shifted as required to meet national needs.

"We have not made a decision upon which of these courses we will embark." (Testimony; *NASA Auth. Hearings*, 32-3, 148)

- Performance of Apollo Command Module (Cm), launched Feb. 26 from ETR on AS-201 flight, was successful enough to obviate any delay in schedule for a 1969 manned lunar expedition, Dr. Joseph Shea, manager of MSC Apollo Spacecraft Office, told Gladwin Hill of the *New York Times*. The Cm, on display at North American Aviation, Inc., in Downey, Calif., remained structurally intact, and the few minor technical problems that had developed during flight were not serious enough to cause any engineering problems in the assembly-line production vehicles for subsequent phases of the testing program, Dr. Shea said. "This was the first major confirmation of design processes and it went quite well." (Hill, *NYT*, 3/12/66, 8)

- Prince Philip of U.K. toured MSC for three hours, accompanied by department heads explaining the apparatus used in spaceflight. With assistance of Astronaut James A. McDivitt, the Prince spent 30 minutes in an astronaut training craft, twice steering it into simulated orbital rendezvous with an Agena rocket. "It was all a fascinating experience," he said later. "When you are inside, everything seems to be out of context." Dr. Robert Gilruth, MSC Director, presented Philip a British flag which had been carried aboard GEMINI VII during its Dec. 4-18 space mission with the intention of presenting it to the Prince during his tour. (MSC *Roundup*, 3/18/66, 8; Waldron, *NYT*, 3/12/66, 8)

March 11: France's report for the Register of the U.N. Committee on the Peaceful Uses of Outer Space of her successful Nov. 26, 1965, launch of A-1 scientific satellite from Hammaguir Range, marked first such report by nation other than U.S. or U.S.S.R. (Text)

March 12: Gemini 8 hatch difficulty which threatened to delay March 15 mission was solved by Astronaut David R. Scott. After KSC technicians had discovered that 65- to 70-lbs. pull pressure was required to close cabin door because of too-thick sealant around hatch edges, Scott worked with hatch-closing mechanism until he was satisfied he would have no difficulty closing the door in space. Officials preferred to have required pull pressure at 40 to 45 lbs., but mission rules would tolerate 80 lbs. If hatch could not be closed properly, astronauts would be unable to repressurize their cabin.

Mission review board of NASA, USAF, and industry officials reaffirmed March 15 launch date after studying all factors of the flight: Titan II and Atlas-Agena boosters, Gemini 8 spacecraft, control centers at KSC and MSC, global tracking network, and recovery forces. (AP, Balt. *Sun* 3/13/66)

- Meteorite crater 1,476 ft. in diameter and with an average depth of 100 ft. —third largest crater on record—was discovered in the western foothills of the Chilean Andes by geological team led by Dr. William A. Cassidy, Columbia Univ. Lamont Geological Observatory, and Joaquin Sanchez, Instituto de Investigaciones de Geologicas de Chile, NSF announced. Crater, which lies in an area 1,000 ft. above sea level accessible only by foot, was discovered while team was investigating a depression shown in an aerial photograph. Crater's age had not been determined. (*Chic. Trib.*, 3/13/66)

- Astronaut Virgil I. Grissom, in Toronto to address the Canadian Aeronautics and Space Institute, said at a news conference that although computers predicted that only one of the first three manned U.S. attempts to land on the moon would be successful, human judgment would perhaps prove them wrong. He said computers had predicted the loss of two of seven astronauts on the Mercury program but none was killed. (AP, *Denver Post*, 3/13/66)

- Soviet biologist Norair Sissakian, believed to have played key role in designing life-support systems for animal and human passengers in spacecraft, died suddenly. His official obituary, signed by Communist Party leader Leonid Brezhnev, Premier Alexey N. Kosygin, President Nikolay Podgorny, former President Anastas I. Mikoyan, and all other members of 12-man ruling presidium of Soviet Communist Party, referred to his important contribution to "the formation and development of a new science—space biology." Sissakian would be buried in Moscow's Novodevichye Cemetery March 14. (Tass, 3/13/66; *Pravda*, 3/14/66, 3)

- Sir Sydney Camm, British aircraft designer who designed the RAF Hurricane fighter used in World War II, the Hunter fighter, the P-1127 Vtol aircraft, and other combat aircraft, died in London. Sir Sydney, honorary fellow and member of council of Royal Aeronautical Society, had received the Society's Gold Medal in 1958 and had been selected to receive the Guggenheim Medal, highest U.S. aeronautical award, in the spring. (*NYT*, 3/14/66, 31)

March 13: NASA Nike-Apache sounding rocket launched from Churchill Research Range was third and last flight [see Feb. 9 and 16] in series of GSFC experiments to measure number and distribution of electrons in energy range of 1-300 kev, which produce visual aurora. Although radar lost track early in flight and peak altitude was not available, rocket performance was near predicted and series was considered completely successful. (NASA Rpt. SRL)
- Formation of Science Advisory Committee, composed of astronomers, biologists, physicists, and geologists from eight universities, to advise NASA on conduct of future space projects was announced by NASA Administrator James E. Webb. Chaired by Dr. Norman F. Ramsey, Harvard Univ. physicist, committee would work directly with Dr. Homer E. Newell, NASA Associate Administrator for Space Science and Applications, but would study future space activities on NASA-wide basis, including manned and unmanned flight programs. In particular, it would review how best to conduct such programs as Voyager planetary spacecraft, including automated biological laboratory, post-Apollo lunar exploration program, and National Space Astronomy Observations—projects recommended by NAS Space Science Board. (NASA Release 66-55)
- U.S. would land a man on the moon before 1970 and before the Russians, predicted Astronaut Walter M. Schirra, Jr., at a news conference held with Astronaut Frank Borman in Sydney, Australia. Schirra said U.S. space program had completed three basic requirements for a manned lunar landing: 14-day spaceflight, ability to rendezvous in space, and controlled reentry into earth's atmosphere. Astronauts Schirra and Borman were visiting Australia on their eight-nation good-will tour of Asia. (Reuters, *NYT*, 3/14/66, 6)
- Gradual and unusually frank disclosures by U.S.S.R. confirming failures of VENUS II and VENUS III missions were discussed by Evert Clark in the *New York Times*. He noted Soviet scientists' admission that VENUS III's trajectory would have missed the planet by 37,500 mi.—passing Venus when neither object would have been visible in Moscow—had its trajectory not been corrected by Dec. 27 radio command, and that VENUS III had failed to eject landing sphere covered with heat-protective material to measure temperature and pressure at planet's surface. (Clark, *NYT*, 3/14/66, 7)
- U.S. physics was facing severe shortage of funds necessary to sustain progress in research and education, concluded NAS Physics Survey Committee in its report *Physics: Survey and Outlook*. Report recommended that physics support by Federal agencies be increased by 1969 to two and one-half times the 1963 $500-million level for physics and astronomy. Assessing U.S. strength in six subfields, report noted for astrophysics "a pressing need for more observational facilities." Based on 1964-65 studies by an 18-member group headed by George E. Pake, Provost of Washington Univ., St. Louis, report stated: "Our strength in observational astrophysics with optical telescopes has long been established with the 200-in. telescope on Mount Palomar, but we have many more bright astrophysicists and astronomers than have access to the two US telescopes most suited for frontier research. Our relative strength will be altered with the implementation of plans for construction of several large telescopes in the Soviet Union. Any nation can, by placing a large telescope in the Southern Hemisphere, assume leadership in the observational astronomy of stellar evolution and cosmology, because

the Magellanic Clouds are the nearest of all external galaxies. The United States has taken the initiative in the expensive but highly promising field of space-based optical and x-ray astronomy.

"In radioastronomy the United States now has an impressive group of major radio telescopes, but the US position is not preeminent. Even the new instruments nearing completion at the California Institute of Technology and the National Radio Astronomy Observatory are inferior to existing instruments in Australia and the Soviet Union and to large new crosstype arrays nearing completion near Sydney and Moscow. The US position in space physics and cosmic radiation is good, with some question whether present conditions permit further strengthening of that position. Research on gravitation is at present not a large sector of research, but the US effort is of very high quality and is being increasingly recognized." (NAS Release; *Science*, 3/18/66, 1363-6; *Physics Today*, 4/66, 23-36)

March 14: Gemini VIII launch would be delayed at least 24 hrs. because of a leak in a unit of the spacecraft's environmental control system and an overflowing fuel tank in the Atlas booster, NASA announced. Launch was reset, NASA announced March 15, for March 16. (Wilford, *NYT*, 3/15/66, 17; Sehlstedt, Balt. *Sun*, 3/15/66; Hines, Wash. *Eve. Star*, 3/14/66, 1)

- Global television relayed by U.S. and U.S.S.R. satellites could be used as a political instrument, threatening the independence of some countries and causing far-reaching changes in world relationships, warned Eurospace, organization of 160 European aerospace industries, in a memorandum released in London. When ordinary home television sets began to receive programs from these satellites "the national independence of developed countries without similar means of expression might be threatened by this persuasive new weapon," memorandum said. Eurospace warning, released three days after an announcement that U.K. and other European nations had joined U.S.-dominated Intelsat consortium to launch system of world-spanning comsats by 1968, urged Europe to play its own major role in space "before it is too late." Intelsat's space communications system would utilize Early Bird-type satellites which Eurospace did not consider "powerful enough to have noteworthy political or economic repercussions." (Reuters, *Wash. Post*, 3/15/66, A16)

- Two meteorologists from U.S. National Environmental Satellite Center were in Geneva to discuss with their Soviet counterparts a cooperative approach on gathering weather information by satellite, including information exchange about meteorological satellite design and operation, *Aviation Week* reported. Full review of U.S.-U.S.S.R. agreement was scheduled for May. (*Av. Wk.*, 3/14/66, 25)

- U.S. might achieve a manned lunar landing by 1968, Dr. Gerard P. Kuiper, director of Univ. of Arizona's Lunar and Planetary Laboratory, told Phoenix *Gazette*. "There is nothing on the moon's surface to prevent a landing and subsequent take-off. The surface is crunchy but solid. It will support the Lem (Lunar Excursion Module), and man can walk on it with relatively little trouble," Kuiper added. (AP, Balt. *Sun*, 3/15/66)

- Investigations of seals' habits might yield valuable information for astronauts exploring the moon, particularly with regard to enduring extremes of cold and lack of oxygen, Dr. Carleton Ray, New York Zoological Society, told the *New York Times*. Ray, who recently returned from Antarctica expedition sponsored by the Society with NSF aid, said that of pertinent interest would be "reasons why seals can endure tremendous cold, sleep under water without breathing, and use what appears to be a remarkable echo-sounding system for navigating in total darkness under thick ice to return to isolated breathing holes needed sporadically to help them remain submerged in a state somewhat akin to semihibernation." (Devlin, *NYT*, 3/14/66, 6)

March 15: U.S. delegate to the Geneva disarmament conference Adrian Fisher urged U.S.S.R. to reconsider President Johnson's offer of a "verified freeze" of the production of missiles and other strategic vehicles for launching nuclear bombs as a "logical first step in the control of the growth of inventories of strategic nuclear vehicles." Soviet delegate Semyon K. Tsarapkin had rejected the U.S. plan March 3. (*NYT*, 3/16/66, 10)

- Compañia Telefónica Nacionale de España (CTNE)—the Spanish national telephone company under NASA contract—awarded ITT contract to establish a ground station terminal complex on Grand Canary Island to support communications for U.S. Apollo mission. Stations, equipped with 42-ft. parabolic antenna reflectors, super-cooled low-noise amplifiers, and 10-kw transmitters, would furnish multichannel telephone and telegraph service. (ITT Release)

- New York Univ. announced completion of New York Univ.-NASA aerospace laboratory in Bronx was scheduled for June 1966. One-story, 15,000-sq.-ft. facility constructed with $582,000 NASA grant would include wind tunnel capable of duplicating speeds of up to 14 times the speed of sound for four seconds; models of supersonic transports, spacecraft, and other advanced aerospace designs would be tested there. (*NYT*, 3/16/66, 55)

- A laser to function as a ground-based device to recharge an orbiting satellite's batteries was suggested by Dr. Charles H. Townes, MIT, and Dr. Arthur L. Schawlow, Stanford Univ., co-inventors of the laser, in a paper before the Optical Society of America's meeting in Washington, D.C. Report predicted exploitation of the laser would grow into a billion-dollar industry by 1970. (Clark, *NYT*, 3/16/66, 45M)

- Employment in aerospace industry would increase to 1,266,000 by June 1966—a gain of 94,000 or eight per cent from September 1965, reported Karl G. Harr, Jr., president of Aerospace Industries Assn. Aircraft employment was expected to show greatest increase, with missile and space employment remaining relatively stable. (*Aerospace News*, 3/15/66)

March 15-16: AAS-NSC Fourth Goddard Memorial Symposium—"Space Age in the Fiscal Year 2001"—was held in Washington, D.C.

Dr. Robert C. Seamans, Jr., NASA Deputy Administrator, delivering the first Annual Dr. Robert H. Goddard Lecture, stressed the importance of the contributions of men such as Dr. Robert H. Goddard and Dr. Hugh L. Dryden to the dramatic development of the space program: "The history of NASA reflects much of Dr. Dryden's understanding of how progress can be made and what the conditions are that produce effective responses to the challenges being faced. A pivotal concept that has guided NASA administration has been that of the relation of the research and development project to many rapidly growing disciplines of science and technology. A project serves a larger purpose than its own defined immediate ends. A project is a disciplined and organized effort directed to a specific objective; one of its principal characteristics is that it has a schedule. As such a project provides a creative and driving force in the total achievement of science, engineering, and technology . . . ." (Text)

Wesley A. Kuhrt, United Aircraft Research Laboratories, predicted interplanetary and translunar tourist travel for FY 2001: "A six-day round trip to the moon via 35-passenger, nuclear-powered 'clipper' ships would cost each adventurer $10,000. For an 18-month trip to Venus, the passenger would pay $32,000. Steerage class to Mars, a two-year undertaking, would cost $35,000. But for $70,000 a passenger could ride the Mars Express to and from the red planet in 11 months. . . ." All fares assumed the Government would underwrite the cost of developing the spacecraft, Kuhrt said. Other predictions made at the symposium included teams of superhuman astronauts bred over several generations as a result of present studies in genetics and molecular basis of life; communications between ordinary persons in their homes and anyone, anywhere; versatile robots to perform human tasks; cosmic cargo transported at $10 per pound; network of satellites to link everyone in the world so they could speak in their native languages via rapid automatic translating computers; and man-shaped robots sending back three-dimensional pictures to human masters hundreds of miles away. Harvard astronomer Carl Sagan speculated that the trip from other galaxies had already been made and that earth had already been visited from space at least once. (Text)

Rep. Joseph E. Karth (D-Minn.) told a luncheon audience that "we must lay out basic objectives far enough ahead that they will serve as relatively fixed goals—or in non-technical terms, as dreams. Such an early dream of man was to fly like a bird; later, we gradually translated this dream into making it easy for everyone to fly in an aircraft. We have these dreams and they affect our progress strongly. . . ."

To meet these ends, Karth recommended forming a "Hoover-type" Commission to review planning and decision-making system in the area of science and technology. In addition he cited the need for "frequent, systematic, large-scale surveys and analysis of science and technology from the near-term to the distant frontiers of knowledge. . . ." (Text)

In a prediction of what the "Space Age Society" would be like in 2001, NASC Executive Secretary Dr. Edward C. Welsh told the symposium: "The major development over the coming years will be the speed of change. We already have great difficulty maintaining our perspective regarding the change which has taken place during the last 8 or 10 years. Well, that rate of change is like the tortoise pace compared with what is coming in the future. . . .

"Some with fertile imaginations foresee the time when there will be large scale emigration from earth to growing settlements on the moon and the planets. While I see the great growth in the ability to travel to such distant spots safely and speedily, I see little likelihood of any sizable movement of population even though nuclear reactors would furnish much of the competence to make conditions in such places both livable and productive. My reason for not predicting such movement is due to the simple belief that living conditions are going to be so improved here on earth that few would want to leave. . . .

"As for space travel, we can certainly let our imaginations soar, as we are now only in the infancy of growth. The space child has barely learned to walk. The future with its space travel being almost as common as airplane travel is today is yet to come . . . but it will come. Interplanetary trips, at speeds many times what we now attain, will take place in craft as reusable as the jet plane is today, with at least as many passengers aboard. . . .

"One more thing I can predict with great confidence. The debate on the worthwhileness of space technology and space exploration—still continuing on the part of those of little vision—will have disappeared completely within a decade or so. The benefits will have become so obvious and the potential so exciting that even the most near-sighted will be aware of the rich space harvest. . . ." (Text; Weil, *Wash. Post*, 3/16/66, A10; Hoffmann, *N.Y. Her. Trib.*, 3/16/66)

Importance of materials science and space vehicle technology was discussed by Milton B. Ames, Jr., Director of Space Vehicle Research and Technology, NASA Hq. OART. "The structural designer of future launch vehicles will face many new and challenging problems. Launch vehicles . . . [will be] larger, complex, and more costly. If we are to reuse these launch vehicles without having to spend large sums of money either to bring them back or to refurbish them after recovery, major advances will be required in structural and materials technology." Development of new materials and methods, such as "superinsulations" to prevent heat leakage; stronger shielding materials to protect spacecraft from meteoroids and intense radiation; advanced lifting-body designs for greater reentry precision; and folding spacecraft parts which could be reassembled in space, would play a major role in U.S. space program before the year 2000, he said. (Text)

Dr. Charles S. Sheldon II, NASC staff member, discussed overall economic outlook for space age up to FY 2001. Sheldon noted that U.S. space program took about one per cent of gross national product and estimated that "Soviet Union is putting about two per cent of their GNP into space efforts . . . .The close tie with military rocketry has speeded space development in these recent years, but it has also saddled space technology with an expendable ammunition philosophy rather than aiming at the lower operating costs of a recoverable transport philosophy. . . .

"Today we probably are headed toward an age of conservation, and one of deliberate recycling . . . the task of the remainder of this century may include the development of closed ecological systems for long-duration space flight, and on a different scale, heroic measures to recycle waste water and to reprocess industrial and consumer wastes into reusable products. . . .

"If our urbanized society requires closed ecological systems even here on Earth, we can, in effect, air condition whole cities."

Discussing weather modification and climate control, he said: ". . . whether we bend nature itself, or create new sealed urban units, each almost self-sufficient in its ability to control its temperature, clean and recycle its air and water, and put all other wastes to productive use, some such heroic measures will be needed as this century runs out." (Text)

March 16: Gemini VIII rendezvous and docking mission—world's first space docking—began at ETR with launch of 7,000-lb. Agena Target Vehicle (Gatv) by Atlas booster at 10:00 a.m. EST, followed 100 min. later by launch of GEMINI VIII by two-stage Titan II booster with Astronauts Neil

A. Armstrong, command pilot, and David R. Scott, pilot. Primary objectives of the mission—sixth manned flight in the Gemini series—were rendezvous and docking of GEMINI VIII spacecraft with GATV and performance of extravehicular activity (Eva) for one orbit.

GATV was injected into near-circular orbit with 186-mi. (299.5-km.) apogee and 184-mi. (296.3-km.) perigee; GEMINI VIII entered initial orbit with 128.8-mi. (206.4-km.) apogee and 74.9-mi. (119.5-km.) perigee. Near third apogee, Armstrong fired aft thrusters, circularizing spacecraft's orbit at 168-mi. (270-km.) altitude. Trailing GATV by 139 mi., Armstrong received intermittent lock-on between GEMINI VIII radar and GATV's companion transponder. After sighting GATV at 66-mi. range at 04:12 GET, Armstrong fired thrusters and Scott read out computer data until 05:15 GET, when they switched from catch-up to rendezvous mode. At 150 ft. from GATV, Armstrong reported he had matched velocity for brief station-keeping exercise. Conditions at that point were so favorable that Armstrong passed under GATV and came up 80 ft. in front of it, where he was able to read some data on GATV's status display.

Successful docking was confirmed at 6:33 GET. Plan had been to remain docked with GATV through 7½-hr. sleep period until beginning Eva, but at approximately 7:00 GET spacecraft-GATV combination began to roll and yaw violently. Unable to stabilize joined spacecraft, Armstrong fired GEMINI VIII's thrusters to undock and maneuvered away from the GATV. Once freed, however, GEMINI VIII began to roll even more rapidly; roll, pitch, and yaw rates increased, approaching one revolution per min. For three minutes, the astronauts fought to control GEMINI VIII; finally, with OAMS turned off, Armstrong fired his Reentry Control System (Rcs) rockets in final attempt to counteract violent roll. This maneuver was successful and eight minutes later, at 7:25:30 GET, the spacecraft was stabilized. Within a matter of minutes, the problem was isolated to the spacecraft's No. 8 thruster, which had fired intermittently while GEMINI VIII was docked with GATV and stuck open until Armstrong managed to turn off OAMS. It was later determined that a short circuit was probable cause of thruster's malfunction. Premature use of reentry rockets forced Gemini flight director John D. Hodge to order an emergency landing, with recovery planned in the western Pacific during seventh revolution. (Safety rules of the Gemini program required immediate landing of any spacecraft once its Rcs squibs were blown; if the flight were allowed to continue, danger of Rcs fuel depletion would jeopardize firing of Rcs rockets, vital to safe return of spacecraft and crew.)

Spacecraft landed 500 mi. west of Okinawa in regular emergency recovery area at approximately 10:23 p.m. EST—10 hrs. 42 min. after launch. Landing was only three miles from intended impact point. Within minutes USAF frogmen parachuted from USAF C-54 rescue aircraft and placed flotation collar around capsule. At 1:30 a.m. EST capsule and astronauts were hoisted aboard recovery ship U.S.S. *Leonard F. Mason*.

Despite early termination, GEMINI VIII mission accomplished one of two primary objectives—rendezvous and docking. Original flight plan had specified four GEMINI VIII-GATV dockings and two hrs. 15 min. of extravehicular activity by Astronaut Scott. Five scientific, four technological, and one medical experiment were to have been conducted. Flight had been scheduled to last 44 revolutions—approximately 73 hrs.

(NASA Proj. Off.; NASA Release 66-52;Wilford, *NYT*, 3/17/66, 1;Waldron, *NYT*, 3/17/66, 20; Hines, Wash. *Eve. Star*, 3/17/66, 1; Simons,*Wash. Post*, 3/17/66, 1; *Av. Wk.*, 3/31/66, 30-32)

March 16: Astronauts David Scott and Neil Armstrong awaiting recovery in the Pacific Ocean after their GEMINI VIII space flight.

March 16: President Johnson issued statement after GEMINI VIII carrying Neil Armstrong and David Scott had splashed down safely following difficulties encountered in flight: "They are disappointed that their mission could not continue, but Gemini 8 accomplished the first docking in space—a major step on the course we have set. The information they have acquired will help us to perfect the spacecrafts that will carry us even further.

"From their skill and strength we all take heart, knowing that the personal qualities of the astronauts and their colleagues will ultimately prevail in the conquest of space. We are very proud of them." (*Pres. Doc.*, 3/21/66, 400)

March 16: COSMOS CX, launched Feb. 22 with dogs Veterok and Ugolyek aboard, landed safely in U.S.S.R. Tass reported that though "the state of the animals after landing is good," both dogs had suffered heart beat irregularities that had become more pronounced toward the end of the 22-day flight. Respiration rate had changed insignificantly. During initial period of weightlessness, some disruption of coordination had been revealed in "a rocking of the head." One of the dogs had been "subjected to all the complex of irritants to indicate the reaction and functional ability" of the cardiovascular system; the other was used for control and comparison. Electronic sensors had provided continuous checks on blood pressure, pulse and respiration rates, and other data. Dogs had been fed a paste injected into their stomachs by a pneumatic device. Spacecraft had carried 300 thermoluminescent dosimeters to measure radiation doses in relation to shielding and ten sets of nuclear emulsions to determine the composition of radiation. Scientific information obtained from the flight, which took the animals through the Van Allen radiation belts once during each of COSMOS CX's 330 revolutions, "required special processing after which final conclusions about the state of the animals and inferences about plans for future studies will be drawn." (Tass, 3/16/66; *Av. Wk.*, 3/28/66, 28)

- President Johnson received 1966 Robert H. Goddard Memorial Trophy Award of the National Space Club at White House for "his efforts in shaping the space program from its infancy while a U.S. senator and as chairman of the National Space Council while Vice President and for providing the leadership for this nation's preeminence in space as President," and affirmed: "We intend to land the first man on the surface of the moon and we intend to do this in the decade of the sixties."

  He also said: "I think if I were writing my own epitaph this morning I perhaps would prefer to be remembered, for the period I was Vice President, more by this one thing than any other. President Kennedy asked me to attempt to organize and give direction to the Space Agency. I interviewed about 19 men through the United States and finally selected Jim Webb. I prevailed upon him to come and begin this undertaking.

  "Within 15 minutes from the time I called the President originally, Jim Webb was out here being announced as the new Space Administrator. I have been very proud of that announcement ever since.

  "Your modesty, humbleness, your great executive ability, your great courage, your Marine spirit and determination have made us the envy of the world, Mr. Webb, and we salute you publicly.

  "Dr. Goddard was a great prophet, a true prophet. To some it seems almost incredible that a year before Lindbergh had ever flown the Atlantic he was dreaming and working to take us up into the stars. Like so many prophets, he was long without honor in his own country, but he never lost faith because his faith was founded on fact. He could turn aside the rebuffs of the shortsighted; he could laugh at the jokes of his detractors.

"I just wish today that he could be here with us to observe what we have just seen. I don't think he would be the least surprised by the progress we have made. I think he would be very proud." (*Pres. Doc.*, 3/21/66, 399; *SBD*, 3/8/66, 49; *CR*, 3/21/66, A1621)

March 16: The late Dr. Robert Hutchings Goddard launched the first liquid-fueled rocket from Auburn, Mass., 40 yrs. ago. Annually on the anniversary of the first flight, Goddard is honored by the National Space Club at a memorial banquet in Washington, D.C. In 1962, Vice President Johnson predicted that within 25 yrs. there would be a permanent base on the moon. Vice President Hubert Humphrey, delivering the 1966 address, emphasized that the space budget must be tightly planned: "I feel the necessity for cost consciousness. This is the need, to put it another way, of getting the most space for the tax dollar. These are times when we must exert high discipline in public expenditure and our space program cannot be exempt from that discipline." Humphrey said he could foresee "dramatic achievements ahead" in space exploration, notably: (1) exploration of the lurain and possible establishment of one or more permanent bases there; (2) development of "a whole family of earth-orbiting stations, supplied by regular ferry services"; (3) development of "recoverable and reusable launch vehicles" and maneuverable spacecraft; (4) building of spaceports around the country; (5) improvement of propulsion methods so that planetary trips could be made in a week or less; (6) "launching of unmanned probes to every part of the solar system—and perhaps manned planetary expeditions as well." (Text; *CR*, 5/4/66)

National Space Club Press Award "in recognition of his exceptional photographic documentation of the manned space flight program" went to NASA photographer William P. Taub; Astronautics Engineer Award was given to Gemini program manager Charles W. Mathews "for his contribution to the nation's manned space flight program as a key engineering pioneer to investigate the feasibility of manned space flight, culminating in Project Mercury...and in the successful Gemini program, as well as in the establishment and refinement of the Apollo configuration"; Nelson P. Jackson Aerospace Award for "making a major advance in the state of the art of space systems" was presented jointly to MSC, USAF Space Systems Div., McDonnell Aircraft Corp., Martin Co., Aerojet-General Corp., General Electric Co., IBM Corp., and Westinghouse Electric Corp.; Robert H. Goddard Historical Essay Award was made to Airman 2/C Frank H. Winter for "Danish Rocketry in the 19th Century"; Robert H. Goddard Scholarship ($1,500 to university of recipient's choice) "in recognition of the need for higher scientific education of American youth of exceptional merit" was awarded Benjamin N. Early, a senior at Howard Univ., Washington, D.C. (Text; Program)

- NASA turned ESSA II (Environmental Survey Satellite) over to ESSA for operation in accordance with June 30, 1964, agreement between NASA and Dept. of Commerce. NASA had supervised design and construction of the satellite, conducted launch operations (Feb. 28), and tracked spacecraft after orbit was achieved; Dept. of Commerce had handled mangement and funding. (NASA Release 66-62)
- Breadboard version of Nerva (Nuclear Engine for Rocket Vehicle Appli-

cation) was tested by NASA and AEC at Nuclear Rocket Development Station, Jackass Flats, Nev. Control positioning error in initial ignition led to automatic flow shutdown. In 18-min. test following second, low-pressure start, reactor reached design power of 1,090 mw and exhaust temperature reached 3,630°F maximum. Test was fourth in series which began Feb. 3 and marked system's eighth and ninth start. (*Atomic Energy Programs, 1966,* 186)

March 16: Smithsonian Institution had photographed reflections of GEMINI VIII and GATV from tracking station 20 mi. north of Johannesburg on spacecraft's first orbit. Photographs were described by station manager as being "not of the space vehicles themselves, but of the sunshine reflected by them. They will appear on the film as dots." This was reportedly first time such photographs had been taken. (AP, *NYT,* 3/18/66, 8)

- Rep. Mark Andrews (R-N.D.) introduced a resolution calling for NSF to make recommendations for wider geographic distribution of Federal funds for research and development. (*CR,* 3/16/66, 5755)

March 17: U.S.S.R. launched COSMOS CXII with scientific instruments aboard for continued space research, Tass announced. Orbital parameters: apogee, 565 km. (350.9 mi.); perigee, 214 km. (132.9 mi.); period, 92.1 min.; inclination, 72°. All systems were reported to be functioning normally. (Tass, 3/17/66)

- NASA selected Bendix Corp. for negotiation of a $17-million cost-plus-incentive-fee contract to manufacture four Apollo Lunar Surface Experiments Packages (Alsep). The 170-lb. packages containing scientific instruments to measure lunar surface characteristics and atmosphere would be carried to moon in Lunar Excursion Module (Lem) on initial Apollo landing missions and would transmit data back to earth for six months to one year after crew's departure from lunar surface. Contract would be managed by MSC. (NASA Release 66-63)

- Heinz Kaminski, chief of the Bochum Observatory, West Germany, said the emergency abort March 16 of the GEMINI VIII mission showed that control systems onboard the spacecraft as well as on the ground were versatile enough to cope with even the most serious trouble: "The Americans now have realistic data on space emergencies and. . .the mission has been very successful. The emergency has been a blessing in disguise." (Reuters, *Chic. Trib.,* 3/18/66)

- Tass told the Soviet people that GEMINI VIII had achieved history's first docking in space, and added: "But its program was carried out only partly. Its main elements—more practice in linking up, a lengthy stay by an astronaut outside his ship and the maneuvering of the ship and astronaut—remained unsolved." (UPI, *NYT,* 3/19/66, 41)

- Taped conversations between Astronaut Neil A. Armstrong and ground controllers disclosing the drama of the emergency but revealing nothing of the difficulty that had caused emergency landing March 16 of GEMINI VIII were released at MSC press conference. There was no indication of panic aboard the spacecraft during motion. In an "almost matter-of-fact tone," Armstrong had reported to MSC: "Well, we consider the problem serious. We are toppling end-over-end. . . . We cannot turn anything off." It was announced there would be no public discussion of their mission by Armstrong and Scott for at least nine days during which they would rest and undergo debriefing at KSC.

MSC Director Robert R. Gilruth announced that preparations for the

May 26 Gemini IX launching would continue "just as if nothing had happened to GEMINI VIII.... While we are expending every effort to find the cause of the GEMINI VIII occurrence, preparations will move ahead for Gemini IX so when the results of this flight become available, we'll be as far along as possible with Gemini IX." (Waldron, *NYT*, 3/18/66, 20; Simons, *Wash. Post*, 3/18/66; AP, *Wash. Post*, 3/18/66)

March 17: Problems that beset eight of the U.S.'s 12 manned space flights were listed by AP: July 21, 1961, before any American had achieved orbit, Astronaut Virgil I. Grissom had had to swim for his life when the hatch blew off the LIBERTY BELL 7 after it had landed at sea; Feb. 20, 1962, Astronaut John H. Glenn had had to take partial manual control of FRIENDSHIP 7 when a small jet thruster failed; May 24, 1962, Astronaut M. Scott Carpenter was forced to use the horizon when a short circuit cut off the device which told him attitude of AURORA 7; May 15-16, 1963, drinking water leaked into the cabin of FAITH 7, and ground stations, fearing the automatic reentry system might have been short circuited, directed Astronaut L. Gordon Cooper to fly the spacecraft manually in reentry—the first astronaut to do so; June 4-7, 1965, Astronauts James A. McDivitt and Edward H. White II, because of malfunctioning computers, controlled GEMINI IV reentry on instructions from the ground; Aug. 21-29, 1965, Astronauts L. Gordon Cooper and Charles Conrad, Jr., flew GEMINI V through 120 revolutions after it seemed a fuel pressure problem might force them down after six; Oct. 25, 1965, Gemini VI mission was canceled when the Agena target vehicle exploded before going into orbit; March 16, 1966, Astronauts Neil A. Armstrong and David R. Scott brought GEMINI VIII under control from a spin that had reached at least one rpm and landed it almost exactly in the planned emergency zone. (AP, *Wash. Eve. Star*, 3/17/66)

- Press commented on the GEMINI VIII mission.

New York Times: "The docking maneuver thus successfully pioneered is of the utmost importance. It is through combining the loads brought into space by individual rockets that the great structures can be built that will be necessary for man's activities in the cosmos in the years ahead. Tomorrow's stations in space will serve as laboratories for astronomers and other scientists, as terminals for transshipment of passengers and freight between short-range and long-range rockets, and unfortunately—if the future resembles the past—also as military bases reflecting in the sky the tensions and hostilities of earth. But whatever purpose they serve, these islands in space will have been made possible by repeated use of the same maneuver that Astronauts Armstrong and Scott performed for the first time last night."

Washington Post: "Space feats have become so commonplace and have been carried out with such precision that we tend to forget about the hazards. The abrupt but fortunately safe return of Gemini 8 is a timely reminder that space is still a frontier and that the exploration remains at best an inexact science.

"That Astronauts Armstrong and Scott emerged unharmed from their awesome experience with their space capsule out of control is a tribute not only to their own cool judgment but also to the extraordinary competence of Space Agency officials. Americans listening to reports of the emergency must have applauded collectively when the crewmen made a successful reentry. They survived a predicament in space which, had it been on a highway, would have meant disaster."

Washington *Evening Star*: "In addition to its docking achievement, the mission was a success in the sense of demonstrating the care with which emergency devices were built into the space ship and the thoroughness of the planning for rescue operations in event of a forced landing. It is possible, indeed, that more can be learned from a mission that goes wrong than from one which is successful all the way.

"Incidentally, there is a commentary on our times in the fact that while the suspense-filled space story was unfolding, Batman was being shown on television. The ABC network interrupted the program with special news announcements, and was rewarded with more than 1,000 phone calls from complaining viewers." (*NYT*, 3/17/66, 36M; *Wash. Post*, 3/18/66, A20; *Wash. Eve. Star*, 3/17/66, A16)

March 17: Discussing the urgency of NASA's deciding "what to do in space after the first Apollo astronauts return from the moon about 1970," William Hines said in the Washington *Evening Star*: ". . . it is hard to forget a hint, dropped . . . by a leading Russian space official several years ago. The Soviet Union, this functionary suggested, might very well celebrate the 50th anniversary of its great revolution by having cosmonauts plant a flag on the lunar surface. That deadline—if it is a deadline—is now less than 20 months away.

"But suppose the Soviet Union is not racing us to the moon, or for that matter suppose that she is. In the present state of planning, the U.S. with the moon in its grasp would be very much like a dog chasing an automobile—what does he do with it if he catches it?" (Hines, *Wash. Eve. Star*, 3/17/66)

- USA-USAF-USN X-22A V/Stol airplane, designed and built by Bell Aerosystems Co. under Navy contract, completed 10-min. first flight at Niagara Falls (N. Y.) International Airport. The four-engine, dual-tandem, ducted-propeller craft made four takeoffs and landings, reaching 25-ft. maximum allowable altitude. While hovering, Bell test pilots Stanley J. Kakol and Paul Miller, Jr., executed a 180° counterclockwise turn and reported that X-22A responded to "all control commands." After planned 225-hr. flight test series, Bell would deliver their two X-22A research aircraft to a tri-service group at Patuxent Naval Air Test Center, Md., for further evaluation. (Bell Releases 3/17/66, 4/4/66-20)
- Dr. S. Bhagavantam, scientific adviser to Indian Defense Ministry, reported at a news conference that India had developed a "limited guided missile." He said that tests of the missile, which could be fired from aircraft, had been completed and arrangements were being made to produce it for Indian armed forces. (AP, *NYT*, 3/19/66, 3)
- Eighth anniversary of launching of VANGUARD I, second U.S. earth satellite. (NASA Historian)

March 18: Astronauts Armstrong and Scott landed safely in Honolulu after KC-135 jet aircraft carrying them from Okinawa to KSC lost oil pressure in its number two engine about 800 mi. east of Hawaii. Armstrong and Scott had earlier received an enthusiastic reception—with brass band—when they disembarked at Okinawa from recovery ship U.S.S. *Leonard F. Mason* for flight to Honolulu. (UPI, *Wash. Post*, 3/19/66, A4; AP, *NYT*, 3/18/66, 21)

- USAF launched two unidentified satellites with Atlas-Agena D booster from WTR. (*U.S. Aeron. & Space Act.*, 1966, 149)
- Dr. Donald F. Hornig, Director of Office of Science and Technology and

science adviser to the President, in memorandum to President Johnson said: "In October 1965 I convened an ad hoc Jet Aircraft Noise Panel ... I am pleased to be able to present to you the report of this Panel entitled 'Alleviation of Jet Aircraft Noise Near Airports.' " Principal recommendations were: (1) initiation of Federally-supported studies of expected scope of noise problem through 1975 and of public and private programs needed to combat the problem; (2) creation of a Federal task force to analyze the problem; (3) development of valid, broadly applicable standards of noise measurement; (4) pursuit of definitive technical study to reduce noise levels; (5) establishment of task force to investigate methods for Federal participation in program for compatible land utilization near airports; (6) initiation of program to modify operating procedures and techniques that would reduce engine noise without compromising safety. (*Pres. Doc.*, 3/21/66, 404)

March 18: First anniversary of the VOSKHOD II mission when Soviet Cosmonaut Aleksey A. Leonov became the first man to float in space. Interviewed on Radio Moscow, Leonov said his space walk had caused the spacecraft to begin rotating unexpectedly: "Down on earth we had trained for every emergency, but not for this since we had thought my weight of 176 pounds would have no effect on a six-ton ship. Yet, my weight caused the ship to rotate."

Pravda disclosed that VOSKHOD II, carrying Leonov and Col. Pavel I. Belyayev, had landed outside the target area in a snowbank near the city of Perm in the Ural Mountains: "Searchers took more than 24 hours to reach the cosmonauts and another 24 hours and a 12-mile ski trip before the men could be lifted out of the forest by helicopters." (Tass, 3/18/66; *Wash. Post*, 3/19/66, A4; UPI, *NYT*, 2/21/66)

- Catalog identifying and locating every known star in the heavens down to the ninth magnitude, providing astronomers with instant reference to the entire sky, was published by GPO. Compiled by Smithsonian Astrophysical Observatory under NASA grant, 2,700-page, four-volume catalog unified 50 different sources to provide position, photographic and visual magnitude, spectral type, and other essential data for 250,000 heavenly bodies. (Sci. Serv., *NYT*, 5/5/66, 21)

- Speaking in Denver at Colorado Women's College, Dr. Nancy G. Roman, NASA Chief of Astronomy Programs, OSSA, said a primary reason for lack of women in space was that few had the scientific training. Even NASA's 300 women professionals were few compared to their number in other agencies, she said. Young women interested in a career in space science were advised to concentrate first on the basics of their preferred area "because the fundamentals of the science used in space study are no different from those in more traditional fields. Equally important is to learn how to think, to face and solve problems and how to find the facts we need, not just remember them .... Space as a career depends on the individual who's considering it. It is glamorous— and it's also dirty work." Dr. Roman said she was sure there would be women astronauts "if manned space flights continue as we expect them to." (von Ende, *Denver Post*, 3/19/66)

- Douglas Aircraft Co. conducted successful acceptance test of fourth flight Saturn S-IVB (204) at its Sacramento Test Center. Stage was fired for about 455 sec. (*Marshall Star*, 3/23/66, 7)

March 19: Statement from MSC on GEMINI VIII mission said that "a short in the circuits controlling a spacecraft yaw thrustor has been pinned down as the probable cause of the difficulty which cut short the flight of GEMINI VIII." MSC Director Dr. Robert R. Gilruth said data ruled out any possibility of pilot error and that "in fact, the crew demonstrated remarkable piloting skill in overcoming this serious problem and bringing the spacecraft to a safe landing." NASA Associate Director for Manned Space Flight Dr. George E. Mueller, who had participated in review of data, joined Dr. Gilruth in saluting Astronauts Armstrong and Scott and commended the evaluation team for "coming up with the answer" in less than 72 hrs. following the incident. "This will give us time to study the Gemini 9 configuration and make any adjustments necessary in order to press forward with our Gemini flight test program."

16-mm. color films taken by an automatic camera mounted on Armstrong's window in the GEMINI VIII were made public by NASA. Films showed maneuvers of GEMINI VIII in docking with GATV and gyrations of the spacecraft after the malfunction. (Text, *NYT*, 3/21/66, 19; AP, Balt. *Sun*, 3/21/66)

March 19: Astronauts Neil A. Armstrong and David R. Scott arrived at KSC to begin an intensive three-day conference with Gemini project officials to attempt to reconstruct events leading to their loss of control of the GEMINI VIII spacecraft March 16. Scott told a crowd at the airstrip that "the lift off, launching, rendezvous, and docking were really tremendous. We were really looking forward to the whole mission." Armstrong agreed, adding: "We had a magnificent flight in the first seven hours. It was a magnificent launch—without a doubt one of the best there's ever been." (AP, *NYT*, 3/20/66, 82)

- Rep. John W. Wydler (R-N.Y.) said in a speech before the American Society of Tool & Manufacturing Engineers in Westbury, N.Y.: "Two years ago I asked Administrator Webb for NASA's future space plans and was told that none existed. I pointed out then that such plans were essential. Last year, we received assurance that the presentations of the fiscal year 1967 budget would reveal such plans. But none has been issued. Instead, NASA Administrator Webb has told us 'the future of the NASA program lies largely in the decision for fiscal year 1968.' . . .

"Our space program began in an atmosphere of perplexity and fear. Our adversaries in Russia had chosen a new arena for the struggle of international power and prestige. They threatened to achieve impressive capabilities which we would be unable to match. But, characteristically, the United States rose to the occasion in one of the biggest sustained technological efforts ever seen. Today, the U.S.S.R. is denied many of the options for surprising space spectaculars which they once had. From month to month, one country or the other may seem to be leading the so-called space race, but our proven capability shows that we are not to be placed at any military or strategic disadvantage.

"Therefore it is time again for public opinion to assert itself. What do we want to do with this hard won space capability?" (Text)

- USAF successfully launched Nike-Javelin combination carrying a payload of plume measurement instrumentation seconds after launch of its target—an Atlas-D ICBM—from Vandenberg AFB. Launch was first in High Altitude Background and Signal to Noise (Hitab) program in which data on infrared radiation and emission characteristics associated with missile exhaust plumes and other exhaust products were telemetered back to Vandenberg telemetry station. (UPI, *Chic. Trib.*, 3/20/66, 22; UPI, *NYT*, 3/19/66, 82; UPI, *Chic. Trib.*, 3/25/66)

March 20: Tokyo Univ. Space and Aeronautics Institute announced it had launched a two-stage meteorological sounding rocket, last in a series of 15, from test center in Uchinoura, Kyushu—600 mi. southwest of Tokyo. (AP, *NYT*, 3/21/66, 27)

- Astronomers at Bochum Observatory, West Germany, reported they had sighted a giant new sunspot at the approximate center of the solar disc. Visible to the naked eye in early morning or just before sunset, sunspot had 88,000-mi. dia. and was expected to affect weather conditions and radio communications on earth during the next few days. Sunspot was unexpected since 1966 had been designated "year of the quiet sun" and scientists had expected only a minimum of solar disturbances. (UPI, *Chic. Trib.*, 3/21/66)
- Pratt & Whitney Div., United Aircraft Corp., announced that JTF-17A engine developed under FAA supersonic transport engine competition would be tested at company's West Palm Beach, Fla., center. Twin-spool turbofan engine would develop 50,000-lbs. thrust—more than twice amount generated by current commercial engines—by burning fuel in a full-length fan exit duct. (P&W; *NYT*, 3/20/66, 83)
- Benefits from atomic test-ban treaty signed in July 1963 by U.S. and U.S.S.R. were reported. Fallout from strontium 90, which might cause bone cancer or leukemia, was decreasing; iodine 131, believed harmful to the thyroid gland, had virtually disappeared from the milk supply three years ago. (AP, Crowley, *Wash. Post*, 3/20/66, A6)
- Proposal by Dr. Herbert Friedman of the Naval Research Lab. that a scanning system be mounted in a lunar crater to report to earth the precise location of x-ray sources as they appeared over the crater's rim because of the moon's rotation, was discussed by Walter Sullivan in the *New York Times*. Recent rocket observations [see Mar. 2] had disclosed objects, some of them otherwise invisible, that strongly emitted x rays—part of the spectrum which cannot penetrate earth's atmosphere. "A systematic survey of the sky in this wavelength," Sullivan said, would "open an entirely new window on the universe." (Sullivan, *NYT*, 3/20/66, 8E)
- Commentary in the Washington *Sunday Star*: "One of the saddest lessons of Gemini 8 is that bureaucrats will be bureaucrats. . . . NASA officials suppressed for about 18 hours tape recordings of ground-to-orbit conversations under circumstances that strongly suggest their only reason for the suppression was to make sure nothing embarrassing was on the tapes.

  "For as long as the United States has been active in space, it has cultivated an image of openness in contrast to the Russians' obsession with secrecy. The overnight gag on tape recordings—which turned out to be legitimate items of news interest—only served to tarnish this image of candor." (*Wash. Sun. Star*, 3/20/66, G1)
- Endorsement of suggestion in Jan. 9 open letter to President Johnson that he invite "a Russian cosmonaut to join an American astronaut for a space ride in a two-man capsule" was reported by letter's author, *Parade* editor Jess Gorkin. Newspapers throughout the world had publicized the idea and written favorable editorials. *Le Figaro* (Paris) had given it front page coverage. Melbourne (Australia) *Herald* had said the idea had "great merit."

  Asahi Shimbun (Tokyo): "As long as America and Russia accelerate space development separately, there will be much money wasted. Both countries will probably suspect that the other party is engaged in space development for military purposes. . . . If the United States and the Soviet Union cooperate and 'all the families of the earth' pitch in to further space development, war on earth will gradually become absurd."

New York *World Telegram*: "What do we have to lose? If we succeed and the two men orbit the earth together the psychological effect will be tremendous. If Russia turns down the idea, we will be on sounder moral grounds." (Gorkin, *Parade*, 3/20/66)

During week of March 20: Several hundred sightings of unidentified flying objects (Ufo's) near Ann Arbor, Mich., were reported. Rep. Weston Vivian (D-Mich.) said he would ask DOD to make an investigation. (UPI, *NYT*, 3/22/66, 19; UPI, *NYT*, 3/23/66, 22; Hines, *Wash. Eve. Star*, 3/23/66, A1; UPI, *Wash. Post*, 3/25/66, A20; AP, *Wash. Eve. Star*, 3/25/66, A1)

March 21: U.S.S.R. launched COSMOS CXIII with scientific instruments aboard for investigation of outer space, Tass announced. Orbital parameters: apogee, 327 km. (203 mi.); perigee, 210 km. (130 mi.); period, 89.6 min.; inclination, 65°. All systems were reported to be functioning normally. (Tass, 3/21/66)

- NASA announced final results of Feb. 23 test firing of 260-in.-dia. solid propellant rocket motor by prime contractor Aerojet-General Corp. in Dade County, Fla.: maximum chamber pressure was recorded 40 sec. into test, with peak thrust level of more than 3.5 million lbs.; motor burned 114 sec. near peak thrust before tapering off; total useful thrust time was 129.9 sec.—only .1 sec. more than engineering predictions; maximum chamber pressure was 601 lbs. psi compared with predicted 604 lbs. psi. Only difficulty was delayed operation of quenching system which caused charring damage to nozzle expansion cone. Program was managed by LRC. (NASA Release 66-61)

- Twelve astronauts were selected for future NASA missions and two others assigned earlier were shifted to a different mission. Prime crewmen for the Apollo earth-orbital mission scheduled in the first quarter of 1967 were Virgil I. Grissom, Edward H. White II, and Roger B. Chaffee; backup crew was James A. McDivitt, David R. Scott, and Russell L. Schweickart, NASA civilian employee. Prime crewmen for the Gemini XI mission scheduled in the last quarter of 1966 were command pilot Charles Conrad, Jr., and pilot Richard F. Gordon, Jr.; backup crew was Neil A. Armstrong, command pilot, and William A. Anders, pilot.

  James A. Lovell, Jr., and Edwin E. Aldrin, Jr., backup crew for the Gemini XI mission, were reassigned as backup crew for Gemini IX. Original Gemini IX backups Thomas P. Stafford and Eugene A. Cernan became prime crewmen for that mission after the Feb. 28 deaths of Astronauts Elliot M. See, Jr., and Charles A. Bassett II. (NASA Release 66-67)

- NASA and DOD signed memorandum of agreement establishing Manned Space Flight Experiments Board (MSFEB) to coordinate scientific, technological or nonmilitary experiments on DOD and NASA flight missions. Board would recommend approval or disapproval, assignment, and relative priority of experiments submitted, after screening, by sponsoring NASA or DOD Program Offices. Dr. George E. Mueller, NASA Associate Administrator for Manned Space Flight, would act as chairman; other members would be Dr. Homer E. Newell, NASA Associate Administrator for Space Science and Applications; Dr. Mac C. Adams, NASA Associate Administrator for Advanced Research and Technology; Daniel J. Fink, DOD Deputy Director for Strategic and Space Systems; and Gen. Bernard A. Schriever, AFSC Commander. (Text; NASA NMI 1154.4, 4/27/66)

March 21: NASA Nike-Apache sounding rocket launched from Andoeya, Norway, carried scientific payload to estimated 133-km. (82.6-mi.) altitude. Experiment, conducted by GSFC and Norwegian Defense Research Establishment (NDRE), was designed to obtain measurements at high spectral resolution of D-region composition, using a new spectrometer based on ion cyclotron resolution principles. Instrumentation functioned properly, and clamshell nose cone ejected at programed 55 sec. (NASA Rpt. SRL)

- Secretary of Commerce John T. Connor announced appointment of Dr. George S. Benton, Chairman of Dept. of Mechanics at Johns Hopkins Univ., as director of ESSA's Institutes for Environmental Research. Dr. Benton, an authority in atmospheric, oceanographic, and hydrological sciences, would assume his new duties in July. (Dept. of Commerce Release 66-60)

- Merger of NASA Apollo Applications program with the USAF Manned Orbiting Laboratory (MOL) was urged in a report by the House Committee on Government Operations' Military Operations Subcommittee: "Inasmuch as both programs are still research and development projects without definitive operational missions, there is reason to expect that with earnest efforts both agencies could get together on a joint program incorporating both unique and similar experiments of each agency.... Such a merger should be effected within the existing scale of priorities which accords to the military experiments greater urgency." (H.R. No. 1340, 46)

- President Johnson announced appointment of five new members to four-year terms on President's Science Advisory Committee: Dr. Ivan L. Bennett, director of dept. of pathology, Johns Hopkins Univ.; Dr. Sidney D. Drell, professor of physics, Stanford Univ.; William R. Hewlett, president, Hewlett-Packard Co.; Dr. Charles P. Slichter, professor of physics, Univ. of Illinois; and Dr. Charles H. Townes, provost, MIT. (*Pres. Doc.*, 3/28/66, 421).

- Rocket Propulsion Research Laboratory at AFSC's Arnold Engineering Development Center was completed. (AEDC)

- Lockheed-California Co. revealed that F-104 Starfighter had taken off from a Zero Length Launch (Zell) platform in 1963 at Edwards AFB, using a Rocketdyne booster rocket in combination with aircraft's General Electric J-79 engine. After 80-min. flight, Lockheed's pilot Ed Brown said of no-runway takeoff: "It was one of the easiest... I ever made." The secret tests, conducted under contract to the German Air Force, had been made public following Luftwaffe announcement that new series of Super Starfighter Zell tests would be carried out at Lagerlechfeld Base, near Munich. (Lockheed Release, 3/21/66)

- A space rescue service, "preferably under cooperative auspices of the United States and the Soviet Union," was urged in a *New York Times* editorial: "Its function would be to send space craft, on very short notice, to aid a space vessel marooned in orbit and incapable of returning to earth under its own power. In some cases the need might be for additional rocket fuel or replacement parts; in others the primary function might be to save lives by taking aboard the crew of a disabled capsule.

". . . with relatively little development, tow-trucks of the cosmos may be a regular part of the safety arrangements for each new step on the road to the moon." (*NYT*, 3/21/66, 30M)

March 21: Reporting on a January Thiokol Chemical Corp.-sponsored Trendex poll on public opinion of the U.S. space program, William Coughlin wrote in *Missiles and Rockets*: "Some 77% of those interviewed now favor the program as against only 20% opposed to it. This is a dramatic gain from 1963, when only 59% were in favor of the lunar landing and 39% were opposed. . . .

"In the four cities in which the latest poll was conducted—Cincinnati, Syracuse, Pasadena and Des Moines—the public appears to be far ahead of the Administration in its assessment of the importance of military space. The response was: defense and security, 38%; scientific and non-military benefit, 35%; and opportunities for economic growth and employment, 18%." (Coughlin, *M&R*, 3/21/66, 46)

- GEMINI VIII Astronauts Neil A. Armstrong and David R. Scott returned to their families in Houston after debriefing at KSC. (AP, *NYT*, 3/22/66, 18)

March 22: President Johnson requested Congressional approval of amendment adding $200 million to pending request for $80 million in FY 1967 budget for FAA's Sst development program. In a letter transmitting BOB amendment to Speaker of the House John W. McCormack, President Johnson said funds would be used "to complete the design competition, for expanded economic and sonic boom studies and for the start of prototype construction." Work carried out with these funds would "bring us closer to a supersonic transport which is safe for the passenger, superior in performance to any comparable aircraft and profitable for both the airlines and the manufacturers." (Text, *Pres. Doc.*, 3/28/66, 431)

- Construction of a $3.5-billion radiotelescope of 10,000 steerable dish antennas was proposed by Dr. Bernard M. Oliver, outgoing president of IEEE, at the Institute's convention in New York. It could gather 1,000 times more radio energy than the largest radiotelescopes now in use and could detect radio signals emitted in space 200 light yrs. away. Dr. Oliver said the telescope should be built by NASA as a "natural adjunct to the manned and unmanned exploration of space." (*NYT*, 3/23/66, 38)

- AEC announced selection of Martin Co. for an estimated $10-million contract to develop Snap-29—nuclear power source for short-lived earth orbital space missions. Generator would be planned to produce 400 watts of power—nearly ten times the output of any radioisotope space power source now being developed by AEC. Weighing about 400 lbs., it would be fueled with plutonium 210 and would draw power from decay of the radioactive fuel without mechanical generators. (AEC Release J-69; Martin Co.)

- Heat shields on the unmanned Apollo spacecraft that underwent reentry test Feb. 26 had developed three "hot spots," NASA announced. Problem would be corrected with improved insulating compound. No delay in Apollo program was anticipated. (AP, Wash. *Eve. Star*, 3/23/66, A29)

- USAF announced Dr. J. Allen Hynek, chairman of Northwestern Univ.'s Dearborn Observatory and scientific consultant to USAF's Project Blue

Book, would investigate reported Ufo sightings in the Ann Arbor, Mich., area. (UPI, *NYT*, 2/23/66, 22)

March 22-25: Air Force Assn. held its 20th annual convention in Dallas, Tex.

USAF Chief of Staff Gen. John P. McConnell told the meeting that U.S. and U.S.S.R. had enough long-range nuclear weapons to make a nuclear war between them extremely costly for either side, but that a stalemate did not exist between them: "It follows that each country has its own nuclear threshold, that is, the point or level of restraint above which it may no longer be deterred from war. This threshold is by no means static but dynamic, as it is determined by a number of factors which are both variable and unpredictable." McConnell said the strategic superiority of USAF during the past 20 yrs. had kept the nuclear threshold of all potential aggressors at a safe level. (Text)

USAF's first priority—continued maintenance of strategic superiority over any potential enemy—required a mixed force of missiles and manned bombers, Secretary of the Air Force Dr. Harold Brown told the convention. "We will continue to need manned bombers in our strategic forces as far ahead as any of us can see, and I think that should include at least the entire decade of the 1970s.

". . . There is no doubt about the technical feasibility of advanced manned bombers of all sorts of designs. The questions which have to be answered . . . are how we propose to *use* such a weapon system and how it fits in to the missions of assured destruction and damage limitation. . . . This issue is under intensive study by the Air Staff. . . ." (Text)

A Citation of Honor—Air Force Assn.'s highest award—was presented to B/G Joseph S. Bleymaier for "outstanding management achievements while directing the Air Force's development program for the Titan III-C space booster," and to SSD's 655th Aerospace Test Wing at Patrick AFB, for "successful missile and space launches, including six Gemini launches which placed ten astronauts in orbit and achieved the first manned rendezvous in space." (AFSC Release 48.66)

March 23: House Science and Astronautics Committee's Space Science and Applications Subcommittee recommended cancellation of NASA's proposed 1967 Venus mission and proposed that $42 million of otherwise programed money be used in the Voyager program to speed exploration of Mars—crashlanding a probe on that planet in 1969. Subcommittee action would go to full Science and Astronautics Committee in form of report on FY 1967 budget authorization for NASA's OSSA. Such reprograming action was new in space effort. Subcommittee Chairman Joseph E. Karth (D-Minn.) said: "Congress is for the first time asserting its prerogatives, using its judgement, setting priorities of its own." (Wash. *Eve. Star*, 3/23/66, 6)

• President Johnson, in World Meteorological Day statement, said: "This day symbolizes for us—and for all mankind—a new dawn of hope for a better, safer, and more meaningful life. It looks to the time when all our science and technology, and all the wonders of the space age, will give us the power of which man has always dreamed—not the power of one nation over another, but the power of the human race over the forces of nature.

"The instrument of this program is the World Meteorological Organization—a specialized agency of the United Nations with a membership of 127 countries. Through the World Meteorological Organization, the concept of a World Weather Watch is now taking shape. On this occasion, I am proud to say that the United States strongly supports international cooperation in this vital field." (*Pres. Doc.*, 3/28/66, 439)

March 23: NASA Nike-Tomahawk sounding rocket launched from Churchill Research Range carried instrumented payload to estimated 180-mi. (290-km.) altitude. GSFC experiment measured magnetic and electric fields and low-energy electrons and protons in a visible aurora. Rocket aspect measurement and evaluation of a xenon flashing light for trajectory determination were secondary objectives. (NASA Rpt. SRL)

- Lowell T. Wingert, AT&T vice president, wrote a letter to ComSatCorp urging discussions on development of a large-capacity communications satellite for "long haul domestic communications of all types," to be owned and operated by ComSatCorp with "equitable access to domestic common carriers" in return for "bearing their share of ComSatCorp's charges." He called for ComSatCorp and other interested communications users "to undertake a joint study promptly with the hope of placing this newest type of communications facility into early domestic use." (*NYT*, 3/25/66, 48; AT&T Release)

- First captive flight of M2-F2 lifting body vehicle was made from Edwards AFB to check out systems in flight while M2-F2 remained mated to B-52 launch aircraft. (NASA Release 66-89)

March 24: Initial tests of first liquid-hydrogen and liquid-oxygen turbopumps for NASA's 1.5-million-lb.-thrust M-1 rocket engine technology program had been successfully completed at LRC. Pumps, driven by 200,000-hp gas generator, were largest ever built for handling propellants for high-energy rocket engine. (LRC Release 66-10)

- A $50-million, 18-mo. flight research program involving AFSC's supersonic XB-70 experimental aircraft was scheduled to begin in spring 1966, AFSC announced. Joint USAF-NASA effort would acquire experimental data on flight dynamics, flight loads, structures, propulsion and air induction, and environmental factors. Overall direction would be by AFSC's Aeronautical Systems Div. (ASD); technical direction would be by NASA. (AFSC Release 15.66)

- First of fleet of nine, 135-ft. NASA barges for transport of liquid hydrogen and liquid oxygen had been placed in service on inland waterways between New Orleans and MSFC's Mississippi Test Facility, Air Products and Chemicals, Inc., announced. Company had designed double-walled cryogenic tanks for mounting on modified existing hulls according to requirements of NASA, U.S. Army Corps of Engineers, and Coast Guard. The three 250,000-gal.-capacity hydrogen barges and six 105,000-gal.-capacity oxygen barges could carry "the largest quantities of the two liquids ever transported in individual vessels." Docking facilities in test stand area at MTF would permit pumping of liquids directly into rocket fuel tanks. (AP&C Release, 3/24/66)

- NASA Nike-Cajun meteorological sounding rocket launched from Sonmiani, Pakistan, carried 25-grenade payload to estimated 84-mi. (135-km.) altitude. 20 grenades successfully ejected and exploded in experiment, conducted for British National Space Research Committee and Pakistan Space and Upper Atmosphere Research Committee, to obtain wind, temperature, and other meteorological data at 31–93-mi. (50–150-km.) altitudes. All grenade flashes were photographed by K-24 ballistic camera network; 20th explosion occurred at about 54 mi. (87 km.). (NASA Rpt. SRL)

March 24: NASA would negotiate two incentive-fee contracts totaling $315 million with Boeing Co. and North American Aviation, Inc., for procurement of additional Saturn V 1st (S-IC) stages and F-1 rocket engines: Boeing Co. would supply five S-IC stages, costing $165 million; North American Aviation, Inc., would handle production, support, and sustaining engineering of 33 F-1 engines for $150 million. (NASA Release 66-69; *WSJ*, 3/25/66, 10)

- GEMINI VIII's GATV had developed a "pretty serious problem," MSC Agena systems engineer Melvin Brooks revealed. The trouble, apparently in a yaw hydraulics system, had developed when mission controllers decided to fire the rocket system by remote control to shift the orbit from 186-mi. (299.5-km.) apogee and 184-mi. (296.3-km.) perigee to circular orbit at 253-mi. (407.33-km.) altitude—leaving it as a rendezvous target for the flight of future Gemini spacecraft. Rocket reached 330-mi. (531-km.) and then 442-mi. (701.6-km.) altitudes. Brooks called these "pretty serious offheadings, serious misalignments." Eventually, the desired orbit of 253 mi. (407.33 km.) was obtained. GATV was expected to drop gradually to orbit of about 184 mi. (296.3 km.).

    GEMINI VIII flight director John D. Hodge said short circuit that had locked a yaw thruster on the Gemini instrument module into wide-open position, forcing emergency landing of the spacecraft March 16, had not been found. (UPI, *NYT*, 3/25/66; *WSJ*, 3/25/66, 1)

March 25: In answer to charge that the "rush of Jodrell Bank Observatory to distribute Luna IX pictures and of scientists both in the U.S. and U.K. to comment on the pictures can only be considered a breach of ethics [see Feb. 4 and 5]," Sir Bernard Lovell said in an open letter in *Science*: "We did not 'rush' to distribute the Luna IX pictures in advance of the Russians. Our pictures came off the facsimile machine several hours after the Russians had convened a large press conference in Moscow to show the pictures which Luna IX had transmitted the previous night. Why they did not do so, or publish them in *Izvestia* the following morning remains a mystery. As for commenting on the pictures before the Russians, my only remarks were that the lunik seemed to be resting on a hard surface with little evidence of dust, and these comments had previously been made by scientists in Moscow. Any suggestion that we published a scientific analysis of the results before the Russians is absurd." (*Science*, 3/25/66, 1477)

- Series of power tests for breadboard version of Nerva (Nuclear Engine for Rocket Vehicle Application) was completed by NASA and AEC at Nuclear Rocket Development Station, Jackass Flats, Nev. Final 16-min. test studied duration capability at approximately design power and temperature—1,130 mw and 3,715°F. Engine system had been started 10 times during series [see Feb. 3, 11, March 3, and 16] and had operated for 110 min., including 29 min. at nominal full power. (*Atomic Energy Programs, 1966*, 186)

- USAF successfully carried out second launch from Vandenberg AFB in High Altitude Background and Signal to Noise (Hitab) program to investigate and measure infrared radiation and emission characteristics associated with missile exhaust plumes and other exhaust products. Nike-Javelin combination carrying payload of plume measurement was launched seconds after launch of its target, a Titan II ICBM, and data were telemetered back to Vandenberg station. Preliminary reports indicated that missile was on target for preselected Pacific target area. (UPI, *NYT*, 3/26/66, 27; UPI, *Chic. Trib.*, 3/25/66)

March 25: USAF launched unidentified satellite from WTR with Scout booster. (*U.S. Aeron. & Space Act., 1966*, 149)

- Low-level penetration flight of 1,201.8 mi.—longest to date—was made by a USAF F-111A variable-sweep-wing fighter from Edwards AFB to General Dynamics Corp.'s Fort Worth, Tex., plant. L/Col. R. C. W. Blessley, director of F-111 joint test force at Edwards, flew three-hour nonstop mission at an even 1,000 ft. above uneven terrain varying from 500-ft. to more than 10,000-ft. altitude. (Gen. Dyn. Corp. Release)

- Provisional analysis of LUNA IX's photographs by Gerard P. Kuiper, Robert G. Strom, Rudolph S. Le Poole, and Ewen A. Whitaker of the Univ. of Arizona's Lunar and Planetary Laboratory appeared in *Science*. Referring to theories that spacecraft landing on the lurain might encounter electrostatic dust problems, the scientists said: "There is no evidence that this process is important since the Luna IX components visible on the record appear in a very clean condition. Also, the camera pictures are clear and sharp and no stray light appears above the horizon. Apparently the optics remained clean upon exposure a few minutes after landing. The cleanness of the lunar surface may also explain the absence of a visible dust cloud from the impacts of Rangers VI, VII, VIII, and IX." (*Science*, 3/25/66, 1561-63)

- Dr. J. Allen Hynek, Northwestern Univ. astrophysicist and scientific consultant to USAF's Project Blue Book, discussed at a press conference the findings of his investigation of reported Ufo sightings in Hillsdale and Dexter, Mich., near Ann Arbor: "The majority of observers ... reported only lights. They have not described an object. Both sightings were in swampy areas—a most unlikely place for a visit from outer space. The glow was localized here. It could have been due to the release of variable quantities of marsh gas." Rotting vegetation in swamps produces the gas, Dr. Hynek said, "which can be trapped by ice and winter conditions. When a spring thaw occurs, the gas may be released in some quantity, producing flames which could be yellow, red, or blue-green." In the Hillsdale case, he added, the sighting might have been assisted by youths "playing pranks with flares." Hynek's conclusions were disputed by several of those who had made the sightings. (AP, Wash. *Eve. Star*, 3/26/66, A3; Rugaber, *NYT*, 3/26/66, 31)

- Rep. Gerald R. Ford (R-Mich.) said he would ask the House Armed Services and Science and Astronautics Committees to consider "full-blown" investigations of unidentified flying objects. Several hundred people in Michigan had reported mysterious lights during the past week. Ford said that if swamp gas had caused the lights as investigators suggested, "USAF should have no hesitancy in explaining that to a committee." (Clark, *NYT*, 3/27/66, 61)

- Detection of low exposure ages in four iron meteorites—from 4 million to 50 million years—was reported in *Science* by Dr. James C. Cobb, Brookhaven (N.Y.) National Laboratory geochemist. Radioactivity measurements of 18 iron meteorites revealed four with unusually low amounts of stable isotope argon-38, compared to normal amounts of radioactive product argon-39. Both "are produced by interactions of cosmic rays with the meteorite while it is in space." Dr. Cobb's calculation of exposure age, based on ratio between the two isotopes, "provided definite evidence of an overlap . . . between the chondrites and the iron meteorites." Iron meteorites previously had been thought to range in age from 120 to 2,200 million years, except for Braunau meteorite estimated at 4.5 million years. (*Science*, 3/25/66, 1524)

March 25: Evidence that Soviet scientists and engineers had acquired a determining voice in decision-making was presented to the Midwest Slavic Conference by Colgate Univ. professor Albert Parry. He cited a report of a plenary session of the Moscow Communist Party Central Committee in which surprise was expressed that the capital's 700 scientific establishments, with more than 350,000 scientists, engineers, and technicians, were escaping party control. Prof. Parry said he saw in Siberia's new science city in Novosibirsk the final and irrefutable proof of the emergence of the Soviet scientist as a dominating force in society: "The experience in Siberia's science city in collecting and integrating so many divergent institutes and laboratories in one place; the role of computers in cutting across disciplines . . . all this is sufficient proof that Russia's science and technology cannot be so readily or lastingly thwarted by the party hierarchs from free and easy cross-fertilization and thus from one more step to independence and political influence." (Handler, *NYT*, 3/27/66, 10)

- W. Cameron Roberts, Jr., director of Atlantic Research Corp.'s Production Div., died of a kidney ailment. Roberts, an expert on solid-propellant and meteorological rocketry, had conducted extensive research for U.S. upper-air and space programs, and was co-inventor of Arcas—one of world's most widely used meteorological rocket systems. (*Wash. Post*, 3/27/66, B10)

March 26: GEMINI VIII Astronauts Neil A. Armstrong and David R. Scott received Exceptional Service Medals from NASA Deputy Administrator Dr. Robert C. Seamans, Jr., in ceremonies at MSC. Armstrong received a quality step increase in his Civil Service pay, and Major Scott (USAF), promotion to lieutenant colonel by President Johnson.

At a televised MSC news conference, Armstrong said he and Scott developed some "anxiety" but that they never for a moment thought they could not return the spacecraft to earth. Both Scott and MSC Director Robert R. Gilruth said that if the trouble had occurred over a ground tracking station, mission might have been salvaged. Malfunction would have been localized immediately and the information relayed to the astronauts. As it was, they had had no idea what was wrong or even whether the trouble was with the GATV/GEMINI VIII or the spacecraft alone. (Waldron, *NYT*, 3/27/66, 64L; Hines, Wash. *Eve. Star*, 3/27/66, A1; Simons, *Wash. Post*, 3/27/66, A1)

- Secretary of Agriculture Orville Freeman announced that the Department had awarded a contract to NASA for use of remote-sensing equipment in large-area surveys. Photographic and other instrumentation carried in spacecraft would identify types and conditions of soil; detect plant diseases in early stages; determine a crop's vigor and agents which could cause loss of water; indicate whether soils were suitable for certain crops; pinpoint moisture and salt content; and detect conditions such as drought and insect infestation. (DA Release 918-66)

March 27: NASA Nike-Apache meteorological sounding rocket launched from Sonmiani, Pakistan, carried 25-grenade payload and dummy trimethyl-aluminum (TMA) canister weighing 25 lbs. to estimated

118-mi. (190-km.) altitude. 23 grenade flashes were photographed by K-24 ballistic camera network in experiment conducted for British National Space Research Committee and Pakistan Space and Upper Atmosphere Research Committee to obtain wind, temperature, and other meteorological data between 31-93 mi. (50-150 km.). (NASA Rpt. SRL)

March 27: FAA lost its first customer for supersonic transport (Sst), paid Pan American Grace Airways (Panagra) $200,000 refund requested by company on Feb. 25. Panagra was one of eight U.S. and 14 foreign customers which had deposited money to reserve delivery positions for Sst. (AP, *NYT*, 3/28/66, 57)

- Possibility that a random or deliberate radio signal caused GATV's attitude control thrusters to fire during GEMINI VIII docking period March 16 was being studied by electromagnetic interference experts, Richard Lewis reported in a *Denver Post* article. "So far in the space age," Lewis said, "there has been no record of any radio-controlled space vehicle being manipulated by any power other than the one that launched it." Other explanations were that a signal sent to GATV to turn on its tape recorder might have activated the thrusters, or that a docking jar had caused intermittent short circuit. (Lewis, *Denver Post*, 3/27/66)

- Soviet aircraft designer Col. Alexander Aykovlev said in *Krasnaya Zvezda* article that U.S.S.R. was planning a supersonic passenger and cargo aircraft with cruising speed of 1,560–1,900 mph. (Reuters, *Chic. Trib.*, 3/28/66)

- Light aircraft of polyester plastic had been built and successfully tested by Wassmer Aviation (France), AP reported. Called the WA-50, four-seater aircraft had a 150-hp engine, could attain 180 mph, and weighed about one metric ton when loaded. (AP, Wash. *Eve. Star*, 3/27/66)

March 27–30: FCC Chairman E. William Henry warned the broadcasting industry that the day might come when direct-broadcasting satellites would outmode "overnight" local radio and television stations. Speaking to the National Assn. of Broadcasters in Chicago, Ill., he said direct-broadcast satellites were still only "a tiny cloud on the horizon," and there remained "mountains of unsolved problems. . . . But smaller clouds have often traversed higher peaks and released a torrential downpour of new technology and new service on the ground below. The resultant floods have swept aside the older order." (*WSJ*, 3/30/66, 5)

ComSatCorp president Joseph V. Charyk told the NAB convention that communication by satellites would recast the "nature of the world: It may profoundly influence the means of distribution to local stations throughout this country. It may establish new arrangements and new patterns that will have a profound effect on the national television and radio picture." Charyk said this development was likely to result from satellite linkage with metropolitan communications centers. Linkage would include "color television and stereophonic FM radio, aural and visual telephone service, high-speed facsimile data and newspapers, library reference, theatre and transportation booking services, access to computer facilities, shopping and banking services of all types, and centralized charging and billing." (Text)

March 28: First fully successful launch and air-snatch recovery of the Air-Launched Air-Recoverable Rocket (Alarr) was made over WSMR. A modified MB-1 Genie carrying instrumented payload for upper atmosphere research, Alarr was launched from F-4C Phantom II jet fighter at 44,000-ft. altitude. Rocket reached 150,000-ft. altitude and 180-lb. payload immediately separated. C-130B transport operating from Edwards AFB made successful mid-air recovery of payload at 5,000 ft. in second pass at target. Test, conducted for Kirtland AFB, was to develop new technique for high-altitude space probes: launch aircraft's speed would serve as first stage of a rocket—eliminating need for a launch pad; descending payload would be recovered to prevent damage or loss. (AP, Balt. *Sun*, 3/29/66; *M&R*, 4/4/66, 8)

March 28: Speech by Rep. Henry S. Reuss (D-Wisc.) on need for research and development on socioeconomic problems such as transportation, construction, and waste management was inserted in the *Congressional Record*: "We should not allow heavy concentration of research and development in the three fields of space, defense, and atomic energy to blind us to the potential for research and development in civilian technology. . . .

"But much of the growth and success of these three . . . 'giants' must be explained in terms of the remarkable men associated with them. . . .

"We think of Admiral Rickover and the nuclear submarine. Of Jim Webb and the space program. . . . Of the personal sponsorship of Presidents Kennedy and Johnson for our manned landing on the moon. . . ." (*CR*, 3/28/66, 6529)

- Maj. Donald E. Keyhoe (USMC, Ret.), director of the National Investigations Committee on Aerial Phenomena, demanded at a press conference in Washington, D.C., that Congressional inquiry be made into Ufo's and USAF's handling of investigations of reported sightings. NICAP had long charged USAF with suppression of evidence and censorship. (Carmach, *Wash. Daily News*, 3/29/66; Young, *N.Y. J/Amer.*, 3/29/66, 1)

- Redeye, USA's shoulder-fired antiaircraft missile system designed for forward-area defense against low-flying aircraft, had successfully completed Arctic phase of service testing at Ft. Greely, Alaska. (Gen. Dyn. Corp. Release)

March 28–30: AIAA-sponsored "Stepping Stones to Mars" meeting was held in Baltimore. Vice President Hubert H. Humphrey, in a speech read by NASC Executive Secretary Edward C. Welsh, called for international cooperation in man's efforts to land on Mars: ". . . it is indeed worth considering the possibility of having man's first voyage to Mars become a truly international undertaking. We in this country are eager to share the thrills and the benefits of space exploration with other nations. This enterprise is of such tremendous, such breath-taking scope that there is ample opportunity for many nations to contribute to it, each in its own way. There are many imaginative and ingenious scientists and engineers in other countries, and we welcome their ideas on how Mars may best be reached and explored." (Text)

Dr. Ernst Stuhlinger, Director of MSFC Research Projects Laboratory, told the conference that "all technologies are sufficiently developed to make flight [to Mars] in 1986 appear possible." Five Saturn V launch vehicles would carry the components of electrically powered Mars spacecraft and of Nerva II-powered nuclear earth-escape stage into 300-mi.-altitude earth orbit. After assembly and checkout, nuclear stage would accelerate electrically powered spacecraft to escape velocity. Flight time to Mars on a direct transfer trajectory would be 145 days;

280 days would be required to return to earth. Spacecraft would enter high earth orbit outside Van Allen radiation belt and astronauts would be brought back to earth by commuter rocket. Dr. Stuhlinger said: "Of the key developments still necessary, the electrical power system undoubtedly requires the major technological and funding effort." Coauthors contributing to the paper presented by Stuhlinger were Joseph C. King and Dr. Russell D. Shelton of MSFC Research Projects Laboratory and Gordon R. Woodcock, MSFC Advanced Systems Office. (MSFC Release 66-67)

LRC scientist J. Reece Roth described an electric rocket life-support system that would supply interplanetary crews with meals which, when metabolized, would serve as fuel for spacecraft's engine. According to Roth's concept, some or all of the propellant, in the form of food, water, and oxygen, would be metabolized by crew to carbon dioxide, water, and solid wastes. CO_2 and water could then be used as propellant for an electric engine while solid residues would be small enough to be stored or discarded. Propellant reserve could be used for food, navigational corrections, and radiation shielding—enhancing safety of the crew. Weight savings from integrated propulsion and life-support system would be so great, Roth said, that a somewhat less efficient thruster could be used with no overall loss in performance. (LRC Release 66-12)

Human experience most similar to one-to-three year space flight was life on submarine, according to Dr. S. B. Sells, Director, Institute of Behavioral Research, Texas Christian Univ. Dr. Sells presented results of NASA-sponsored study designed to formulate problems of "group organization, structure, and interpersonal interaction of crew members in the environmental circumstances of a typical space mission." Fifty-six social system characteristics of extended space flight—such as motivation, status, background, training, and command structure of crew, as well as conditions of confinement, isolation, and environmental hazards—were compared with those of 11 other social systems. (Text)

NASA Deputy Associate Administrator for Space Science and Applications Edgar M. Cortright said: "Thousands of dedicated people have laboriously laid ... stepping stones to Mars. This nation is ready and able to step up the exploration of Mars, Venus, and the rest of the solar system. Just when such a program will get a go ahead depends on many factors. This coming year will be a crucial one for the decision makers.

"Of one thing I am absolutely certain: comprehensive exploration of the planets is inevitable. Who will be leading this exploration seems less certain; but we can if we choose." (Text)

March 29: Launch of NASA Atlas-Centaur AC-8 rocket, on mission to perform first double-ignition of a hydrogen engine in space, was aborted at ETR seconds before liftoff when a power cord failed to disconnect from the booster. Failure automatically stopped the Atlas engines from firing and countdown was halted at two seconds. No new launch date was set. (NASA Proj. Off.; AP, Balt. *Sun*, 3/30/66)

• ComSatCorp, on behalf of Intelsat—the international telecommunications satellite consortium—issued requests for proposals (rfp's) for 210-lb. synchronous satellite capable of providing "airplane-to-ground-to-airplane" communications with two-way channels. Satellite would supplement high-density North Atlantic communications and "permit orderly expansion of the aeronautical service on a global basis." (ComSatCorp Release)

March 29: Wilhelm Forster Observatory (West Germany) reported that on 127th orbit, Soviet spacecraft COSMOS CXIII returned to earth and soft-landed at Baikonur, Kazakhstan. (AP, *Wash. Post*, 3/30/66, A14)
- U.S. Committee on Extension to the Standard Atmosphere (COESA) announced plans to publish new volume, *U.S. Standard Atmosphere Supplements*, 1966, which would tabulate and detail properties of earth's atmosphere at all altitudes up to 600 mi.—information necessary to investigate atmosphere's effect on aerospace vehicle operation and design. Sponsored by NASA, ESSA, and USAF Cambridge Research Lab., publication would be based on data gathered through tracking of artificial earth satellites and instrumented rockets and satellites. (NASA Release 66-72)
- Proposed crew transfer tunnel connecting a Gemini spacecraft and a Mol vehicle was tested at Arnold Engineering Development Center. (AEDC)

March 30: USAF successfully launched the OV1-IV and OV1-V scientific satellites into circular orbits from Vandenberg AFB with Atlas-Agena D booster. OV1-IV satellite bore photosynthetic organisms to determine effects of zero-g and four types of thermal control coating samples to test durability and resistance to heat and radiation. Each of two thermally insulated sample holders on satellite carried two control surfaces and four circular samples, slightly less than an inch in diameter. Protected during launch, coatings would be exposed to space environment and thermistor measurements of erosion, corrosion, and changes in color telemetered during anticipated year in orbit. OV1-V measured optical radiation of earth, background, and space. (AFSC, RTD Release 37.66; *M&R*, 4/4/66, 8; *U.S. Aeron. & Space Act.*, 1966, 149)
- Attempt to launch NASA's Orbiting Astronomical Observatory (Oao) was halted by failure of one of the three engines in the Atlas booster to reach full 60,000-lbs. thrust upon firing. Launch was rescheduled for April 2, later for April 8. (UPI, *Phil. Inq.*, 3/31/66; UPI, *NYT*, 4/1/66, 15)
- USAF launched unidentified satellite from WTR with Thor-Agena D booster. (*U.S. Aeron. & Space Act.*, 1966, 149)
- Addressing the AAS in Hampton, Va., NASA Associate Administrator for Space Science and Applications Dr. Homer E. Newell said: "... as a science, space astronomy simply cannot thrive in separation from ground-based astronomy. In fact, the phrase 'space astronomy' is a disquieting one. The discipline is astronomy, the objective of which, broadly speaking, is to advance our understanding of the universe in which we live. The word 'space' in space astronomy refers simply to the fact that sounding rockets, satellites, and space probes are new tools that the astronomer can use in wrestling with the difficult problems of astronomical observation. In perspective, the astronomical community has before it the task of using these new tools in effective combination with ground-based observing so as to best advance the science of astronomy."

Dr. Nancy G. Roman, NASA Chief of Astronomy Programs, OSSA, said: "So far, all space astronomy has been automated. However, as the manned space program matures, plans are being made to incorporate the flexibility of manned operation to the program to an increasing extent.... We recognize that even simple operations by man could greatly extend both the lifetime and the versatility of an OAO, and therefore have been looking into the possibilities of using man in such maintenance activities. Looking further ahead, we also realize that eventually astronomers will want larger telescopes in space than are possible with the current series of OAO's. Therefore we have had a study conducted on the possibility of building, mounting, and using a telescope of approximately 120 inches in space. The problems are far from trivial, but they also do not appear insurmountable, given the resources necessary for the job." (Texts)

March 30: ComSatCorp asked FCC for authority to build $5-million earth station near Moorefield, W.Va., to supplement ComSatCorp-operated facility at Andover, Me. Station would be in operation by late 1967, handling all types of commercial communications and supporting two Atlantic satellites, EARLY BIRD I, and the proposed Apollo system. (ComSatCorp Release)

- Secretary of the Air Force Dr. Harold Brown, testifying before Senate Armed Services Committee and Subcommittee on DOD of Senate Appropriations Committee, said USAF was "working on" increasing procurements directly with prime contractors, but with Government-furnished equipment. He added: "There comes a point, however, where the trouble that the Government buys itself by furnishing equipment is more than the money it saves . . ." (Testimony; Senate, *Military Procurement Auth. for FY 1967*, 907)

- Dr. J. Allen Hynek's report on investigation of Ufo sightings in Michigan received commentary in the *Christian Science Monitor*: "Many saucer sightings have been merely glimpses of familiar things seen under unfamiliar circumstances. In this case, Dr. Hynek says it was marsh gas fire. But he also has said there is an impressive and constantly growing list of well-documented sightings that cannot yet be explained. . . . It is time for the scientific community to conduct a thorough and objective study of the 'unexplainables.' " (*CSM*, 3/30/66, A1)

March 31: U.S.S.R. launched LUNA X unmanned spacecraft toward the moon "to test a system insuring the setting up of an artificial moon satellite with the aim of exploring near-lunar outer space and also testing systems installed aboard for placing the station in a selenocentric or near lunar orbit," Tass announced. Spacecraft was following a trajectory close to the calculated one and onboard equipment was functioning normally. (*Pravda*, 4/1/66, 1, USS-T Trans.; Tass, 3/31/66; Loory, *Wash. Post*, 4/1/66, A3)

- Fifth Saturn IB booster (S-IB-5) was successfully captive-fired by Chrysler Corp. for 2½ min.—longest duration firing—at NASA Marshall Space Flight Center. Stage would be shipped to Michoud Assembly Facility for post-static-test checks and would be used in checking out the Apollo spacecraft, perfecting rendezvous and docking techniques, and training astronauts in preparation for manned lunar landing. (MSFC Release 66-69)

- B/G J. C. Maxwell (USAF), FAA deputy administrator for supersonic development, told a Washington meeting of Aerospace Industries Association's Public Relations Advisory Committee that total cost of delivering the first American Sst for airline service may exceed four billion dollars. He explained: "The SST program has been, and still remains, a high risk program. That fact, combined with the sheer magnitude of the development cost, is why the Government is participating in the program to develop an aircraft aimed solely at a civil market." Adding that the cost estimate was "considerably more than the total net worth of the major aircraft manufacturers combined," Maxwell said: "Our philosophy for financing the SST program beyond the prototype construction and flight test phase is to encourage industry and the financial community to assume as much responsibility and risk as practicable. We eagerly look foward to the day when we can turn the whole program over to private enterprise. Conceivably, this could occur when the prototype has flown and proven out the design. If all goes well—and the work performed by the airframe and engine manufacturers during the past nine months gives us cause for optimism—the SST prototype will take to the air in late 1969 or early in 1970 to begin an intensive period of testing. At about this time, production development will get underway with the first production model rolling off the line early in 1973 followed by type certification and entry into commercial service by 1974." Although the British-French Concorde was expected to be in airline service by 1971, Maxwell said this three-year timelag would be offset by the U.S. Sst's advantages: (1) cruising speed of mach 2.7, compared with Concorde's 2.2; (2) 250-passenger capacity, compared with Concorde's anticipated 140 passengers; and (3) growth potential of titanium construction contrasted with volume-limited aluminum of Concorde. (Text)

March 31: Pratt & Whitney Div.'s entry in Sst engine competition—the JTF-17A—was operated for the first time at P&W's Florida Research and Development Center, W. Palm Beach, Fla. Twin-spool turbofan "was started and operated for the programmed time." (P&W Release, 4/1/66)

- FAA had submitted to aircraft manufacturers a new supersonic transport (Sst) financing plan, reported William Chapman in the *Washington Post*. Manufacturers, who claimed they could not afford to pay 25% of expected $1-billion cost of next building stage, would be able to contribute a lighter share of development costs than they had been paying under the 75-25% cost-sharing plan arranged by President Kennedy. (Chapman, *Wash. Post*, 3/31/66, H8)

Australia's Minister of Supply Denham Henty announced that U.S., U.K., and Australia would cooperate in experiments with U.S. Redstone rockets at Woomera Range to study effects of rocket reentry into the atmosphere. (Reuters, *Wash. Post*, 4/1/66, A20; Reuters, *NYT*, 4/1/66, 24)

- Japanese Transportation Ministry announced a seven-year program to establish system of navigation satellites to be launched in 1972. A new ministry, the Satellite Navigation Research Office, would begin to develop electronic equipment, ground stations, and control systems as soon as the government's budget for FY 1966 was approved by the current session of the Diet. R&D program would be carried out entirely by Japanese scientists. (*NYT*, 4/1/66, 59M)

- Growing "impatience, mistrust, and hostility of news media and congressmen toward NASA" was reported by William Hines in the Washington *Evening Star*: "Newsmen have long contended that the initials NASA stand for 'Never A Straight Answer,' and on at least one occasion a high-ranking official was publicly called to account by a reporter for consistent telling of inconsistent stories.... If the situation continues unaltered, the space administration may soon replace the State Department and the Central Intelligence Agency at the bottom of the capital's credibility list." (Hines, Wash. *Eve. Star*, 3/31/66, A14)

- NAS President Frederick Seitz told Charter Assembly of the Aerospace Science and Technology Branch of the Research Society of America, Bolling AFB, he believed the scientific community would pursue the interests of science far more effectively if it recognized its unique opportunity to make the space program scientifically valuable by lending support to those components "which do have a significant bearing on scientific knowledge." He believed the knowledge we could gain of the solar system through the space program represented "a major contribution to our heritage of knowledge, fully as important as the knowledge of the earth we have gained in the last 500 years through the exploration which began with Henry the Navigator's systematic explorations of the coast of Africa." (Seitz, *Science*, 6/24/66)

March 31: Col. Donald Boyer Phillips (USA, Ret.), military aviation pioneer, died of pneumonia at age 74 in Alexandria, Va. Colonel Phillips, native of Washington, D.C., had designed, built, and flown his first glider in 1905. When he retired in 1947, he had flown in nearly 90 types of planes. (*Wash. Post*, 4/4/66, B2)

During March: Study of Vertically Incident Cosmic-Ray Trajectories Using Sixth Degree Simulations of the Geomagnetic Field—consisting largely of computer-generated tables of cosmic-ray trajectories at more than 130 locations around the world—was issued by AFCRL. Since cosmic ray trajectories are directly influenced by the configuration of the magnetic field, which varies diurnally, seasonally, and with solar-cycle variations, it was possible to derive information on the changing magnetic field by observing cosmic-ray events. Also, because calculations of cosmic-ray rigidities are equally valid for solar protons, calculation techniques developed under this study could be used to evaluate the trajectories of these solar particles as well. (OAR *Res. Rev.*, 3/66)

- Thirtieth anniversary of Jet Propulsion Laboratory of Cal Tech. In March 1936 Dr. Theodore von Kármán approved initiation of rocket propulsion studies by Frank J. Malina with the assistance of John W. Parsons and Edward S. Forman at the Guggenheim Aeronautical Laboratory. (NASA Historian)
- Seven French-built Centaure rockets carrying experiments for study of polar auroras were successfully launched in February and March by ESRO from base at Andoeya, Norway. (*M&R*, 4/11/66, 7)

April 1966

April 1: Vice President Hubert Humphrey, speaking to a group of *Life* magazine executives visiting KSC, remarked: ". . . the space program has added greater meaning to our lives. I have talked to a great many Americans recently about the benefits of the space program, and I have found that my fellow Americans get a thrill and a sense of accomplishment and satisfaction out of this great undertaking of our nation.

"Our young people have a new frontier to explore. And this comes at a time when we thought all the frontiers were gone. They have new and exciting places to go out in the universe, new ideas to bring home and enrich their lives, new challenges to hone their character.

"Yesterday our horizon was limited by the globe of earth, and today, as a result of this new frontier, our horizons are as limitless as the universe itself." (Text)

- NASA announced transfer of project management of the RL-10—first hydrogen-fueled rocket engine—from MSFC to LRC and of Headquarters program authority from NASA OMSF to OSSA. RL-10, which had powered Saturn I's 2nd stage (S-IV) in six successful flights prior to conclusion of MSFC's Saturn I program July 30, 1965, also powered Centaur upper stage launch vehicle, managed by LRC. (NASA Release 66-74)
- Soviet Defense Minister Rodion Y. Malinovsky told Communist Party Congress meeting in Moscow that Soviet "anti-aircraft defense means assure the reliable destruction of any aircraft and many enemy rockets . . . calmly and confidently we are guarding the peaceful toil of our people especially now that the construction of our blue belt defense has been completed. . . ." U.S. military experts believed the "blue belt" defense referred to by Malinovsky was the Soviet nuclear submarine force. (Gwertzman, *Wash. Eve. Star,* 4/7/66, A3)
- Astronomers in six European countries—Denmark, Sweden, France, the Netherlands, Belgium, and West Germany—had agreed to share the expense of constructing $7-million observatory on Lasilla, a mountain peak several hundred miles north of Santiago, Chile. (Reuters, *NYT,* 4/2/66, 27; Reuters, *Wash. Post,* 4/2/66, A16)
- Launch of U.S.S.R.'s LUNA X spacecraft indicated Soviet goals and techniques for reaching moon were similar to those of U.S., according to William Hines in Washington *Evening Star,* and raised question whether U.S.S.R. would attempt manned lunar landing in November 1967 to mark 50th anniversary of Bolshevik revolution. Lack of evidence that Russians had developed a large booster comparable to U.S.'s Saturn V suggested they would use earth orbit rendezvous method to reach moon, Hines said. (Wash. *Eve. Star,* 4/1/66, A1)
- ComSatCorp's third annual report revealed that EARLY BIRD I, operating since June 28, 1965, as the first commercial communications satellite,

had earned $2,138,577 as of Dec. 31, 1965. Report omitted any corporate earnings statement, explaining that "with Comsat's singular structure—and its newness—we are not yet able to present the corporate financial statements in a conventional mode." With FCC approval, "the Corporation has adopted the accounting practice of deferring all revenues and expenditures until the commencement of full operations." Financial statement showed, however, that $9,677,000 had been spent for "the construction of the Early Bird satellites, the cost of the launch in April [6,] 1965 and expenditures for satellite system equipment." By the end of 1965, 75 of EARLY BIRD I's 240 two-way voice channels had been leased—64 by AT&T, nine by Candian Overseas Telecommunication Corp., and one each by ITT and RCA. Corporation also earned $8,083,745 in income from temporary investments, compared with $4,312,714 in 1964. Report noted that membership of International Telecommunications Satellite Consortium (Intelsat) had grown from 20 to 48 countries during 1965, and cited plans to launch two improved satellites in late summer of 1966 "to provide communications support" for NASA's Apollo program. (Text)

During week of April 1: Pulsed laser beam communications system which could penetrate cloud of plasma or charged gases surrounding spacecraft during reentry and avoid communications blackout was patented by General Electric Co. engineers Dr. Kiyo Timiyasu and James R. Whitten. Principal use of invention would be in line-of-sight communications between earth stations and spacecraft, but two land stations not in line of sight also could exchange information by using cloud reflections. (Jones, *NYT*, 4/2/66, 32)

April 2: 124,000-mi.-long cluster of sunspots which could affect weather and radio and television signals was detected by West Germany's Bochum Observatory. (AP, *Wash. Post*, 4/3/66, B7)

• Gatv test program—Project "Sure-Fire"—was completed at Arnold Engineering Development Center; priority testing began when Gatv failure Oct. 25, 1965, caused canceling of Gemini VI mission. (AEDC)

April 3: Sir Bernard Lovell, director of Jodrell Bank Experimental Station, told the press that LUNA X spacecraft had gone into orbit around the moon, but that "all may not be well" with it. He suggested that LUNA X was tumbling uncontrollably in orbit around moon at distance of 300 or 400 mi.; this theory evolved after Jodrell Bank detected a series of maneuvers beginning at 12:50 p.m. EST—15 min. after Jodrell Bank had first picked up LUNA X's signals. Frequency of the signals had begun to decrease, indicating the probe was under the moon's gravitational pull at a distance of about 2,000 mi.; signals then faded out entirely, suggesting that the retrorockets had been fired and the probe had landed. At that point Lovell said: "We think Luna 10 landed at 1:44 P.M. [EST]." Minutes later, however, signals again picked up and the situation became "extremely obscure." He concluded: "At last hearing our signals could be interpreted as coming from a tumbling spacecraft. We find it difficult to believe that it is meant to be tumbling."

U.S.S.R. had announced LUNA X's lunar-orbit mission after March 31 launch. Sir Bernard's theory came a day before Soviet announcement of mission results [see April 4]. (*NYT*, 4/4/66, 9; AP, Balt. *Sun*, 4/4/66; *Wash. Post*, 4/4/66, A1)

April 4: U.S.S.R.'s LUNA X unmanned spacecraft had achieved lunar orbit (1,000-km. [621-mi.] apolune, 350-km. [217-mi.] perilune, 3-hr. period) at 1:44 p.m. EST April 3 to become world's first artificial moon satellite, U.S.S.R. announced at 23rd Communist Party Congress meeting in the Kremlin. Accomplishment was dramatized by direct broadcast from LUNA X of the Communist anthem, "Internationale."

Tass released details of LUNA X's structure and flight. 3,530-lb. spacecraft consisted of two main parts: moon satellite section and engine facilities section. 540-lb. satellite was an airtight package containing radio equipment, telemetry system, equipment for studying the moon and near-lunar space, heat-regulation and power supply systems, and small jet engines to stabilize the flight.

Approximately 8,000 km. (4,969 mi.) from the lunar surface, spacecraft was oriented so that jet of the engine was directed against direction of its movement. At approximately 1:44 p.m. EST retrorockets fired automatically, slowing spacecraft to 1.3 km. (.8 mi.) per second and putting it in lunar orbit. Twenty minutes after switching off retrorockets, earth command was sent to separate instrumented lunar satellite from engine facilities section, and first radio communications session was held. Equipment was functioning normally and radio communications were stable. There was no suggestion that photos of the lunar surface were being transmitted to earth, or that LUNA X was "tumbling" [see April 3]. (Tass, 4/4/66; AP, Wash. *Eve. Star*, 4/4/66, A6; AP, *Wash. Post*, 4/5/66, A12)

April 4: House Committee on Science and Astronautics voted to report $4,986,864,150 NASA authorization bill—$25,135,580 less than $5,012,000,000 NASA had requested for FY 1967. (NASA LAR V/57-58)

- Nike-Apache launched from NASA Wallops Station carried 59-lb. instrumented payload to 108-mi. (174-km.) altitude and impacted 67 mi. downrange in the Atlantic. Conducted for ESSA and Graduate Research Center of the Southwest, Dallas, the experiment sought to obtain ionospheric electron density profile measurements using a plasma resonance probe and to obtain data concerning sporadic E phenomenon in the valley region between the E- and F-layers of the ionosphere. Data were telemetered to ground stations during flight. Experiment was first of three to compare techniques for measuring characteristics of upper atmosphere and ionosphere. (Wallops Release 66-16; NASA Rpt. SRL)

- NASA announced a change in sequence of two Apollo/Saturn IB launches scheduled for late 1966: AS-202 mission was rescheduled to follow AS-203 mission to "provide additional time for checkout of the Apollo spacecraft to be flown in . . . AS-202." AS-203 would test Saturn V 3rd stage (S-IVB) and verify that orbital operations features of the liquid hydrogen propulsion system were satisfactory. (NASA Release 66-78)

- NASA announced selection of 19 pilots, to join 31-member NASA astronaut team in early May: Vance D. Brand, civilian; Lt. John S. Bull (USN); Maj. Gerald P. Carr (USMC); Capt. Charles M. Duke, Jr. (USAF); Capt. Joe H. Engle (USAF); LCdr. Ronald E. Evans (USN); Maj. Edward G. Givens, Jr. (USAF); Fred W. Haise, Jr., civilian; Maj. James B. Irwin (USAF); Dr. Don L. Lind, civilian; Capt. Jack R. Lousma (USMC); Lt. Thomas K. Mattingly (USN); Lt. Bruce McCandless II (USN); LCdr. Edgar D. Mitchell (USN); Maj. William R. Pogue (USAF); Capt. Stuart A. Roosa (USAF); John L. Swigert, Jr., civilian; LCdr. Paul J. Weitz (USN); and Capt. Alfred M. Worden (USAF). During recruitment, which began Sept. 10, 1965, NASA received 351 applications of which 159 met basic requirements. (NASA Release 66-67)

April 4: Defense Electronics, Inc., received a $2,380,000 contract from USAF for telemetry equipment for Apollo airborne tracking aircraft. Firm would supply C-135 trackers with prediction playback monitors and other equipment to provide communications link to Apollo spacecraft prior to insertion into lunar trajectory. (*Wash. Post*, 4/5/66, C7)

- FAA Administrator William F. McKee announced that Dr. Raymond L. Bisplinghoff would serve as a consultant on FAA research and development efforts—particularly in supersonic transport (Sst) program—while continuing to serve as Special Assistant to NASA Administrator. (FAA Release 66-30)

- Extent of the impact of Bureau of the Budget's new directive (A-76), stating that "in some instances . . . it is in the national interest for the government to provide directly for products and services it uses . . ." would be realized this month, reported George C. Wilson in *Aviation Week*. BOB official George C. Mullins told Wilson that the definitive standards in A-76 would facilitate a more consistent Federal contracting policy: "The high sounding but unspecific language in 60-2 [previous directive] was subject to different interpretations by various federal agencies." Mullins added: "We agree that the private enterprise system should be preserved. But there is such a thing as a national objective. By acknowledging this in the new directive, we feel we are basing our policy on a much stronger foundation. Preserving the free enterprise system is one good reason" for contracting with private industry, "but it is not the only reason." (Wilson, *Av. Wk.*, 4/4/66, 26)

- UCLA dedicated new center for space sciences, Schlichter Hall, made possible by $2-million NASA grant. Six-story building, named for Dr. Louis Byrne Schlichter, Prof. Emeritus of Geophysics and Director of UCLA Institute of Geophysics from 1947 to 1962, would contain 30 large laboratories and 50 research and administrative offices.

    Dr. T. L. K. Smull, Special Assistant to the NASA Administrator, said in a dedication speech: "NASA thinks that this is a singularly important type of activity as many of the problems confronting NASA in the space program reach across disciplinary lines and require the combined efforts of chemists, physicists, mathematicians, engineers, biologists, psychologists and even, on occasion, political scientists, social scientists and economists." Other speakers were Dr. Thomas Gold, director of Cornell Univ. Center for Radiophysics and Space Research; Rep. Alphonzo Bell (R-Calif.) of House Committee on Science and Astronautics; UCLA Chancellor Franklin D. Murphy; Dr. Schlichter; Prof. Gordon J. F. MacDonald; and Nobel Laureate Willard F. Libby, director of UCLA Institute of Geophysics and Planetary Physics. (UCLA Release, 4/3/66; Text)

- First direct measurement of minimum amount of energy required to initiate simple chemical reaction was announced by Cal Tech, where graduate student John Michael White had shown that one third of an electron volt of energy was necessary to split a hydrogen molecule and link a deuterium atom with one of the hydrogens. AEC-supported study had been suggested to White by Dr. Aron Kuppermann, prof. of chemical physics, who said measurement would make it possible to learn "whether bimolecular chemical reactions can be described by the laws of classical mechanics or whether it is necessary to use quantum mechanics." (Cal Tech Release, 4/4/66)

- USAF and NASA signed agreement to establish procedures for joint space program accident investigation. Ad hoc joint boards, consisting of at least three voting members appointed by USAF Director of Aerospace Safety and three by NASA Safety Director, would study particular launch vehicle accidents. Chairman would be from agency with "primary responsibility at time of accident," determined by circumstances of each case. (Text; NASA NMI 1052.74, 4/27/66)

April 4-6: The importance of imagination in shaping the future was emphasized by AFSC Commander Gen. Bernard A. Schriever in a speech before Arnold Air Society's 18th National Conclave, Dallas: "Back in the 1930s we had what we thought were pretty fantastic ideas for the future. We thought that it might be possible to fly a plane up to the sound barrier. There were even people who thought it might be possible to break through it somehow—at least for short periods. A few people had heard about Dr. Robert Goddard and his work with rockets. They knew that he had already launched a rocket several hundred feet into the air. Science writers were speculating that we might be able to get something out into space by the beginning of the next century.

"As it turns out, even the most fantastic thinking 30 years ago has proved to be very conservative. We have already flown planes like the SR-71 at speeds greater than mach 3 for extended periods. The experimental X-15 has flown at mach 6 and reached altitudes of over 50 miles. Our long-range ballistic missiles have ranges of from 5,000 to 9,000 miles. And our Mercury and Gemini astronauts have sped through space for extended periods at speeds better than 17,000 miles per hour. With the Apollo program we are not too many years away from landing a man on the moon." (Text)

Entertainer Bob Hope presented General Schriever the inaugural award of Arnold Air Society's Eugene M. Zuckert Trophy "to recognize and stimulate the growth of professionalism in the United States Air Force." General Schriever was cited for his "outstanding dedication to the task of insuring the development of vitally needed modern weaponry for national defense" and the gathering of "highly qualified technological management teams required to perform this task." (AFSC Release 74.66)

April 5: LUNA X was continuing in circumlunar orbit with 1,017-km. (632-mi.) apolune, 350-km. (217-mi.) perilune, 2-hr. 58-min. period, and 71° inclination, Tass reported. Eighteen radio communications had been held with the satellite "during which a large volume of scientific data had been received." Scientific equipment onboard included: (1) recorder of meteoric particles for investigating meteor showers along earth-moon route, for seeking dust cloud around the moon, and for registering the micrometeorites in circumlunar space; (2) gamma-spectrometer for studying the gamma-radiation of the lunar surface; (3) magnetometer for refining the value for the strength of the moon's magnetic field; (4) equipment for studying solar plasma in circumlunar space; (5) equipment for recording the infrared radiation of the lunar surface; and (6) instrumentation for studying the radiation conditions in the moon's environs. Variations in the satellite's orbital parameters would be used to investigate the moon's gravitational field. (Tass, *Pravda*, 4/6/66, 1, USS-T Trans.)

- NASA had invited 33 firms to submit proposals on a 12-mo. study of Mars and Venus flyby missions in 1975-1982 period using Saturn/Apollo hardware with minimum modifications. Study would emphasize earth escape stages based on modifications of the upper stages of Saturn V launch vehicle (S-II and S-IVB), orbital tanking and assembly operations, launch windows, spacecraft, and probes to be launched from spacecraft to planet's surface. (MSFC Release 66-73)

April 5: NASA MSFC awarded cost-plus-incentive-fee contracts totaling more than $61 million for "engineering, fabrication, and institutional support services in . . . Saturn launch vehicle program" to nine firms: Sperry Rand Corp.; Vitro Corp.; Brown Engineering Co.; Spaco, Inc.; Northrop Corp.; Hayes International Corp.; Management Services, Inc.; Rust Engineering Co.; and RCA Service Co. Awards were first one-year renewals under options of original contracts awarded in 1965. (MSFC Release 66-74)

- Secretary of the Air Force Dr. Harold Brown testified before the House Armed Services Committee that Ufo reports investigated by USAF under Project Blue Book indicated no threat to our national security: "Based upon 10,147 reported sightings from 1947 through 1965 . . . the Air Force has succeeded in identifying 9,501 of these objects. Virtually all of these sightings were derived from subjective human observations and interpolations. The most common of these were astronomical sightings. . . . Other major sources of reported sightings include . . . satellites, mirages, and spurious radar indications. The remaining 646 . . . are those in which the information available does not provide an adequate basis for analysis, or for which the information suggests a hypothesis but the object or phenomenon explaining it cannot be proved to have been there or taken place at that time." Dr. Brown assured committee that USAF would contract a group of scientific experts to conduct an "in depth study" of selected Ufo sightings. (Text; AP, *Wash. Post*, 4/22/66, A4)

- Success of U.S.S.R.'s LUNA IX and LUNA X missions supported two important conclusions, suggested the *New York Times* in an editorial: "The Soviet Union is now well ahead of the United States in scientific exploration of the moon; Moscow's scientists are straining to beat this country in achieving man's first round-trip to the moon. . . .

    "But no matter which nation will be first to conquer the moon, it is important to clarify the legal issues arising from this development. There should be general international agreement that the moon is the property of all mankind and not of any one nation . . . [and] is open for research by scientists of all nations. . . ." (*NYT*, 4/5/66, 36M)

April 6: U.S.S.R. launched COSMOS CXIV with "scientific equipment designed for continuing the outer space research." Orbital parameters: apogee, 374 km. (232 mi.); perigee, 210 km. (130 mi.); period, 90.1 min.; inclination, 73°. Equipment was functioning normally. (*Pravda*, 4/7/66, 10, USS-T Trans.)

- MSFC had purchased 22 additional Saturn IB H-1 rocket engines from North American Aviation, Inc., under $7,634,742 modification to previous contract. In addition to the 205,000-lb.-thrust engines, which would complete engine requirements for Saturn IB/Apollo program, NAA would provide three years of support services. (MSFC Release 66-75)

- Astronauts had made more than 1,000 "flights" in the Rendezvous Docking Simulator and other facilities at LaRC to help develop techniques for the operation of vehicles in space. Developed by LaRC scientists to explore, under controlled laboratory conditions, man's ability to complete a rendezvous in earth or lunar orbit during final 200 ft. of the space-docking maneuver, the simulator had been used extensively by

GEMINI VIII Astronauts Neil A. Armstrong and David R. Scott, who performed world's first space docking March 16. (LaRC Release)

April 6: FAA's seventh annual report to the President and Congress described "significant cost reductions achieved while meeting increasing demands for aeronautical services." Activities for FY 1965 included: (1) advancement of supersonic transport (Sst) program toward initial prototype construction by end of 1966; (2) issuance of new rules requiring greater safety margins during bad weather, higher operational standards for small aircraft, and new standards for passenger evacuation in emergencies; (3) beginning of field tests of a computerized air-traffic-control system; (4) commission of first Distance Measuring Equipment-Instrument Landing System (Dme-Ils) combination to facilitate bad weather approaches; (5) establishment of General Aviation Jet Training Standards Board and issuance of regulation requiring biennial requalification of all flight instructors to assure pilot efficiency; and (6) $20-million savings from cost-reduction program and other substantial savings from increased employee productivity and joint use of equipment and wire communications services with DOD. (FAA Release 66-31)

- DOD announced that USA, USN, and USAF would test British-built P.1127 aircraft—redesignated XV-6A V/Stol by U.S.—to determine aircraft's ability to meet requirements of each individual service. Evaluations, which would be completed by July 31, would be performed at Fort Campbell, Eglin AFB, Patuxent Naval Air Station, and aboard an aircraft carrier. (DOD Release 270-66)

- Dr. Edmund P. Learned, Harvard Univ. professor of business policy, became adviser to FAA Administrator William F. McKee on "economic and financial aspects of the supersonic transport (SST) development program." (FAA; FAA Release 66-38)

- ComSatCorp suggested separate meetings during week of April 18 with AT&T and Western Union to explore use of multipurpose domestic communications satellites and discuss ownership of ground stations. Dr. Joseph V. Charyk, ComSatCorp president, stated: "We believe that there are no inhibitions on Comsat's ability to provide domestic satellite service, and ... we view earth stations as an integral and critical part of a satellite communication system which must be planned and operated in concern with the space segment. Accordingly, we feel that this responsibility must be reflected in ownership." (ComSatCorp Release)

- AFSC achieved more than $1 million in savings or cost avoidance during the first year of its Internal Zero Defects Program, AFSC Commander Gen. Bernard A. Schriever said in an Andrews AFB ceremony where he presented AFSC Zero Defects Achievement Award to the Air Proving Ground Center, Eglin AFB. (AFSC Release 82.60)

- Senegalese President Leopold Sedar Senghor, speaking at a press conference in Dakar, criticized the wealthy nations for spending money on space programs and abandoning the world's poor. He said that was the reason he had never cabled congratulations to heads of state for their countries' space exploits. (Reuters, *Chic. Trib.*, 4/7/66)

- London bookmaker William Hill, who had offered 1,000-to-1 odds against a manned lunar landing by the end of 1968, cut the odds to 10 to 1 after LUNA X entered a lunar orbit. (UPI, *Wash. Daily News*, 4/6/66)

April 7: NASA Atlas-Centaur AC-8 launch vehicle was launched from ETR with dummy Surveyor payload. Centaur stage's failure to achieve successful double ignition in space caused 1,730-lb. dummy Surveyor payload to remain in earth orbit with the Centaur stage rather than enter simulated lunar trajectory as programed. Orbital parameters: apogee, 208 mi. (335 km.); perigee, 113 mi. (182 km.); period, 90 min.; inclination, 31°. Eighty minutes after launch NASA officials at KSC announced that only one of booster's two hydrogen-fueled RL-10 rocket engines had fired the second time. Both engines were to have ignited for approximate 107-sec. burn and inserted payload into lunar trajectory. Later analysis of data indicated the failure was because of a malfunction in Centaur's attitude control system [see May 3]. Purpose of AC-8 mission was to demonstrate Centaur's capability to restart its high-energy engines in the space environment following coast period in earth orbit. Success of this mission was vital to NASA's first operational mission which would attempt to land Surveyor spacecraft on the moon. Centaur program was directed by NASA OSSA and managed by LRC. (NASA Releases 66-58, 66-92; NASA Proj. Off.; *Wash. Post*, 4/8/66, A9; AP, Balt. *Sun*, 4/8/66)

April 7: USAF launched unidentified satellite with Thor-Agena D booster from WTR. (*U.S. Aeron. & Space Act., 1966*, 149)

• New USA OH-6A helicopter, developed by Hughes Aircraft Co., completed a 2,230-mi. flight—longest nonstop helicopter flight in history—from Culver City, Calif., to Ormand Beach, Fla., in 15 hrs. 13 min. Flight broke record set by USN Sikorsky SA-3A helicopter, which flew 2,105 mi. nonstop on March 5, 1965. (*NYT*, 4/11/66, 70; Hughes Aircraft Corp.)

• NASA Apollo Service Module (SM) propulsion engine test phase was completed at Arnold Engineering Development Center. (AEDC)

• NASA Administrator James E. Webb was elected to National Geographic Society's board of trustees. (*NYT*, 4/8/66, 3)

April 8: OAO I Orbiting Astronomical Observatory was launched by NASA from ETR with Atlas-Agena D booster into parking orbit and then into planned circular orbit. Orbital parameters: 505.4-mi. (813.3-km.) apogee; 493.2-mi. (793.7-km.) perigee; 101-min. period; and 35° inclination. Satellite's solar panels and instrumented booms would be deployed April 10.

First of four spacecraft scheduled for NASA's OAO program, OAO I—largest, heaviest, and most electronically complex unmanned spacecraft ever developed by U.S.—weighed 3,900 lbs., carried 10 telescopes, and contained 44,000 separate parts and 30 mi. of electrical wiring. It carried four scientific experiments to provide astronomers with their first sustained look at the universe from above the earth's atmosphere, including Univ. of Wisconsin broad-band ultraviolet telescope package and detection devices to study the ultraviolet, x-ray, and gamma-ray regions of the electromagnetic spectrum. Through its complex ground-command spacecraft attitude control system, OAO I would be aimed at individual objects in space with precision never before attained by an orbiting satellite. Information from experiments would be radioed back to earth in the form of digital data for analysis by experimenters. OAO program was managed by NASA Goddard Space Flight Center. (NASA Release 66-60; *NYT*, 4/9/66, 7; *NYT*, 4/10/66, 10E)

• NASA Nike-Apache sounding rocket launched from NASA Wallops Station reached 99-mi. (159-km.) altitude and impacted 64 mi. downrange in the Atlantic. Objectives of flight were to compare five ionospheric plasma probes using stable ionosphere as a laboratory for evaluating the probes and to check relevant theories of probe operation. (NASA Rpt. SRL)

April 8: FAA established a Noise Abatement Staff to direct the agency's participation in the Government campaign to bring relief to areas disturbed by aircraft noise. Objectives of staff, to be directed by Raymond A. Shepanek (Capt., USNR), former FAA aeronautical inspector, included development of aircraft noise measurement standards, formulation of aircraft noise standards, development of air traffic procedures and aircraft operating procedures to minimize noise, and initiation of research programs to provide for alleviation of the various features of aircraft noise. (FAA Release 66-33)

- Prospect of using a computer to simulate vast changes in weather and climate was discussed by Walter Orr Roberts, director of the National Center for Atmospheric Research, Boulder, Colo., in an editorial in *Science.* Roberts suggested that with data from World Weather Watch—an experiment proposed by the National Academy of Sciences for observing the world's weather continuously in three dimensions—"an adequate computer, and a global mathematical model, a vast array of experiments on weather and climate-modification can be 'performed' by numerical computation rather than in nature. The beneficial consequences can be evaluated and compared to the expected costs. The full effect and the potential hazards can be determined without risk to life and property...." World Weather Watch system was being studied by ESSA and World Meteorological Organization. (Roberts, *Science,* 4/8/66, 159)

- U.S. Government's increased interest in transatlantic cooperation on scientific research was attracting attention in Europe, Victor K. McElheny reported in *Science.* "The most dramatic of the suggestions made recently is that of sending an 'artificial comet' spacecraft toward Jupiter or close to the sun, or to use such a spacecraft to investigate the belt of asteroids between Jupiter and Mars...." McElheny said NASA Administrator James E. Webb was planning to visit West Germany and other European nations to discuss details of the project and to emphasize "that the U.S. is not thinking necessarily in terms of bilateral collaboration with individual countries, of the sort exemplified by current U.S. launching of satellites for Britain, Italy, and France.

    "Where it can, the U.S. prefers to work with multinational European organizations, such as the European Atomic Energy Community (Euratom) or the European Space Research Organization (ESRO) ..." (McElheny, *Science,* 4/8/66, 190-191)

April 9: LUNA X had detected electrons with energies higher than cosmic radiation, Tass reported: "Data have been obtained which may be interpreted as evidence of the existence in the near-lunar space of fluxes of electrons with energies of tens of thousands of electron volts. These fluxes are 70 to 100 times more intense than the cosmic ray background. Possibly, this phenomenon is due to the earth's magnetic 'tail.' Later measurements will permit more positive conclusions on the radiation conditions near the moon." In addition, preliminary data analysis showed that intensity of the moon's magnetic field was "somewhat above the level of magnetic fields in free interplanetary space" and that meteorite "particle density" near the moon was higher than in "interplanetary space."

UPI reported that U.S. scientists, including Dr. James A. Van Allen, State Univ. of Iowa, believed radiation observed by LUNA X was either the earth's comet tail or a similar feature produced by the moon itself. Reported level of radiation would be no threat to manned operations near the moon, they said. (Tass, 4/9/66; UPI, *NYT*, 4/10/66, 79; AP, Wash. *Sun. Star*, 4/10/66, A6)

April 9: Problems of preventing space-probe contamination of other planets were discussed by James L. Haggerty, Jr., in *Journal of the Armed Forces*. He noted remarks of Boeing engineers C. S. Bartholemew and D. C. Porter at the March 28-30 AIAA-AAS meeting in Baltimore that "sterilization and spacecraft reliability are 'competing characteristics.' If you apply enough heat to kill the bacteria, you may also damage the sensitive electronic gear. Alternatively, you can use chemicals such as ethylene oxide, but it must be applied wet and moisture is similarly an 'enemy agent' to electronics equipment. . . . Boeing engineers feel that heat-sterilizable electronics are 'within reach,' but considerably more investigation is required." (Haggerty, *J/Armed Forces*, 4/9/66, 9)

April 10: Main battery system of OAO I Orbiting Astronomical Observatory had failed, NASA officials at KSC announced: "Overheating of the OAO. primary battery, a problem which became critical late yesterday, has resulted in complications leading to degradation of the power supply from all three batteries aboard the spacecraft, and telemetry signals no longer are being received.

"Efforts to overcome the problem have been unsuccessful and the OAO mission appears to have been lost."

Officials added, however, that before satellite had failed it had demonstrated a key maneuver of the OAO program by successfully locking into a series of stars for precise aiming. OAO I had been launched from ETR April 8. (AP, *NYT*, 4/11/66, 71; AP, Wash. *Eve. Star*, 4/11/66, B10; AP, Balt. *Sun*, 4/11/66)

- The 19 astronauts selected by NASA April 4 would be "too late to take part in the first lunar landings . . . [but they] should get into some of the follow-on Apollo flights and take some 60-day space trips," Dr. Charles A. Berry, Chief of Medical Programs, MSC, said in a press interview in Colorado Springs. (Gibney, *Denver Post*, 4/10/66)
- First full-scale checkout version of 365-ft. Saturn V rocket had been assembled at KSC to aid installation of electrical and plumbing connections on three mobile launching towers. (*Marshall Star*, 4/6/66, 1)
- Aviatrix Geraldine Mock—who in 1964 became first woman to make a solo flight around world—set a new world nonstop distance record for women in a 4,550-mi., 31-hr. flight from Honolulu, Hawaii, to Columbus, Ohio, in single-engine Cessna 206. Previous distance flight record of 3,671 mi. had been set by three Russian women in 1938. (*NYT*, 4/12/66, 25; Wash. *Eve. Star*, 4/11/66, A3)
- USNS *Kingsport*, first ocean link in U.S. research in satellite communications, was retired from NASA's networks and returned to DOD. In July 1963 *Kingsport* had participated in world's first demonstration of communications by satellite in synchronous orbit when a message was sent in 45,000-mi. loop from *Kingsport* anchored in Lagos harbor, Nigeria, to SYNCOM II comsat and back. Several days later, she was part of first direct exchange of radio message via satellite, linking Lakehurst, N.J., terminal station with land circuits at Lagos. On Aug. 4, *Kingsport* and Lakehurst transmitted and received via SYNCOM II first satellite ex-

change of news copy and photo facsimile. Several days later, caught in a squall 40 mi. offshore, she became first terminal station to establish satellite communications from the open sea. Two months later, she participated in first trans-ocean press conference by satellite: stationed at Rota, Spain, she connected SYNCOM II communications between United Nations in New York City and NASA Hq. in Washington, D.C., with the International Telecommunications Union Conference in Geneva. *Kingsport* first demonstrated the practicality of communications by satellite during manned space missions March 16, 1966, by relaying voice reports from GEMINI VIII Astronauts Neil A. Armstrong and David R. Scott via SYNCOM II to Hawaii ground station. (NASA Release 66-79)

April 11: NASA Langley Research Center would negotiate $450,000 study contract with Martin Co. for 11-mo. study of "costs, crew size, and complexity of a flight research program using a manned lifting body vehicle." In conducting the study, the firm would be required to consider an HL-10 lifting-body concept capable of carrying one, two, four, six, or eight crew members. HL-10 was considered representative of advanced lifting entry vehicles.

NASA had specified Titan II, Titan III, and Saturn IB as potential launch vehicles. For study purposes, Rogers Dry Lake at Edwards AFB would be a probable landing site. (NASA Release 66-82; LaRC Release)

- Clarence L. Johnson, Lockheed Aircraft Corp. vice president and designer of numerous aircraft, was named recipient of the Air Force Academy's 1966 Thomas D. White National Defense Award for "outstanding contribution to the national defense and security of the United States." (UPI, *Wash. Post*, 4/12/66, A18)

- Development of the world's brightest laser—a light beam 250 billion times brighter than the sun—was announced by Dr. Theodore Maiman, president of Korad Corp., a subsidiary of Union Carbide Corp. Laser had an intensity of 250 trillion watts per square centimeter when focused to a point 1/1,000 of an inch in diameter. (UPI, *Wash. Post*, 4/12/66, A5)

April 12: Cosmonautics Day in U.S.S.R., celebrating fifth anniversary of first manned space flight by Yuri Gagarin, was marked by condemnations of military orientation of U.S. space efforts and predictions of Soviet space successes. Cosmonaut Gagarin, speaking in the Kremlin, said: "American scientists and cosmonauts have carried out a number of interesting experiments. This is to their credit. It is to be deeply regretted, however, that American cosmonautics is increasingly falling under the influence of the Pentagon, which regards outer space as the theater of future military operations." Gagarin pledged "all his energies and knowledge" to help in the lunar mission and said man must make himself at home on the moon "to establish stations which will serve as points of departure for longer space voyages and also as astronomical observatories and scientific laboratories."

Cosmonaut Gherman Titov, speaking in a taped interview on Moscow television, predicted that "builders and assemblers will one day exit into space and put together various parts of space stations, assemble spaceships and stations on the moon, and build various structures."

In a *Pravda* article Marshal Konstantin Vershinin, head of Soviet Air Force, predicted a new manned space mission that could take several cosmonauts through the Van Allen radiation belts. He noted that biomedical data from the COSMOS CX flight would "doubtlessly . . . be used in preparing new manned space flights."

In an interview published in Prague, Cosmonaut Konstantin Feoktistov predicted that man would land on Mars and Venus between 1970 and 1980. (*Pravda*, 4/12/66, 2; *Wash. Post*, 4/11/66, A8; AP, *NYT*, 4/11/66, 37; AP, Wash. *Eve. Star*, 4/11/66, A8; UPI, *NYT*, 4/14/66, 7; *Wash. Post*, 4/13/66, A14; Loory, *N.Y. Her. Trib.*, 4/13/66)

April 12: USAF XB-70 No. 2 flew 20 min. at mach 3—longest to date—and reached 70,000-ft. maximum altitude in 1-hr. 49-min. flight from Edwards AFB. Pilots were North American Aviation, Inc.'s chief test pilot Alvin S. White and Col. Joseph Cotton (USAF). (Flight log; AFFTC PIO)

- U.K.'s Associated Electrical Industries, Ltd., General Electric Co., Ltd., and Plessey Co., Ltd., had joined forces to sell ground terminals at $4 million each to countries that wanted to pick up and relay telephone, telegraph, and television signals from communications satellites, reported Clyde H. Farnsworth in the *New York Times*. The British companies believed there would be market for about 30 terminals in the next two to three years and about 70 to 80 in the next ten years, and said that at least seven nations were already interested in purchasing terminals. (Farnsworth, *NYT*, 4/1/66, 57)

- National Broadcasting Co. (NBC) and ComSatCorp. presented progress reports on their separate feasibility studies on domestic communications satellite system at ComSatCorp-sponsored meeting in Washington, D.C. Proposed systems, which could provide 24-hr. full-color television service and interconnected facilities for radio networks, would each cost about $100 million and would consist of four to six three-channel synchronous satellites, back-up satellites and boosters, ground-to-satellite stations, control stations, and ground receiver terminals. NBC's system would serve only television and radio networks; ComSatCorp's "common carrier" system could be expanded to accommodate all types of communications. Meeting, attended by television network, communications, industrial, manufacturing, and government officials, was first in a series to be held before final plans were submitted to FCC for approval. (NBC Release; ComSatCorp; *Wash. Post*, 4/13/66, D6; Wash. *Eve. Star*, 4/13/66, D12)

- Ability to induce genetically a hibernation state in man that would lessen the body's needs by cooling could be beneficial for long space voyages, suggested Dr. R. R. Chaffee, Univ. of Missouri Space Science Research Center, in a speech before Federation of American Societies for Experimental Biology meeting in Atlantic City. Dr. Chaffee told of genetic experiments in which the ability to hibernate gradually was increased over several generations in some families of Syrian hamsters and reduced in others, indicating that only a few hereditary genes might be involved in the hibernation process. (Carey, *Wash. Post*, 4/13/66, A3)

April 13: Gamma-ray spectrum of the lunar surface obtained by LUNA X indicated that the moon's crust was similar to the earth's, Tass announced. The satellite had detected lunar gamma rays from two sources: one was "an increased intensity of gamma radiation caused chiefly by interaction of cosmic rays with the surface layer of lunar material"; the other was "natural radioactivity from the lunar rocks associated with uranium, thorium, and potassium.

"Compared with the analogous radioactivity of earth rocks, the observed spectrum comes close to the radioactivity of basic rocks—basalts." (Sullivan, *NYT*, 4/14/66, 1, 21)

April 13: NASA Nike-Tomahawk sounding rocket launched from Churchill Research Range carried GSFC payload to 184-mi. (296-km.) altitude for simultaneous measurements of magnetic and electric fields, low-energy electrons, and low-energy protons in a visible aurora. Secondary objectives were rocket aspect measurement and evaluation of xenon flashing light for trajectory determination. (NASA Rpt. SRL)

- GEMINI VIII Astronaut Neil A. Armstrong and his family were honored with a gala parade on their arrival in Armstrong's hometown of Wapakoneta, Ohio. Ohio's Governor James A. Rhodes announced plans for the construction of a $200,000 airport near Wapakoneta to be named for Armstrong. Sen. Frank J. Lausche (D-Ohio) and Rep. William M. McCulloch (R-Ohio) announced a congressional resolution of tribute to Armstrong and his fellow GEMINI VIII Astronaut David R. Scott. (AP, Balt. *Sun*, 4/14/66; *Wash. Post*, 4/14/66, D16)

- Pan American World Airways, Inc. (Pan Am), would purchase 25 new Boeing 747 jet transports—including twenty-three 490-seat passenger models and two all-cargo models which would carry 214,000 lbs. freight—for over $525 million, announced Pan Am's chairman and chief executive, Juan T. Trippe. Deliveries of order, reported to be largest and most expensive in airline history, would be made between September 1969 and May 1970. The 680,000-lb. aircraft would be powered by new Pratt & Whitney JT9D-1 fan-jet engine capable of delivering 41,000-lbs. thrust and would have 45,000-ft. cruise altitude—nearly one mile higher than current transports—and mach .90 cruise speed—compared with mach .82 for current models—and would provide "safety margins superior to any previous commercial airliner" with multi-gear landing system which would require only 9,150 ft. of takeoff runway. Trippe forecast lower passenger fares and reduced cargo tariffs because of 747's large capacity and efficiency of operation. (Pan Am Release; Bedingfield, *NYT*, 4/14/66, 1; Edwards, *Wash. Post*, 4/14/66, K1)

- AFSC Commander Gen. Bernard A. Schriever was presented the American Legion's General William E. Mitchell Memorial Award—highest honor bestowed by the Legion's Aviator's Post—at a luncheon ceremony in New York's Wings Club. Schriever was cited for his "outstanding contributions to the aerospace science and military posture of the United States." (AFSC Release 76.66)

April 14: Two space cooperation agreements were confirmed by Spanish Foreign Minister Fernando Castiella and Secretary of State Dean Rusk in a State Dept. ceremony attended by NASA Deputy Administrator Robert C. Seamans, Jr. First agreement provided that NASA and Spain's Instituto Nacional de Tecnica Aeroespacial (INTA) would share in the operation of NASA space tracking and data acquisition station near Madrid which maintains radio contact with unmanned lunar and interplanetary probes and would support manned Apollo mission. Effective from May 1, 1966, to Jan. 28, 1974, and in accordance with government-to-government agreement between Spain and U.S. of Jan. 29, 1964, contract provided that Spanish engineers and technicians would be trained and assigned to tracking, telemetry, communications, and support. Some of INTA's key personnel had already begun special training at Deep Space Facilities, Goldstone, Calif.

Second agreement, part of a continuing Spanish-U.S. program of cooperation in space investigation, provided for joint project to launch four sounding rockets in Spain to study winds and temperatures at high altitudes. Agreement was in form of a memorandum of understanding between INTA and NASA: INTA would establish sounding rocket range, construct grenade-type payloads, provide two Nike-booster rockets, conduct launchings, and analyze and publish the data; NASA would lend range tracking and telemetry equipment and a launcher, provide two Nike-booster rockets, provide prototype payload, and train Spanish personnel as needed. Both agencies would bear the cost of discharging their respective responsibilities. Results of the experiments would be made available to world scientific community. (NASA Release 66-83)

April 14: Two NASA Aerobee 150 sounding rockets were launched from WSMR: first reached 114-mi. (183-km.) altitude and measured intensity of ultraviolet dayglow between 1160 A and 1400 A in the zenith, in the nadir, and on the horizon; second reached 120-mi. (192-km.) altitude and measured the profile and intensity of the solar Lyman-alpha line. (NASA Rpt. SRL)

- NASA's Apollo Applications program would be hampered by lack of payloads unless Congress granted additional funds in FY 1968 budget, NASA Administrator James E. Webb told press conference at Colorado Women's College. NASA's efforts to obtain appropriations for post-Apollo projects had been hindered by rising costs of the Vietnamese war and Congressional discontent with NASA's increasing administrative costs. Webb said the U.S. was behind in the space race "except for rendezvous and docking, and the Ranger mission. . . . Russia has widened the gap during the past two years, launching two Proton payloads of 27,500 pounds each—which is heavier than we can lift—recovering payloads of 10,000 to 15,000 pounds, and launching a three-man spacecraft. . . ." Questioned about the House Government Operations Committee's suggestion that NASA abandon the Apollo Applications program and participate in USAF's Manned Orbiting Laboratory program, Webb said he did not believe a "complete, common use" of facilities was possible, noting that many of the countries in which the U.S. had tracking facilities would not cooperate if installations were used for military projects. (*Denver Post*, 4/15/66)

- First simulated exploration of the moon had shown that two astronauts could live and intelligently perform tasks for at least 18 days in cabin with 117 cu. ft. of living-working space. Test was performed in support of Apollo Applications program by two MSFC engineers in Minneapolis [see Feb. 20, March 10]. (MSFC Release 66-80)

- Recent Soviet space accomplishments indicated that a manned space spectacular was imminent, suggested William Hines in the Washington *Evening Star*. He noted LUNA IX's soft-landing on the moon Feb. 3, LUNA X's launch into lunar orbit March 31, and the 22-day flight of two dogs in COSMOS CX Feb. 22-March 16. (Hines, Wash. *Eve. Star*, 4/14/66, A12)

- Secretary of the Air Force Dr. Harold Brown issued memorandum approving recommendations of panel appointed to investigate USAF relations with "not-for-profit" corporations and ordering that fees paid to the corporations be substantially reduced. In report, issued Feb. 1, panel had said that nonprofit corporations "... continue to be an indispensible factor in developing and acquiring complex aerospace systems and in assuring their function through command and control," but that "guidelines... point to a fee substantially lower than at present." (Text; AP, Wash. *Sun. Star*, 4/17/66, A10; *Wash. Post*, 4/17/66, A14)

April 14: Inter-American Development Bank awarded Page Communications, Inc., $250,000 contract to "determine the most economic locations for ground stations in South America which would be linked to the satellite system, and the means of connection with local telecommunications systems." Study would be carried out with ComSatCorp cooperation. (IDB Release; *Wash. Post*, 4/16/66, C6)

- In a lecture at Rice Univ., Dr. Henry Faul of the Graduate Research Center of the Southwest's geosciences division insisted on the terrestrial origin of tektites in opposition to lunar-origin theory of many experts: "The fatal objection to the theory that tektites are of lunar origin is the fact that there is no known mechanism of flight that can explain how something that hits the moon, generates a lot of heat and is melted into a jet stream, is somehow dumped in the widely varying patterns and areas where tektites have been found on earth." (Justice, *Houston Post*, 4/19/66)

April 15: Representatives of Intelsat (International Telecommunications Satellite Consortium) toured NASA Marshall Space Flight Center and received briefing on MSFC programs and research projects. (MSFC Release 66-77)

- NASA's Procurement Office reorganized into two main divisions: Policy and Review, and Contract Management. George J. Vecchietti, Director of Procurement, named Paul A. Barron, formerly GSA Assistant General Counsel for Regulations and General Law, Assistant Director of Procurement for Policy and Review. William P. Risso, formerly Special Assistant to Director of Procurement, was named Assistant Director of Procurement for Contract Management. Vecchietti also announced establishment of Procurement Surveys Div., with Harvey M. Kennedy, Jr., Director, and Staff Operations Div., with Harold E. Pryor, Director. (NASA Release 66-81)

- U.S.S.R. had constructed two stations in the United Arab Republic and Mali to photograph artificial earth satellites, Tass reported. Stations had been equipped by U.S.S.R. and were being operated by Soviet, U.A.R., and Mali specialists. (Reuters, *NYT*, 4/16/66, 13)

- Vice President Hubert Humphrey said in address at the Air Force Academy, Colorado Springs, Colo.:

    "It was seventy years ago this May that Samuel Pierpont Langley, Secretary of the Smithsonian Institution, launched his sixteen-foot, steam-powered model airplane off the Potomac River and flew a half a mile in the incredible time of 90 seconds....

    "Seventy years—in the larger scheme of history—is not a very long time.

    "But though those seventy years are but the average length of a full American life, they reach back into an incredibly distant world—a world that has receded from us by a quantum of change that no other period in the vast sweep of human history can surpass.

    "One measure of that change is flight itself.

    "I flew here this morning from Washington—nonstop—in 3 hours and 20 minutes.

"Had I made the flight from Washington to Colorado Springs in Langley's plane, it would have required 127 days—and we would have had to make 2,987 fuel stops along the way. . . ." (Text)

April 15: USAF had accepted and funded at cost of $18.5 million more than one fourth of the unsolicited proposals it had received from industrial and scientific community during first half of FY 1966. The 319 proposals accepted promised needed technological advances in operational aerospace systems. (AFSC Release 71.66)

- GEMINI VI command pilot Walter M. Schirra, Jr., received an honorary Doctor of Aeronautical Engineering degree from Lafayette College, Easton, Pa., during convocation commemorating 100 yrs. of science and engineering instruction on the campus. (MSC *Roundup*, 4/29/66, 1; UPI, *Wash. Post*, 4/16/66, D4)

- NASA Administrator James E. Webb, incoming president of American Society for Public Administration (ASPA), discussed responsibilities of administering a "Great Society" at ASPA conference in Washington, D.C. Webb said that ". . . with the seething mass of humanity all around the world, not many of them very happy today, we have a strong requirement for effective administration, precision in the use of public power, for prediction, for planned use of limited resources. . . .

  "It seems to me that the public administrator in this period . . . must be a bigger, broader person than has been required in the past because the forces he must work with are so much larger, and yet so much more complex. . . . Somehow he must command the respect and following that will convert criticism from destruction of ideas to clarification and perfection of those ideas. . . . I believe his performance will be measured . . . by the scope and sweep of his vision and adjustment to some forces that will be beyond his ability to prediot or control; also his ability to ride the crest of the wave, that wave of the incoming tide of tomorrow that is now bringing in the fruits of a vital and vigorous society." (Text)

April 16: Preliminary results of flight of U.S.S.R.'s LUNA X, placed in circumlunar orbit April 3 to become moon's first artificial satellite, were announced at Moscow press conference. President of the Soviet Academy of Sciences Mstislav Keldysh said that considerably less braking power had been required to put a spacecraft in orbit around the moon than had been required in a soft landing. This had made it possible to increase the payload in LUNA X and to equip it with instruments for important studies of the moon and the space surrounding it.

Academician Aleksandr Vinogradov, geochemist, said that study of radioactivity of rocks on the lunar surface had indicated the overall intensity of gamma radiation on the lurain was comparable to the intensity of gamma radiation emitted by terrestrial granites [see April 13]. However, part of it was not due to natural radioactivity but to interaction of cosmic rays and lunar matter. Such radiation was absent in terrestrial rocks because of the shielding effect of the earth's atmosphere. Lunar rocks were comparable in natural radioactivity to basalt rocks on earth, indicating that "the processes of the formation of the core of the planets of the earth group apparently have the same mechanism."

Prof. Naum Grigorov said readings of LUNA X's magnetometer indicated presence near the moon of a weak, homogenous, and regular magnetic field. Streams of ions of small energies were registered. Analysis of materials transmitted by LUNA X showed the intensity of particles in the moon's radiation belt was 100,000 times less than in the earth's radiation belts.

Prof. Alexandr Mikhailov, director of the Pulkovo Observatory, said existence of an artificial moon satellite was important not only for physical research, but also for determining more exactly the mass and shape of the moon—information vital for building an exact theory of the motion of the moon and for study of the unevenness of the earth's rotation.

Replying to questions, Academician Keldysh said LUNA X did not carry photocameras but was intended for physical study of near-lunar space and the moon itself—its nature, ionosphere, radiations, and fields. Satellite would orbit the moon for "a few years." Duration could not be predicted since parameters of the moon's gravitational field were not known. Keldysh said that since assembly of large spacecraft in orbit was an important step toward the conquest of outer space, cosmonauts were preparing for the solution of this task. Referring to the problem of manned flights to the moon, he stressed that the most difficult task to be solved was return of the spacecraft into earth's atmosphere. Keldysh said dogs Veterok's and Ugolyek's passages through radiation belts in COSMOS CX had had no apparently serious effect on them but that prolonged observations were necessary. (Tass, 4/18/66)

April 16: NASA Assistant Administrator for International Affairs Arnold W. Frutkin addressed the American Academy of Political and Social Science in Philadelphia on international significance of space program: ". . . the United States has welcomed direct foreign participation in space research and exploration. The U.S. has contributed its boosters to launch six satellites which were conceived, engineered, instrumented and funded by cooperating foreign countries. Eight more such satellites are to be launched by us under existing agreements and, at this date, other agreements are in early prospect. We have opened our own satellites to foreign experimenters. Some twenty experiments proposed by scientists abroad have been selected on their merits for flight on our satellites. These foreign experiments are contributed to the program; we contribute space in the satellite and its support systems and retrieve the data for the experimenter. Even wider cooperation—with twenty different countries—is achieved through programs which utilize small, nonorbiting rockets to obtain data for which satellites do not commend themselves. In all of these projects, we have shared the tasks and costs in literal cooperation without exporting a dollar." (Text)

April 17: New evidence supporting theory that Venus' environment could sustain life was published by Johns Hopkins Univ. astrophysicists William Plummer and John Strong in *Astrophysical Journal*. Past measurements of microwave emissions from Venus had indicated very high surface temperatures. Plummer and Strong concluded, on the basis of recent detailed measurements from high-altitude balloon observations by Cal Tech radioastronomers, that possibly "30 percent of the observed microwaves in the past had no bearing on surface temperatures but rather emanated from electrical discharges within the cloudy atmosphere of Venus." They estimated Venus' temperature ranges from 580°F at equator to 9°F at poles and said there are extensive regions—greater than earth's land area—"where man would find the temperature comfortable." (*NYT*, 4/18/66, 4; *Wash. Post*, 4/18/66, A7)

April 17: Ufo was sighted by hundreds of people in Ohio and Pennsylvania, including two deputy sheriffs who followed it for 85 mi. USAF later attributed sightings to combination of a satellite and the planet Venus. (UPI, *NYT,* 4/18/66, 4; *Wash. Post,* 4/18/66, A7; UPI, *NYT,* 4/24/66, 95; Wash. *Eve. Star,* 4/23/66, A2)

April 18: ComSatCorp filed with FCC a $31,985,000 fixed-price contract with TRW Systems, Inc., for research, development, and production of six advanced synchronous satellites for global commercial satellite service. Cylindrical, 240-lb. satellites would have 1,200 two-way voice channels and five-year operational life and would handle all types of communications. (ComSatCorp Release)

- Supersonic transport and rocket flights in the next few decades could raise the average surface temperature of the earth by almost half a degree by spewing tons of chemical pollutants into the atmosphere, said Dr. Gordon J. F. MacDonald, UCLA planetary and space physicist, in an interview with the *Los Angeles Times.* Heavy burning of coal, oil, and gas introduced by the industrial revolution in the 19th century and the automotive revolution in the 20th had increased carbon dioxide in the atmosphere by about 15 per cent, MacDonald estimated. The carbon dioxide created a layer of gas that allowed heat from the sun to reach the earth but stopped it from escaping when it was re-radiated back from the earth. This was the "greenhouse effect," which raised the temperature below the layer and lowered the temperature above the layer. "Four hundred supersonic aircraft flying four flights a day on intercontinental routes, would affect surface temperatures as much as the whole industrial and automotive revolutions have to date," MacDonald said. (Getze, *L.A. Times,* 4/18/66)

- Striking members of International Union of Electrical Workers (AFL-CIO) picketed all gates of Kennedy Space Center, NASA, in wage dispute with United Technology Center, builder of solid-fuel boosters for USAF Titan III launch vehicle. UTC had about 175 workers at KSC, one-third of whom were IUEW members. Some 300 of a total force of 366 construction workers honored the picket lines and failed to show up for work on the Saturn V mobile service structure and Vertical Assembly Building (Vab). Strike was first at KSC during 1966, but the ninth in 27 mos. (AP, Wash. *Eve. Star,* 4/18/66, A1; AP, *NYT,* 4/19/66, 18)

- A new nuclear particle—the most massive and most stable yet found—had been discovered by physicists at Argonne National Laboratory (ANL). Three and a half times as heavy as the proton, the particle was "a 'nucleon resonance' known simply as N*3245." The number "3245" stood for 3,245 million electron volts, the amount of its energy. According to Dr. Alan D. Krisch of Univ. of Michigan, nucleon resonances "may be 'excited' or highly energetic states of protons and neutrons ... and can be produced only in accelerators. They are not observed in nature." Krisch, along with John R. O'Fallon, Keith Ruddick, and Steven W. Kormanyos of Univ. of Michigan, and Lazarus G. Ratner of ANL, made the discovery using Argonne's 13.5-billion electron volt (Bev) zero gradient synchrotron. They said this "abnormally stable" particle has a lifetime of one ten-thousandth of a millionth of a millionth of a millionth of a second, which is "longer than that of any of its family of nuclear particles." Krisch suggested N*3245 may have "an unusually high spin which prevents it from decaying easily."

Discovery was reported in *Physical Review Letters*. (*Phys. Rev. Letters*, 4/18/66, 709; Amer. Inst. of Physics Release, 4/18/66)

April 18: Capt. Robert F. Freitag (USN, Ret.), Director of Field Center Development, NASA OMSF, speaking in Washington, D.C., before the Community Development Conference of the 16th District of Texas, said that the magnitude and importance of the investment in the space program dictate that informed citizens from all walks of life participate in a national dialogue on decisions involved: "We in NASA do not feel that the planning of our space programs is the prerogative of NASA alone. . . . The budget for the coming fiscal year permits NASA to hold open the option for a program to procure additional flight vehicles beyond those now programmed, so as to employ the Apollo hardware, facilities and capabilities at least through 1971. If we do not exercise this option in the decision for the 1968 budget, we will have to begin a phase-down of the manned space flight activities and the 'mothballing' of some of our facilities. At a small fraction of the initial cost we can continue and we can expand our operations in space for the next ten years and more. These operations will have tremendous implications for our national security and for our position of world leadership, in addition to the benefits of scientific advancement and the betterment of man's life on Earth. . . . In the area of national security, our expenditures in space probably represent the cheapest insurance policy the nation can buy for the future." (Text)

April 18, 19: Press commented on $525-million order from Pan American World Airways, Inc. (Pan Am), for 25 Boeing 747 aircraft [see April 13]: Order had cast doubt on market value for Anglo-French Concorde SST, according to Andrew Wilson in *Washington Post*. BOAC Chairman Sir Giles Guthrie had said his corporation would meet with Boeing representatives during week of April 18 to discuss purchase of about six new jets. 490-seat Boeing 747 would reduce operating costs by as much as 37 per cent, and Pan Am order, Wilson said, "makes it inevitable that B.O.A.C. will follow suit." (*Wash. Eve. Star*, 4/15/66, C-3; Wilson, London *Observer*; *Wash. Post*, 4/19/66, C7)

"The 747 concept promises to be as revolutionary in relation to the current generation of jets as the initial 707s were to the piston-powered transports they superseded," *Aviation Week* editorialized.

"Biggest technical advancement will come from the new generation of high bypass ratio, high temperature, large turbofans typified by the Pratt & Whitney Aircraft JT9D. This new gas turbine cycle promises as much improvement in load-carrying capacity and operating economy as the original turbofan gained over the straight turbojet. It is the real root of the economic and technical revolution embodied in the 747 concept." (Hotz, *Av. Wk.*, 4/18/66, 21)

April 19: BOAC grounded four Boeing 707 jets after precautionary inspection had revealed hairline cracks around bolt holes in tail assemblies. Following BOAC announcement, FAA disclosed that 22 Boeing 707 and 720 jetliners, all three or more years old, were under repair for tail assembly cracks. (*NYT*, 4/20/66, 1; Ingraham, *NYT*, 4/21/66, 66M; *NYT*, 4/22/66, 65; Toth, *Wash. Post*, 4/24/66, M6)

- USAF launched unidentified satellite from WTR with Atlas-Agena D booster. (*U.S. Aeron. & Space Act., 1966*, 150)
- House Committee on Science and Astronautics filed its report on H.R. 1441, the $4.986-billion NASA budget authorization. (NASA LAR V/65-66)

April 19: International Association of Machinists and Boeing Co. settled dispute on seniority issues under Federal mediation, averting strike of 50,000 Boeing employees, including those working on Saturn V boosters at Kennedy Space Center, NASA. (UPI, *NYT*, 4/20/66, 48; UPI, *Wash. Post*, 4/20/66, A6)

- Eastern Airlines reserved two delivery positions for U.S. supersonic transport (Sst); transaction raised total number of reserved delivery positions to 96 and number of companies holding reservations to 22. (FAA Release 66-36)

April 19-20: American Geophysical Union held 47th annual meeting in Washington, D.C. Dr. George P. Woollard, AGU president and director of Univ. of Hawaii's Geophysics Research Center, in opening speech, urged intensive study of earth's interior. As an example of complexities of the planet brought to light by detailed knowledge, he noted that satellite observations, leading at first to the view of earth as "pear-shaped," had been supplemented by new information on local variations. "We will probably wind up," Woollard said, "with a warty, oblate spheroid." (Sullivan, *NYT*, 4/20/66, 47M)

Univ. of Chicago astronomer Edward Anders told AGU that organic compounds in certain meteorites were result of "natural chemical processes," not "products of living organisms from afar." Hydrocarbon formation, Anders said, "will happen whenever carbon monoxide, hydrogen and meteoritic dust cool on a rapid time scale." John Wood of Smithsonian's Cambridge Astrophysical Observatory said that some meteorites were "'truly primordial' matter, formed along with the rest of the solar system from a solar nebula." (*Wash. Post*, 4/21/66, A3)

Experimenters presented preliminary reports on NASA's PIONEER VI spacecraft at AGU meeting. Initial data gathered by 140-lb. spacecraft, launched into heliocentric orbit Dec. 16, 1965, indicated that: (1) low-energy cosmic radiation generated in solar flares traveled in well-defined streams coming from varying directions; (2) electron density in interplanetary space, measured by Stanford Univ. radio propagation detector, was nearly 10 per cu. cm.; (3) solar wind contained not only hydrogen and doubly-charged helium ions but also a small number of singly-charged helium ions; and (4) interplanetary magnetic field had structures interpreted as filamentary "stringers" associated with solar coronal filaments seen during eclipses. (NASA Release 66-87)

April 19-21: Nine NASA astronauts visited MSFC for briefing on Saturn IB launch vehicles to be used in initial manned Apollo missions. Briefing—including tours, hardware descriptions, design philosophy, structural and propulsion tests, and flight data from Feb. 26 AS-201 flight—was attended by Astronauts Virgil I. Grissom, James A. McDivitt, David R. Scott, Russell Schweickart, Edward H. White II, Frank Borman, Walter M. Schirra, Jr., Roger Chaffee, and Walter Cunningham. (MSFC Release 66-79)

April 20: COSMOS CXV scientific satellite was launched into earth orbit by U.S.S.R. "to continue space research," Tass reported. Satellite was said to have an apogee of 294 km. (183 mi.); perigee of 190 km. (118 mi.); period, 89.3 min.; and inclination, 65°. Equipment was functioning normally. (Tass, 4/20/66)

- International Union of Electrical Workers acceded to Government request to remove pickets at KSC from entrance gates used by building trades workers. Electricians had struck for higher wages April 18 after expiration of contract with United Technology Center. (*NYT*, 4/21/66, 27)

April 20: H. Julian Allen, Director of NASA Ames Research Center, was among 27 elected to National Academy of Engineering "in recognition of their outstanding contributions to engineering theory and practice or to the pioneering of new and developing fields of technology." (NAE Release, 4/20/66)

- GSFC turned over all Tiros photographs to the Environmental Science Services Administration (ESSA) for archival purposes. (GSFC Historian)
- BOAC announced grounding of two more Boeing 707 jetliners for repair of tail assembly cracks; four 707's had been grounded April 19. (Wash. *Eve. Star*, 4/20/66, A10)

April 21: NASA Aerobee 150 sounding rocket launched from WSMR reached 96-mi. (154-km.) altitude in Johns Hopkins Univ. experiment designed to observe ultraviolet emissions from Venus. Scattered light produced by solar illumination of ITT startracker optics prevented acquisition and lock-on of target Venus, but good data were obtained on Lyman-alpha. (NASA Rpt. SRL)

- Senate Armed Services Committee Chairman Richard B. Russell (D-Ga.) announced unanimous vote to increase by $167.9 million DOD's FY 1967 budget "to provide 'long-lead-time' production items for the Nike-X antimissile missile system." Senator Russell said Committee's decision had been influenced by information reported April 21 in *Washington Post* on Soviet missile-defense buildup and "by the unanimous recommendation of the Joint Chiefs of Staff to go ahead, during recent closed-door testimony." Administration budget had already carried $447 million for missile defense. (Rosenfeld, *Wash. Post*, 4/21/66, A1; Norris, *Wash. Post*, 4/22/66, A4; Senate Rpt. 1136, 4/25/66)
- NASA Administrator James E. Webb praised development of the Nerva nuclear rocket engine following a tour of AEC's Las Vegas, Nev., test site. He observed the nuclear engine was capable of producing longer-duration flights than engine systems in current use. (UPI, *Houston Chron.*, 4/22/66)
- NASA awarded $50,000, 60-day fixed-price contracts to Douglas Aircraft Corp., McDonnell Aircraft Corp., and Grumman Aircraft Engineering Corp. to perform definition and preliminary design studies and evaluate plan to make spent Saturn V S-IVB stage hydrogen tank habitable for manned space missions of up to 30 days. Contracts would be managed by MSFC. (MSFC Release 66-83)
- Immediate need for an emergency space rescue squad was underlined by the curtailed March 16 GEMINI VIII mission and by electrical malfunction aboard OAO I successfully orbited April 8, William Hines asserted in the Washington *Evening Star*. "Without such capability," Hines suggested, "large sums of money and vast amounts of scientific-technological effort are being dissipated, and it is only a matter of time before lives will be placed in jeopardy. . . . Where manned missions are concerned, a rescue capability is not merely desirable but imperative. . . . In the next five years somewhere between 25 and 30 missions involving about 75 astronauts will be flown. Durations will range from three to 45 days, and mission objectives will reach as far as the moon's surface. At the most conservative estimate there are probably a dozen chances for something to go seriously wrong on a typical mission. To put it another way, there are somewhere between 300 and 500 accidents waiting to

happen in the next five years—any one of them potentially tragic." (Hines, Wash. *Eve. Star*, 4/21/66, A14)

April 21: U.S.S.R. announced annual Lenin prizes. Winners in science included anonymous space experts responsible for LUNA IX and LUNA X missions. LUNA IX made the first softlanding on the moon Feb. 3; LUNA X became the first manmade satellite of the moon April 3. Prize commemorated Lenin's birthday and was considered highest in the Soviet Union. (AP, *Wash. Post*, 4/22/66)

April 22: AFSC Commander Gen. Bernard A. Schriever, who directed the development of USAF's strategic missile force, would retire this summer, the White House announced. General Schriever, retiring for personal reasons, would be replaced Aug. 31 by L/G James Ferguson, USAF Deputy Chief of Staff for Research and Development. (AP, *NYT*, 4/23/66, 5; Norris, *Wash. Post*, 4/23/66, A2)

- USAF OV3-I satellite carrying AFCRL-instrumented 152-lb. payload was launched into polar orbit from Vandenberg AFB with Scout booster (SLV-1A). Reported orbital parameters were: apogee, 3,557 mi. (5,724 km.); perigee, 219 mi. (352 km.); period, 152 min.; inclination, 82°. Objectives were measurement of charged particle distribution in earth's magnetic field and acquisition of radiation intensity data for comparison with AFCRL satellite measurements in 1960-1963. Previously scheduled for April 19, launch had been postponed after detection of malfunction in launch vehicle's ground support system. (OAR Release; *U.S. Aeron. & Space Act., 1966*, 150)

- ComSatCorp Chairman of the Board James McCormack said ComSatCorp was contacting communications and broadcasting companies to get largest possible consensus on costs and technological feasibility of domestic multipurpose communications satellite system, and would present plans to FCC by Aug. 1. (AP, *Wash. Post*, 4/23/66, E7; Bishop, *WSJ*, 4/25/66, 32)

- LUNA X completed its 150th orbit of the moon and marked Lenin's birthday by beaming back to earth the Communist anthem, "Internationale." (UPI, *NYT*, 4/23/66, 5)

- Graduate Research Center of the Southwest president Dr. Lloyd V. Berkner addressed MSC employees. Outlining postlunar landing exploration, Dr. Berkner said: "The very broad strategic objectives for the exploration of the solar system are clearly outlined in the National Academy of Sciences Summer Study that was done last summer.... It is urgent and imperative that this study be adopted, or something like it, as our major basic strategy beyond the moon. The first job is to adopt a national strategy. This is not just a decision of NASA; it is a decision of the administration, of the Congress, of the American people.

"And second it is urgent and imperative that this study be translated by joint actions of NASA, the National Academy of Sciences and American scientists into specific tactical plans for a specific program with dollars attached to it. For example, in my opinion, it makes little sense to undertake the first Martian landing without some precedent exploratory steps. We must first orbit and map the planet and ascertain where a landing would be profitable.

"To do the planetary exploration job effectively, we now need a tactical program of specific flights on specific dates with the instrumentation necessary to plan to accomplish those tasks. I remind you that what we are doing today was planned in 1961. And yet, when we get to our planetary program in the 1970's, we don't have corresponding specific plans." He reemphasized the urgency by saying: "If we are to achieve in the planetary exploration of the future in an intelligent way, in light of the objectives of the Space Act, the time has come to get our scientists started now. Your vehicle program is already ahead of us." Dr. Berkner said that the long-range program should meet certain specifications: "(1) step-by-step advancement of experiments in the proper order; (2) use of the right vehicles to carry out the experiments; (3) effective use of Saturn capabilities, from unmanned missions to the ultimate manned landing on Mars and perhaps Venus, and (4) proper phasing-in of smaller vehicles with the larger ones, and reasonable time/cost programming." (MSC *Roundup*, 5/27/66, 3)

April 22-23: JPL Director Dr. William H. Pickering received American Philosophical Society's Magellanic Gold Medal at society's annual meeting in Philadelphia "for his achievements in the fields of astronomy and navigation." (*NYT*, 4/24/66, 83)

AEC Chairman Dr. Glenn T. Seaborg told APS that a technological society must give the humanities economic freedom in order to assure development as a "human" society. Discussing role of National Foundation on the Arts and Humanities, established in September 1965, Seaborg said: "The new support of the arts and humanities by the Federal Government should help break down some of the artificial barriers we have created between the world of science and technology and that of the arts and humanities.... though science may have a pervading and ascending influence in our lives today, there cannot be any clearcut division between science and nonscience in interdisciplinary civilization which a livable future world will require." (*CR*, 5/4/66, 3-4)

April 23: Saturn V 2nd stage test vehicle (S-II-T) was successfully captive-fired for 15 sec. at MSFC's Mississippi Test Facility in first test of flight-weight, all systems S-II stage. 33-ft.-dia., 81-ft.-long stage—largest and most powerful liquid-oxygen/liquid-hydrogen stage known—developed 1-million-lbs. thrust from five J-2 engines. Test also marked first operational use of MTF. (MSFC Release 66-84)

- Russian space dogs Veterok and Ugolyek, in orbit Feb. 22-March 16 aboard COSMOS CX, appeared on Moscow television with Soviet physician-cosmonaut Dr. Boris Yegorov, who told viewers that "the dogs had undergone many tests and were found quite normal." The dogs, who has passed through Van Allen radiation belts during each of 330 revolutions, would be adopted as pets by workers at Russia's space center, Dr. Yegorov said. (Tass, 4/23/66)

- Materials Research Center at Rensselaer Polytechnic Institute—$1.6 million, NASA-sponsored facility—and adjoining Engineering Science Research Building—built with Rensselaer funds and matching NSF grant—were dedicated at Troy, N.Y. Principal speaker at ceremony, which climaxed two-day "Man and Materials" symposium, was NASA Administrator James E. Webb. Rensselaer had received first grant under NASA program for establishment of interdisciplinary centers at selected universities. (RPI Release, 4/2/66)

- In an article in *Krasnaya Zvezda*, retired Soviet M/G of Aviation B. Teplinskiy said: "In 1961 ... at the [U.S.] Defense Department, a special command for developing weapons systems for the air forces was created. Control of military space plans was assigned it.... It was

April 23: NASA Mississippi Test Facility conducts its first test firing, a 15-sec. test of the Saturn V S-II stage.

at that time that Lt. Gen. Bernard Schriever, who then headed the Air Force Systems Command and is a well-known fanatic advocate of the militarization of outer space, stated: 'Our military space requirements cannot be fully satisfied by the civilian programs now being worked out or planned.' . . .

"Not all people agreed with this. Prominent American scientist Dr. Hugh Dryden declared shortly after his appointment as NASA deputy administrator: 'I have learned—I would like to say I regret to have learned—that it is assumed in some circles that if our outer space work . . . has no military potential we will not receive corresponding financial support from any future administrations or congresses." (*Krasnaya Zvezda*, 4/23/66)

April 24: NASA Administrator James E. Webb, in an article written for AP, discussed national power in space: "The question that arises is this: in terms of national power, is there going to be a real contest for the control of this new environment? Is there going to be a shared mastery or an effort at domination?

"The danger is clear in having such a very great power available to only one nation, and especially to the Communist nations who have a strong, determined drive to dominate the world, to enforce their views on others. . . .

"The U.S. space program gives us a mastery of the space environment. It gives us the knowledge which permits us to judge what is in our own interest. It gives us the assurance that we are going to be at the decision table when the big decisions of the future are made. It also gives us the image of a 'can do' nation, the image of a nation that is not going to focus on pure commercialism or fail to respond to a great challenge. It gives us also the power to exploit space." (Text)

April 24: Jet aircraft operating from Washington National Airport. (Wash. *Sun. Star*, 4/24/66, A1)

- U.S.S.R. began series of carrier rocket tests in two areas of the Pacific west and northwest of Midway Island. (AP, *NYT*, 7/5/66, 28)

April 25: Third MOLNIYA I satellite was orbited by U.S.S.R. to assist "development and further improvement of a satellite radio and TV communication system (SOVCOMSAT)," Tass announced. Orbital data: apogee, 39,500 km. (24,545 mi.); perigee, 499 km. (310 mi.); period, 11 hrs. 50 min.; inclination, 64.5°. Equipment included broadcasting system "as well as instruments of a command-measuring set, an orientation system, orbital correction devices, and power supply" for "further development of onboard systems, satellite equipment, and onground tracking facilities." (Tass, *Pravda*, 4/26/66, 6, USS-T Trans.)

- Five-man Observatory Class Spacecraft Review Board had been appointed by Dr. Homer E. Newell, NASA Associate Administrator for Space Science and Applications. Robert F. Garbarini, NASA Deputy Associate Administrator for Space Science and Applications (Engineering), would head the board, established to "study all phases of design, development, testing and space operations procedures of the orbiting observatory spacecraft." Other Review Board members were: F. John Bailey, Chief, Flight Safety Office, MSC; Jack N. James, Director for Lunar and Planetary Programs, JPL; Dr. Albert J. Kelley, Deputy Director, ERC; and Francis B. Smith, Assistant Director, LaRC. (NASA Release 66-91)

- Tokyo Univ.'s Institute of Space and Aviation announced successful launch of 240-lb. meteorological sounding rocket from Uchinoura launching range in Kyushu, southern Japan. Single-stage rocket reached 40.8-mi. (65.7-km.) altitude in 2 min. 4 sec. Data were telemetered to ground station. (*Wash. Post*, 4/22/66, B2)

- NASA would ask $1 billion for Apollo Applications programing in FY 1968 to prevent a lapse in manned space flight by offsetting reduction in expenditures and aerospace industry manpower for Project Apollo, William Normyle reported in *Aviation Week*. He continued: "NASA was permitted by the Administration to request only $41.9 million for Apollo Applications in Fiscal 1967. Some of NASA's advanced planners argued for $500 million in the Fiscal 1967 budget now under consideration, but the agency went to the Bureau of the Budget with a request for only $270 million, which was reduced to the $41.9 million." (Normyle, *Av. Wk.*, 4/25/66, 30)

- Object described as green and yellow with a long fiery tail—later determined by scientists to be a meteor traveling northwest at 80- to 100-mi. altitude—was sighted by thousands of people in eastern U.S. and southeastern Canada. Scientists estimated that meteor, which was observed at around 8:15 p.m. EST, might have ranged in size from size of a football to several hundred pounds, and had probably disintegrated between Albany, N.Y., and Montreal. (*Wash. Post*, 4/26/66, A1; *NYT*, 4/27/66, 24)

April 25: Referring to NASA-DOD decision of 1964 to exchange experts in each other's programs, *Aviation Week* reported the score to date: 184 USAF, 78 Army, 25 Navy, and six USMC officers assigned to NASA; transferred to DOD was one NASA official, Dr. Michael Yarymovich, now technical director of USAF's Manned Orbiting Laboratory (Mol) program. (*Av. Wk.*, 4/25/66, 25)

- Federal mediators in San Francisco announced interim agreement in wage dispute between International Union of Electrical Workers and United Technology Center. Striking electricians would return to work by April 28; negotiations for a new contract would continue until May 20. AP dispatches from KSC reported union, which had limited picketing to one gate April 20, had resumed picketing all five gates, charging "bad faith on the part of UTC." (UPI, *Wash. Post*, 4/26/66, A6; AP, *NYT*, 4/26/66, 14)

- Two controversial theories about the earth's crust to be tested by the Mohole geological project were reviewed in *Science*. One theory held that material flows slowly but constantly up from deep in the earth and then spreads out across the ocean floors. Second theory was that the earth's magnetic field reverses periodically so that successive bands of rock would have alternate polarization, depending on the time the molten material had frozen and locked in its magnetism. Drills would attempt to probe different rock bands to determine if their charges were different and when they were formed. First field trials were scheduled for early 1968. (*Science*, 4/25/66)

- Dr. Wallace J. Eckert, Columbia Univ. prof. of celestial mechanics and director of IBM's Watson Laboratory, received National Academy of Sciences' 19th James Craig Watson Medal "for his pioneering contributions to the scientific application of electronic computers and to the theory of the motion of the moon." (NAS Release, 3/29/66)

April 25-28: American Physical Society met in Washington, D.C. Geophysicist S. K. Runcorn, Univ. of Newcastle-upon-Tyne, U.K., suggested that Jupiter's great red spot may be "the top of a 200-mi.-high column of 'stagnant air' carried along" by a meteorite crater in the planet's hydrogen surface. This view would contradict a widely held notion that the spot is associated with a high mountain. "We don't believe in mountain ranges on Jupiter, at Newcastle," Runcorn said, explaining that Jupiter's crust would not support a mountain. He noted that the large surface feature connected with the stagnant air column might be a depression caused by impact of an asteroid or "a moonlet." Runcorn said explanation of red spot as column of gases supported theory of Jupiter's fluid metallic hydrogen core. (APS Release, 4/20/66; Simons, *Wash. Post*, 4/27/66, A10; Hines, *Wash. Eve. Star*, 4/26/66, A3)

Dr. Harold Brown, Secretary of the Air Force, examined management role of scientifically trained people and invited their participation in the "governmental decision-making process." Brown, a former physicist, attributed rising influence of technocrats "to the increasing technical content of modern life" and to "the less narrowly technical orientation of the technically trained." He noted transition from part-time to full-time scientific advisers in the late 1950's as "one of the major lasting consequences of the Soviet Sputnik achievement." (Text)

NASA's OGO I, launched Sept. 4, 1964, and EXPLORER XXVIII (IMP-C), launched May 29, 1965, had yielded data essential to evolution of an accurate physical description of the universe, according to paper on low-energy cosmic rays presented at APS meeting by Dr. Donald E. Hagge of NASA Goddard Space Flight Center. Hagge noted satellite measurements and recent "quiet sun year" had given physicists "an unprecedented opportunity... for meaningful cosmic ray research." High-apogee satellites permitted measurement of low-energy particles "not directly observable on the surface of the earth" due to protective atmosphere and influence of earth's magnetic field. Minimum of "solar modulation"—suppression of galactic particles by sun's magnetic field upon attempted penetration of solar system—meant that "particles observed during 1965 most nearly represent the true situation outside the solar system." (GSFC Release, 3/27/66)

J. A. Simpson of Enrico Fermi Institute, Univ. of Chicago, reported these satellite experiments had revealed "new features in the cosmic radiation regarding the energy distribution, the chemical and isotopic composition of the cosmic rays at hitherto unattainable energies." Origin of "ultra-low-energy nuclei," which reached vicinity of earth during 1965 period of minimum solar modulation, may be supernovae, Simpson suggested. (Fermi Institute Release, 4/27/66)

Simpson and J. J. O'Gallagher, also of Fermi Institute, presented results of study of solar flare particle propagation based on simultaneous observations by MARINER IV space probe and Imp satellites. Evidence showed that "protons are confined by interplanetary magnetic fields to narrow channels extending outward from the sun" and "particles in these channels are carried around the sun by magnetic fields which co-rotate with the sun." (Fermi Institute Release, 4/29/66)

Paul J. Coleman, Jr., of UCLA, presented results of "a recently completed statistical analysis of records of the magnetic field and plasma velocity, obtained in late 1962 with instruments aboard the spacecraft MARINER II," including "experimental evidence that [hydromagnetic] waves do exist in interplanetary space." Discussing properties of solar wind, Coleman said MARINER II's measurements had indicated average temperature of positive ions in plasma was about 200,000°K. He noted that its high temperature and low density "suggest that the plasma is completely ionized." (APS Release, 4/26/66)

Discussing importance of their NASA-sponsored stellar wind study, S. H. Lam of Princeton Univ. and G. Sandri of Aeronautical Research Associates of Princeton, Inc., said: "The cosmic radiation that reached the Earth, or any satellite that has been sent into space by man, has been attenuated, or as the specialists say 'modulated,' by the solar wind. Therefore, we cannot measure directly the constitution of the cosmic radiation as it exists in interstellar space unaffected by the solar wind. Yet, to be able to understand the origin of the cosmic radiation, we must know its true constitution in interstellar space. A proper understanding of the solar wind will greatly facilitate this difficult task." Lam and Sandri presented electrostatic model for stellar winds at APS meeting. (APS Release, 4/29/66)

Discovery of first celestial body, "Cygnus GR-1," identifiable as a "point source of high energy gamma radiation," was announced to APS by Univ. of Rochester physicists J. G. M. Duthie, Roland W. Cobb, and Joseph Stewart. Discovery of source, which coincided with a region already identified as an x-ray source known as "Cygnus XR-1," climaxed five years of research sponsored by NASA and NSF. (Univ. of Rochester Release, 4/29/66)

Describing the Alfven Propulsion Engine (Ape) in a paper on hydromagnetic propulsion and drag in space vehicles, M. A. Ruderman, New York Univ., said: "A source of electrical power (nuclear, or solar panels) sufficient to power one hundred common light bulbs could raise the altitude of a 5-ton satellite about 100 miles in a week of operation. By allowing the motor to act as a generator the orbital energy of this same satellite can be converted into electrical energy, contributing sufficient drag to lower the satellite altitude by almost 15 miles per day." (APS Release, 4/26/66)

Cal Tech physicist H. Victor Neher described latitude survey of cosmic rays: "Once we satisfied ourselves that instruments sent by balloons up to altitudes of 140,000 feet, near the earth's magnetic poles, gave answers directly comparable to those found by Mariners II and IV, it became less important to depend on spacecraft for our data. This was good news because we have a flexibility and control of balloon flights that are not possible with spacecraft." (APS Release, 4/28/66)

April 25-28: Radiotelescopes trained on radio galaxies might discover the quark, which had been proposed as the most elementary particle, Purdue Univ. physicists C. S. Shen and T. K. Kuo theorized to APS. Suggested two years ago by Dr. Murray Gell-Mann of Cal Tech, the quark was described by Shen as "a very, very elegant, beautiful theory. There are so many elementary particles. Physicists are all confused as to how they are constructed. If they are all constructed from quarks, it is very simple." (Clark, *NYT*, 4/26/66, 28; Wash. *Eve. Star*, 4/29/66, A14)

April 26: First test transmissions of long-range radio and television communications between Moscow and the Far East via third MOLNIYA I comsat, launched April 25, were successful. Pictures were of good quality. Tass said the satellite would be used to "further refine the system of long-range, two-way television and telephone-telegraph radio communications and pilot operations." (Tass, 4/27/66)

- COSMOS CXVI scientific satellite was launched into earth orbit by U.S.S.R. "to continue outer space research," Tass announced. Satellite was said to have an apogee of 478 km. (297 mi.); perigee of 294 km. (183 mi.); period, 92 min.; and inclination, 48° 25 min. Equipment was functioning normally. (Tass, *Krasnaya Zvezda*, 4/28/66, 1, USS-T Trans.)

- NASA Nike-Apache meteorological sounding rocket launched from Sonmiani, Pakistan, carried acoustic-grenade payload to 118-mi. (190-km.) altitude in experiment conducted for Pakistan Space and Upper Atmosphere Research Committee (SUPARCO) and British National Space Research Committee. Grenades did not eject, but data were obtained from trimethyl-aluminum (TMA) cloud release and spectrum photography. (NASA Rpt. SRL)

- ComSatCorp requested FCC permission to construct a $6-million high-capacity earth station at St. Croix, Virgin Islands, to "serve the expanding needs of the Caribbean area, and provide for reliable service between the United States and Puerto Rico, South America, Europe and Asia."

  In a separate request, ComSatCorp petitioned FCC to deny AT&T's and ITT Communications, Inc.-Virgin Islands' applications to build and operate a cable between U.S. mainland and St. Thomas, Virgin Islands. (ComSatCorp Release)

- USAF selected Martin Co. to develop SV-5P manned lifting-body vehicle to explore flight characteristics and atmospheric maneuverability of wingless lifting bodies. Initially, the 5,000-lb., 24-ft. wingless vehicle would be dropped from a B-52 bomber at 40,000- to 50,000-ft. altitude at 500 mph for powerless, gliding flight, landing at Edwards AFB at 120 to 150 mph; in later powered tests, it would accelerate to mach 2 speeds at 100,000-ft. altitude and then maneuver to landing. Airfoil shape of vehicle, which resembled "a delta shaped porpoise with vertical fins," provided lift normally derived from wings. Future vehicles, developed as part of AFSC's Piloted Low Speed Test (Pilot) project, would be launched to ferry supplies and crews between earth and orbiting space stations. (DOD Release 343-66; Wilford, *NYT*, 4/27/66, 22)

April 26: NASA Convair 990 jet aircraft began ARC-managed, preliminary flights in support of NASA's Nimbus C weather satellite scheduled for launch from WTR May 13. Four GSFC experiments installed in jet would test new spacecraft sensors for weather measurement and collect data at altitudes over 40,000 ft. for comparison with data obtained by Nimbus spacecraft over the same areas. (NASA Release 66-107)

- DDR&E Deputy Director (Strategic & Space Systems) Daniel J. Fink, in an address before the National Space Club in Washington, D.C., said: "We in the Department of Defense have always looked upon space less as a matter of adventure and more as a matter of necessity; we explore its potential not because it is 'there' but because we have needs that are 'here.' It is not that we have identified fundamentally new missions since the early days of our space effort. We have, however, begun to recognize new applications for space-based systems, particularly in the tactical field. The opportunities are tremendous—and sometimes so are the problems. We are no longer as enamored with the 'how' of developing a satellite system. We now take a hard look at the 'why' of choosing one approach to solving a military need over another—and this most assuredly includes competing earth-based systems as well. We have come to recognize that our launch vehicle capability coupled with better sensors permits us to do more missions from synchronous orbit. We also are beginning to sense that there is a high potential for combining operational functions in a multiple-purpose satellite where common sensor packages and orbital parameters exist." (DOD Release 341-66)

- Editorial comment on retirement of AFSC Commander Gen. Bernard A. Schriever: "Few men had a larger hand ... in developing the scientific 'mix' of manned aircraft and missiles that is today's U.S. Air Force.
 "From the day he was put in charge of our ballistic missile program in 1954 until today he has been the key man in Air Force research and development.
 "Only last August he was given what is probably the single most important scientific assignment in Air Force history—developing the military Manned Orbiting Laboratory (MOL). . . ." (*Wash. Daily News*, 4/26/66)

- LaRC would test parachutes for possible landing of instrumented unmanned capsules on Mars beginning in summer 1966, in support of NASA's Voyager planetary exploration project. Parachutes would be carried by balloons and sounding rockets to 130,000-ft. altitude—where thin earth atmosphere simulates Martian atmosphere—and would then be deployed behind test units accelerated to mach 1.2 in level flight [see Aug. 30]. (NASA Release 66-90; LaRC Release)

April 27: Meteorological sounding rocket launched from Chamical, Argentina, was first in series of launches planned for NASA-Argentine Space Commission (CNIE) cooperative program. CNIE had scheduled 13 launches for 1966 as part of Inter-American Experimental Meteorological Rocket Network (EXAMETNET), in which Argentina, Brazil, and U.S. would cooperate to establish in the Western Hemisphere a north-south chain of stations for coordinated sounding rocket launchings to obtain experimental data on weather patterns. Brazil launched first rocket under EXAMETNET program from Natal, Brazil, Jan. 12. (NASA Release 66-103)

- JPL 100-kw. radar was flown successfully on NASA Convair 990 at 3-6-mi. (5–10-km.) altitude in preliminary test for rocket flight [see May 9]. (JPL Release, 6/26/66)
- Third MOLNIYA I communications satellite, launched into elliptical orbit by U.S.S.R. April 25, transmitted television broadcast from Vladivostok. Transmission of televised May Day celebration from Moscow to Vladivostok via satellite had been scheduled for May 1. All equipment was functioning normally. (*Pravda*, 4/28/66, 4, USS-T Trans.)
- Soviet press reported that Belka and Strelka, dog passengers on U.S.S.R.'s SPUTNIK V in orbit more than 24 hrs. Aug. 19-20, 1960, were alive anp well. (AP, *Wash. Post*, 4/28/66, B7)
- Army Corps of Engineers had established Extra-Terrestrial Research Agency, headed by geologist Alice S. Allen, to coordinate studies of lunar surface in anticipation of moon exploration requirements. Comprehensive study of such surface properties as soil density, light reflectivity, and thermal and electrical conductivity, Corps of Engineers believed, would permit deductions about problems of "walking, digging, and moving of surface loads . . . providing shelter, water supply and necessary construction materials and tools." (Henry, Wash. *Eve. Star*, 4/27/66, C3)
- Properly managed, the value of the returns from Government investment in research and technology outweighed by "orders of magnitude" the investment itself, Dr. Chalmers W. Sherwin, deputy director (research and technology), Defense Department Research and Engineering, told a meeting in Chicago jointly sponsored by the Patent Law Assn. of Chicago and the Chicago Assn. of Commerce and Industry. Sherwin said the value of R&D had been confirmed by "Project Hindsight," a study covering such systems as the Mark 46 torpedoes, Minuteman II, Hound Dog, Polaris, Sergeant, and Lance missiles. "We are not sure whether good ideas attract 'good' (that is, flexible) money, or if innovative organizations just 'happen' to always have such money. Available technology money simply must be spread all over the place in little pockets near the need. It is not the ivory towers which need flexible money the most, it is rather the organizations heavily involved in real problems—particularly in the early stages of development of new systems." (Text)
- A surface-based deep-sea search and retrieval system was described by its inventor, Willard N. Bascom, president of Ocean Science and Engineering, Inc., in an interview with William Hines of Washington *Evening Star*. Bascom, holder of a patent on "SEARCH", said his system could be ready in a year and that "total cost of a 'SEARCH' vessel ready for operation anywhere in the world would not exceed $\$$₃ million." He cited "presently unrecoverable losses of spacecraft and experimental weapons as an urgent justification for developing a deep-sea retrieval system,"

and said that he had offered "SEARCH" to the Navy. "Similar proposals have been submitted to the Air Force and NASA," Bascom said. His system, an outgrowth of the Mohole deep-sea drilling project, would use scanning devices and grappling tools suspended from a converted cargo ship. (Hines, Wash. *Eve. Star*, 4/27/66, A7)

April 27: FAA had total of 9,566 U.S. airports, heliports, and seaplane bases on record at end of 1965, 76 more than 1964 tally. Texas, with 846 landing facilities, led other states. (FAA Release 66-43)

- Dr. Eric A. Walker, president of Pennsylvania State Univ. and chairman of the National Science Board, was elected president of the National Academy of Engineering. (NAE News)

  Dr. Vannevar Bush, Honorary Chairman of the Corporation, MIT, received the National Academy of Engineering's first Founders' Medal at 2nd annual NAE meeting in Washington, D.C. Dr. Bush, former president of Carnegie Institution of Washington, director of Office of Scientific Research and Development during World War II, and NACA Chairman 1939-41, was cited as "the engineer who mobilized this country's scientific and technical resources for war and created the blueprint for the post-war cooperation of science and government." (NAE Release, 4/19/66)

- About 25,000 acres at KSC had been set aside as a national wildlife refuge, the National Geographic Society revealed. Cape Kennedy and nearby Cocoa, Fla., were especially rich in bird life; a record total of 204 species had been tabulated by the Audubon Society. (*NYT*, 4/27/66, 52)

April 28: NASA issued a fact sheet on U.S. international space programs which listed criteria applied to determine value and acceptability of cooperative arrangements in space activity: (1) cooperative project must have scientific validity and mutual interest; (2) cooperative project must be conducted openly, with the scientific results being made freely available to the world scientific community; (3) each participant must accept financial responsibility for its own contribution to the project.

Since their inception, NASA's international programs had involved 71 countries or separate jurisdictions and had included satellite projects, individual experiments flown on NASA satellites, sounding rocket investigations, and a variety of programs in which foreign ground-based observations were coordinated with flight projects.

U.S. had concluded bilateral agreements providing for launch of 14 scientific satellites instrumented and, in all but two cases, engineered abroad. Six of these satellites had already been successfully launched (one each for France and Italy and two each for Canada and U.K.). Other satellites would be contributed by the European Space Research Organization (ESRO), U.K., Canada, Italy, and Germany.

U.S. had also opened its own satellites, manned and unmanned, to participation by foreign experimenters. Sixteen such experiments had been accepted for flight.

Some 17 countries had entered into agreements with NASA for one or more cooperative sounding rocket projects. Such investigations had been initiated by both U.S. and foreign proposals. They had accounted for roughly 150 launchings in a five-year period with rockets ranging widely in size and type, from boosted Darts and Arcas to Nike combinations, Aerobees, Shotputs, British Skylarks, and French Centaures.

Ten cooperating countries made substantial contributions to experimental communications satellite work by constructing and operating overseas ground terminals used for TV and radio demonstrations. Conventional weather observations in 42 countries helped interpretation of cloud-cover photographs from weather satellites.

A 1962 agreement with U.S.S.R. had resulted in: (1) one-way communications demonstrations using ECHO II, (2) some exchange of ground-based magnetic field data in anticipation of an agreed exchange of satellite data, and (3) exchange of conventional weather data over a special communications channel set up between Washington and Moscow on a shared-cost basis pending the availability of satellite data from U.S.S.R. (Text)

April 28: European Launcher Development Organization (ELDO) ended three-day, seven-nation Inter-ministerial Conference in Paris without reaching decisive conclusions. Issues involved were: (1) Western Europe's independent space role competitive with U.S. and U.S.S.R.; (2) extent of collaboration to keep pace with technological advance; and (3) continuation of initial ELDO project, three-stage Europa-1 rocket. Cost estimate had more than doubled to $420 million, and recent proposal to move Europa-1 into space communications would add $60 million. Another meeting was scheduled for June. (Mooney, *NYT*, 4/29/66, 8; UPI, *Wash. Post*, 4/27/66, A15; *NYT*, 4/28/66, 24)

- Fire in altitude chamber used to simulate environment of Apollo Command Module (CM) at AiResearch Manufacturing Div. Torrance (Calif.) Facility severely damaged Apollo Environmental Control System, which had undergone 480 hrs. 37 min. of a 500-hr.-duration qualification test. Investigating board concluded that probable cause of ignition had been failure of electrical heater tape installed as part of the test set-up, and that some "ground test equipment and materials used . . . were not suitable for application in the vacuum and 5 psia 100% oxygen environments." Board recommended improvement in selection of "some materials used in the Environmental Control System and Apollo Command Module" and pointed to a "potential fire hazard from arcing or direct short circuits." (Senate, Hearing, *Apollo Accident*, Pt. 1, 2/7/67, 32-4)

- NASA Aerobee 150 sounding rocket launched from WSMR carried NRL-instrumented payload to 109-mi. (175-km.) altitude in experiment to photograph solar corona and disc and to measure solar disc variations and total solar flux in Lyman-alpha. Performance of rocket and instrumentation was considered excellent. (NASA Rpt. SRL)

- U.S. Army parachutists made first manned flight tests of NASA-developed parawing at Army's John F. Kennedy Center for Special Warfare, Fort Bragg. Concept of maneuverable, parachute-like, flexible wing had been originated by LaRC scientist Francis M. Rogallo and Gertrude S. Rogallo in 1948, and on July 18, 1963, NASA had awarded $35,000 to the inventors in recognition of their contribution to aeronautics. Parawing's gliding and maneuvering capability results from aerodynamic lift developed in flight; conventional parachutes are drag devices and do not develop lift. Roughly triangular and about 400 sq. ft. in area, parawings used in tests had been built to LaRC specifications and could be folded, packed, and deployed like parachutes. Army Aviation Materiel Laboratories (AVLABS) had recommended evaluation of parawings for personnel use, and NASA engineers and observers assisted Army with technical consultation. (NASA Release 66-106)

- International Air Transport Assn. (IATA) director Knut Hammarskjold announced that most of IATA's member airlines would accept U.S. State

Dept. demands for changes in Warsaw Convention—an international treaty which limited liability of airline involved in international-flight crash to $8,300 per passenger. U.S. had demanded increase in liability to $75,000 per passenger and an absolute liability system under which carriers would be held liable for damages under all circumstances—even in cases of sabotage. (IATA News 12; Long, *NYT*, 4/28/66, 7; *NYT*, 4/29/66, 42)

April 28: *Izvestia* described how the Communist Party anthem was transmitted from LUNA X to earth: "Before the launching, an electronic generator was installed in the artificial moon satellite to radiate in a definite sequence electric oscillations of the frequencies necessary for reproducing selected fragments of the melody. . . . The use of semi conductors made it possible to substantially reduce the weight and volume of the set and considerably increase the range of transmissions. The 'music' signals of the set installed in the Luna 10 could be received even from the neighborhood of Venus.

"The increase in range and the overcoming of cosmic noise were achieved by means of special narrow band-pass filters. The cycle of functioning of the instrument reproducing 'Internationale' consists of two parts: generating in a certain sequence signals of various frequencies —10 seconds—and creating intervals before the repetition of the melody—45 seconds. The instrument is switched on by a signal from the earth." (Tass, *Izvestia*, 4/28/66)

- *New York Times* reported hint by Leonid I. Dubrovin, chief of 11th U.S.S.R. Antarctic expedition, broadcast by radio from Mirny Antarctic station, "that the Soviet Union planned to use the Antarctic as a testing ground for manned landings on the moon." Dubrovin had said fierce Antarctic conditions would test man's adaptability and scientists could learn much there "that would help man to survive in the desolate conditions of the moon." (*NYT*, 5/1/66, 15)

April 29: One of the world's largest and most sensitive automatic space tracking and telemetry antennas was officially dedicated at Goldstone, Calif. Newest facility in NASA's Deep Space Network (DSN), 210-ft. dish antenna was U.S.'s largest fully steerable antenna and world's largest built for research by spacecraft. Operated by JPL, $14-million antenna would be used to track future Mariners and other spacecraft to Mars, Venus, and even Pluto on the outer reaches of the solar system. Dedication ceremonies, hosted by JPL Director Dr. William H. Pickering, included addresses by Sen. Clinton P. Anderson (D-N. Mex.), chairman of Aeronautical and Space Sciences Committee, and Rep. George P. Miller (D-Calif.), chairman of Science and Astronautics Committee. NASA Deputy Administrator Dr. Robert C. Seamans, Jr., and Edmond C. Buckley, NASA Associate Administrator for Tracking and Data Acquisition, also participated. (NASA Release 66-88; MSC *Roundup*, 4/29/66)

- NASA had selected three firms—Computer Div. of Control Data Corp., Industrial Electronics Div. of General Electric Co., and the Univac Div. of Sperry Rand Corp.—for competitive negotiations leading to an estimated $20 million contract to provide a new computing system at MSFC. (MSFC Release 66-89)

- McDonnell Aircraft Corp. received a $2,071,882 cost-plus-incentive-fee contract from AFSC for work on Gemini spacecraft heat shield for Manned Orbiting Laboratory (Mol). (DOD Release 356-66)

April 29: NASA's 210-ft.-diameter Deep Space Network antenna, dedicated at Goldstone, Calif., April 29.

April 30: Test pilots saved one of two U.S. $500 million XB-70 research aircraft during test flight from Edwards AFB by short circuiting two terminals with a paper clamp. When a jammed door prevented hydraulic extension of nose landing gear, Col. Joseph Cotton (USAF), co-pilot with North American Aviation, Inc.'s chief test pilot, Alvin S. White, removed paper clamp from his brief case and touched it to two pin terminals in a small relay unit, following instructions radioed by NAA ground engineers. Gear extended, and although main gear burst into flame as

185-ft., delta-winged aircraft landed, NAA reported damage was light. (*Wash. Post*, 5/3/66, A1; *NYT*, 5/4/66, 23)

April 30: President Johnson issued Executive Order 11277 designating International Telecommunications Satellite Consortium (Intelsat) as an organization "entitled to enjoy, from and after August 20, 1964, all of the privileges, exemptions, and immunities provided by Section 4(a)" of the International Organizations Immunities Act. (Text, *Pres. Doc.*, 5/9/66, 600)

- LUNA X completed 218th orbit of moon after a total of 120 radio communications periods, Tass announced. Equipment was functioning normally, and radio link was said to be stable. (Tass, 4/30/66)
- GISS marked fifth anniversary with lecture by Dr. Harold C. Urey, prof. of chemistry at large, Univ. of California, La Jolla. Reception preceding lecture had honored Dr. Urey's 73rd birthday on April 28. (*Goddard News*, 5/30/66, 8)

During April: Dr. Raymond L. Bisplinghoff, Special Assistant to NASA Administrator James E. Webb and AIAA President, sought to define the role of a professional and of a professional society in an editorial in *Astronautics & Aeronautics*.

". . . in my reading of the history of science and technology, I have been impressed by the key role of individuals who related science and technology to social needs. I would suggest that they are the professionals, and that a true profession deliberately exercises science and technology to meet social needs.

"Unlike most professional and technical societies, AIAA composes a wide variety of engineers and physical and life scientists. Yet definite social concerns bind us in a single profession—*society's expressed need to transport goods and people by atmospheric flight and to explore space.* . . ." (*A&A*, 4/66, 30-31)

- Three U.S. airlines—all claiming to be Nation's oldest—celebrated their 40th anniversaries: United Airlines, April 6; Western Airlines, April 17; and Trans World Airlines (TWA), April 17. (UPI, *NYT*, 4/3/66)
- Study of buoyant Venus probes that could avoid the high temperatures close to the planet's surface by staying in static equilibrium at higher and cooler altitudes was published in *Journal of Spacecraft and Rockets*. Report, by F. R. Gross of the Goodyear Aerospace Corp., said probes would be launched by Saturn IB boosters, have isotensoid shape, and trail behind the instrument package. To withstand reentry heating, the balloons would be constructed from high-temperature fabrics or films now under development. To permit use of conventional instrumentation, probes would be designed to reach equilibrium at altitudes where the ambient temperature was 160° F. Weight analysis had shown that in the minimum temperature atmosphere, worthwhile biological and atmospheric experiments could be performed. (*J/Spacecraft*, 4/66, 582-7)
- CAB study revealed that in 153 U.S. air carrier accidents between 1955 and 1964 caused by or resulting in fire, 297 of 1,955 fatalities would not have occurred had fire been prevented. 1,628 persons died of impact forces, and 30 for other reasons. All fire victims were involved in 13 of the accidents. Report recommended strengthening of structures to withstand impact, improvement of onboard and ground fire-fighting and rescue equipment, continued improvement of exit facilities, development of emergency communication systems, and more detailed passenger briefings. (CAB, BOSP, 7-6-3)

During April: Men and women engaged in the national space program should reverse defensive and apologetic attitudes toward the commitment to space exploration, according to Aerospace Industries Assn. President Karl G. Harr, Jr., addressing American Astronautical Society annual meeting in San Diego. "If we were the only nation in the world that had the capability to proceed into this new dimension, we should do it," Harr said.

"We should do it as part of a national renaissance on all fronts; for a great society . . . cannot confine its attention to certain challenges selected under past conditions, but must . . . move comprehensively to accept the challenge of the future.

". . . As the huge challenge of space generates and consumes and then regenerates our material and human resources, so does it widen the scope of the environment in which our citizens will exist and make their contribution in the years ahead." (*Aerospace*, 4/66, 7-9)

- Commentary on numerous reports of Ufo sightings:

    Sir Bernard Lovell, director of Britain's Jodrell Bank Experimental Station, said Ufo's were "purely American phenomena." What people really see "is most likely weather balloons, meteorites, fireballs, re-entry of nose cones or other space debris. No trained observer or astronomer has ever reported such a sighting."

    Philadelphia *Evening Bulletin:* "Sir Bernard Lovell, the eminent British astronomer who runs the Jodrell Bank space observatory, believes that seeing UFOS . . . is strictly an American phenomenon.

    "Maybe he is right, but IF Americans are seeing UFOs, and IF there are visitations from outer space, it may be that the terrestrial visitors believe that only America is worth investigating." (Phil. *Eve. Bull.*, 4/26/66)

    Rev. Francis J. Heydon, professor of astronomy at Georgetown Univ., said that people "are seeing things," but that reports of flying saucers would "doubtless increase. The phenomenon of flying saucers is 100 percent imagination."

    Roscoe Drummond, *Washington Post:* "History is littered with examples of the most eminent scientists who were dead certain that things couldn't be done . . . and wrote with great displays of scientific evidence to prove that they couldn't be wrong. We need more objective investigation than we have been getting. The scientific unbelievers may be right—but they could be totally wrong. They have been before."

    Howard Simons, *Washington Post:* ". . . scientists point out that even traveling at the speed of light, which is 186,000 miles per second, whoever it is that is buzzing America in a flying saucer or saucers would have to have left the nearest galaxy 50,000 years ago." (AP, Wash. *Eve. Star*, 4/25/66, A2; Wash. *Eve. Star*, 4/18/66; Drummond, *Wash. Post*, 4/6/66, A25; Simons, *Wash. Post*, 4/24/66, A25)

May 1966

May 1: NASA launched series of four Nike-Cajun meteorological sounding rockets within four hours from Natal, Brazil; Point Barrow, Alaska; Churchill Research Range; and NASA Wallops Station. GSFC acoustic-grenade experiment was designed to obtain temperature, pressure, density, and wind data at 22–59-mi. (35–95-km.) altitudes during transition from wintertime westerly wind circulation to summertime easterly pattern. Natal launch was first in cooperative Brazil-U.S. project under agreement signed Nov. 15, 1965, by NASA and Brazilian Space Activities Commission (CNAE). (NASA Release 66-101; NASA Rpt. SRL)

- Javelin sounding rocket was successfully launched by USAF Cambridge Research Lab. from Churchill Research Range, with payload designed to measure total and component magnetic field values and primary electron and proton fluxes during a magnetic-absorption event. (M&R, 5/16/66, 10)
- Parachutist Nick Piantanida suffered critical brain damage from oxygen cutoff and rapid decompression during emergency descent in a balloon gondola near Worthington, Minn. Gondola was electrically severed at 57,000-ft. altitude in response to Piantanida's emergency cry, but parachute did not appreciably slow its descent until about 20,000 ft. USAF specialist later reported that pressure suit had failed, resulting in loss of oxygen.

  Piantanida, in his third attempt to break world's free-fall record, had planned to jump from 124,000 ft., free-fall to 7,000 ft., and then parachute to earth, proving that trained parachutists could free-fall from 100,000 ft. without health impairment and without stabilizing devices. (UPI, *NYT*, 5/2/66, 40; *Wash. Post*, 5/2/66)
- Moscow's May Day parade featured no new military weapons for first time in three years. Soviet commentators emphasized increased mobility of missiles and claimed that antiaircraft rockets had "radioelectronic devices [which] home the rockets on their targets, even if they take evasive action or use radio jamming devices as a cover." (*Wash. Post*, 5/2/66, A1, A14)
- NASA's 31 veteran astronauts would lose over $6,000 annually from sale of their personal stories if 19 new astronauts joined them in contract with Time, Inc., and Field Enterprises Educational Corp., AP reported. Under terms of contract, Time, Inc., and Field paid $520,000 annually for equal division among "participating astronauts," in return for exclusive rights to all unofficial personal stories and photos. (AP, *Wash. Post*, 5/1/66, N8)

May 1: Australian mobile space tracking station soon would begin seven-week, nonstop operation on Thursday Island off Cape York Peninsula, according to Australian News and Information Bureau announcement. Minister of Supply Denham Henty had said mobile station, part of NASA's Anna geodetic satellite project for measurement of size and shape of the earth, would transmit data to U.S., via a fixed station near Adelaide, for analysis at Johns Hopkins Univ.'s Applied Physics Laboratory. (*NYT*, 5/1/66, 70)

May 2: Resources of U.S. space program were minimal "while the Russians apparently are increasing theirs," Dr. George E. Mueller, NASA Associate Administrator for Manned Space Flight, told a news conference at MTF. "Competition in space is not lessening with time, but rather is increasing. What is needed is a clear definition of where the nation is going in space after the moon is explored." (Russell, *Miami Her.*, 5/3/66)

- Six-week NASA-sponsored exploration trip to search for plant life in Chile's Atacama Desert—area resembling Martian terrain because of its extreme aridity—was begun by Richard W. Davies, JPL Advanced Studies Section, and two other JPL scientists. Explorers, who planned to send back 1,000 lbs. of sandy soil for analysis, would concentrate on high plateau near Calama, 80 mi. inland and 5,000 to 8,000 ft. above sea level. Similar trip last year had yielded a microflora in top four inches of sand. (JPL Release)

- A Soviet victory in the race to achieve first manned lunar landing "would be a worldwide propaganda disgrace because we ourselves chose the target . . . [and] a crushing political blow at home to the President linked with it, which the present White House incumbent has been from the inception," wrote William J. Coughlin in *Missiles and Rockets*. He suggested a possible solution: ". . . extend the U.S. national goal prior to a manned lunar landing by either the U.S. or the Soviet Union. Then, if the Russians were the first to arrive on the Moon, we would be in a position to gracefully acknowledge their achievement while pointing out that to us the Moon was but a way-station en route to a more distant objective. If we accomplished the first landing, we would also have the benefits of victory. But we would not be so exposed in the event of a defeat. . . .

    "Without establishment of a higher goal and a greater challenge, U.S. space capability will dissipate. NASA efforts to prevent this by getting a worthwhile Apollo Applications Program under way have not met with notable success. This provides a second good reason for lifting our national goal. . . ." (Coughlin, *M&R*, 5/2/66, 46)

During week of May 2: MSC officials had not pinpointed exact cause, but had isolated probable cause of faulty thruster which prematurely terminated GEMINI VIII mission March 16, *Missiles and Rockets* reported. Corrective action would affect all these components on Gemini 9 and later spacecraft. Gemini program manager Charles W. Mathews emphasized that thruster and its associated hardware and electronics were contained in adapter module abandoned by crew in orbit prior to reentry. "If we can believe telemetry data, we can isolate the cause of the problem to either spacecraft wiring or thruster valve solenoids," Mathews said. Specific trouble spot within wiring or valve solenoids would be difficult to determine without testing actual equipment. (*M&R*, 5/9/66, 17)

May 2-4: AIAA met in Washington, D.C. Satellite Educational and Informational Television (SEIT)—concept of combining television and communications satellites to raise standards of living in less-advanced nations—was discussed by Vincent P. Rock, director of Communications Central. He emphasized lack of literacy requirement for television instruction and noted that it could (1) "transmit skills and knowledge required for modernization of productive activity; (2) transmit fantasy which helps a person understand the role and techniques of association required for participation in a modern society; and (3) transmit information about events and goods required for action in the modern world."

Following his speech, Rock told the press that a demonstration SEIT system would cost about $30,000,000, excluding ground receiving equipment, and could be ready in three years. Transmissions from satellites could either be received on six-foot "dish" antennas and relayed to individual receiving sets or relayed directly to home sets. (Sehlstedt, Balt. *Sun*, 5/3/66)

James D. O'Connell, White House Director of Telecommunications Management, warned of frequency shortage for satellite communications by 1970 unless new techniques were devised. O'Connell urged aerospace engineers who started "this communications explosion" to "consider all ideas and concepts" that would squeeze more talk into existing broadcast frequencies. (Clark, *NYT*, 5/4/66, 23)

U.S. should assume active world leadership in direct satellite-to-home radio and television broadcasts which offer tremendous potential for mass education and mass propaganda, suggested Columbia Univ. law professor Richard N. Gardner in a panel discussion. Warning that several nations had already asked international bodies to outlaw general broadcasting from space for fear of "unilateral penetration" or unwanted propaganda, Gardner urged: (1) the Administration to proceed immediately in developing direct broadcast satellites; (2) the organization of a pilot demonstration project, perhaps in India, with U.N. endorsement; (3) the involvement of AID and other international lending agencies in educational use of space communications; and (4) the reassurance of friendly nations that comsats would not be used for propaganda.

Murray L. Schwartz, Univ. of Calif. School of Law, told panel that satellites were one more example of a technology that might be moving too fast for existing social institutions. A "knowledge explosion" helped by space communications would accelerate homogenization of society and "one of the great problems is to make sure the homogenization is not too great," he said. (Clark, *NYT*, 5/5/66, 17; Simons, *Wash. Post*, 5/5/66, A1)

May 3: House passed NASA authorization bill for FY 1967 (H.R. 14324) totaling $4,986,864,150 as follows: $4,248,235,000 for research and development; $94,419,000 for construction of facilities; and $644,210,150 for administrative operations. NASA had requested $5.012 billion.

During debate preceding bill's passage, Rep. James G. Fulton (R-Pa.) offered amendments, subsequently rejected, to eliminate $41.9-million funding for Apollo Applications program and $9.1-million funding for construction of lunar sample receiving laboratory at MSC; to reduce advanced missions by $5 million; and to cancel $10 million of prior authorizations for Facility Planning and Design. Rep. Fulton also offered, but later withdrew, an amendment to establish an Inspector General's Office in NASA.

Commenting on NASA's $5.012 billion request for FY 1967, Rep. Fulton said: "[It] generally represents a tight program in terms of the allocation of resources to the programs that have been regularly approved by Congress as parts of the Nation's effort in aeronautics and space.

"At a level under $4.9 billion, I do not believe NASA can restructure its Apollo program moon landing in a form that would not cast the most serious doubt as to its ability to meet the objective of the manned lunar landing within this decade.

"At $4.8 billion the moon landing objective could not be met. Below this level of the budget, it would become clear that the Nation had abandoned the concept of developing a well-rounded total space capability directed toward achieving space preeminence. . . ."

Rep. John Wydler (R-N.Y.), proposing an amendment to earmark $20 million of the total R&D funds for development work on the problem of aircraft noise, suggested: ". . . it is incumbent upon us to bring this [space] program closer to the people and show them some way in which it can benefit them. We often hear there is going to be a fallout as a result of this program—some kind of fallout that will affect the civilian population beneficially. Here is an opportunity to give them some fallout directly that will be of benefit to them. . . ." Rep. Wydler's amendment was defeated. (*CR*, 5/3/66, 9189-9223)

May 3: Malfunction in Atlas-Centaur launch vehicle's attitude control system was responsible for Centaur stage's failure to achieve double ignition in space during April 7 AC-8 mission from ETR, NASA announced. Preliminary analysis indicated that one or more small attitude control jets on Centaur stage apparently used excessive amounts of hydrogen-peroxide gas, causing early depletion of system's fuel supply. Exact cause of excessive use had not yet been determined. Preparations for AC-10 mission, which would launch dummy Surveyor payload on direct ascent to the moon, continued on schedule at ETR. (NASA Release 66-92)

- Gemini 9 spacecraft was mechanically mated to top of Titan II booster at ETR in preparation for May 17 launch after 48-hr. delay because of tiny leak in jet-powered maneuvering unit stored behind spacecraft. NASA officials said delay would not change launch date. (AP, *Phil. Eve. Bull.*, 5/4/66; AP, Wash. *Eve. Star*, 5/3/66, A4)

- NASA launched four Nike-Cajun sounding rockets carrying acoustic grenade payloads within four hours from Natal, Brazil; Point Barrow, Alaska; Churchill Research Range; and NASA Wallops Station. GSFC experiment was designed to obtain temperature, pressure, density, and wind data at 22–59-mi. (35–95-km.) altitudes during seasonal wind shift from westerly to easterly circulation. First group of Nike-Cajuns had been launched from same sites May 1 as shift to summertime easterly pattern began. (NASA Rpt. SRL)

- Long-range HC-130H cargo aircraft equipped with Fulton Recovery System successfully snatched Capt. Gerald T. LyVere (USAF) from Rogers Dry Lake at AFFTC in first test of new ground-to-air recovery system for rescuing downed fliers and astronauts. At 400-ft. altitude aircraft clamped onto balloon-suspended nylon cable attached to harness dropped earlier by parachute, pulled Captain LyVere up behind aircraft, and reeled him over rear ramp. In later demonstration, Col. Allison Brooks (USAF) and A-3C Ronald Doll (USAF) were recovered simultaneously in side-by-side two-man pickup using double harness. Lockheed-built HC-130H, which would become operational June 30, could fly missions to more than 2,000 mi. from its base, loiter at sea level, and rescue up to five persons in multiple pickups from land or sea. For NASA projects, aircraft would carry complex UHF spacecraft reentry tracker equipment to locate reentering spacecraft. (AFFTC Release)

May 3: Rep. Lester L. Wolff (D-N.Y.) put in *Congressional Record* a telegram from Charles H. Ruby, president of International Air Line Pilots Association, opposing $75,000 absolute liability award to survivors of victims of international air accidents. Proposal, Ruby said, would invite sabotage ". . . and would offer . . . criminals the incentive of an automatic payoff without requiring them to buy insurance." (*CR*, 5/5/66, A2474)

- Air Transport Assn. (ATA) noted major developments in air transportation in 1966 annual report, "Facts and Figures": (1) airlines accounted for 59% of all inter-city common carrier services compared to 14% in 1950; (2) sustained growth rate of 16.2% over past three years outdistanced all other major U.S. industries; (3) improved efficiency and economy of operations enabled airlines to absorb steady advances in wage rates, material, and equipment prices; (4) heavy airline investment in modernization and improvement permitted reduction in average fares and eased inflationary pressures; and (5) more than 85% of record profits were reinvested in expansion and improvement program. (ATA Release 39)
- Thailand and Malaysia joined Intelsat—International Telecommunications Satellite Consortium headed by ComSatCorp—bringing total number of member nations to 50. ComSatCorp said it "warmly welcomes this aspect of the development of the consortium and the intention of other countries to increase Intelsat membership even further in the near future." Yugoslavia and Romania had also shown interest in Intelsat and were considering joining. (ComSatCorp Release; Clark, *NYT*, 5/3/66, 9)
- Criticism of Government research and development programs in Richard J. Barber's *The Politics of Research* was cited in a review by Donald Mintz for the Washington *Evening Star:* (1) DOD, NASA, and AEC contributions —90 per cent of total R&D support—indicated emphasis on defense and space at the expense of social and behavioral sciences; (2) 70 per cent of R&D funds was directed to development; most of remainder was spent on applied, not basic, research; (3) Government agencies had given little attention to consequences of their undertakings. (Mintz, Wash. *Eve. Star*, 5/3/66, A10)
- Former Vice President Richard Nixon told a Republican fund-raising dinner in Houston that U.S. should divert some of its space exploration funds to the Vietnam war effort. (AP, Wash. *Eve. Star*, 5/4/66, A36)
- Minuteman II had entered operational service with Strategic Air Command, with delivery of first 50-missile squadron at Wing 6, Grand Forks AFB. (*M/S Daily*, 5/3/66)

May 4: NASA selected Control Data Corp. and IBM for competitive negotiations of a $15-million contract to furnish a large-scale digital computer complex at LaRC. (NASA Release 66-102)

- Soviet physicist Peter Kapitsa, who had received the 1966 Rutherford Prize for Physics May 3 in London, told a press conference that U.S.-U.S.S.R. rivalry was good, stimulated research, and produced variety of solutions to different problems. Kapitsa said he thought the Soviet space program was ahead of U.S., but could not prove it. (*Wash. Post*, 5/5/66, A20)
- NASA had selected ARINC Research Corp. and ITT/Federal Electric Corp. for competitive negotiations leading to $1,750,000, one-year, cost-plus-

incentive-fee contract to provide technical support for Saturn launch vehicle reliability program at MSFC. (NASA Release 66-104)

May 4: A Space Sciences Division in the Engineering and Development Directorate, a consolidated medical directorate, and title changes for key management positions became effective at MSC. Medical Research and Operations Directorate would be headed by Dr. Charles A. Berry, formerly Chief of Center Medical Programs. Medical directorate would include Biomedical Research Office, Medical Operations Office, and Occupational and Environmental Medicine Office. Addition of medical directorate brought the number of Center directorates to five. The four Assistant Directors would be known as Directors: Director of Administration, Wesley L. Hjornevik; Director of Flight Crew Operations, Donald K. Slayton; Director of Engineering and Development, Dr. Maxime A. Faget; and Director of Flight Operations, Christopher C. Kraft, Jr. (MSCI 1130.1; MSC *Roundup,* 5/27/66, 1,2)

- Astronauts returning to earth after long space flights in sterile environment might have to take bacteria pills, suggested Dr. T. D. Luckey, Univ. of Missouri Medical School, at American Society for Microbiology convention in Los Angeles. Dr. Luckey suggested that astronauts' sterile air, food, and water eventually could dangerously reduce the amount of bacteria in their systems, resulting in bacterial illness when they returned to earth. Pills could restore normal degree of immunity to terrestrial bacteria. (AP, Wash. *Eve. Star,* 5/4/66, A12)

- Time intervals of less than a billionth of a second were measured for first time in a laboratory by Dr. Arnold Shastak, Office of Naval Research. Dr. Shastak reported that such intervals might be significant in determining the frequency of electromagnetic waves and in research into structure and distribution of particles in atomic and molecular systems. (Henry, Wash. *Eve. Star,* 5/4/66, A27)

- Aviation/Space Writers' Assn. named William Hines of the Washington *Evening Star* recipient of the Robert S. Ball Memorial Award for excellent writing. Cited for his continuing coverage of NASA's MARINER IV Mars probe, Hines would receive a trophy and $300 at the Association's annual meeting in New York May 24-26. (AP, *NYT,* 5/5/66, 6; AP, *Wash. Post,* 5/5/66, B2)

- JPL Director Dr. William H. Pickering received 1965 Spirit of St. Louis Medal for "outstanding service to the U.S. in directing the research, design development, and successful operation of the Ranger unmanned lunar reconnaissance spacecraft and the Mariner interplanetary spacecraft." Presentation, by ASME and the City of St. Louis, was made at National Transportation Symposium in San Francisco. (*Av. Wk.,* 5/16/66, 23)

May 5: Independent Offices Appropriations Bill H.R. 14921 for FY 1967 was reported to the House and scheduled for floor action May 10. Appropriations Committee recommended $4,245,000,000 for NASA research and development, compared to $4,248,235,000 authorized by House; $75,000,000 for construction of facilities, compared to $94,419,000 authorized by House; and $630,000,000 for administrative operations, compared to $644,210,150 authorized by House. Total recommendation was $4,950,000,000, compared to $4,986,864,150 authorized by House. (Text; NASA LAR V/75, 5/5/66)

- NASA Administrator James E. Webb appointed Dr. John F. Clark as Director of GSFC. Dr. Clark, formerly Deputy Associate Administrator for Space

Science and Applications (Science), had been Acting Director of GSFC since July 1965. (NASA Release 66-105)

May 5: Sen. Milton R. Young (R-N.D.) introduced bill (H.R. 3309) to authorize prototype construction of supersonic transport (Sst) financed by Government-guaranteed bonds under Supersonic Transport Authority. Proposed procedure would: (1) eliminate necessity of Government's appropriating up to $1.5 billion for project; (2) provide management guidelines and sound basis for determining industry's share in development costs; (3) make use of development-phase assets in production phase; and (4) permit contractors and public to participate as stockholders in corporation to produce and sell Sst. (*CR*, 5/5/66, 9396)

- NASA's first manned Apollo flight might be made in 1966 instead of 1967, as originally scheduled, predicted Dr. Joseph F. Shea, manager of NASA Apollo Spacecraft Office at MSC, in a speech at NAA's plant in Downey, Calif., to press group touring manufacturing and test sites for Apollo spacecraft and Saturn boosters. "The Apollo program is about on schedule in time, cost and performance. We have found the performance of the hardware some better than we had expected, on the basis of tests. All testing for the first manned flight probably will be completed by the end of June.

    "I hope we can make the first manned flight this year, although it is not scheduled until next year. I think that the first lunar excursion module will be sent up next year.

    "I am sure personally that our first lunar landing will be done comfortably before the end of the decade."

    C. R. Able, group vice president of Douglas Aircraft Co.'s Missile and Space Systems Div., outlined to the group plan for converting empty hydrogen tank from Saturn V's 3rd stage into manned space capsule. For conversion, series of metal balloons containing liquid oxygen would be added to original orbited payload, to serve later for breathing and for combination with hydrogen to operate fuel cells for producing electricity. An Apollo or Gemini spacecraft would dock with empty 3rd stage, and astronauts would emerge to remove end cover plate in end of the tank, after purging it of any hydrogen remaining from launch. An airlock, carried aloft as part of original payload or launched later, would be attached to opening left by removal of cover plate and tank would be repressurized with oxygen from metal balloons to provide astronauts with "shirtsleeve environment." Tank walls would have about 100 shallow threaded holes for mounting work benches, furniture, or other convenient items, presumably launched separately. Able emphasized that purpose of stage conversion—to offer astronauts relief from cramped quarters and extended weightlessness—did not duplicate function of USAF's Manned Orbiting Laboratory (Mol). Douglas Aircraft Co., McDonnell Aircraft Corp., and Grumman Aircraft Engineering Corp. each had $50,000 NASA contracts to study use of booster cases. (Wallin, *St. Louis Post-Dispatch*, 5/5/66; UPI, *Miami Her.*, 5/6/66)

- Technology and automation derived from U.S. space effort could not raise social, educational, moral, or religious values, Dr. Edward C. Welsh, Executive Secretary of NASC, told Churchman's Club in Baltimore, but "science and technology do increase the opportunity to make improvements. Where there is more leisure, more accumulated knowledge, and greater understanding, there is also increased opportunity to improve character and raise human values. Certainly, the opportunity which such progress gives toward the elimination of war is an impressive justification for the entire space program." (Text)

May 5: Small hydrogen-oxygen engine originally developed for Saturn S-IV rocket stage was being adapted to serve as heater for expanding helium gas used in repressurizing S-IVB propellant tanks in space. Burner, providing an 800-lb. payload gain through weight reduction, would make it possible to meet pressure requirements by using supercold helium from bottles immersed in liquid hydrogen inside fuel tank, and would serve as possible source of low-thrust ullage control power during weightless flight. (MSFC Release 66-92)

- Small semi-pressure weather balloon designed to drift indefinitely entered its fourth orbit of the earth, establishing world's record for flight duration, reported Walter Sullivan in the *New York Times*. Balloon was one of a series launched March 30 from Christchurch, New Zealand, as part of Southern Hemisphere experiment to examine feasibility of Global Horizontal Sounding Technique (Ghost). Ghost—plan to use satellites to locate and read out information from approximately 10,000 balloons floating around earth at constant altitudes—would express state of entire earth's atmosphere in numbers which would then be processed by computer to achieve reliable forecasts for two weeks ahead. Participants in Ghost project were National Center of Atmospheric Research, ESSA, and New Zealand Weather Service. (Sullivan, *NYT*, 5/5/66, 16)

- Bell Aerosystems Co. announced plans to develop 25-ton air-cushion vehicle (ACV) capable of carrying 90 passengers or 12 tons of cargo. Capable of amphibious operations, hovercraft would be 56 ft. long, 16½ ft. high, and have cruising speed of at least 60 mph, and could be available within one and one half to two years. (Bell Aerosystems Release 30)

- AFSC Commander Gen. Bernard A. Schriever received Henry H. Arnold Gold Medal for "distinguished service to augment progress and readiness in air and space" and for "distinguished service in defense of our country" at American Ordnance Assn. meeting in Washington, D.C. In his acceptance speech, General Schriever quoted General Arnold's 1938 prediction: " 'Planes of the future may have telescopic wings, which, once in flight, can be foreshortened, telescoped, or pulled in . . . thus greatly accelerating the forward progress of the vessel of reduced size and decreased air resistance.' This description . . . anticipated the principle which is employed today on the F-111 sweepwing fighter. . . ." (Text)

May 6: U.S.S.R. launched COSMOS CXVII carrying scientific instruments for continued study of outer space. Orbital parameters: apogee, 308 km. (191 mi.); perigee, 207 km. (129 mi.); period, 89.5 min.; inclination, 65°. Instruments were reportedly functioning normally. (*Pravda*, 5/7/66, USS-T Trans.)

- NASA research pilot John B. McKay made successful emergency landing on Delamar Dry Lake, Nev., when engine of X-15 No. 1 aircraft shut down prematurely—probably because ammonia fuel pump ruptured—32 sec. after aircraft had been dropped from B-52 bomber at 45,000-ft. altitude. McKay escaped injury, and X-15 suffered only minor damage. (Edwards AFB Memorandum; X-15 Proj. Off.)

- First of uprated J-2 rocket engines designed to power upper stages of Saturn IB and Saturn V launch vehicles had been delivered to MSFC for static firing. NAA's Rocketdyne Div. had increased engine's maximum thrust to 230,000 lbs. by strengthening turbine wheels of turbopumps and modifying engine control system. (MSFC Release 66-91)

May 6: Dr. Athelstan F. Spilhaus, dean of Univ. of Minnesota Institute of Technology, received patent for an astronomical clock to keep track of the movements of the sun, moon, stars, and tides; provide a perpetual calendar; and show local and world time. Dr. Spilhaus said his clock could aid astronomers and navigators. (Jones, *NYT*, 5/7/66, 35)

- NASA had selected Computer Sciences Corp. and General Electric Computer Div. for competitive negotiations leading to a one-year, $6-million, cost-plus-incentive-fee contract to provide support services to MSFC's Computation Laboratory. (MSFC Release 66-96)

- Two rocket engines which had key roles in U.S. space program—RL-10 and H-1 engines—were being given to Smithsonian Institution by MSFC for National Air Museum's display on the history of rocketry. (MSFC Release 66-95)

May 6-7: Former Astronaut John H. Glenn attended annual meeting of the Transatlantic Council, Boy Scouts of America, in Garmisch, Germany, and participated in ceremonies marking German-American Friendship Week. While in Europe, Glenn, in his capacity as NASA consultant, would speak on U.S. space program to educational, scientific, and technical groups in Germany, England, Sweden, Finland, Norway, Denmark, Belgium, France, and Switzerland. (NASA Release 66-100)

May 7: President Johnson announced U.S. would seek a treaty through the United Nations "laying down rules and procedures for the exploration of celestial bodies." He listed as "essential elements" of the proposed treaty: (1) the moon and other celestial bodies should be free for exploration and use by all countries; (2) no country should be permitted a claim of sovereignty; (3) there should be freedom of scientific investigation, and all countries should cooperate in scientific activities relating to celestial bodies; (4) efforts should be made to avoid harmful contamination; (5) astronauts of one country should give any necessary help to astronauts of another country; (6) no country should be permitted to station weapons of mass destruction on a celestial body; and (7) weapon tests and military maneuvers should be forbidden. (Text; *Pres. Doc.*, 5/16/66, 622; *NYT*, 5/8/66, 66)

- Editorial comment on President Johnson's proposal:

"... In addition to rules governing space exploration a body of law is needed to govern the exploitation of useful resources that may conceivably be found on other solar bodies. Omission of this topic from the first international treaty on space law could produce future difficulties. But the President may well have felt that any present attempt to reach agreement on this delicate matter might prolong negotiations beyond the time of the first manned lunar landing. ... There is no time to lose in getting international agreement to and ratification of this essential document." (*NYT*, 5/9/66, 36)

"... Someday such a treaty may save us from a new sort of far-out interplanetary imperialism. As revived by Mr. Johnson, however, the proposal has to do with a problem that will develop sooner and closer to home. After all between now and 1969 or 1970, Americans are going to be lofted to the moon's surface, and so are Russians. This is a prospect full of very great historic importance. It lends special force to the President's call for a renewed effort ... to negotiate an understanding under which the always unpredictable Russians will sign a treaty worthy of the name...." (Wash. *Eve. Star*, 5/10/66, A6)

"The first manned landing on the moon seems likely before 1970. The implications are immense, regardless of whether the first one is American or Russian. President Johnson's proposal for an international compact to govern this and outer space explorations makes much sense.

"...Perhaps a willingness by Moscow to entertain a treaty on space may also open the door to discussions about pooling some of the costs, as the late President Kennedy proposed." (*Boston Globe*, 5/11/66)

"... Probably the most succinct argument against a comprehensive treaty of this nature is, 'Who needs it?' Monopolization of the lunar surface by military units and establishment of terror weapons on the moon or in orbit appear to be neither immediate dangers nor remote possibilities. The bugaboo of an orbiting multimegaton H-bomb seems, on close examination, to be more trouble than it would be worth....

"There is no slightest indication that the Russians propose to lay claim to the moon if their cosmonauts get there first, any more than we do if we win the race. The precedent followed by both countries in Antarctica—nearest thing to the moon on earth—is neither to assert nor to recognize territorial claims. This was U.S. and Soviet policy before the Antarctic Treaty was signed in 1959...." (Hines, Wash. *Eve. Star*, 5/12/66, A20)

May 7: Secretary of Defense Robert S. McNamara's contention that Nike-X antimissile missile system was incapable of defending U.S. against Soviet attack and should be kept in R&D stage, received support from Society of American Scientists' executive board. In a statement opposing Senate vote providing $168 million toward deployment of Nike-X system, the SAS board said: "These expenditures, truly enormous over extended periods, would be the hallmark of a frightened, not a great society...." (*Wash. Post*, 5/8/66, P14)

May 8: GEMINI VII Astronaut Frank Borman told a British television audience in London that the space vehicle, not man, was limiting the length of manned space flight. "We are finding that man is very adaptable and that if you provide him with the normal comforts he expects on earth, he can function as well at zero gravity as he can on earth." (AP, Wash. *Eve. Star*, 5/9/66, A2)

- Lockheed Aircraft Corp. chairman Courtland S. Gross described flight in proposed U.S. supersonic transport (Sst) in an interview for *This Week*: "It will make the world a lot smaller. Its near two-thousand-mile-an-hour speed will carry passengers to London and Paris in two and a half hours. With room for 255 to 266 passengers, five abreast, it will be a 222-foot walk from the tail to the pilot's cabin.

"But the biggest thrill will be cruising at 70,000 to 80,000 feet above the earth. At that height you will enjoy the kind of stars and sky seen heretofore only by U-2 pilots and astronauts—pinpoint icicles suspended in the blue-black velvet backdrop of space. Most of the atmosphere is below the supersonic transport's cruising altitude, so that at night, on the horizon, there will be the bright blue band first reported by the astronauts—light radiating from the other side of the earth. When flying the polar routes around the world, passengers will enjoy spectacular electrical displays from the ionization of gases in the upper atmosphere—great shimmering curtains of color that flicker through the sky." (*This Week*, 5/8/66)

May 9: NASA Aerobee 150 sounding rocket launched from WSMR carried JPL 100-kw. radar to 102-mi. (164-km.) altitude in experiment designed to gather data on behavior of radar echoes as a function of altitude for use in developing planetary terrain sounding technique. GSFC-managed flight marked first successful functioning of high-powered radar in space. Similar radar had been flown successfully on NASA Convair 990 April 27. (JPL Release, 6/26/66; JPL, *Lab-Oratory*, 6/66, 10)

- U.S.S.R.'s LUNA X completed its 284th orbit around moon and transmitted Communist anthem, "Internationale," on command from Soviet center of deep space communications. Instruments were functioning normally. (*Pravda*, 5/10/66, 1, USS-T Trans.)

- House passed bill authorizing appropriations for AEC during FY 1967 totaling $2,259,958,000: $1,964,128,000 for operating expenses and $295,830,000 for plant and capital equipment. Rep. James G. Fulton (R-Pa.) urged that the Appropriations Committee consider the possibility of studying lunar materials in AEC radiation laboratories rather than building new NASA Lunar Sample Receiving Laboratory at MSC. (NASA LAR V/76)

- Commenting on President Johnson's proposed U.N. treaty banning sovereignty claims on the moon, Rep. George P. Miller (D-Calif.) told the House: "The President has taken a very progressive and forward step . . . in trying to establish a barrier against the use of the moon or outer space as a place of contest for the nations of the world. He has asked that these areas be internationalized and we follow in space exploration that principle which has been established already with respect to the use of nuclear weapons.

    "This is a great step forward by the President, and we in the House should appreciate it, and commend him for it." (*CR*, 5/9/66)

- Text of Soviet Academy of Sciences' August 1965 proposal to International Astronomical Union (IAU) for an international effort to search for signals from other worlds was made public in Vienna. Proposal, which implied that peculiar objects such as quasars represented the kind of beacon a civilization with extraordinary capability might build, recommended two kinds of searches. First would look for narrow-band signals which a "civilization with the meager energy resources of earth" could send to another star. Concentration of the energy into narrow range of wavelengths would ensure signals' survival for hundreds of light years. Search would be confined to 21-centimeter wavelength in the spectrum emitted throughout space by hydrogen.

    Second search would be for signals spanning much of the spectrum which a super civilization with the energy resources of a star would generate. Such signals would be difficult both to decode and recognize as artificial. Proposal, which recommended round-the-clock observations for three to five years, would be discussed at IAU General Assembly in Prague Aug. 22-31, 1967. (Sullivan, *NYT*, 5/10/66, 19)

- NASA Deputy Administrator Robert C. Seamans, Jr., said NASA would request $6 billion for FY 1968, including $900 million for Apollo Applications and $100 million for Voyager, *Aviation Week* reported. Seamans' statement that "Apollo Applications will fill the post-Apollo gap and that there is no need to define another goal now" was said to conflict with the recent contention of Dr. George E. Mueller, NASA Associate Administrator for Manned Space Flight, that a clear definition of the Nation's post-Apollo goal was necessary. (*Av. Wk.*, 5/9/66, 25)

May 9: AFSC Commander Gen. Bernard A. Schriever, at Armed Forces Management Assn., in Washington, D.C., predicted a management revolution commensurate with the explosive advance of technology and the pressing nature of military requirements. "I think that the biggest challenge to management in the years ahead will be to find ways to encourage and use the full creative potential of individuals. . . . We must create the atmosphere where the unique human capacity for innovation and judgment can be fully developed and employed. . . ." (Text)

- Communist China conducted "a nuclear explosion that contained thermonuclear material" over its western areas, Hsinhua, Chinese Communist press agency, reported. U.S. officials in Washington, D.C., surmised that explosion indicated China was on the verge of developing a thermonuclear or hydrogen bomb. (AP, *NYT*, 5/10/66, 4; Topping, *NYT*, 5/10/66, 1)
- Industry bids to AFSC for study of new integrated satellite system as follow-on to the Midas early warning satellite would be due June 24, *Aviation Week* reported. Proposal request included studies for sensor payload and spacecraft. Primary sensors would be optical, covering the infrared, ultraviolet, and visible spectra. (*Av. Wk.*, 5/9/66, 23)
- USAF announced it would award contracts to "individuals within the scientific community" to conduct "prompt, in depth investigation of selected Ufo reports." Funds for study, which had been recommended by Air Force Scientific Advisory Board in March, would be requested from FY 1967 and FY 1968 budgets. (DOD Release 388-66)
- Results of survey by DMS, Inc., an aerospace market research firm, indicated that nearly twice as many aircraft would be built from 1966 through 1975 in the non-Communist world as were produced during the previous decade as a result of continued increases in airline passenger traffic, expansion of "private" flying, and pressing need to replace obsolete military aircraft. This would amount to about 260,000 aircraft for total investment of $100 billion. (*DMS Aerospace News*)

May 10: House passed Independent Offices Appropriation Bill (H.R. 14921) for FY 1967 appropriating $4.95 billion for NASA. Amounts were those recommended by Appropriations Committee May 5. An amendment proposed by Rep. H. R. Gross (R-Iowa) to deny funds for Lunar Sample Receiving Laboratory at MSC was defeated by a voice vote. (NASA LAR V/77)

- First perfect dual countdown rehearsal in Gemini series conducted at KSC in preparation for May 17 Gemini IX mission. (AP, *NYT*, 5/11/66, 16; AP, Wash. *Eve. Star*, 5/11/66, A6; *Wash. Post*, 5/11/66, A7)
- Accelerated programs for launching communications satellites had improved prospects of earning profit in 1967, reported ComSatCorp president Dr. Joseph V. Charyk at annual shareholders' meeting in Washington, D.C. Predicting a 100-fold increase in ComSatCorp's revenue potential during five-year development plan, Dr. Charyk said: ". . . we have taken firm steps for the development of even larger and more versatile satellites for deployment on a global basis in 1968. These spacecraft will weigh approximately 240 pounds in orbit and will have a capacity for 1,200 two-way telephone circuits or four television channels." He also noted ComSatCorp had proposed to ESSA that it conduct joint feasibility study with IBM to develop satellites to collect, process, and disseminate weather data on a global basis. (Vartan, *NYT*, 5/11/66, 65; Fouquet, *Wash. Post*, 5/11/66, E6)

May 10: Alex P. Aven, owner of an Oklahoma City, Okla., petroleum consulting firm, was sworn in by NASA Administrator James E. Webb as consultant to the Administrator on management. (NASA Release 66-112)

May 10-13: NASA successfully launched 18 Arcas and Hasp meteorological sounding rockets to 200,000-ft. altitudes at intervals ranging from four minutes to six hours to investigate daily cycle of wind and temperature variations in upper atmosphere, compare results of various rocket and balloon measuring systems, and study effects of radiation from the sun on meteorological instruments. Instrumented payloads descending by parachute transmitted atmospheric temperature, density, and pressure data, and ten specially instrumented high-altitude weather balloons, interspersed among rocket firings, transmitted information as they rose to 130,000 ft. Radar tracked both balloon and rocket payloads to determine speed and direction of high-altitude winds. Launches were managed by NASA Langley Research Center and sponsored by NASA, USA, USN, USAF, and ESSA. (NASA Release 66-118; Wallops Release 66-25)

May 10-19: 1966 Committee on Space Research (COSPAR) meeting was held in Vienna.

National Academy of Sciences delegate Dr. Richard W. Porter, summarizing activities of 1965 U.S. space program, said a manned lunar landing by 1970 would be a "difficult goal" but U.S. had made "substantial progress."

Dr. Porter presented an analysis of data transmitted by NASA MARINER IV Mars probe which appeared to support theory that earth sometimes has comet-like tail stretching past orbit of Mars. On Feb. 5, 1965, a peculiarly timed solar eruption sent series of intense proton clouds into space, five of which were detected within 48 hrs. by MARINER IV and two other U.S. satellites in near-earth orbit. Second and third clouds were reported by MARINER IV three to five hours prior to their detection near earth, possibly because solar eruption had blown earth's tail to enormous length, forcing two clouds to detour through tail in order to penetrate earth's magnetic field. (Sullivan, *NYT*, 5/11/66, 16; Toth, *Wash. Post*, 5/12/66, A14)

Chief Soviet delegate Anatoli A. Blagonravov summarized 1965 U.S.S.R. space program, and said LUNA X, which entered lunar orbit April 3, had recorded radiation pattern from lunar surface similar to that from basaltic part of earth's crust. Apparently lunar surface exhibited no differentiation equivalent to granite on earth, indicating that earth and moon were created separately. LUNA X had found level of moon's magnetic field barely above that of interplanetary space; small increase was attributed to moon's passing through earth's magnetic tail at time of measurement. Implication was that moon is cold throughout, since liquid metal core would produce large magnetic field. Report suggested that new minerals would be found on the moon because of unusual chemical reactions caused by its extreme temperature range, the impacts of micrometeorites, and rains of high-energy radiation. (Sullivan, *NYT*, 5/11/66, 16; Toth, *Wash. Post*, 5/12/66, A14)

Soviet delegate A. G. Prishchep, in a joint paper with colleague V. I. Vashkov, revealed that U.S.S.R. planetary probes were assembled in clean rooms and sterilized with two-fold process involving moist heat to prevent dryness and a gas mixture, harmless to the skin, which might have important surgical applications. Gas mixture contained 50% ethylene oxide—extremely poisonous substance which is dangerously explosive when combined with oxygen—diluted with 40% methyl bromide for safety reasons.

General Electric scientists H. G. Lorsch and M. G. Koesterer reported that current sterilization cycles varied from treatment at 320° for three hours to baking at 200°F for 14 days. They argued that no more than 10 billion microbes should exist within a spacecraft prior to beginning of sterilization if reasonable chance of killing virtually all of them was to be assured. (Sullivan, *NYT*, 5/12/66, 79; Beller, *M&R*, 5/16/66, 17-18)

Drs. Herbert Friedman and R. W. Kreplin, Naval Research Laboratory, reported that solar radiation level would probably exceed highest ever recorded in 100 yrs. during 1969-1970—time of first projected manned lunar landing. Basing their prediction on measurements of solar x-ray emissions taken by EXPLORER XXX (IQSY Solar Explorer) satellite, Friedman and Kreplin explained that x-rays, which could not penetrate earth's atmosphere, warned of approaching sunspots before they could be seen. Solar activities in the form of sunspots and gaseous flares occurred in 11-12 yr. cycles, but there was also a gross cycle which reaches its greatest intensity every 80 to 100 yrs. (Sullivan, *NYT*, 5/13/66, 19; Toth, *Wash. Post*, 5/13/66, A12)

Dr. Bessel Kok, Research Institute for Advanced Studies, described a small instrument which could determine the existence of life when landing on another planet. Designed on the assumption that even the simplest forms of life use water and such common chemical compounds as phosphates and nitrates, device would trace the transfer of nonradioactive isotope—Oxygen 18—from these chemicals to water. (*Wash. Post*, 5/13/66, A12)

Reports on little-publicized efforts were presented to the meeting: (1) Italy's platform for launching San Marco earth satellite at sea was in African port of Mombasa and would be towed shortly to site where U.S. Scout booster would launch satellite into equatorial orbit; (2) Pakistan was conducting monthly firings of rockets that injected copper chaff into stratosphere to disclose winds in that region over subtropics; (3) U.S.S.R. had fired 150 meteorological rockets—almost half from shipboard—and U.S. was firing about 120 monthly; (4) Romania had used orbital data from Washington and Cambridge, Mass., to photograph ECHO I and ECHO II comsats in coordination with Soviet expeditionary stations in Egypt and Mali; (5) Sweden, Finland, Russia, and East Europeans had joined in observing orbital flight changes of ECHO I and ECHO II attributable to density variations in wisps of air at satellite altitude; and (6) East German scientists reported their conclusions on upper air structure derived from real-time transmissions from NASA EXPLORER XXII satellite. It was estimated that 60 stations in various parts of the world were monitoring satellites that sent out "blind" data. (Sullivan, *NYT*, 5/15/66, 10E; Sullivan, *NYT*, 5/16/66, 88)

Dr. A. I. Lebedensky, Soviet Academy of Sciences, reported that LUNA IX photos revealed lunar surface was steadily eroding, at least in some places. He suggested that erosion could be caused by solar winds or impact of micrometeorites. Lebedensky pointed out two peculiar features on LUNA IX's photos which supported his theory. First was series of streaks, resembling mineral veins in earth's bedrock, on floor of crater where LUNA IX had landed. It had previously been assumed that moon's surface was buried deep in debris churned up by meteorite impacts. Second was presence of rock-like objects on small pedestals similar to those on earth where soft but rock-strewn surface had been eroded. Stones, which ranged in size from one foot to less than one inch across and whose pedestals appeared as high as they were wide, "were the most unexpected and important result of the flight," Lebedensky said. "Our conclusion is that the stone is harder than the ground around it and screens the ground underneath from the erosion effects of solar winds and micrometeorites." Lebedensky admitted the possibility that erosion was peculiar to the crater or a particular region of the moon, but saw no obvious reason why this should be true. (Sullivan, *NYT*, 5/17/66, 28; Toth, *Wash. Post*, 5/17/66, A17)

Radio occultation measurements from MARINER IV's July 14, 1965, Mars flyby indicated that regions on the surface of Mars might vary in height by three miles (five kilometers), JPL scientist Dr. Arvydas J. Kliore reported. (JPL Release)

Venus was rotating clockwise only once every 243 days, but during that period it seemed to be synchronized with earth, Dr. Richard Goldstein, chief of JPL Communications Systems Research, told the meeting. Dr. Goldstein also reported that radar probes of winter 1965 when Venus was nearest earth indicated prominences with the ability to depolarize microwaves. "They may well be mountain ranges," Dr. Goldstein said, "although large fields of boulders would also depolarize microwaves. On earth, of course, such areas usually indicate the presence of nearby mountains." (JPL Release)

Chief Soviet delegate Anatoli Blagonravov, addressing the annual COSPAR press conference, said France and U.S.S.R. had been negotiating —within the framework of COSPAR—on cooperation in space research which might extend to incorporation of French instruments in Soviet satellites "if this should be asked for." Asked if he could imagine similar close cooperation between U.S. and U.S.S.R., Blagonravov said there was no need for it; if the situation arose, it would be better for them to cooperate in processing data obtained from their own satellites. (Reuters, *Wash. Post*, 5/20/66, A5)

May 11: NASA's Surveyor spacecraft made first successful soft-landing under its own power in test at Holloman AFB. 36-sec. descent of spacecraft, which had been carried by balloon to 1,000-ft. altitude, was slowed from velocity of 55 fps to 5 fps at point of impact by its rocket engine. Test's success was necessary prerequisite to May 30 Surveyor launch from ETR for soft-landing on the moon. (AP, *NYT*, 5/12/66, 39; AP, Wash. *Eve. Star*, 5/12/66, A16)

- U.S.S.R. launched COSMOS CXVIII carrying scientific instruments for continued space research into circular orbit at 640-km. (398-mi.) altitude with 97.1-min. period and 65° inclination. Instruments were functioning normally. (Tass, 5/11/66)

- NASA was negotiating two $300,000 contracts with SDS Data Systems and Perkin-Elmer Corp. for "design, development, and fabrication of two-gas atmosphere sensor system" for future manned spacecraft operating for 45 days or longer. New system would give an accurate measurement of the mixed gases and information on presence of carbon dioxide and water vapor. By measuring partial pressure of each component present, system would determine correct ratio required for controlling total atmosphere pressure. Manned spacecraft currently used a single-gas system operating on oxygen. Longer missions lasting 45 days or more might require a two-gas system composed of oxygen and either nitrogen, helium, or another inert gas for crew health and safety. Contracts would be managed by LaRC. (NASA Release 66-113)

May 11: USN successfully conducted first guided firing of Phoenix air-to-air guided missile over Pacific Missile Range. Missile, developed by Hughes Aircraft Co., was fired from USN A-3 aircraft against drone target. All test objectives were met. (USN Release)

- NASA selected Dynatronics, Inc., to provide 18 additional pulse code modulated (Pcm) telemetry decommutator systems for Project Apollo communications under $2.3-million fixed-price contract. Systems would receive, identify, sort, and prepare "real time" and "onboard stored" data. (NASA Release 66-114)

- NASA would negotiate $800,000 contract with RCA for design study of improved Tiros Operational Satellite (Tos) for national operational weather satellite program. Proposed 500-lb., cylindrical satellite would be 5 ft. high and 4½ ft. in diameter—compared with current 300-lb. satellites 2 ft. high, 3½ ft. in diameter—and would incorporate camera system now carried in two operational spacecraft. Stabilized by flywheel and magnetic attitude control system, satellite would always point to earth, and its infrared radiometers would scan entire world's cloud cover at night. (NASA Release 66-115)

- ITT, RCA, and Western Union International requested FCC permission to build a $6-million ground station near Woodland, Ga., or at Moorefield, W.Va. The three carriers and AT&T also asked FCC to deny ComSatCorp's request to build station at Moorefield. (*WSJ*, 5/12/66, 1)

May 12: Senate Committee on Aeronautical and Space Sciences recommended NASA FY 1967 authorization of $5,008,000,000—$58,000,000 more than House appropriation but $4,000,000 less than NASA request. Recommendations by Committee included restoration of 1967 Mariner-Venus mission and deletion of amounts added by House for other flight projects in lunar and planetary programs; concurrence with NASA request of $3.5 million for 260-in. solid fuel rocket motor project; reduction of amount requested for Lunar Sample Receiving Laboratory by $1 million; and restoration of $14,689,850 to administrative operation, but separation of amount approved into two categories: personnel compensation, and benefits and other expenses. (NASA LAR V/78-79)

- Lee R. Scherer, manager of NASA's Lunar Orbiter program, in interview with Seattle *Times* reporter Robert L. Twiss, expressed confidence that the Boeing-built spacecraft could help resolve many unanswered questions about the moon. With funding and NASA approval, Lunar Orbiters equipped with devices that could be deployed to the moon could be launched in 1969, yielding valuable scientific information, Scherer said. "We could gain substantial information about other areas of the moon than the site chosen for manned landing." (Twiss, Seattle *Times*, 5/12/66)

- West Germany's first satellite would be launched in 1968 with U.S. Scout booster to study inner radiation belts and aurora borealis, Science Ministry in Bonn announced. Orbit would have 2,898-km. (1,800-mi.) apogee and 258-km. (160-mi.) perigee. Nine West German firms were participating in joint NASA-West Germany Science Ministry project. (Reuters, *Wash. Post*, 5/13/66, A23; Reuters, *NYT*, 5/13/66, 12)

- Results of nationwide Gallup Survey on Ufo sightings: "More than five million Americans claim to have seen something they believed to be a 'flying saucer.' And about ten times as many people—nearly half of the

U.S. adult civilian population—believe that these frequently reported flying objects, while not necessarily 'saucers,' are real and not just a figment of the imagination." (Gallup, *Wash. Post*, 5/12/66, K7)

May 13: 156-in.-dia., glass-reinforced, plastic-cased, submerged-nozzle solid motor reached full thrust of 300,000 lbs. in static firing at Thiokol Chemical Corp.'s Wasatch Div. Nozzle had expansion ratio of 34:1 and directed exhaust into 83-ft.-long diffuser tube to simulate high-altitude atmospheric conditions. Motor was one of two units being built by Thiokol under contract to AFSC Space Systems Div. to demonstrate feasibility of a high-mass-fraction, high-performance, upper-stage 156-in. solid motor. (*Av. Wk.*, 5/23/66, 29)

- Apollo Spacecraft Lem Adapter (Sla) was transported by helicopter from Tulsa to MSC for series of vibration and acoustics tests in conjunction with other Apollo spacecraft modules and a Saturn IB instrument unit. (MSC *Roundup*, 5/27/66, 1)
- Electrical fire broke out in basement of unoccupied room at JPL and briefly interrupted communications with NASA PIONEER VI interplanetary probe. Cause of fire had not yet been determined. (AP, *Wash. Post*, 5/14/66, A4)
- World's first commercial model of nuclear power generator was marketed by Martin Co. for $63,000. Designed to produce 25 w of electricity for at least five years without refueling, 3,000-lb. device converted heat from radioactive strontium-90 into electricity. (Wilford, *NYT*, 5/13/66, 55)
- AFSC awarded Martin Marietta Corp. $3,115,200 increment to existing contract to study Manned Orbiting Laboratory (Mol) compatibility requirements for Titan III program. Work would be done in Denver, Colo. (DOD Release 413-66)
- NASA was negotiating $35,000 contract with RCA for continued research on simplified process for manufacturing microelectric printed circuits by printing patterns of paste or slurry through screens onto a ceramic wafer. Utilizing new transistor developed jointly by NASA and RCA, process would enable transistor formerly inserted by hand to be printed onto the wafer simultaneously with resistors and capacitors and permit several connections to be formed at one time. Contract would be managed by LaRC. (NASA Release 66-109)
- "Moonwalker"—device developed by Space General Corp. for astronauts to travel on the moon—had been adapted for crippled children. Built under HEW grant, 90-lb. machine, which walked on eight mechanical legs, could negotiate sand or mud, climb and descend stairs, step up a normal street curb, and climb grades of 15° or more. (Wash. *Eve. Star*, 5/13/66, A1; *Brevard Sentinel*, 5/13/66)

May 14: Gemini 10 spacecraft, manufactured by McDonnell Aircraft Corp., was loaded aboard USAF transport plane in St. Louis for flight to KSC in preparation for launch in summer 1966. (UPI, *Wash. Post*, 5/15/66, A2)

- Potential of U.S. space program was described by Capt. Robert F. Freitag (USN, Ret.), NASA Director of Field Center Development, OMSF, in a speech before Michigan Assn. of the Professions in Detroit. "Man's exploration of space, the moon and the planets is potentially the greatest inspirational venture in the history of man, and one in which the entire world can share. Thus, the exploration of the space frontier can well provide the seed for new flowering—sparking a Twentieth Century Renaissance, stirring man's imagination, ennobling his ambitions, and igniting his spirit with fresh hope and excitement. As the European Renaissance was sparked by the discovery of new continents and seas, so a new Age of Discovery in many arenas may well be awakened by man's venture into space. . . ." (Text)

May 14: USAF launched two unidentified satellites from WTR with Atlas-Agena D booster. (*U.S. Aeron. & Space Act., 1966,* 150)

May 15: NASA NIMBUS II (Nimbus C) meteorological satellite was launched from WTR with Thrust-Augmented Thor-Agena B booster into near-polar orbit with 730-mi. (1,175-km.) apogee, 678-mi. (1,091-km.) perigee, 108-min. period, and 80° inclination. Flight plan had called for 600-mi.-altitude orbit.

NIMBUS II, three-axis earth-stabilized R&D weather satellite, would flight-test basic technology essential to meteorological satellites and instrumentation necessary to study atmospheric structure. It would also extend meteorological observation to region of electromagnetic spectrum not previously covered. Satellite carried Advanced Vidicon Camera System (Avcs) and Automatic Picture Transmission System (Apt), both of which operate during daylight portion of each orbit; Medium Resolution Infrared Radiometer (Mrir) to store global medium resolution data on earth's heat balance; and High Resolution Infrared Radiometer (Hrir) to provide high-resolution nighttime cloud-cover photos and cloud-top temperatures. Addition of Hrir data to Apt system was initial demonstration of stored and direct readout day-night cloud coverage capability. 3,000 weather photos would be taken daily by 912-lb. satellite and transmitted to any of 150 Apt stations around the world—eight of which had been modified to obtain nighttime cloud cover. Second in NASA's Nimbus series of meteorological satellites, NIMBUS II was managed by GSFC under overall direction of NASA OSSA. (NASA Proj. Off.; O'Toole, *NYT,* 5/16/66, 10)

- Boosted Arcas sounding rocket was launched from deck of USNS *Range Recoverer* near Koroni Beach, Greece, as test for flight series to obtain data on electron and ion density distribution in ionospheric D region during solar eclipse May 20. No data were obtained because 2nd stage failed to ignite, but problem was corrected to ensure successful flights during eclipse. (NASA Rpt. SRL)

- Gemini IX Astronauts Thomas P. Stafford and Eugene A. Cernan passed their final major physical examinations at KSC. Mission review board confirmed May 17 launch date. (Hines, *Wash. Eve. Star,* 5/15/66, A1)

- Manufacturers of rocket components, faced with possible work shortage because of NASA's delay in formulating plans for post-Apollo missions, were making "frank pitches to the U.S. Space Agency—and presumably to the Defense Department—for more business," wrote Alvin B. Webb in the *Washington Post.* Webb noted that "Chrysler is expected to begin building the last of its Saturn 1-B boosters by the end of this year. The same applies to Douglas production of S4B upper stages for the Saturn 5.

"Boeing, under the present schedule, probably will begin to run out of work on the Saturn first stage in late 1967." (Webb, *Wash. Post,* 5/16/66, D9)

May 16: NIMBUS II meteorological satellite was reported "working beautifully" by GSFC. Among the thousands of photographs transmitted were those of the Gemini IX recovery area. The 912-lb. spacecraft had been launched May 15. (UPI, *NYT,* 5/17/66, 77)

May 16: Live commercial TV coverage of Gemini IX recovery would be provided under an agreement between FCC, ITT, USN, NASA, and ComSatCorp, *Aviation Week* reported. ITT would operate a station on the U.S.S. *Wasp* in the Atlantic Ocean recovery force. Signals would be transmitted to the U.S. via EARLY BIRD I comsat. (*Av. Wk.*, 5/16/66, 33)

- Soviet space doctors' examination of dogs Ugolyek and Veterok after their 22-day flight in COSMOS CX—during which they had been harnessed and force-fed through stomach tubes—indicated that prolonged stay in space produced functional changes in animals which increased upon return to earth and then gradually disappeared. X-rays of dogs' skeletal systems showed that calcium had been "washed out." Muscle mass had changed, preventing full resumption of normal motor activity until eight or 10 days after return. High pulse rates and dehydration—findings common to both U.S.S.R. and U.S. manned experiments—were observed, and on fourth to fifth day after dogs' return there was "rather considerable speedup" of erythrocyte sedimentation reaction and an increase in number of leucocytes in blood, both apparently in response to normal gravity. In summary, doctors said that while functional changes did occur during flight, they were not irreversible. Whether or not zero gravity could change functional state of organism to extent that it could not return to normal was not yet known. (Tass, 5/16/66, USS-T Trans.; *M&R*, 5/23/66, 12)

- Sen. Karl Mundt (R-S.D.) informed Senate that Subcommittee on Government Research expected to make specific recommendations to improve Federally-financed research and development program, for which almost $16 billion was being expended during FY 1966. Sen. Mundt, ranking minority member of the subcommittee, said primary task of group was to determine and enunciate a more definitive, longer-ranging science policy. He said U.S. had reached the crossroads, where "it has become painfully obvious that we cannot continue to support all who demand support, but must be selective in the allocation of research funds." This indicated that one of the most pressing questions was how to direct NASA efforts, with Apollo program reaching its goal and no follow-on approved: "NASA now is on the threshold of placing a man on the Moon within this decade. After this what? This question concerns the leaders of this fine NASA organization as well as its highly trained, scientific and technical staff. Where do we turn? Shall the space program continue on its present course, or shall it be broadened to better serve the everyday needs of the American people? This is a priority crossroads. . . ." (*CR*, 5/16/66, 10212)

- NASA awarded Lockheed Missiles and Space Co. and RCA $90,000, six-month contracts to study characteristics for proposed Orbiting Data Relay System (Odrs) which would eliminate gaps in communications and provide continuous contact with spacecraft. Studies would probe kind of synchronous satellite and ground facilities required to relay data from spacecraft in earth orbit to NASA's mission control centers at GSFC, JPL, and MSC. (NASA Release 66-119)

- ComSatCorp had negotiated $441,412 contract—largest ever negotiated by ComSatCorp with a foreign country—with Nippon Electric Co., Ltd., Tokyo, Japan, for multiplex subsystem equipment for earth stations at Oahu, Hawaii, and Brewster Flat, Wash., which would be completed by 1967. System would modulate signals received from communications carriers for transmission over satellite links, demodulate signals received from satellite for relay to land distribution systems, and provide continuous monitoring for quality of all signals being passed through the system. (ComSatCorp Release)

May 16: Possible space developments in 1970's were listed by Dr. Edward C. Welsh, NASC Executive Secretary, in a letter to Rep. Olin E. Teague (D-Tex.), chairman of House Committee on Science and Astronautics' Subcommittee on NASA Oversight: (1) improved propulsion, including nuclear; (2) merging of aeronautics and astronautics in development of lifting bodies and winged spacecraft with maneuverable reentry capability; (3) lunar surface exploration and possible establishment of lunar bases; (4) manned earth-orbiting space stations; (5) unmanned probes within solar system and manned expeditions to planets "whenever that becomes promising and practicable"; (6) voice and TV direct communications from satellites to home receivers "throughout large sections of the world," and advances in operations and research on other applied satellites. (Text)

- Discussing "problems of prosperity" confronting airlines in the jet age, Robert Hotz said in *Aviation Week* that growing number of air travelers meant larger segment of U.S. population was concerned with "how much of this increasing airline prosperity is being shared with the traveling public in better service and lower fares. . . ." Noting that CAB chairman Charles Murphy had been urging airlines for about one year to "revamp their fare structure" in their own way to avoid Federal intervention, Hotz said airlines could either "take a vigorous industry approach" that would satisfy customers and retain corporate solvency or "continue their traditional piston-powered attitudes and have an angry public goad federal legislators and regulators into drastic solutions that could leave both the airlines and their customers in a worse muddle than ever." (Hotz, *Av. Wk.*, 5/16/66, 21)

- AFSC had completed 110 successful tests of Bak-11 (Barrier Arresting Component)—device to stop aircraft on runway during emergency—at Air Force Flight Test Center (AFFTC), Edwards AFB. Device, developed by Research, Inc., consisted of an arresting cable which engaged main gear struts when actuated by aircraft's wheels rolling over switches mounted on the runway. (AFSC Release 46. 66)

May 16-27: U.S. sponsored communications satellite seminar to provide new and developing countries with basic knowledge and practical information on economic and technical requirements for earth stations. Seminar, conducted under auspices of International Telecommunications Union, included lectures, panel discussions, informal exchanges, and equipment exhibits by 17 countries. Talks were in English with simultaneous interpretations in Spanish and French. Representatives of 44 nations attended. (State Dept. Release 108)

May 17: Gemini IX mission was scrubbed when Atlas booster failed to place Gemini Agena Target Vehicle (Gatv) in planned 185-mi. (298-km.) circular orbit. Launch of Gemini 9 with Command Pilot Thomas Stafford and Pilot Eugene Cernan was to have followed 99 min. after Atlas-Agena launch; Gemini spacecraft was to rendezvous and dock with Gatv to evaluate extravehicular life support and maneuvering equipment and procedures.

Simultaneous countdown of the Gemini Launch Vehicle/Spacecraft and the Atlas-Agena had progressed smoothly and the Atlas launch was normal until 121 sec. after liftoff. Then, said the Flight Safety Review

Board: "The Atlas No. 2 booster engine swiveled to an extreme hardover position about 10 sec. before booster engine cutoff. The other booster engine and the sustainer engine, acting under autopilot control, continued to work to counter the asymmetrical thrust.

"After booster separation, the vehicle continued the flight under sustained thrust but at a down angle. It had also rolled to a position where ground guidance could not lock on or reacquire. Signals to shut down the sustainer engine and inhibit ignition of the Agena engine were sent and acted upon by the vehicle. Agena separated on schedule and both vehicles plunged into the sea."

Gemini IX-A mission was scheduled for May 31—and later rescheduled for June 1—using Augmented Target Docking Adapter (Atda) as target vehicle. (NASA Releases 66-97, 66-124; MSC *Roundup*, 5/27/66, 1; Wash. *Eve. Star*, 5/17/66, A1, A6; Lewis, *Wash. Post*, 5/18/66, A1, A6; *NYT*, 5/18/66, 1, 26)

May 17: In its 17th annual report, the U.N. Commission to Study the Organization of Peace said the General Assembly should "declare the title of the international community" to the high seas and outer space. Six reasons were cited: (1) to avoid conflicts over them between nations; (2) to keep their resources from being wasted; (3) to prevent their military use; (4) to prevent their contamination; (5) to ensure that all nations benefit from their resources; and (6) to provide the U.N. with an independent source of income.

Report continued: "Member nations are not willing to pay assessments adequate for the task they expect the United Nations to perform. The revenues of the United Nations must therefore be augmented by independent sources of income. No one can estimate now what the income of the United Nations might be from its granting licenses for the exploitation of the resources of the seas and the revenues which should accrue to it from outer space communications. It is estimated, however, that the amount of money to be realized certainly should make an important contribution to the United Nations budget. Furthermore, it should help pay for . . . technical assistance to the developing states." (AP, *Chic. Trib.*, 5/18/66; Oatis, *Wash. Post*, 5/19/66, G5)

- 150-sec. static firing of Saturn S-II-T, an all-systems version of Saturn V booster's 2nd stage, was conducted at NASA Mississippi Test Facility. In captive ground test of Nation's largest, most powerful hydrogen-oxygen engine (200,000 lbs. thrust), all major test objectives were met: 1,100 measurements were taken; gimbaling of 4 of the 5 engines—which in flight would provide stability and control to the stage—was performed. (MSFC Release 66-108)

- U.S.S.R.'s LUNA X spacecraft, in lunar orbit since April 3, had completed 349 revolutions, covered distance of over 5 million km. (3.1 million mi.), and transmitted data for 178 communications periods with all instruments functioning normally. (Tass, 5/17/66)

- U.S.S.R. had developed Polaris-type missiles which could be fired by remote control from containers planted under the sea, diplomatic sources reported. (UPI, *NYT*, 5/18/66)

- Intensive investigation by GSFC and contractors had not yet determined causes of OAO I failure April 10. OSSA Review Board of Observatory-class Spacecraft would continue analysis of failure as a primary objective. (NASA Proj. Off.)

May 18: L/Col. Robert A. Rushworth flew X-15 No. 2 for 81 sec. in test to measure stability and control of X-15 with ventral fin attached. He reached 3,689 mph (mach 5.43) and 99,000-ft. altitude. Flight was 158th in NASA/USAF X-15 research program. (FRC; X-15 Proj. Off.)

- In letter to Congress transmitting report on the Nation's oceanographic work, President Johnson urged that funds be voted to continue work on Project Mohole. House Appropriations Committee had denied (May 5) $19.7-million Presidential request for NSF work on Mohole in FY 1967. Total cost was estimated as $110 million; about $25 million had already been spent. President Johnson wrote: "The Mohole Project will provide the answer to many basic questions about the earth's crust and the origin of ocean basins. It will teach us how to drill in the ocean depths—the prelude to the future exploitation of resources at the bottom of the sea." (*Pres. Doc.*, 5/23/66, 661; AP, *NYT*, 5/19/66, 23; *Wash. Post*, 5/19/66, A6)

- U.S.S.R.'s third MOLNIYA I comsat, launched April 25, successfully carried out "first experiment in observing the earth and taking television pictures of our planet" when it transmitted photographs from 30,000- to 40,000-km. (18,634- to 24,845-mi.) altitude. Lenses and light filters of television camera, focused on earth by means of autonomous focusing system, were changed to obtain images of earth on different scales and to observe earth's surface elements under different illumination. (Tass, 5/18/66)

- Summing up the effect of May 17 loss of Gemini IX mission's Agena Target Vehicle, William Hines said in the Washington *Evening Star:* ". . . it is obvious that at the very least the American space effort has lost three weeks of precious time and possibly $25 million in money—the cost of the lost hardware plus overhead. . . .

    "One thing should be clear from the experience of Gemini 9.

    "There was nothing of omission or commission that can be charged against NASA in the failure of the Agena's Atlas launching rocket. If the hackneyed phrase 'random failure' ever was properly used, it was in describing yesterday's trouble.

    "The Atlas, as a standard space launching vehicle, has had 49 successes in 52 attempts over the last three years—a better success record than the over-all American space program has enjoyed in any year since its inception in 1957." (Hines, Wash. *Eve. Star*, 5/18/66, A8)

- Eight photoelectric photometer-telescopes were flown from NASA Wallops Sta. on Aerobee 150A sounding rocket to 131-mi. (211-km.) altitude to measure stellar ultraviolet radiation which did not penetrate earth's atmosphere and could not be observed from the ground. Readings on light in four spectral bands were telemetered to ground stations during the 7½-min. flight conducted for Univ. of Wisconsin's Space Astronomy Lab. (NASA Releases 66-30, 66-123; NASA Rpt. SRL)

- Editorial comment on May 17 scrubbing of Gemini IX mission: *New York Times:* "For the third time in the past few months, a space experiment involving the Atlas-Agena rocket has run into trouble. Late last year a sequence of events similar to yesterday's led to a postponement of the original Gemini rendezvous experiment, while in March of this year, Gemini 8 nearly met catastrophe when it was docked with an Agena target vehicle. There may well be no connection between these three separate incidents; but since the Atlas-Agena has been a highly reliable and successful instrument in other aspects of the nation's space program, its emergence as the weak link in the Gemini series poses something of a mystery. The setback caused by last year's failure to orbit an Agena target vehicle was made up for by the brilliant improvisation that permitted Geminis 6 and 7 to rendezvous last December. Warned by that earlier experience, space officials were better prepared for this latest setback. The substitute target rocket they have on hand may permit much of the originally planned Gemini 9 experiment to take place early next month. But for the moment at least, a shadow has again been thrown over this country's lunar program." (*NYT*, 5/18/66)

Washington Post: "We shall do well to prepare for many more disappointments before man reaches the moon, including the possibility of serious accident. What remains essential in the competition is that considerations of human safety not be sacrificed to too feverish a schedule." (*Wash. Post*, 5/18/66)

Washington *Evening Star:* "The simple fact is that the United States has been incredibly lucky that no catastrophic accident has yet taken place during a manned space flight. The further fact is that the nation cannot logically expect this luck to last forever. Progress in space, as with nearly all human progress, will come at the cost of men's lives. And the nation should be prepared for the day when our luck runs out; prepared to accept tragedy as the inevitable concomitant of so ambitious a voyage of discovery." (Wash. *Eve. Star*, 5/18/66, A18)

May 18: Lockheed test pilot Ed Brown flew American-designed, German-built F-104G Starfighter to 400 mph in 8 sec. from Lagerfeld Base, Munich, Germany. Rocket-assisted takeoff from Zero Length Launch (Zell) platform at 30° angle demonstrated aircraft's capability to carry out combat operations from hidden areas without airfields. (*New York News*, 5/19/66; Lockheed Aircraft Corp.)

- Unidentified USAF satellite was launched from Vandenberg AFB by a Scout booster. (*U.S. Aeron. & Space Act., 1966*, 150)

May 19: Interplanetary spacecraft PIONEER VI, launched by NASA Dec. 16, 1965, reached perihelion—75.7 million mi. from sun—at 10 p.m. EDT. Among findings reported during probe's 260 million mi. of travel: exact measurement of average numbers of particles in interplanetary space; fact that high-energy particles thrown out by solar flares (principal radiation hazard to man) traveled in well-defined streams which were twisted about one another; and solar wind did not travel in straight lines. PIONEER VI had identified new ion (singly ionized helium) in solar wind, had found magnetic-field effects on solar wind temperatures, and discovered that sun's magnetic field occurred in filament-like "stringers" extending out from its corona. Spacecraft had revealed that regions on sun produced many solar flares and high solar wind velocities, as reported by MARINER IV. During its 154 days in solar orbit, spacecraft had transmitted to earth 340 million readings of 3,000 separate scientific measurements and sent back 3 million readings of 100 individual engineering measurements. Specifications called for operating life of six months (complete on June 16, 1966), but it could be extended if larger antennas were available to hear PIONEER VI as it moved progressively away and ahead of the earth on its smaller orbit around the sun. (NASA Release 66-122)

- XB-70A—only operational aircraft that approximated size, weight, and speed of proposed supersonic transport (Sst)—successfully made sustained 2,000-mph (mach 3) flight at 70,000-ft. altitude for 32 min. from

Edwards AFB. Although air friction on exterior surfaces caused heat buildup of 620°F, crew inside cockpit enjoyed 75°F environment, proving aircraft's skin and internal systems could withstand this temperature range. Flight accomplished major goal in USAF's XB-70 test program and proved that sustained "triple sonic" speed was possible for larger aircraft. (AFFTC Release 66-4-7)

May 19: Supersonic transport (Sst) would "grow one foot in length" when it reached its normal cruising speed of 1,800 mph—result of metal's expansion and contraction during transition between subsonic and supersonic flight—reported AP. "At altitudes for subsonic flight, an SST will be exposed to temperatures as low as 65 degrees below zero.

May 19: XB-70 research aircraft, shown here in an earlier flight. Later this year NASA would join USAF in using XB-70 to investigate sonic booms and other high-speed flight phenomena.

But in 17 minutes it can reach supersonic altitudes where the temperature on the outside of the fuselage will be 450 degrees." (AP, Phil. Eve. Bull., 5/19/66)

• Cuba and U.S.S.R. had signed agreement to improve Cuba's meteorological service and hurricane detection system, reported the *New York Times.* Agreement provided for "meteorological laboratory" equipped with Soviet radar and instruments which would enable Cuba to receive weather data from all over the world and to exchange aerological soundings with the Soviet Union. (*NYT*, 5/19/66, 8)

May 19: William Hines charged in Washington *Evening Star* that publicity given to proposed U.S. supersonic transport (Sst) was ". . . part of a propaganda barrage aimed at the American taxpayer, who in the years ahead will be called upon to underwrite deficits clearly predictable in building and operating a costly fleet of airplanes capable of spanning the Atlantic in two hours.

"The SST program has been publicized since the program's inception in 1963 with the same combination of gobbledygook and public-relations hokum used so successfully in merchandising the moon program. Technological challenge, the asserted psychological-spiritual imperative, and the added spur of an international 'race'—all are cited as reasons compelling development of a 2,000-mile-an-hour super-airliner.

"Actually the SST, when built and placed in operation, will be no more a free-enterprise product than the Saturn-5 moon rocket is. It will be built to government specifications, by contractors selected by the government, with development costs largely if not entirely underwritten by the government." (Hines, Wash. *Eve. Star*, 5/19/66)

May 20: Test version of Surveyor spacecraft designed for unmanned soft-landing on moon made smooth 900-ft. descent from a balloon over Holloman AFB in successful final test prior to scheduled May 30 flight. Three liquid-fueled engines, operating on information from two radar systems aboard spacecraft, slowed it from descent speed of about 50 mph to about 3.5 mph for landing. Test, second successful Surveyor descent without parachutes, had been delayed more than two hours by strong winds and optical problems. (AP, Wash. *Eve. Star*, 5/20/66, A1; AP, *Chic. Trib.*, 5/21/66)

- Saturn V all-systems 2nd stage (S-II) was successfully fired at MTF for 355 sec.—its full duration. The big stage developed 1-million-lbs. thrust from five hydrogen-and-oxygen-powered J-2 engines. (MSFC Release 66-116)

- Radio astronomy experiment designed to measure average intensity of cosmic radio noise originating outside the solar system was launched from NASA Wallops Station by four-stage Javelin rocket on 20-min. ballistic flight with 570-mi. (918-km.) apogee. Measurements were made at frequencies which could not be accurately detected by ground-based receivers. Data would aid study of galaxy's formation. (Wallops Release 66-31)

- Five successful sounding rocket launchings were conducted from the deck of USNS *Range Recoverer* during eclipse of sun in Greece, NASA announced. Launchings of the Boosted Arcas rockets, part of a cooperative project of the Greek National Committee for Space Research and NASA, were planned to investigate ionization of the upper amosphere during solar eclipse. (NASA Release 66-32; NASA Rpt. SRL)

- USN announced successful completion of flight tests of V/Stol XC-142A aircraft aboard U.S.S. *Bennington*. Tests included 44 short and six vertical takeoffs at speeds of up to 35 mph backward and as much as 400 mph forward at 25,000-ft. altitude. (DOD Release 443-66)

- In a press conference, NIMBUS II project officials at GSFC said the meteorological satellite was "exceeding all expectations" with the amount and quality of data it was returning to earth. Launched into nearly circular orbit May 15, satellite could "produce in 12 hrs. as much refined weather data as took six months to get out of earlier satellites." (Wash. *Eve. Star*, 5/20/66, A24)

May 20: U.S. and Greece launch Boosted Arcas sounding rockets from USNS *Range Recoverer* near Koroni Beach, Greece.

May 20: Checks of distributions of Martian and lunar crater diameters indicated that the visible surface of Mars was 2.2-to-3 $\times 10^9$ years old, Alan B. Binder, Univ. of Arizona, reported in *Science.* This result implied that in the early history of Mars large-scale subaerial erosion occurred. Of 69 Martian craters with diameters greater than 10 km., 13% had central peaks. This compared favorably with the frequency (11.7%) of central peaks among lunar craters and may indicate the peaks resulted from the impact mechanism rather than post-impact volcanism. Well-defined system of lineaments shown in MARINER IV photographs may indicate that Mars had lost appreciable angular momentum during its history. (Binder, *Science*, 5/20/66, 1053-5)

- NASA Aerobee 150 sounding rocket launched from WSMR carried grazing incidence telescopes and bandpass filters to estimated 107-mi. (172-km.) altitude in GSFC experiment to obtain solar x-ray photographs. Another GSFC-instrumented Aerobee 150 was launched from WSMR eight hours later to 121-mi. (195-km.) altitude to obtain solar spectral data from 2–60 A. Both rockets and their experiments performed satisfactorily. (NASA Rpt. SRL)
- NASA Goddard Space Flight Center held an "open house," including movies, live demonstrations, and exhibits. (GSFC Release G-7-66)

May 21: Boosted Arcas sounding rocket, last in series launched during solar eclipse from USNS *Range Recoverer* near Koroni Beach, Greece, carried GSFC experiment to obtain data on electron and ion density distribution in ionospheric D region. (NASA Rpt. SRL)

May 22: 15-lb. pigtailed monkey, connected to intricate battery of miniaturized instruments, would be launched into earth orbit in 1967 to provide U.S. scientists with 30-day study of his health, JPL announced. Monkey—an astronaut stand-in—would be part of the Biosatellite series projected by NASA. Particularly meaningful to NASA's space biology program would be urinalysis tests, showing changes in calcium and other primate urine constituents, which should indicate how astronaut's metabolism might vary during prolonged space flights. ARC was in charge of the program. (JPL Release)

- U.S. Army claimed 21 world records for its new light observation helicopter, OH-6A. NAA-supervised flights at Edwards AFB resulted in 12 speed records, and three each for distance, climbing, and sustained altitude. Records had been submitted to Fédération Aeronautique Internationale in Paris for official approval. (*Chic. Trib.*, 5/23/66, 16)

May 23: Debate began in Senate on authorization of NASA budget for FY 1967. Sen. Margaret Chase Smith (R-Me.) said: "It is important to note that the pressures creating the great surges in progress—whether it be in science, technology, exploration, or economic and social development—come from outside forces and considerations. . . .

"It is these forcing functions which mainly set the pace of progress and development. Space is no exception. Indeed, the great surge in space research and exploration is a classic example of this historical pattern. It was these forcing functions which set in motion the revolution in science and technology which has brought us to the era we know as the Space Age. And it is these forcing functions which today and tomorrow will determine our future course and our future policies in space. . . .

"The decisions which confront us today are those which will determine whether the history of the last half century will repeat itself a few years hence, and whether we will once again experience a bitter awakening to the fact that others have seized the initiative in the more advanced space missions of the future. Indeed, in today's cold war arena, the question boils down to whether this Nation can afford such a luxury, in terms of our very survival. For it is clear that future leadership in this scientific and technological competition—which has such great economic, military, and political significance—will not be easily held, nor would it be easily regained once lost." (*CR*, 5/23/66, 10681-2)

May 23: Important information on one of several x-ray sources, observable only from space, had been obtained by experiments on March 8 sounding rocket flight from WSMR, NASA announced. Initial results from data analyzed revealed that the angular size of the x-ray source in the constellation Scorpius, Sco X-1, could not exceed 20 arc-seconds. This discovery disagreed with widely held view that this source represented an ancient nearby (Milky Way) supernova. It strengthened current speculation that the sources in Scorpio and many of the other x-ray sources are an entirely new class of celestial objects characterized by small angular size which are either undetectable or star-like in visible light or radio emission. The detail with which the angular size of this x-ray source was measured was more than 20 times greater than any previous measurement [see June 17-18]. (NASA Release 66-125)

- USAF launched unidentified satellite from WTR with Thor-Agena D booster. (*U.S. Aeron. & Space Act., 1966*, 150)
- NASA Administrator James E. Webb announced appointment of William B. Rieke, NASA Assistant Administrator for Industry Affairs and former president of Lockheed Aircraft International, Inc., as Acting Assistant Administrator for Administration. Mr. Rieke would assume new duties beginning May 30, replacing John D. Young who resigned to become an assistant to BOB Director. (NASA Release 66-131; NASA Ann.)
- NASA semiannual procurement report revealed continuing sharp upsurge in use of incentive contracts. During six-month period, NASA had awarded 41 new incentive contracts and converted two cost-plus-fixed-fee contracts to incentive awards. Target value of all incentive contracts as of Dec. 31, 1965, was $2.184 billion—more than three times the total at the end of first half of FY 1965. (Text)
- A NASA Contract Administration Services Representative's office—headed by Donald R. Mulholland, Deputy Director of NASA's Western Operations Office—had been established at headquarters of AFSC's Air Force Contract Management Div. (AFCMD), Los Angeles, Calif. New office, which completed network of representatives' offices in major DOD centers, would "provide liaison between AFCMD and NASA elements on contract administration services support. . . ." (NASA Hq. *Bull.*, 5/23/66, 6)
- International Galabert Astronautical Prize was awarded jointly to MSFC Director Dr. Wernher von Braun "for the whole of his work in the United States, particularly on the Saturn project," and to French space technicians Jean-Pierre Causse, of the Centre National d'Etudes Spatiales (CNES), and Roger Chevalier, of the Société pour l'Etude et la Réalisation d'Engins Balistiques (SEREB), "as symbols of the French space effort." The $4,000 prize would be presented at October meeting of International Astronautical Congress in Paris. (French Embassy; Reuters, *NYT*, 5/24/66, 11)

May 23: Britain's W. J. Strang, deputy technical director of the Anglo-French Concorde Sst project, dropped all references to the Concorde's being "the world's first supersonic transport" from the prepared text of an address before the Aviation/Space Writers' Assn. in New York. When asked about this omission, Bernard A. Darrieux of France's Sud Aviation Co. replied that French aviation experts who had visited the U.S.S.R. and who were informed of Soviet plans said there was a "strong possibility the U.S.S.R. may be first." (Norris, *Wash. Post,* 5/24/66, A28)

- Astronaut Scott Carpenter decided not to attend Second International Congress on Oceanography in Moscow after learning that the U.S. ocean research vessel U.S.S. *Silas Bent* had been denied permission to dock at Leningrad in connection with the conference. Carpenter, who had participated in the Sealab II experiment in living and working underwater, would have attended as an aquanaut and would have been the first astronaut to visit Moscow. NAS had designated *Silas Bent* to respond to Soviet invitation for oceanographic ships to visit Soviet ports during Congress so delegates could visit them but Soviet Foreign Ministry had refused permission. (AP, Balt. *Sun,* 5/24/66)
- NASA Apollo Lunar Excursion Module (Lem) descent engine, Phase II, test began at Arnold Engineering Development Center. (AEDC)

May 23-25: Harsh indictment by A. V. Cleaver, chief engineer for rockets at Rolls-Royce, of European politicians and civil servants involved in the European Launcher Development Organization (ELDO) opened the sixth annual European Space Symposium at Brighton, U.K.: "I believe the future of ELDO is extremely bleak, unless Europe as a whole and the United Kingdom in particular, adopts an entirely fresh approach to its problems.... Europe could have made a better showing if her politicians had been of equal caliber to her technical men, insofar as their grasp of 20th century realities was concerned.... The lesson is this: at any time, the leading nations are those preeminent in technology, usually in the technology of transport." Cleaver noted that the European nations in ELDO had, among them, about 120% of the U.S.'s population and more than 50% of the U.S. gross national product, and should be doing far more than the present one-twentieth of the U.S. space effort. (Coleman, *Av. Wk.,* 5/30/66, 32)

- Optimistic assessment of prospects for exploring Mars during the next decade and finding some form of life there was presented by scientists at the AAS-sponsored symposium on "The Search for Extraterrestrial Life" in Anaheim, Calif. Carlos De Moraes, director of planetary program studies for Martin Co., visualized a series of space flights from 1969 to 1979 and culminating in the landing of two 3,500-lb. "space buses." First flight, in 1969, would be of an 850-lb. spacecraft that would provide televised mapping of the Mars surface and take measurements of the atmosphere. After a series of intermediary flights, the space buses would land in 1979 and release both automated biological laboratories and roving vehicles that would explore and gather samples at points away from the immediate landing area.

  Dr. Norman H. Horowitz of JPL noted that "any form of life, however primitive, found on Mars would be of immense scientific value." He said that no matter how bleak the picture, the possibilities of life on Mars should be explored to the fullest because that planet appeared to be the likeliest site for life beyond the earth.

GE engineers P. G. Thome and Ernest J. Merz described giant clamshell under development that would be part of a system launched by a Saturn V booster in 1973. The thick disc-shaped object would open like a giant mollusk when landed on Mars, releasing devices that would start gathering information on atmosphere, weather, soil, and life. Information would be telemetered back to earth. Disc could be powered by radioisotope-fueled generator that could keep it in operation for up to six months. (Texts)

Speculating about meetings with intelligent beings from or on another planet, Harold D. Lasswell, Yale Univ. professor of law and political science, asked: "Would earthlings unite or wage interstellar war? Would we have to kow-tow to a more advanced civilization, or dominate an inferior one? Suppose they had no more, or even less, feelings of love and responsibility? Might meeting with an advanced society push man further toward creating a culture of frantic fun?" (AP, Balt. *Sun*, 5/26/66)

May 24: Senate unanimously passed by voice vote the NASA FY 1967 authorization bill of $5,008,000,000—$121,135,850 more than the sum voted by the House but $4 million less than President Johnson had requested. In speech on Senate floor during passage of bill, Sen. Spessard L. Holland (D-Fla.) said that if the U.S. failed to continue development of solid propellants, it would fail in the one thing it "should be doing to catch up with the Soviet Union in the field of space effort." Senate bill included only the $3.5 million requested by the Administration for one additional firing of a half-length 260-in. solid rocket motor. In other debate, Sen. Joseph S. Clark (D-Pa.) said he felt that national priorities were "substantially out of order when the Senate, in the course of 5 or 10 minutes, passes a $5 billion authorization bill to send a man to the moon," and that he hoped the space effort could be put on an international cooperation and not a competitive basis. Clark was joined in his remarks against the bill by Sen. J. William Fulbright (D-Ark.), who said such action was "typical, today, of any bill relating either to outer space or the military . . . in my view, it is an outrageous distortion of priorities." (*CR*, 5/24/66, 10715-20)

- NASA Aerobee 150 launched from WSMR reached 82-mi. (132-km.) altitude in GSFC-Princeton Univ. experiment designed to study ultraviolet radiation of the star Zeta Ophiuchus. Burnthrough in rocket's combustion chamber at 50.1 sec. caused underperformance, and payload did not observe primary target. Most data were lost through long exposure of payload film to direct sunlight before recovery. (NASA Rpt. SRL)

- U.S.S.R. launched COSMOS CIX carrying scientific instruments for continued space research. Orbital parameters: apogee, 1,305 km. (810.5 mi.); perigee, 219 km. (136 mi.); period, 99.8 min.; inclination, 48.5°. Instruments were functioning normally. (*Pravda*, 5/25/66, 2, USS-T Trans.)

- First Europa 1 booster for European Launcher Development Organization (ELDO) was successfully launched from Woomera Range although malfunction in an impact indicator at a downrange radar station led to early reentry. Booster, composed of British Blue Streak first stage and dummy French and German second and third stages, was programed for 143-sec. flight. Range safety officer cut flight short by eight seconds when radar indicated Europa 1 was heading off course to the left.

There was no malfunction in the rocket itself. Reduction of data later showed booster had been on a correct course at 30.3-mi. (48.8-km.) altitude. Hawker Siddeley Dynamics officials estimated test was 90% successful, and that "objectives were achieved." Europa 1 was sponsored by U.K., France, Australia, Italy, West Germany, the Netherlands, and Belgium. (UPI, *Wash. Post*, 5/24/66, A12; *NYT*, 5/25/66, 22; *Av. Wk.*, 5/30/66, 32)

May 24: NASA and West Germany's Ministry for Scientific Research had signed cooperative agreement to investigate the physics of comets, interplanetary medium, and earth's magnetosphere. Initial experiments, to be launched in fall 1966, would release vaporized metal in upper atmosphere, creating artificial ion clouds which would be observed from ground. Results would be reviewed to determine desirability of conducting similar experiments on larger vehicle at a distance of several earth radii. Under agreement, BMwF would provide rocket payloads, supplementary cameras, and photometric equipment. NASA would furnish one Javelin and one Nike-Tomahawk rocket, launching range, support facilities, and equipment for optical observations. (NASA Release 66-121)

- Former Astronaut John H. Glenn, Jr., received Great Silver Medal of Paris from Mayor Albert Chavanac for being "one of the pioneers of space, who was able to give a human aspect to the experiment he attempted so successfully." (AP, *NYT*, 5/25/66, 33)

May 24-25: Representatives of NASA and its major spacecraft and launch vehicle contractors attended Saturn Manufacturing Review meeting at MSFC's Michoud Assembly Facility. Meeting, sponsored annually by MSFC, was held to "promote the interchange of space vehicle manufacturing technology." (MSFC Release 6-115)

May 24-26: Aviation/Space Writers' Assn. held annual meeting in New York.

AFSC Commander Gen. Bernard A. Schriever, in luncheon address, cited three factors which could restrain full advancement of technology: "The first of these constraints is the relative silence of our opponents. A few years ago the headlines were full of Soviet threats of orbital bombs, missiles that could 'hit a fly in the sky,' and similar weapons. Today, by contrast, the Soviets are not boasting about their new weapon developments.... The second constraint is the argument made in some quarters that U.S. military research and development could be 'provocative' to our opponents. This argument overlooks the fact that our very existence as a free society is 'provocative' to our opponents, because it contradicts their theory of history.... The third constraint is the immediate need for large funds to support our forces in Vietnam...." To overcome these constraints, General Schriever said, "we must harness our management experience and technological knowledge to serve the national purpose," in an effort involving every "segment of our national life...." (Text)

Dr. George E. Mueller, NASA Associate Administrator for Manned Space Flight, discussed implications of 1968 space budget decisions. He said U.S. today was "*not* ahead in this deadly serious competition," and "strong and increasing effort" would be required "to prevent the Soviets from forging ahead as the unchallenged leader in space." (Text)

Nike X system had been improved so that it could not only destroy hostile satellites but also provide defensive cover for the entire Nation, reported Dr. Oswald H. Lange, chief scientist for the Nike X project office at Redstone Arsenal, Ala. He described the Nike X system as employing two interceptors—the long-range Zeus and short-range, high acceleration Sprint. (*NYT*, 5/26/66, 4)

R. I. Mitchell, vice-president of Lockheed Aircraft Corp.'s JetStar program, said that Lockheed would produce advanced version of four-engine JetStar business aircraft capable of taking off on shorter runways and carrying more cargo and passengers than previous models. New Dash 8 JetStar, powered by Pratt & Whitney JT12A-8 engine, would reduce runway requirements from 6,425 ft. to 5,550 ft. and would have 1,000 fpm faster rate of climb. (Lockheed Release)

Two space rescue systems being tested for possible use by USAF were discussed by Harold L. Bloom of General Electric Co.'s Re-Entry Systems Dept.: (1) "Emergency cocoon"—an inflatable shelter to maintain astronaut in a "livable environment" while awaiting rescue—would use thin silicone rubber membrane in dacron and Mylar skin to provide CO_2 and water vapor control; (2) one-man reentry vehicle for self-rescue—"Manned Orbital Operations Safety Equipment (MOOSE)"—would consist of foldable heat shield bonded to Mylar bag. Astronaut would zip himself into MOOSE bag, leave spacecraft, and initiate polyurethane foaming process; foam would insulate and support him during reentry oriented with small rocket in an attitude/de-orbit package. Form-fitting couch, designed to be traveling at less than 30 feet per second at the time of impact, would also double as a life raft. (Text; Wilford, *NYT*, 5/8/66, 66)

May 25: NASA successfully launched EXPLORER XXXII (AE-B) aeronomy satellite from ETR with three-stage Delta booster. The 2nd stage burned 8 sec. longer than planned—failing to cut off and running to propellant depletion—and boosted the satellite into a higher orbit than planned. Orbital parameters: apogee, 1,688 mi. (2,717.7 km.), as opposed to planned 750 mi.; perigee, 180 mi. (289.8 km.); period, 116 min.; inclination to equator, 65°. Launch was third time Delta 2nd stage had failed to shut down on time and sent 3rd stage and payload to higher altitudes.

Built by GSFC, and the last of the original satellites outlined by NASA when it was established in 1958, the 495-lb., 35-in.- dia. EXPLORER XXXII was designed to investigate temperatures, composition, densities, and pressure in the upper atmosphere and their diurnal, seasonal, and annual variations on a global basis. NASA later announced that adjustment of onboard programing of spacecraft's sensors had compensated for higher apogee and useful scientific measurements would be taken that were not planned originally. All eight experiments onboard were functioning "as expected" and spacecraft was spin stabilized at the required 30 rpm.

First aeronomy satellite, EXPLORER XVII, was launched April 3, 1963. Results included first direct measurement of neutral helium; first *in situ* measurements of concentrations of neutral atmospheric constituents at satellite altitudes as a function of time and solar activity; first data revealing difference by a factor of two (approx.) in value for atmospheric density as determined from onboard sensors and from changes in satellite's orbit; first detailed description of diurnal and latitudinal behavior of summer ionosphere near altitude of F_2 maximum over eastern U.S. at time of solar maximum. Results had indicated more specialized measurements were needed to reveal basic atmospheric processes. EXPLORER XXXII incorporated many improvements aimed at achieving this objective. (NASA Releases 66-96, 66-139; AP, *Wash. Post,* 5/26/66, E17; *WSJ,* 5/26/66, 1)

May 25: Fifth anniversary of President John F. Kennedy's call before a joint session of Congress to undertake a manned lunar landing in this decade.

- Referring to President Kennedy's challenge five years ago "to achieving the goal, before this decade is out, of landing a man on the moon and returning him safely to earth," William Hines commented in the Washington *Evening Star* on the progress made:"Of course, Kennedy did not 'invent' space. Nor did he, in a single 1,100-word exhortation to Congress, create all the conditions needed to make a lunar program possible. The United States had quite a respectable space effort going for several years before Kennedy spoke, and indeed the first American astronaut had flown a brief suborbital mission three weeks earlier.

  "The foundations on which Kennedy could build a moon program were there. What he did—with advice and urging from many advisers including the then Vice President Johnson and the late 'grand old man of space,' Hugh L. Dryden—was set a goal. . . .

  "The call to leadership in space came at a particularly troublous time in the new Kennedy administration's development. In the month just past, April 1961, the United States had suffered two tremendous setbacks in what had been called 'the battle for men's minds.' On April 12 Cosmonaut Yuri Gagarin orbited the earth—a full 10 months, as it turned out, before Astronaut John Glenn would match his feat. And on April 19 the bottom fell out of the ill-conceived Bay of Pigs adventure to overthrow Fidel Castro in Cuba." (Hines, Wash. *Eve. Star,* 5/25/66, A1, A5)

- First full-scale Apollo/Saturn V booster-spacecraft combination rolled out at NASA Kennedy Space Center exactly five years after President Kennedy committed U.S. to manned lunar landing by 1969. Designated AS-500-F, the 365-ft., 500,000-lb. facilities vehicle was moved from Vehicle Assembly Building (Vab) on 3,000-ton, diesel-powered, steel-link-tread crawler transporter to Pad A for use to verify launch facilities, train launch crews, and develop test and checkout procedures. (MSFC Release 66-114)

- USAF Flight Safety Review Board announced it had pinpointed cause of the Atlas booster failure during the aborted Gemini IX mission May 17: gimbaling of the booster engine which forced the Atlas-Agena into a hardover trajectory was caused by a short circuit in the electrical command system of the engine pitch servo valve. Although several other failures could have caused a similar occurrence, the Board said, only this one could have caused the exact set of data received from the launch vehicle. Malfunction was called a random failure, but to ensure that it would not happen again, new tests would be made on the Atlas and electrical connections in all new boosters would be x-rayed.

  NASA scheduled the mission, redesignated Gemini IX-A, for June 1. (UPI, *Wash. Post,* 5/26/66, E17; *NYT,* 5/26/66; NASA Proj. Off.)

- U.S.S.R. announced successful completion of rocket tests in the Pacific to test equipment for spacecraft landings at sea. Another series of tests, which had begun April 24, was being continued to improve booster rockets necessary to propel manned spacecraft to moon. (Tass, 5/25/66, USS-T Trans.)

May 25: NASA had selected Boeing Co. for negotiations on $5-million, three-year contract to provide "technology for high-power solar arrays" for manned and unmanned space missions. Program—part of research effort to increase amount of power for each pound of overall power-generating system—would include design and fabrication of nonflight experimental model of a 1,250-sq.-ft. deployable solar panel assembly which would have 12.5-kw. output if completely covered with electrically active solar cells. Completed model would be tested under simulated space environment conditions and would serve as test model in solving major problems of deploying large-area structures, compact packaging, and design of thin solar cells and lightweight material. Contract would be managed by JPL. (NASA Release 66-133)

- In a press interview, Capt. Charles Oglesby (USAF) of the North American Air Defense Command (NORAD) tracking center in Moorestown, N.J., said scanners were checking on 1,100 items in orbit, including functioning payloads and debris. 95% of items in orbit were from the U.S. Of payloads, the U.S. had 167; U.S.S.R., 43; U.K. and Canada, two each; and France, three. (AP, Balt. *Sun*, 5/26/66)

- USAF awarded Ling-Temco-Vought (LTV) Aerospace Corp. a $353,000 contract to develop improved techniques and materials for "rapid landing sites" for V/Stol aircraft and helicopters. Contract called for quick-setting, resinous materials that could be sprayed on ground by unskilled persons in remote areas where it would be too costly and time-consuming to build conventional landing sites and permanent shelter facilities. (AFSC Release 96.66)

- Eastern Airlines president Floyd D. Hall announced company had bought options on two French-U.K. Concorde Sst's for service on Pacific routes and to Denver and Seattle. Eastern previously had taken options on two U.S. Sst's. (*WSJ*, 5/25/66, 14)

May 25-27: Twenty astronauts toured MSFC laboratories and test facilities and received briefings on Saturn IB and Saturn V launch vehicles. (MSFC Release 66-99)

May 26: MARINER IV spacecraft, which took the world's first close-up pictures of Mars July 14, 1965, was again in contact with earth, reporting on the space environment and its own operating performance after 18 mos. of flight. NASA said telemetry from MARINER IV, received over a 197.5-million-mi. radio link with the Deep Space Network's Goldstone Tracking Station, indicated that all spacecraft systems were operating properly. Launched Nov. 28, 1964, MARINER IV had exceeded its design life by more than 100% and gave indications it might function until 1968. (NASA Release 66-135; JPL Release)

- President Johnson issued executive order directing Federal agencies to "provide leadership" in prevention, control, and abatement of air pollution. (*Pres. Doc.*, 5/30/66, 696)

- A superior glass-covered, super-blue solar cell for space use had been developed by scientists at LRC. "Super-blue" referred to cell's extremely high sensitivity to blue light—the higher frequency rays of sunlight that reach only the top surface of the solar cell. Since experiments had indicated that radiation damage was most severe in cell's response to low-frequency red light, program to develop cells ultraresponsive to blue light was begun and the thin, lightweight, super-blue cell evolved. (LRC Release 66-25)

May 26: USAF awarded Lockheed Missiles and Space Co. $12,916,000 contract for Agena rocket launch services from April 1966 to September 1967. Work would be at Vandenberg AFB. (DOD Release 462-66)

May 27: Dr. Walter A. Radius, career Foreign Service officer who had been on detail to NASA from State Dept. since 1963, was appointed special assistant to NASA Assistant Administrator for Policy Analysis, Breene M. Kerr. Dr. Radius would assume new duties beginning June 6. (NASA Release 66-129; PIO)

- NASA selected Lockheed Missiles and Space Co. and Martin Co. for negotiations of parallel one-year $1-million fixed-price study contracts on integration of experiments and experiments support equipment in space vehicles and spacecraft for manned Apollo Applications (Aa) missions. Companies would define experiment integration work (payload integration in Apollo lunar module, Saturn launch vehicle instrument unit, and S-IVB stages of Saturn IB and Saturn V vehicles.) Contracts would be managed by MSFC. (NASA Release 66-137)

- Breene M. Kerr, NASA Assistant Administrator for Policy Analysis, announced appointment of Dr. Irwin P. Halpern as Director, Policy Analysis Staff, Office of the Assistant Administrator for Policy Analysis. Dr. Halpern, a CIA specialist on Soviet and Chinese Communist political and military affairs and doctrine, would assume new duties beginning June 5. (NASA Release 66-130)

- New F-4J Phantom jet, developed by McDonnell Aircraft Corp. for carrier use by USN and USMC, made its first public flight at Lambert-St. Louis Municipal Airport. Aircraft had higher maximum speed, greater range, higher combat ceiling, shorter takeoff distance, and greater combat capabilities than other Phantom models. (McDonnell Release)

- Memorandum of understanding for cooperative meteorological project called "Eole" was signed by NASA and French Centre Nationale d'Etudes Spatiales (CNES). To gather data on atmospheric circulation, Eole project would use network of constant level balloons—which would drift with the wind and act as tracers of air masses—and an earth-orbiting satellite—which would record pressure and temperature data telemetered from balloons for later transmission to ground stations. CNES would be responsible for "development and launching of the balloons and their payloads and for the design, fabrication and testing of the proposed satellite." NASA would provide Scout rocket, handle launching, and train personnel. (NASA Release 66-156)

May 28: Second-stage ground test model of three-stage Saturn V booster exploded while being removed from test stand at MTF, injuring five persons—none seriously. Stage, which had five hydrogen-oxygen J-2 engines capable of generating 1 million lbs. thrust, had been tested May 25 in ground firing which had been terminated after 195 sec. when hydrogen line leak caused automatic cutoff. At time of explosion, technicians were trying to determine cause of hydrogen leak. No hydrogen was in tank when explosion occurred; cause of explosion had not been determined. (UPI, *Wash. Post*, 5/28/66, A18; Wilford, *NYT*, 5/30/66, 22)

- Gemini IX-A was officially cleared for launching June 1 on three-day mission that would include rendezvous, docking, and extravehicular activity. After hearing reports from the astronauts, launching pad crews, and the worldwide tracking network, William C. Schneider, mission director, declared Gemini IX-A "ready to go." (*NYT*, 5/29/66, 58; UPI, *Wash. Post*, 5/28/66, A1)

May 28: NASA Associate Administrator for Manned Space Flight Dr. George E. Mueller, speaking at meeting of Texas Radiation Advisory Board and Texas Atomic Energy Commission in Freeport, Texas, cited reasons that it was vital the U.S. be pre-eminent in space: "First, consider the national security, or peace-keeping, aspects.

"Although our space program is a peaceful endeavor, it cannot help but have a profound effect on our future military position. Although there is no military space force in being in any nation's arsenal, it would be disastrous for the United States to lack the basic understanding, the basic technology, and the basic engineering which would be required if an aggressor should choose to make space a battlefield. . . .

"In the international arena, our space program—and Manned Space Flight in particular—may be considered as a measure of our ability to compete with a formidable rival, and as a criterion of our ability to maintain technological eminence.

"The influence of our scientific and technological progress and prowess is and has been one of the deciding factors in keeping the Cold War peace over the past 20 years. . . . Should we fall behind in the area of space technology, we would jeopardize our national interests, on earth as well as in space." (Text)

May 29: The Nation was facing a crisis in space planning, NASA Administrator James E. Webb said in a press interview at NASA Hq. The question was what to do with the Apollo project after men had landed on the moon. The answer would determine whether the U.S. remained a "major spacefaring nation," and must be found within a year, he said—long before the first lunar landing was made. Search for the answer had been postponed several times because the problem seemed remote and budgets were tight. "It is extremely important not to think you can postpone the decision again. I think it is imperative to have a thoroughgoing national debate on whether we want to go past the point of no return." If the Nation did not mind being second to the moon or chose to "dismantle the Apollo machine without recovering its great investment," he would accept that, he said. But he did not believe that either the Congress or the public had consciously reached these decisions. He feared they just had not thought about them. (Clark, *NYT*, 5/30/66, 1, 22)

• Soviet aircraft designer Andrey N. Tupolev described U.S.S.R.'s Tu-144 supersonic airliner in Soviet trade-union newspaper *Trud*. Four-engine aluminum aircraft would carry 121 passengers at 1,500 mph (compared with U.S. supersonic transport's 250 passengers at 2,000 mph); speed had been deliberately limited to allow construction "of highly durable aluminum alloys," Tupolev said. Tu-144 would be needle-nosed with small delta wing far back on fuselage and four jet engines in single container under rear section. Aircraft would be operational in "the nearest future"—possibly by 1970. (Anderson, *NYT*, 5/30/66, 36C)

May 29-June 4: Mapping the moon for man's first lunar landing was discussed by leading scientists at an 11-nation International Conference on Selenodesy at Manchester, England. (Reuters, *NYT*, 5/30/66, 22)

May 30-June 1: NASA's SURVEYOR I was successfully launched from ETR by Atlas-Centaur (AC-10) booster on 63-hr. direct ascent lunar transfer trajectory. Spacecraft—carrying survey television system and instrumentation to measure lunar surface bearing strength, temperatures, and radar reflectivity—was first in series of seven designed to prove out design, develop technology of lunar soft-landing, and provide basic scientific and engineering data in support of Project Apollo. It would softland on the moon June 2.

After SURVEYOR I separated from Atlas, spacecraft pointed its solar panels toward the sun to power its equipment; at five hours GET, sensing devices fixed on the star Canopus for cruise attitude stabilization.

SURVEYOR I spacecraft successfully completed mid-course correction of flight path on electronic commands relayed from antenna near JPL. Only flaw in launching occurred within one hour GET, when radio signals indicated that one of two antennas on spacecraft might have failed to deploy. JPL's Surveyor project manager Robert J. Parks said the problem was "not going to preclude a possibility of getting full success out of this mission since one antenna is sufficient to receive radioed commands unless the spacecraft gets in a position where the working antenna is out of range of earth." The other antenna was properly deployed and working perfectly.

Surveyor program was under direction of OSSA Lunar and Planetary Program Div.; project management was assigned to JPL; prime contractor for spacecraft development and design was Hughes Aircraft Co. Atlas-Centaur launch vehicle was under direction of LRC. (NASA Release 66-127; Wilford, *NYT*, 5/31/66, 1, 33; Simons, *Wash. Post*, 5/31/66, A1)

May 30: Planned expansion of the global communications satellite network of the 50-nation Intelsat was being postponed two years as a result of pressure by the FCC on ComSatCorp. ComSatCorp was majority shareholder and served as manager for Intelsat. Basic question raised by FCC was whether the expansion was too ambitious for the traffic that would be available. Threatened with FCC veto or modification of its application to construct six 1,500-channel satellites for launch in 1968, ComSatCorp notified FCC that deployment of the follow-on 6,000-channel multipurpose satellite had been postponed from 1969-70 to 1972 or later. (*Av. Wk.*, 5/30/66, 31)

- U.S.S.R.'s LUNA X spacecraft, which entered selenocentric orbit April 3, had stopped sending signals after 219 transmissions, Tass announced. (AP, *Wash. Eve. Star*, 6/2/66, A1)
- The "breakthrough" in the air cargo business was discussed by Robert Hotz in *Aviation Week:* "There is no doubt that air cargo is on the move. Airline cargo performance in 1965 surpassed forecasts by astonishing margins. Domestic cargo soared 25% last year in contrast to the airline predictions of 16%. International cargo rocketed 46%—with at least one carrier recording a 65% gain—compared with forecasts of a 25% increase. Many airlines are now watching their projected cargo goals for 1970 become a reality in 1966.

"Stuart Tipton, president of the Air Transport Assn. and always a conservative spokesman for the airline industry, cited figures in a recent speech that forecast 5.2 billion freight ton miles annually by 1970 and 15.3 billion by 1975. Mr. Tipton noted that, although these levels compared with 1.7 billion freight ton miles flown in 1965, the forecasts for a decade hence might well 'be very conservative.'" (Hotz, *Av. Wk.*, 5/30/66, 21)

May 30: Astronauts Thomas P. Stafford and Eugene A. Cernan, the Gemini IX-A crew, practiced space rendezvous procedures at KSC in preparation for their scheduled June 1 mission. (*NYT*, 5/31/66, 33)

- "The First Soft Step," program discussing Project Surveyor and showing flight operations facilities, was broadcast on six Alabama Educational Television Network stations. Program, hosted by MIT science reporter John Fitch, was first in 13-week "Science Reporter" series which had been taped on location at four NASA centers and nine contractor plants. Second program, to be telecast June 6, would be titled "Landing on the Moon" and would concern NASA's Lunar Excursion Module. (MSFC Release 66-117)
- Dr. Donald P. Burcham, JPL Voyager program manager, told *Aviation Week* in an interview that heavy funding for the Voyager Mars program was anticipated with FY 1968 funds if the Vietnam war did not create new funding problems. (*Av. Wk.*, 5/30/66, 40)
- Adm. William F. Boone (USN, Ret.), Assistant NASA Administrator for Defense Affairs, told *Missiles and Rockets* that under new agreement between DOD and NASA—expected to be signed in about two months—LRC would modify a liquid hydrogen J-85 jet engine with 200,000–500,000 lbs. thrust for use in air-breathing technology development. As its contribution to the joint program, USAF would provide F-106 aircraft for flight tests of engine. (*M&R*, 5/30/66, 31)

May 31: Soviet proposal for keeping moon free of weapons was presented to U.N. Secretary General U Thant by Ambassador Nikolay T. Fedorenko, head of the Soviet delegation to the U.N. Four principles were suggested as basis for international agreement: "1—The moon and other celestial bodies should be free for exploration and use by all states without any discrimination. All states enjoy freedom of the scientific exploration of the moon and other celestial bodies on an equal basis and in accordance with the basic principles of international law.

"2—The moon and other celestial bodies should be used for peaceful purposes only. No military bases and installations including installations of nuclear and other weapons of mass destruction should be stationed on the moon and other celestial bodies.

"3—The exploration and use of the moon and other celestial bodies should be carried out for the benefit and in the interest of all mankind. They are not subject to appropriation or any territorial claim.

"4—In the course of the exploration of the moon and other celestial bodies, the states proceed from the principles of cooperation and mutual assistance and carry on their activities with due account of the respected interests of other states for the purpose of maintaining international peace and security." Proposal paralleled treaty suggested by President Johnson May 7. (Marder, *Wash. Post*, 6/1/66, A1)

- NASA released preliminary report of board investigating explosion of Saturn V 2nd stage (S-II-F) ground test version May 28 at MTF. Board chairman Dr. Kurt H. Debus, KSC Director, said that during checkout tests after five successful static firings, hydrogen tank pressure sensing line had been disconnected. Failure to reconnect line had caused tank rupture when pressurized with helium beyond design limits. (*Marshall Star*, 6/1/66, 1)
- NAS issued final report of Space Science Board exobiology study, undertaken in summer 1964 at NASA's request, including postscript examining results of MARINER IV's July 1965 Mars flyby in relation to study's recom-

mendation that Mars exploration be assigned "highest scientific priority." Report, *Biology and the Exploration of Mars*, was released with companion anthology, *Extraterrestrial Life*, which contained 2,000-entry bibliography. Advance summary of report had been released Apr. 26, 1965. (NAS-NRC-NAE *News Report*, 6-7; NAS-NRC Release, 5/31/66)

May 31: ComSatCorp Chairman James McCormack told Washington Society of Investment Analysts that ComSatCorp faced two issues of national and international significance: (1) "Responsibility for U.S. earth stations, as between ComSat and the other U.S. carriers" and (2) "Designation of the entities which may deal directly with Comsat...." He pointed out that present international agreements would be renegotiated in 1969, and that some participants were "dubious about Comsat's continuing ability as the consortium's manager to spend their money in their interests while being subject to the detailed regulation" of the FCC. (Text)

- Need for detection instruments to warn Sst pilots of excessive radiation during solar flares was described in papers by Sidney Teweles and Raymond M. McInturff of ESSA and Robert F. Jones of British Meteorological Office at Geneva meeting of executive committee of World Meteorological Organization. ESSA scientists said descent of plane from cruising altitude of about 60,000 ft. to 40,000 ft. might be necessary "three or four times a year during the half of the 11-year sunspot cycle when the greatest flares occur." (*NYT*, 6/1/66, 77)
- Gen. William Hugh Blanchard, USAF Vice Chief of Staff, died at 50 after suffering heart attack at Pentagon meeting. General Blanchard was credited with landing first B-29 in China in March 1944. (*NYT*, 6/1/66, 43)

During May: Secretary of the Air Force Dr. Harold Brown said in an article written for *NATO's Fifteen Nations* that one of his principal tasks had been to find the most effective way of integrating technical and nontechnical advice. Of contributions scientists had made to the process of defense policy-making, he said: "First, the scientist, operating at the frontiers of knowledge, may have anticipated many political and military problems posed by scientific advance. Second, military necessity made the scientist a leader in applying sophisticated rational and mathematical analysis to the solution of some problems—especially resource allocation —which have been generated by scientific advance.

"At the same time it must be acknowledged that advice from scientists had been far from infallible, especially where the advice touches upon aspects of a larger problem on which the specialist is less expert, and in some cases, may not even be fully informed." (*NATO's Fifteen Nations*, 4-5/66, 18)

- Dr. Von R. Eshleman of Stanford Univ., speaking at International Scientific Radio Union meeting in Washington, D.C., urged greater interplay between "those concerned with the design and communications of spacecraft and those who decide what scientific experiments will be aboard." He cited success of NASA's MARINER IV mission in use of telemetry signals to learn that Martian atmospheric pressure was 1% that of earth's as example of accomplishment through cooperation in bistatic radar studies. (*Science News*, 5/14/66, 366)
- Three Boeing engineers had invented a jet engine inlet which could "choke" supersonic transport engine-compressor whine during approaches to airports, thereby avoiding airport-area nuisance. Noise-

choking feature was in addition to inlet's primary advantage—its ability to provide a smooth and even flow of air to a jet engine at mach 2.7, or 1,900 mph, the planned cruising speed of the U.S. Sst. (Boeing Release)

During May: Postulating that "many members of the House Space Committee are determined to become full partners with NASA in the development of basic objectives and policy," Space Science Subcommittee Chairman Rep. Joseph E. Karth (D-Minn.) cited recent approach by Subcommittee in reallocation of funds during hearings on NASA FY 1967 authorization: "The Congress makes resources available for our national space program in the name of the American people. Therefore Congress believes that it should have a voice in the allocation of these resources. This principle was reflected in the actions taken by the Subcommittee." Karth was writing in *Astronautics & Aeronautics.* (Karth, *A&A,* 5/66, 26-8)

- Revolution in military communications "to which we are now fully committed and which we are actively pursuing," was discussed by William Beecher in *Astronautics & Aeronautics*: "At this writing, the Pentagon plans a May launch of the first eight of a series of 14 to 22 repeater satellites in its Initial Defense Communications Satellite Project (Idcsp). The first eight, riding piggyback on a Titan III-C test booster, will be deposited into random orbits about 21,000 mi. up. Within 60 days thereafter, another six repeater satellites together with two gravity-gradient experimental satellites will be similarly orbited. If all 14 of the communications satellites function perfectly, sufficient coverage will be provided to obviate the need for a third shot of eight, Pentagon experts say.

  "Although meant primarily to perfect the technology of space communications, after a relatively short test-only period, this Idcsp would be made available for high-priority worldwide military traffic—for instance, important communications between Washington and Saigon. . . .

  "Then, in 1969 or 1970, the Defense Department hopes to put up a fully operational system, called the Advanced Defense Communications Satellite Project (Adcsp). As John S. Foster, Jr., Director of Defense Research and Engineering, described it in recent Congressional testimony, 'This system will take advantage of advances in both booster and communications technology which will permit the establishment of an economical, highly reliable system capable of providing many secure, jam-resistant, long-range communications circuits to support vital security operations wherever they occur.' " (Beecher, *A&A,* 5/66, 10)

- New NASA-developed technique for making biomedical checks on test pilots was being used at Univ. of Kansas Medical Center. Method, which replaced bulky sensors with bare wires acting as their own electrodes, permitted doctors to send instrumented patient home after office visit and record biomedical data while patient resumed normal activities. Wires were applied in three-minute operation. Data obtained in this way—rather than while patient was on operating table—were considered more meaningful by some physicians. Technique was conceived by Dr. James A. Roman, chief of biomedical research at NASA Flight Research Center, and introduced to the Univ. of Kansas by Midwest Research Institute. MRI was under contract to NASA's Technology Utilization Div. in its program of transferring aerospace technology to the public. (NASA Release 66-116)

- B/G J. C. Maxwell, director of FAA Sst development, discussed supersonic air travel in *Aerospace*:

". . . The national SST program, though small (from a dollar point of view) when compared to the national space program, nevertheless may well have a more immediate impact. The challenge inherent in producing an economically attractive supersonic aircraft has already identified and created demands for improved manufacturing techniques, simple yet reliable electronic and communications equipment, advancements in extrusion and machine tooling techniques, more efficient (and less costly) fuels and lubricants and many other needs.

"The economic aspects are more easily definable.

"Concerning the potential market for the SST, varying estimates have been made based on the expected growth of air transportation over the next 20 years and beyond. . . .

"Conservatively speaking, we now look for a three-fold increase in long-haul revenue passenger miles flown by the airlines of the world by 1980 and a five-fold increase by 1990. SSTs could carry almost half of this traffic. . . .

"Obviously, the SST program will have a significant impact on the U.S. balance-of-trade position. Based upon past experience along with current encouraging signs, we expect that the export market for a U.S.-made SST will be about half of the total production rate. Over a 20-year period, this could result in a gold inflow approaching $10 billion." (*Aerospace*, 5/66)

During May: Space research programs of the U.S.S.R. and the U.S. were opening the way for solving earth's future overpopulation problem and mankind's future raw material needs, Soviet Prof. E. Kolman wrote in *Mirovaya Ekonomika i Mezhdunarodnyye Otnosheniye*, a publication of the Soviet Academy of Sciences. He referred to estimates that in 400 yrs. the earth would have a population of about 900 billion people, equivalent to about 10,000 people per square mile. In such a situation, this planet would not be able to give all its people even elementary living space, let alone food. Therefore, he said, "escape into cosmic space . . . will become inevitable." Professor Kolman mentioned the stimulating impact of the space program on progress in science and technology and the likelihood of developments useful in more mundane areas of human activity. He cited the use of meteorological and communications satellites as examples of terrestrially useful byproducts of space research. (*NYT*, 5/22/66, 85)

- Four experimental ramjet missiles launched by France's Office National d'Etudes et de Recherches Aérospatiales (ONERA) reached speeds of nearly 4,600 fps between 40,000- and 115,000-ft. altitudes. Missiles were part of a 10-flight program code-named STATALTEX, to generate data applicable to eventual development of an atmospheric booster that could reach mach 5. (*M&R*, 5/23/66, 9)

June 1966

June 1: Augmented Target Docking Adapter (Atda) for Gemini IX-A mission was successfully launched by Atlas booster from ETR into planned 185-mi. (298-km.) circular orbit, but officials were not certain glass-fiber shroud protecting docking apparatus had been jettisoned—a maneuver essential for docking with Gemini 9 spacecraft. Gemini IX-Titan II launch was postponed until June 3 when malfunction in electronics data-processing equipment blocked guidance signals being sent from KSC launch control center to Gemini spacecraft at T−3. NASA officials said postponement would not change objectives of the mission —extravehicular activity and rendezvous and docking with Atda—but would cause some revisions in timing of events. (Wilford, *NYT*, 6/2/66, 1)

- Saturn V 3rd stage (S-IVB) was flown aboard Super Guppy aircraft from Douglas Aircraft Co.'s Huntington Beach, Calif., facility to company's Sacramento Test Center for static testing. The 33,000-lb. stage, 59 ft. long and 21½ ft. in diameter, was second Saturn V upper stage to arrive at Douglas test site. (MSFC Release 66-126)

- General Accounting Office released report to Congress charging that NASA had wasted $2.5 million on Surveyor program by continuing development of 11 types of scientific instruments for SURVEYOR I after 1962, when weight problems had forced their deletion from planned payload. NASA comments in report's appendix included letter from Deputy Associate Administrator Earl D. Hilburn stating: "Development of these experiments was not halted ... to assure their availability for not only later Surveyor flights but other NASA missions as well." He called continuation "prudent in the light of development lead time for scientific instrumentation." SURVEYOR I had been launched May 30 from ETR on lunar transfer trajectory. (Text; Blair, *NYT*, 6/1/66, 26; *Av. Wk.*, 6/6/66, 25)

- JPL scientists Richard W. Davies and Roy E. Cameron completed month-long expedition in Atacama Desert in Northern Chile and announced in Santiago that analysis of soil samples taken at altitudes up to 20,000 ft. would test theory that "more microorganisms are found under the surface than on it." If so, Davies said, NASA would "redesign its Mars exploration vehicles to look for subsurface life as well as surface life." Atacama region so resembled supposed surface of Mars, they said, that JPL team would attempt further studies in area, called by some the driest on earth. (AP, Balt. *Sun*, 6/2/66)

- William L. Green, Jr., executive assistant in NASA's Office of Public Affairs and former USIA press officer, was appointed NASA Deputy Assistant Administrator for Public Affairs. (NASA Release 66-140; AF)

June 1: Three-month NASA-USAF program to measure reactions of structures and people to varying types of sonic booms and jet engine noise began at AFFTC. Program—in which over 100 residents would participate as test subjects—would provide data for analysis and later application to supersonic transport development. (AFFTC Release 66-5-6)
- Three college students set record by wearing fully pressurized spacesuits continuously for five days in research test conducted for NASA at Aerospace Medical Research Laboratories, Wright-Patterson AFB. Experiment—part of 90-day study to test food requirements for future manned space flights and emergency food comcepts for lunar exploration—broke 24-hr. record set at Aerospace Medical Div., Brooks AFB. (AP, *NYT*, 6/2/66, 30)
- Astronaut Frank Borman, a 1950 graduate of West Point, gave Academy cadets a souvenir from space—the flag of the cadet company he had commanded—carried on his Dec. 4-18, 1965, GEMINI VII mission. (UPI, *NYT*, 6/2/66, 4)

June 2-14: NASA SURVEYOR I became first U.S. spacecraft to softland on the moon when it touched down in the Ocean of Storms and began transmitting the first of its more than 10,000 clear and detailed television pictures to JPL Deep Space Facilities, Goldstone, Calif.

Landing sequence began 2,000 mi. above moon when, traveling at 6,000 mph, SURVEYOR I shifted its normal cruising attitude to position main retrorocket. Triggered by radar ejected at 52-mi. altitude, main retrorocket burned until 37,000-ft. altitude, slowing craft to 400 mph. After retrorocket burnout, smaller guidance rockets ignited, slowing SURVEYOR I to 3½ mph at 14-ft. altitude. Craft then free-fell to lunar surface at 8 mph. Landing—so precise that spacecraft's three footpads touched lunar surface within 19 millisec. of each other—confirmed that lunar surface could support Apollo Lunar Excursion Module (Lem) and proved concept of automatically decelerating a spacecraft from 6,000-mph speed to touchdown speed of 3½ mph.

Photos transmitted by SURVEYOR I before its camera was secured after sunset of first lunar day (June 14) revealed nearly level lunar surface littered with rock fragments [see During June]. "Rubble scattered over the surface is probably a fairly general characteristic all over the moon," NASA Surveyor Project Scientist Dr. Leonard Jaffe told JPL press conference.

Pictures taken June 4 during seven two-second firings of nitrogen gas from spacecraft's attitude control jets showed no disturbance of the lunar surface. In addition to wide- and narrow-angle lurain photographs, SURVEYOR I obtained pictures of the stars Sirius and Canopus to determine spacecraft's exact location.

Halting communication with SURVEYOR I June 14 to allow spacecraft to conserve battery energy throughout lunar night (June 14-29), JPL announced: "It is very difficult to predict if Surveyor's batteries will withstand the −260°F cold. If Surveyor survives, the engineers estimate it will be several days after sunrise before the batteries thaw out" [see June 28-29].

Considerably more complex than Soviet LUNA IX which softlanded on the moon Feb. 3, SURVEYOR I apparatus remained a closed intact system up until touchdown. A tripod-legged arrangement of cameras, antennas, and other equipment which used solar cells to convert sun's energy into electric power for operation on lunar surface, SURVEYOR I

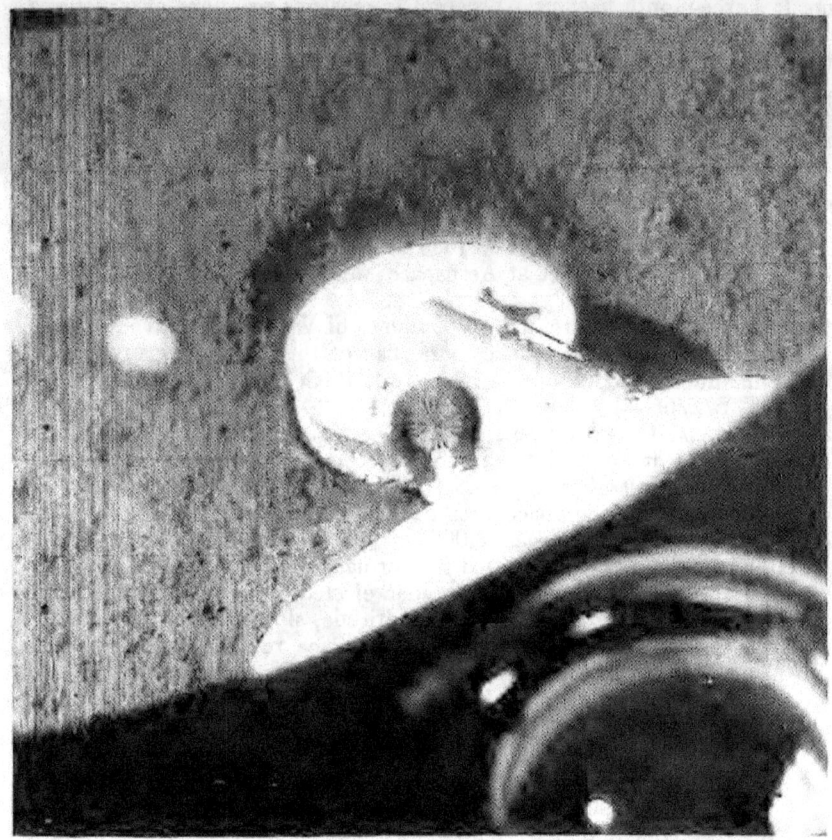

June 2-14: SURVEYOR I returned this photograph of its footpad resting on the surface of the moon.

had a landing system which used radars, a computer, and four rocket engines to adjust its own speed, attitude, and flight path. LUNA IX was crash-proof, uncontrolled photographic instrument package jettisoned by its carrier rocket immediately before impact, which free-fell under pull of lunar gravity. It had transmitted only 12 photos.

The near-flawless performance of SURVEYOR I, an engineering model, far exceeded JPL officials' expectations. "This, in my opinion, puts the Surveyor program ahead a year," Robert F. Garbarini, NASA Deputy Associate Administrator for Space Science and Applications for Engineering, told JPL press conference. "We thought it would take three or four flights to get the craft operational." (NASA Proj. Off.; NASA Release 66-127; JPL Transcript; Spivak, *WSJ*, 6/3/66, 2; *WSJ*, 6/15/66, 1; *Shreveport Times*, 6/3/66, 12A; Clark, *NYT*, 6/3/66, 1, 22; 6/4/66, 12; 6/5/66, 1; 6/6/66, 40M; AP, *NYT*, 17; Hines, *Wash. Eve. Star*, 6/5/66, C3; AP, *Wash. Eve. Star*, 6/6/66, A9; *Wash. Eve. Star*, 6/15/66, A6; Berman, *Wash. Post*, 6/5/66, A14; AP, *Wash. Post*, 6/15/66, A10)

June 2: President Johnson's statement on SURVEYOR I's successful soft-landing on the moon: "Overnight the eyes of Surveyor I have become the eyes of the world on the Moon. Another exciting chapter in the peaceful exploration of the universe is open for men to read and share....

"The odds against achieving full success in such a difficult mission on the first flight were understandably great. This moment of triumph for all who have participated in the Surveyor project has been well earned, for back of Surveyor's perfect performance on this first flight are years of hard work, painstaking care, and brilliant engineering.

"Today our Nation salutes the highly deserving team of scientists, engineers, technicians, and managers—in Government, industry, and the universities—who have a vision and the skills to pursue it successfully.

"As the day approaches when men land on the Moon, it is of the greatest importance that we agree to exchange openly all information that could affect their safety and welfare. It is equally important that we preserve these regions for peaceful, scientific activities. I welcome the constructive approach of the Soviet statement of May 31 on a treaty I have proposed to cover these matters and hope that progress can be made rapidly." (*Pres. Doc.*, 6/6/66, 728)

- NASA SURVEYOR I's successful soft-landing on the moon received worldwide acclaim.

  Daily Telegraph, London: "There is magic of a kind in the latest achievement of American exploration in placing their electronic Surveyor almost directly on its target on the moon.... The technological power behind the enterprise of putting a man on the moon is mounting with visible acceleration. But the human element remains most vital."

  Le Monde, Paris: "In succeeding with their first attempt at a soft landing on the moon only three months after the Soviets, the United States has shown that they are on the point of catching up with the Soviets in the race to the moon. There is only one experiment left in which the United States is behind the Soviet Union: the orbiting around the moon of a lunar probe similar to Luna-10. This lag is not very great, since the first trial of the Lunar Orbiters is set for next month."

  Il Tempo, Rome: "We think that the best judgment, even if purely a prioristic, of the Surveyor is that given by the Soviet Government, which, the day after the departure of the lunar probe, practically adhered to President Johnson's initiative of March 7, tending to a recognition of the internationalization of the moon and other extraterritorial bodies."

  Tass: "The success of the Surveyor is an important part of the American space program and Soviet scientists think very highly of it." (*NYT*, 6/5/66, E13)

  Congratulatory cable from Soviet President Nikolay V. Podgorny to President Johnson: "On the occasion of the successful soft-landing of the Surveyor I spacecraft on the moon, accept ... our congratulations for the American specialists who ensured the success of this flight." (Reuters, *NYT*, 6/5/66, 80)

- Successful overlapping operation of PEGASUS I, II, and III meteoroid-detection satellites—launched Feb. 16, May 25, and July 30, 1965, respectively—had provided full year of data which defined "more precisely than ever before" near-earth meteoroid environment, NASA reported. Data—including cumulative count of 1,000 "hits" on target plate thickness, identification and location of specific panel hit, attitude of spacecraft with respect to earth and sun at impact time, and time of penetration—confirmed "protective adequacy" of Apollo manned lunar landing spacecraft against meteoroids; provided designers of future spacecraft with guidelines on probability of meteoroid penetrations; and improved mapping of South Atlantic anomaly by measuring electron radiation density. (NASA Release 66-134)

June 2: V/Adm. H. G. Rickover (USN), in speech at Athens meeting of Royal National Foundation, noted benefits to humanity in science and warned of dangers inherent in misuse of technology. He urged that technological capability be applied selectively to enhance human welfare and conserve natural resources, rather than to "submit meekly to whatever is technically feasible." (Text, *CR*, 6/21/66, 13114-17)

June 3-6: NASA's GEMINI IX spacecraft, with Astronauts Thomas P. Stafford as command pilot and Eugene A. Cernan, pilot, was successfully launched from ETR by Titan II booster on GEMINI IX-A mission to evaluate extravehicular life-support and maneuvering equipment and procedures, and to rendezvous and dock with Augmented Target Docking Adapter (ATDA), launched from ETR June 1. Initial orbit: 174-mi. (280-km.) apogee; 99-mi. (159-km.) perigee; 90-min. period; and 30° inclination. At T-3 ground communications system malfunction which had postponed June 1 launch recurred, but alternate procedures proved satisfactory and countdown continued.

At 49 min. GET Stafford executed first course correction to position GEMINI IX for rendezvous, firing thruster rockets to raise perigee to 134 mi. After two more orbital maneuvers, spacecraft was within 25 ft. of target. Stafford, sighting ATDA, confirmed suspicions that its shroud had not jettisoned, and docking would be impossible. He radioed ground control: "It looks like an angry alligator out there." NASA later revealed that shroud had not jettisoned because technicians, following "insufficiently detailed" written instructions, had installed lanyards improperly.

Complying with modified flight plan, astronauts photographed ATDA, executed re-rendezvous maneuver, and executed rendezvous-from-above simulating Lunar Excursion Module (Lem) abort from the moon. After this maneuver, which depleted fuel supply to 11%—only 6% more than minimum reserve—Stafford and Cernan were fatigued and recommended 24-hr. postponement of Cernan's walk in space.

At 11:02 a.m. EST June 5, Cernan opened hatch of depressurized spacecraft, climbed out, and retrieved micrometeoroid impact detector attached to side. He moved to full length of 25-ft. tether to take photos. After one hour, Cernan returned to Adapter Section to don Astronaut Maneuvering Unit (Amu)—a task which required "four or five times more work than anticipated." His faceplate visor began clouding over, presumably because environmental control system was not absorbing moisture quickly enough. Cernan's limited visibility and discovery that Amu's radio transmissions were garbled forced Stafford to recall him to spacecraft. He had been scheduled to disconnect from GEMINI IX's oxygen supply and maneuver in Amu to 150 ft. from spacecraft.

Reentry on June 6 was normal. At 10:00 a.m. EST, in the 46th revolution, GEMINI IX impacted 345 mi. east of Cape Kennedy—less

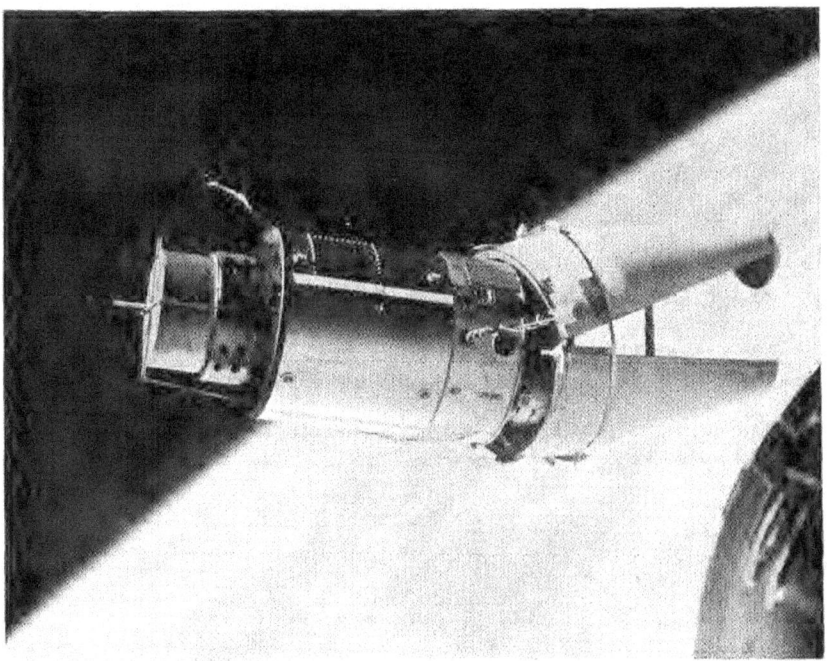

June 3-6: Augmented Target Docking Adapter with unreleased shroud, photographed by crew of GEMINI IX.

than two miles from target—in the most accurate landing to date, after total mission duration of 72 hrs. 21 min. Splashdown and recovery were carried on live television via EARLY BIRD I comsat. Within minutes, USN pararescue team, dropped from helicopters, placed flotation collar around spacecraft. At 10:45 a.m. EST, spacecraft carrying astronauts was hoisted onboard U.S.S. *Wasp.*

Accomplishments of GEMINI IX-A mission were summarized by Dr. George E. Mueller, NASA Associate Administrator for Manned Space Flight, at MSC news conference: (1) demonstration of three new rendezvous procedures—catching target in only three revolutions, locating it without radar, and approaching it from above rather than below; (2) accurately controlled reentry; (3) longest period of extravehicular activities—over two hours; (4) new information on man's ability to work in space necessary for manned lunar landing and for assembling stations in space; and (5) new information on manned observation of another spacecraft in orbit. Five experiments were successfully performed in accordance with modified flight plan. Concerning mission's disappointments, he said: "We will try to do as much as we can on each of our Gemini flights. We won't always succeed in meeting all of the objectives. But we learned a great deal about the problems that do occur during the course of one of these very difficult space missions." (NASA Proj. Off.; NASA Release 66-97; Wilford, *NYT,* 6/4/66, 1, 10; 6/5/66, 1, 80; 6/6/66, 1, 40; 6/7/66, 1, 34; 6/8/66, 1,29; Hines, Wash. *Eve. Star,* 6/4/66, A1, A2; 6/5/66, A1, A6; 6/6/66,

A1, A4; 6/7/66, A3; Simons, *Wash. Post*, 6/4/66, A1, A8; 6/5/66, A1, A14; 6/6/66, A1, A3; 6/7/66, A1, A7, A21; 6/8/66, A3; *WSJ*, 6/7/66, 2; 6/8/66, 1; *Tech. Wk.*, 6/13/66, 14-15; *Av. Wk.*, 6/13/66, 31-34)

June 3: USAF launched two unidentified satellites from WTR with Atlas-Agena D booster. (*U.S. Aeron. & Space Act., 1966*, 151)

- Lockheed Aircraft Corp. signed agreement with Junkers of Munich, Germany, to "provide technical consultation, perform design work . . . and provide components" for ESRO's Highly Eccentric Orbiting Satellite (Heos)—first to be designed and developed in Germany—to investigate charged particles in space and the effects of magnetic fields on them. Overall contract for satellite, which would be launched by NASA into elliptical orbit in late 1968, was $5.5 million, with Lockheed's services valued at $500,000. (Lockheed Release)

- Accumulating geochronologic evidence which "indicates more and more convincingly that tektites were formed from terrestrial rocks in large meteoritic impacts on the earth" was reported in *Science* by Henry Faul. He said exact process of tektite formation remained a mystery, but that age measurements were beginning to show "where" this formation occurred and could ultimately invalidate the lunar origin hypothesis. (Faul, *Science*, 6/3/66, 1341-5)

- GEMINI IV Astronaut Edward H. White II and GEMINI VII Astronaut Frank Borman received National Aviation Club's Award for Achievement in Washington, D.C. (*Wash. Eve. Star*, 6/3/66)

- Edgar G. Bush, senior technician at GSFC since 1959 and designer of first micro-electronic circuitry used for flight computers, died of a heart ailment. Bush had designed computers for VANGUARD III, Explorer satellites, and Lunar Orbiters. (*Wash. Eve. Star*, 6/7/66, B5)

June 4: Role of NASA in aeronautical research and development was examined in *Policy Planning for Aeronautical Research and Development*, 279-page report released by Sen. Clinton P. Anderson (D-N.Mex.), chairman of Senate Committee on Aeronautical and Space Sciences. Document noted "space budget demands have probably hampered what might have been expected to be a normal growth of the level of effort in aeronautics within the agency" and questioned adequacy of $124 million (about 2%) earmarked for aeronautical R&D out of NASA FY 1967 appropriation of $5.012 billion. Separate budget for aeronautics was suggested as possible solution. (Senate Doc. No. 90, 5/19/66)

- It had not yet been determined whether space fatigue caused any physical or psychological problems, but examining effects of this phenomenon had become a medical goal, Dr. Charles Berry, Chief of Medical Programs, MSC, told the press. Space officials believed exhaustion experienced by GEMINI IX-A Astronauts Thomas Stafford and Eugene Cernan June 3 was caused by lack of sleep during first night of mission and execution of complicated and tedious rendezvous. GEMINI VIII Astronaut Neil Armstrong said fatigue would be an important consideration in planning a lunar flight which might take as long as 14 days: ". . . we will evaluate that particular thing to insure that what we learn from these Gemini flights will be applied to that situation and used." (*NYT*, 6/5/66, 80)

- First seven photos transmitted by SURVEYOR I were donated to Clark Univ. by JPL Director Dr. William H. Pickering as the school broke ground for a $5-million library honoring the late rocket pioneer Dr.

Robert Hutchings Goddard. (UPI, *NYT*, 6/6/66, 42; Berman, *Wash. Post*, 6/5/66, A14)

June 5: NASA appointed Dr. Richard L. Lesher as Assistant Administrator for Technology Utilization, replacing Breene M. Kerr, who remained Assistant Administrator for Policy Analysis. Dr. Lesher's former position of Deputy Assistant Administrator for Technology Utilization was filled by Melvin S. Day, former head of Scientific and Technical Information Div. (NASA Release 66-143)

June 6: NASA's OGO III (Ogo-B) Orbiting Geophysical Observatory was successfully launched from ETR with Atlas-Agena B booster into orbit with 75,874-mi. (122,157-km.) apogee; 183-mi. (295-km.) perigee; 48.6-hr. period; and 31° inclination. Third of seven spacecraft in NASA's Ogo program and first to successfully operate three-axis stabilized in highly elliptical orbit, OGO III weighed 1,135 lbs.; it carried 21 experiments—largest number ever carried by U.S. scientific spacecraft—to study solar wind, solar flares, magnetic field disturbances, radiation belt particles, aurora events, ionization, and variations in atmospheric density. Ogo program was managed by NASA Goddard Space Flight Center. (NASA Proj. Off.; NASA Release 66-132; AP, *NYT*, 6/8/66, 15; AP, *Wash. Post*, 6/7/66, A7)

- GEMINI IX-A Astronauts Thomas P. Stafford and Eugene A. Cernan, resting aboard U.S.S. *Wasp*, received tributes from around the world. President Johnson, telephoning from his Texas ranch, said: "We are very proud of both of you. I have been watching off and on over the weekend and saw your return this morning.

    "You have made all of us more aware of what performance under pressure is all about and that includes courage. . . ."

    Tass science commentator said the Soviet people "give their due" to the courage of the astronauts and congratulate them on their safe return to earth. He praised the flight as an "achievement of American astronautics."

    Jodrell Bank Experimental Station director Sir Bernard Lovell said the success of SURVEYOR I and GEMINI IX-A had made the week one of the greatest for American science. "It seems that the Russians will have to stage a spectacular development in order to recapture the lead which they apparently possessed a week ago." (Kilpatrick, *Wash. Post*, 6/7/66, A8; Balt. *Sun*, 6/7/66)

- Sun-powered laser developed by Dr. C. Gilbert Young of American Optical Co. under USAF sponsorship had been successfully operated in earth's atmosphere, producing one watt of power in continuous, highly intense beam of invisible infrared light. A slender, one-inch-long rod of yttrium-aluminum-garnet crystal (Yag), laser was connected to telescope which trapped, focused, and delivered sun's rays—which replaced conventional lamp and electrical apparatus needed by other lasers for "initial boost"—to crystal rod. Laser's beam could carry messages between spacecraft and could be used as intense satellite beacon. (*AIP News* 6)

- FAA's "Staff Study—General Aviation Occupant Load Factor" report showed that general aviation (nonairline) aircraft in the U.S. carried 39.4 million travelers in 1965, while domestic airlines carried 84.6 million passengers. Estimates were based on information gathered by all of FAA's flight service station facilities during August 1965. (FAA Release 66-54)

June 6: U.N. World Meteorological Organization announced award of annual prize for outstanding work and international collaboration in meteorology to Prof. Tor Bergeron of Sweden. Noted for pioneering new techniques in air-mass analysis, weather forecasting, and physics of precipitation, Professor Bergeron would receive a gold medal and $1,200. (*NYT*, 6/7/66, 12)

- Mrs. Jane Marshall, editor of aeronautics bibliographies and education sourcebooks for National Aerospace Education Council, was named 1965 winner of National Aeronautic Association's Frank G. Brewer Trophy for "her contributions to enlarged aerospace horizons for those who teach our nation's youth." (*Natl. Aeronautic Assn. News*)

June 7: Two-stage Wasp (Weightless Analysis Sounding Probe) sounding rocket carried 1,500-lb. payload to 148-mi. (238-km.) altitude in ballistic trajectory from NASA Wallops Station in experiment to study behavior of liquid hydrogen under near-zero gravity conditions. Payload, consisting of scale model of liquid hydrogen fuel tank, impacted 330 mi. downrange in Atlantic; no recovery was involved. Experiment under weightless conditions, continuing for seven minutes, was recorded by television cameras through transparent lucite tank and was telemetered back to Wallops. Cameras, mounted on five-foot extension booms, photographed the liquid's sloshing motion when small thrusters were activated to show how well baffles positioned the liquid in the tank. (NASA Release 66-147)

- MSFC successfully static-tested S-IC stage of the second flight Saturn V launch vehicle for 125 sec. and recorded 1,200 measurements of stage's performance. The 33-ft.-dia., 135-ft.-long stage developed 7.5 million lbs. thrust from five F-1 engines—four of which were gimbaled during test. This was only captive test planned for this flight stage. (MSFC Release 66-129)

- Dr. Hans-Joachim von Merkatz, member of West German Bundestag, charged in report prepared for delivery before meeting of Assembly of the Western European Union that U.S. was trying to "break up" Western Europe's space program with "tantalizing" offers for joint U.S.-European planetary exploration: "In other words, the exploration of Jupiter could divert us from the essential economic benefits to be derived from space through the exploration of communications satellites. The American approach to date has definitely been aimed at insuring United States monopoly and leadership in this field as in the nuclear field." Western European Union was composed of U.K., France, West Germany, Italy, Belgium, the Netherlands, and Luxembourg. (*NYT*, 6/9/66, 3)

- The Baltimore *Sun* commended four companies for their roles in GEMINI IX-A flight: Martin Co. for building and testing Titan II launch vehicle; Westinghouse Corp. for developing sensitive radar equipment; Bendix Corp. for engineering, operating, and maintaining NASA's tracking system; and ITT Corp. for giving "millions of television viewers . . . all the advantages of a grandstand seat for the dramatic recovery. . . ." (Balt. *Sun*, 6/7/66)

- Editorial in Washington *Evening Star:* "The United States has just passed through its most ambitious and taxing week to date in space. So this is, perhaps a fitting time to ask itself just where it is going in space and why. James E. Webb, the head of the National Aeronautics and Space Administration, has been asking that question for some time now, trying to get an answer from the Administration ... and from Congress. ... The question of whether to pack up or to push on should, as Mr. Webb suggests, be the subject of a major national debate." (Wash. *Eve. Star*, 6/7/66)

June 7: U.K. Foreign Office issued following statement after press reports of Cabinet decision to withdraw from European Launcher Development Organization (ELDO): "For some time the Government has had serious doubts about whether or not it should continue to participate in the ELDO programme. These doubts were centered on financial, technical and economical assessments of the initial programme.

"The Government has concluded after a very careful and detailed consideration of all the factors involved, that the latest proposals for modifying the initial programme still do not constitute a sufficient basis for continuing, and it has so informed its partners." (British Embassy INFORM 195/66, 6/7/66)

June 8: U.S.S.R. launched COSMOS CXX carrying scientific instrumentation for continued space research. Orbital parameters: apogee, 300 km. (186 mi.); perigee, 200 km. (124 mi.); period, 89.4 min.; inclination, 51.8°. Equipment was reported to be functioning normally. (Tass, 6/8/66)

- $500-million XB-70 No. 2 experimental bomber and its F-104 chase plane collided and burned near Barstow, Calif., destroying both aircraft and killing F-104 pilot Joseph A. Walker and XB-70 co-pilot Maj. Carl S. Cross (USAF). Walker, NASA test pilot, held world's records for his 4,104-mph and 354,200-ft.-altitude flights in X-15 rocket research aircraft. XB-70 pilot Al White ejected in seat capsule and parachuted to ground, suffering arm and back injuries. On basis of preliminary USAF reports that F-104 apparently hit two vertical stabilizers on XB-70 at 25,000-ft. altitude during "routine" flight, President Johnson praised pilots who gave "their lives to advancing science and technology" and added "immeasurably to the progress this nation is making in the effort" to advance supersonic flight. USAF later revealed that aircraft were flying in formation as close as 10 ft. with T-38, F-4B, and F-5 aircraft "to allow photographic coverage of aircraft powered by General Electric engines" for publicity purposes. Disclosure prompted AFSC Commander Gen. Bernard Schriever to establish two USAF accident boards and Congress to designate House Appropriations Committee to investigate the disaster. (Edwards AFB Release; *Pres. Doc.*, 6/13/66, 748; *Wash. Post*, 6/14/66, A3)

- Rep. Olin E. Teague (D-Tex.) commented on collision of XB-70 and F-104 aircraft: "This accident must serve as a reminder to the American public and to the world that all of the potential hazards of aircraft and space flight can never be entirely eliminated from our experimental program. The phenomenal success of our manned space flight programs Mercury and Gemini and the remaining flights of Gemini and the flights of Apollo ... share the risk that Walker, White, and Cross faced each time they flew to test the XB-70." (*CR*, 6/8/66)

- Nerva (Nuclear Engine for Rocket Vehicle Application) reactor (NRX-A5) was successfully ground-tested by NASA and AEC at Jackass Flats, Nev. Test, in which reactor was operated at design power of 1,100 mw for 15½ min., was first in series to "obtain additional data on reactor characteristics under extended operating duration" under joint NASA-AEC Rover program. (AEC Release J-151)

June 8: President Johnson had sent SURVEYOR I photos to "Chiefs of State of more than 100 foreign nations . . . and to the world's scientific community," Press Secretary Bill Moyers announced. "This effort is made," the President said, "in order that careful study of the photographs will be possible around the world. We intend that the knowledge we derive from space will be available for the enrichment of our common experience and the advancement of peaceful undertakings in the exploration of space." (*Pres. Doc.*, 6/13/66, 748)

- Sen. John J. Sparkman (D-Ala.), speaking on floor of Senate, pointed out the contributions of small business to Gemini program and SURVEYOR I mission and said they were "irrefutable evidence that there is still a significant place in our economy for the ambitious and talented . . . in new and growing business firms." (*CR*, 6/8/66, 12053)

June 9: USAF launched three satellites with a single Atlas-Agena D booster from WTR—one unidentified satellite, ERS XVI satellite to test metal-to-metal adhesion of space propulsion materials, and USA's SECOR VI geodetic satellite. (*U.S. Aeron. & Space Act., 1966,* 152)

- NASA OSO II Orbiting Solar Observatory, activated by GSFC officials, transmitted scientific and engineering data from four of its eight experiments. Launched from ETR Feb. 3, 1965, OSO II had exceeded its six-month life expectancy by 50% before it was turned off Nov. 3. (GSFC Release G-10-66)

- U.S. had current "edge" over U.S.S.R. in overall space capability but this status had been "recently acquired and is not sufficient to form even the slightest basis for complacency," Dr. Edward C. Welsh, NASC Executive Secretary, told Armed Forces Communications and Electronics Assn. Industrial Luncheon in Washington, D.C. Dr. Welsh analyzed the areas of relative accomplishment: "(1) *Number of spacecraft*. The United States has placed almost 400 spacecraft into earth orbit and on escape missions, while the USSR has put slightly less than 200 on such missions. . . . However . . . Soviets have placed each year a greater total weight of payloads. . . . (2) *Lunar and Interplanetary*. The United States has been far less active but considerably more successful in its interplanetary activities. . . . We have photographed Mars and had a productive fly-by of Venus while the USSR's active interplanetary program has been primarily plagued by failure. In addition, we have taken some 17,000 photographs of the moon and are taking hundreds more by the magnificent Surveyor spacecraft. . . . (3) *Manned Flight* . . . U.S. . . . has a distinct lead over the USSR with about 1500 man flight hours in orbit compared with about 500 . . . with the longest individual flight as well as the longest extra-vehicular activity. Also, the U.S. has had two controlled rendezvous maneuvers, one docking experiment, and considerable manned spacecraft maneuvering experience. So far, the USSR has achieved none of these latter goals. (4) *Space Applications*. The United States has navigation, weather, and communications satellite systems in regular operation, while the USSR is still in the developmental stage in such activities. (5) *Scientific Data* . . . the larger number and greater variety of U.S. scientific payloads, as well as the wider and freer dissemination of the information obtained, may well have added more to the world's store of knowledge. (6) *Propulsion* . . . the USSR has surpassed the United States in the field of propulsion so far as the amount of thrust is concerned [but] this thrust advantage is much less true today as both countries have in operation and under development larger and more advanced propulsion systems...." (Text)

June 9: Changes in Apollo nomenclature were announced by Julian Scheer, NASA Assistant Administrator for Public Affairs, in memorandum from Project Designation Committee: (1) Lunar Excursion Module would be known as "Lunar Module"; (2) Saturn IB booster would be "Uprated Saturn I"; and (3) Saturn V stages (S-II, S-IC, and S-IVB) would be "first, second, and third stages." Technical designations for Saturn V stages, Scheer said, were confusing to the public and would be dropped gradually, except in NASA's internal and technical documentation. (Text)

- State Dept. disclosed that on May 26 U.S.S.R. had given U.S. photos taken by LUNA IX after it softlanded on the moon Feb. 3. (UPI, *NYT*, 6/10/66, 8)
- LaRC engineers Upshur T. Joyner and Walter B. Horne received Society of Automotive Engineers' Arch T. Colwell Merit Award—presented annually to authors of papers "of outstanding technical or professional merit"—for "Pneumatic Tire Hydrolapping and Some Effects on Vehicle Performance." (LaRC Release)

June 10: USAF's OV3-IV research satellite was launched by NASA from Wallops Station by Scout booster. Orbital parameters: 2,933 mi. (4,720 km.); perigee, 398 mi. (640 km.); period, 142 min.; inclination, 41°. The 173-lb. satellite carried plastic spheres of simulated human tissue containing linear energy transfer spectrometers to record effects of space radiation and determine how much radiation damage an astronaut's body could undergo in similar manned flight, how long man could stay in space at one time, and how often he could safely return to space environment. Satellite would also give measurements on Van Allen radiation belts and gauge decay of residual radiation in man-made Starfish Belt. Equipment, expected to relay data for one year, was functioning normally. (Wallops Release 66-35; *U.S. Aeron. & Space Act., 1966,* 152)

- Twenty of 21 scientific experiments carried by NASA's OGO III, launched June 6 from ETR, had been turned on and were operating well. The remaining experiment—an ESSA radio propagation measuring device—would be turned on June 11. (NASA Release 66-149)
- Under Secretary of the Air Force Norman Paul told Rep. George H. Mahon (D-Tex.), chairman of House Defense Appropriations Subcommittee, in closed session that USAF officials were unaware that the XB-70 which crashed June 8 was to be used for commercial publicity purposes: "The Air Force definitely does not put its stamp on this sort of business.... Had we been aware, I think it is safe to say it would not have happened." Representative Mahon said Committee would "insist that procedures be established which will make impossible a recurrence of such misuse of men and military equipment." (Morris, *NYT*, 6/11/66, 1; Simons, *Wash. Post*, 6/24/66, A8)
- GEMINI IX-A Astronauts Thomas Stafford and Eugene Cernan were greeted by their families on their arrival in Houston after four days of debriefing at KSC. (UPI, *NYT*, 6/11/66, 62)
- Lockheed Missiles and Space Co.'s first high-reliability Agena was launched from Vandenberg AFB as upper stage on Atlas booster, *Aviation Week and Space Technology* reported. High-reliability Agena had been under development for USAF Space Systems Div. for over three years in con-

junction with Midas infrared early warning system under Program 461. (*Av. Wk.*, 6/20/66, 31)

June 11: Astronauts Neil Armstrong and James McDivitt were among over 700 persons who attended memorial services in Lancaster, Calif., for NASA test pilot Joseph A. Walker, killed in June 8 midair collision of XB-70 and F-104 Starfighter. (UPI, *Wash. Post*, 6/13/66, A17)

- "When U.S. spacemen go to the moon, they will surely meet Soviet people there," Cosmonaut Yuri Gagarin, visiting Sofia for World Federation of Democratic Youth congress, told Bulgarian newspaper *Narodna Armiya*. (AP, Wash. *Sun. Star*, 6/12/66, A6)

June 12: General Dynamics Corp. announced that U.S.S. *Redstone*, one of three World War II T-2 type tankers being converted by corporation into Project Apollo tracking-communications vessel, had undergone initial sea trials out of Quincy, Mass. (*NYT*, 6/12/66)

June 13: House appointed to House-Senate Conference Committee on NASA FY 1967 authorization Reps. William E. Miller (R-N.Y.), Olin E. Teague (D-Tex.), Joseph E. Karth (D-Minn.), Ken Hechler (D-W. Va.), Emilio Q. Daddario (D-Conn.), Joseph W. Martin (R-Mass.), James G. Fulton (R-Pa.), and Charles A. Mosher (R-Ohio). (NASA LAR V/95-96)

- Accomplishments between June 1 and 7—SURVEYOR I, GEMINI IX-A, OGO III, and Wasp missions—had given NASA "the most rewarding week in its short but action-packed history," wrote William J. Coughlin in *Technology Week*. "Big shots and small, it all added up to an active and highly promising seven-day period for NASA. As someone commented, this was indeed the week that was." (Coughlin, *Tech. Wk.*, 6/13/66, 54)

- V/Adm. Rufus E. Rose (USN, Ret.) resigned as special assistant on policy studies and analyses to NASA Administrator James E. Webb to become an assistant to the president of Allied Research Associates, Inc. Admiral Rose, who joined NASA in 1964 after 44 years USN service, would be involved with "planning, organization, management, and overseas operation" at Allied. (NASA Release 66-148; *Marshall Star*, 6/22/66,3)

- France planned to continue work on large space boosters, either alone or with other interested European nations, despite British decision to withdraw from ELDO's Europa rocket program, *Aviation Week and Space Technology* reported. French officials, open in their bitterness at the British decision, viewed the withdrawal as further proof that U.K. made "a poor European partner." (*Av. Wk.*, 6/13/66, 38)

June 14: Nike-Apache sounding rocket launched from NASA Wallops Station carried 50-lb. instrumented payload to 115-mi. (185-km.) altitude. Conducted by NASA for Univ. of Illinois and GCA Corp., experiment was designed to measure electron density in ionospheric D region. (Wallops Release 66-36)

- Gen. Maxwell D. Taylor (USAF), special consultant to President Johnson, was appointed president of Institute for Defense Analysis replacing Dr. J. P. Ruina, who would return to MIT. IDA also named Dr. Gordon J. F. MacDonald, of UCLA's Dept. of Planetary and Space Sciences, vice president for research. Both appointments would take effect in September. (IDA Release)

- NASA's PIONEER VI interplanetary probe launched from ETR Dec. 16, 1965, completed six-month nominal mission at a range of 54.6 million mi. with all spacecraft systems performing well. Perihelion had been passed May 20, 1966, and aphelion passage would occur Oct. 24. Preliminary results of experiments had been reported Apr. 19 at

American Geophysical Union meeting in Washington, D.C. (NASA Proj. Off.)

June 15: On floor of House, Rep. George P. Miller (D-Calif.) paid tribute to SURVEYOR I and GEMINI IX-A mission successes and said they were possible because of "the unity of purpose that exists between the aerospace industry and the National Aeronautics and Space Administration." (CR, 6/15/66)

- Propulsion technology's short-term potential was overestimated and its long-term potential underestimated, AFSC Commander Gen. Bernard A. Schriever told the Second Propulsion Joint Specialist Conference in Colorado Springs. "This means that we should push technology without trying to tie it to specific system requirements. As a matter of historical record, it can be shown that exploratory and advanced development programs designed to push technology have provided valuable savings in time and money when applied to an engineering development program that encounters serious and unexpected problems. . . ." (Text)

- NASA had selected Control Data Corp. to furnish large-scale computer complex for LaRC under $20-million, fixed-price contract. Complex would utilize computers with wide variety of simulators and increase LaRC's capability and effectiveness in conducting advanced research. (NASA Release 66-152)

- Vice President Hubert H. Humphrey proclaimed Dr. Jocelyn R. Gill "Multiple Sclerosis Woman of the Year" in ceremony at his Washington, D.C., office and presented her with engraved bronze hope chest—symbol of Multiple Sclerosis Society. Dr. Gill, an MS patient for over 20 yrs., was with NASA Manned Flights Experiments Office, OSSA. She formerly had been Chief of In-Flight Sciences, OSSA, and on March 1 had received one of six Federal Women's Awards of 1966. (Wash. Post, 6/16/66, D1)

- NASA Western Operations Office (WOO) was disestablished as NASA field installation. WOO functions had been assigned to two component field activities—Western Support Office (WSO), which would serve "NASA Headquarters and installations programs and projects in Southern California and Nevada," and NASA Office—Downey, responsible for "overall NASA representation with North American Aviation, Inc., and for providing support to field installation projects at the contractor's site." (NASA Hq. Weekly Bull., 7/19/66, 3)

- Boeing Co. revealed new design for its 2707 supersonic transport to be submitted to FAA in SST competition with Lockheed Aircraft Corp. Sept. 6. Major structural changes included: shifting aircraft's four engines from under wings to under tail to provide smoother flow of air to engine and quiet compressor whine; integrating variable-sweep wing with horizontal tail to form single lifting surface during 1,800-mph flight; and increasing length (to 298 ft.) and gross takeoff weight (to 600,000 lbs.) to accommodate 300 passengers. (Boeing Release; Edwards, Wash. Post, 6/16/66, C10)

- Hourly bus tours—for a fee—began at KSC. Announcing the new service, KSC Director Dr. Kurt H. Debus said: "We want to provide the public with as meaningful a tour as can be permitted without interference with mission operational requirements. On Sundays, the public will continue to have the option of the free drive-through, in private vehicles, of the [NASA] center and the Cape Kennedy Air Force Station." (NYT, 5/12/66, 21; Hines, Wash. Sun. Star, 6/13/66, E1)

June 15: GSFC selected Brush Instruments Co. for $900,000 contract to provide 22 Unified S-Band (Usb) Systems Monitors for NASA's Manned Space Flight Tracking Network. Systems, composed of high-speed, finite sensing and recording devices, would provide "immediate reading of critical subsystems performance while simultaneously recording operating events as they occur," pinpoint and record trouble or failure, and serve as diagnostic tool for remedial action. (GSFC Release G-11-66)

June 16: USAF Titan III-C booster launched from ETR inserted seven Initial Defense Communications Satellite Program (Idcsp) repeaters and one gravity-gradient satellite into random, near-synchronous, equatorial orbit, creating nucleus of worldwide military communications system. Apogees varied from 15,911 mi. (25,617 km.) to 16,126 mi. (25,963 km.); perigees from 15,811 mi. (25,456 km.) to 15,829 mi. (25,484 km.). Average period was 22 hrs. 10 min. All seven comsats were reported performing nominally. Satellite to explore feasibility of gravity-gradient stabilization at very high altitudes had successfully deployed its 52-ft.-long booms and was performing as expected.

Powered flight of Titan III-C was close to planned parameters. Transtage and payload were inserted into parking orbit, where first transtage burn made necessary course corrections. Second transtage burn—at 110 min. GET—moved stage and load into transfer orbit. Third transtage burn—at 06:03:12 GET—put satellite dispenser frame and its eight satellites weighing 800 lbs. total into near-synchronous orbit with 15,906-mi. (25,608.6-km.) apogee and 15,801-mi. (25,439.6-km.) perigee. At 06:06:52 GET, satellites were ejected one at a time over approximate 3-min. period into their preselected orbits. Within six minutes, telemetry signals were received and surface-station tracking and communications tests begun; within two hours, circuits were established through repeaters. Launch was fourth consecutive success of Titan III-C and first total success of its transtage.

Fifteen additional comsats would be launched by late 1966 to ensure against failure of one or more satellites. New ground terminal would be installed in Vietnam by midsummer to provide additional command and control circuits. (UPI, *NYT*, 6/17/66, 14; UPI, *Wash. Post*, 6/17/66, A4; *Tech. Wk.*, 6/27/66, 16)

- U.S. gave U.N. proposed treaty on exploration of moon and other celestial bodies and asked that U.N. Outer Space Legal Subcommittee meet July 12 to discuss it. Draft—based on May 7 statement by President Johnson urging treaty to reserve celestial bodies for peaceful purposes—declared such bodies open equally to all countries, subject to claim by none, and off limits to nuclear weapons, weapon testing, military maneuvers, and fortifications. Soviet source said U.S.S.R. had submitted similar treaty earlier in day; text of proposal was not made public [see June 17]. (AP, *Wash. Post*, 6/17/66, A4)

- Two Nike-Cajun meteorological sounding rockets carrying exploding grenade payloads were launched from Point Barrow, Alaska, and Churchill Research Range in a coordinated GSFC experiment to study atmospheric parameters of wind, temperature, pressure, and density as summer seasonal maximum of noctilucent cloud sightings approached. Rockets and instrumentation functioned normally. (NASA Rpt. SRL)

- Mrs. James A. McDivitt gave birth in Houston to a daughter—first child conceived by an American astronaut's wife after her husband had returned from space. Astronaut McDivitt was command pilot on the June

June 16: USAF Titan III-C launch vehicle in preparation for multiple-satellite payload mission from Eastern Test Range.

    4-7, 1965, GEMINI IV mission during which Astronaut Edward H. White II made his 21-min. Eva. (AP, Wash. *Eve. Star*, 6/17/66, A2)

June 17: Washington, D.C., National Airport marked 25th anniversary facing ground congestion problems intensified by initiation of short-haul jet service April 24. FAA Administrator William F. McKee told 13 airlines to restrict flights at National or risk restriction by agency edict. Flight statistics released by FAA June 3 had shown 16,109 more passengers at National in May than April. Dulles International Airport's passenger count had dropped by 14,456. (Golden, Wash. *Eve. Star*, 6/20/66, 1; Eisen, *Wash. Post*, 6/26/66, B1, B7)

June 17: U.S.S.R. launched COSMOS CXXI for continued space research into orbit with 354-km. (220-mi.) apogee, 210-km. (130-mi.) perigee, 72.9° inclination, and 98.9-min. period. Instruments were functioning normally. (Tass, 6/17/66)

- GSFC announced completion of OGO III checkout program. The 1,135-lb. spacecraft, orbited June 6, was "operating well" and 21 experiments onboard had recorded and transmitted more than 3,000 hrs. of data. Cause of continuing problem in receiving signals from range and range-rate tracking beacon—a method for precise determination of position in orbit—was under study, but use of alternate tracking system prevented serious effect on OGO III scientific mission. (GSFC Release; AP, *NYT*, 6/18/66, 12)

- At MSC, NASA Deputy Administrator Dr. Robert C. Seamans, Jr., presented NASA Exceptional Service Medals to GEMINI IX-A Astronauts Thomas P. Stafford and Eugene A. Cernan and to Richard E. Dineen (Col., USAF) who had helped solve problems in the Titan II booster that had launched the spacecraft. Referring to difficulties that occurred during the June 3-6 GEMINI IX-A flight, Seamans said: "We know that if we had been a little sharper, if we had worked a little harder, that this would not have happened.... We are... dissatisfied with ourselves when technical and procedural failures keep us from our goals." He added he had requested review of plans for Gemini X, XI, and XII missions for "tightening up" of procedures. Summarizing Project Gemini's results to date—"at the three quarter mark"—and forecasting its probable total impact on the space effort, Seamans said: "I am convinced that this program will be completed with far greater return to the nation in scientific information, technology, and operational experience than we had originally set out to accomplish when we began the program in 1961." (*Wash. Post*, 6/18/66, A2; Waldron, *NYT*, 6/18/66, 10; Hines, *Wash. Eve. Star*, 6/20/66, A10)

- GEMINI IX-A Astronauts Thomas P. Stafford and Eugene A. Cernan held their first post-mission press conference. Cernan said that during his space walk he could not keep his feet from floating, was unable to use both hands while trying to don backpack containing maneuvering unit, and "had to work continually against the pressure suit.... I was devoting 50 per cent of my work load just to maintain position." This extra effort saturated the environmental control system and led to fogging of his faceplate so that he could not test the maneuvering unit. About the same time, he felt his back becoming "extremely hot." It was later revealed some of the insulation had been ripped from his spacesuit. After recounting his adventures, Cernan said: "I found no new voodoo out there.... You're at home out there ... I'm convinced we can do just about anything we want to do." (*Wash. Post*, 6/18/66, A2; *NYT*, 6/18/66, A10)

- U.S.S.R. made public its draft treaty for preserving universe for peaceful exploration and research in interest of all mankind submitted to U.N. June 16. One difference from U.S. treaty proposal, also submitted to U.N. June 16, concerned settlement of disputes: Soviet plan called for consultation and negotiation among parties involved; U.S. proposal would refer disputes to International Court of Justice. Platon D. Morozov, acting head of Soviet delegation to U.N., asked to have draft circulated as document for inclusion on agenda of 21st session of General Assembly, to open Sept. 20. (Daniell, *NYT*, 6/18/66, 10)

June 17: Gemini XII crew was named. Prime crew: James A. Lovell, Jr., command pilot; Edwin E. Aldrin, Jr., pilot. Backup crew: L. Gordon Cooper, Jr., command pilot; Eugene A. Cernan, pilot. Mission was scheduled for late October or early November 1966. (AP, *NYT*, 6/18/66, 10; AP, Wash. *Eve. Star*, 6/18/66, A2)

- Praising June 16 flight of Titan III-C booster which injected eight satellites into orbit, Secretary of the Air Force Dr. Harold Brown told Pentagon news conference that program cost included $33 million for satellites, $45 million for ground stations, and $10 to $20 million for other system elements. Launch costs were excluded since they were covered by Titan III-C development funding. Satellites were resistant to jamming or interference, but "anyone who puts up an antenna" of the right kind could listen in. What they would hear "may or may not be comprehensible to them" since traffic would be coded. Asked if U.S.S.R. could use satellite system, Dr. Brown replied affirmatively. However, DOD Deputy Director of Defense Research and Engineering Dr. Thomas F. Rogers stated: "We have arranged so that it would be very difficult for them to have assured use of it." (Transcript)

- U.K.'s Cable & Wireless, Ltd., would build three ground stations—at Hong Kong, Bahrein Island, and an undisclosed site—for satellite communications, reported the *Wall Street Journal*. Hong Kong station would cost $7 million and would be operational by 1968. Seventeen firms—including seven U.S. companies—had been asked to bid on the project. (*WSJ*, 6/17/66, 12)

- President Johnson signed into law Marine Resources and Engineering Development Act of 1966, establishing national goals for an expanding national oceanographic effort. Legislation, sponsored by Sen. Warren G. Magnuson (D-Wash.), provided for Cabinet-level National Council of Marine Resources and Engineering Development—and administrative and coordinating organization similar to NASC. Chairman would be Vice President. Bill also provided for 15-member Marine Science, Engineering and Resources Commission to conduct 18-mo. study of U.S. oceanographic capabilities and recommend Government organizational plan for oceanography. (*CR*, 6/2/66, 11490-92; *Wash. Post*, 6/19/66, A17)

June 17-18: Optical identification of x-ray source Sco X-1 was made at Tokyo Observatory. Photoelectric and spectroscopic observations continued at Mt. Wilson and Mt. Palomar Observatories through July; data indicated source might be "an uncatalogued old nova." Optical search had been based on information obtained from NASA-American Science and Engineering, Inc.-MIT experiment flown on Aerobee 150 sounding rocket launched March 8 from WSMR. (*Astrophys. J.*, 10/66, 316-21)

June 18: Discussing the "cost to land men on the moon" in *Journal of the Armed Forces*, James J. Haggerty, Jr., said late NASA Deputy Administrator Dr. Hugh L. Dryden, "without much detailed information on which to base an assessment," had estimated $20 billion. Now that work on all major contracts was well advanced, NASA had made new estimate that validated original figure: total cost would be $22.7 billion with possibility of adjustment that would bring it down almost to $20 billion. (Haggerty, *J/Armed Forces*, 6/18/66, 8)

- Development of new solid-fuel improved capability missile (Icm) that could "penetrate even the tightest Soviet defense" was being urged by USAF, Pentagon sources revealed. Missile would also be able to propel several times the payload of present Minuteman and Polaris missiles toward targets in U.S.S.R., could be perfected in about five years, and would ensure U.S. strike capability even after surprise attack. (Beecher, *NYT*, 6/20/66, 1)

June 19: TIROS VII meteorological satellite had operated successfully in orbit three years, traveled more than 445,000,000 mi., and had taken 123,000 pictures of cloud formations, storms, hurricanes, and typhoons with two wide-angle vidicon TV cameras. Orbited from AMR (now ETR) June 19, 1963, it tracked major hurricanes and provided information aiding Ranger, Mariner, and Gemini missions. (AP, Balt. *Sun*, 6/19/66

- Calculations by computer at Astronomisches Rechen Institut, Heidelberg, Germany, and Yale Univ. Observatory had helped astronomers rediscover Tempel-Tuttle Comet responsible for Leonid meteor showers each November. Comet had been found 100 yrs. ago and not seen again until summer 1965. Meteor shower was expected to be spectacular in November 1966; it would occur close to time of comet's return to vicinity of earth. (Sci. Serv., *NYT*, 6/19/66, 26)
- Discussing military satellite as "necessity, not adventure," article in *Houston Post* said DOD had reoriented space research toward "development of a many-purpose military satellite." DOD Deputy Director of Office of Defense Research and Engineering Daniel J. Fink was quoted as saying proposed vehicle could: (1) locate military units precisely; (2) speed military communications; (3) carry weather observation and reporting equipment; (4) furnish early warning against nuclear attack; and (5) detect unannounced nuclear tests. (*Houston Post*, 6/19/66)
- NASA's "new policy of calling a spade a spade—and a failure a failure—in discussing the results of its manned operations in space" was noted by William Hines in the Washington *Evening Star*: "NASA has been subjected to harsh criticism—some of it bordering on ridicule—for the past disinclination of its officials to admit . . . that things occasionally had gone wrong. . . . Seamans' outspoken evaluation [June 17] of the Gemini 9A flight was regarded here as a signal to other NASA officials to lay the facts on the line when things do not go exactly according to plan." (Hines, Wash. *Eve. Star*, 6/20/66, A10)
- Stages in development of facilities at KSC for Saturn/Apollo launches were described in *Aviation Week* editorial by Robert Hotz. Now that the mission was "moving so close to realization," he said, bold and far-reaching decisions were needed "to insure in the post-Apollo decade the full utilization of these Apollo launch facilities and the superb technical team that operates them." Citing "dramatic progress" made by U.S. in first decade of space age, Hotz added: ". . . it is hard to understand why the highest level of national leadership now shrinks from making the decisions today that are necessary to insure this nation's capability for tomorrow." (Hotz, *Av. Wk.*, 6/20/66, 21)
- Martin Marietta Corp. was receiving $7,622,000 fixed-price contract from AFSC for work on Titan III booster. (DOD Release 536-66)
- Proposed compromise by ELDO reducing U.K.'s share of financing was reported in *Aviation Week*. If accepted at July 7-8 ELDO ministerial conference in Paris, financial participation was expected to be: U.K.—27% of future ELDO costs, a reduction from the 38.79% initially agreed upon in 1962-63; West Germany—27%, increased from 22.01%; France, 25%, increased from 23.93%; Italy, 12%, increased from 9.78%; Netherlands and Belgium would share 9%, increased from Belgium's 2.85% and Netherlands' 2.64%. (*Av. Wk.*, 6/20/66, 30)

June 21: NASA awarded Philco Corp. $3.5-million contract for four tri-axis magnetometers and supporting equipment to be left on moon by Apollo astronauts. One of seven geophysical instruments to be carried in Alsep (Apollo Lunar Surface Experiments Package), 14.5-lb. magnetometer would determine presence or absence of lunar magnetic field, measure magnetic penetrability of moon, study interplanetary magnetic field and its diffusion to moon, and detect electrical currents in solar wind above lunar surface. (ARC Release 66-9)

- USAF launched unidentified satellite with Thor-Agena D rocket from WTR. (*U.S. Aeron. & Space Act., 1966,* 153)
- ComSatCorp president Dr. Joseph V. Charyk, speaking at National Space Club luncheon in Washington, D.C., disclosed plans for a $7- to $9-million laboratory to be built by ComSatCorp near Washington, D.C. Facility would employ about 200 people and would be completed by mid-1968. (Clark, *NYT,* 6/22/66, 12)
- $3,000 block of magnesium used as vibration test fixture for Lunar Excursion Module's propellant-gaging system brought $28.14 in Los Angeles junk yard. Two trash collectors had diverted block from Apollo project into their disposal truck while emptying trash cans at Giannini Corp., Duarte, Calif. They were booked on suspicion of grand larceny. (UPI, *Wash. Post,* 6/24/66, A6)
- Development of chemical laser requiring no external energy was advancing so rapidly it would "probably be achieved within a year or two." Prediction was made by Dr. George C. Pimental, professor of chemistry at Univ. of California, Berkeley, at AFOSR-sponsored seminar on "Science Frontiers" at Albuquerque, N. Mex. (Sullivan, *NYT,* 6/21/66, 12)

June 22: U.S. and U.S.S.R.—which had submitted to U.N. (June 16) similar drafts for a treaty banning military use of celestial bodies—agreed to meeting of legal experts of Committee on the Peaceful Uses of Outer Space in Geneva July 12 to arrange compromise between drafts. U.S.S.R. had originally asked that proposals be included in agenda of 21st session of U.N. General Assembly, which would open Sept. 20. (Daniell, *NYT,* 6/23/66, 1)

- Two Nike-Cajun sounding rocket launches, from Point Barrow, Alaska, and Churchill Research Range, were coordinated in GSFC experiment to obtain atmospheric data as time of summer maximum for noctilucent cloud sightings approached. Rockets and instrumentation functioned satisfactorily. (NASA Rpt. SRL)
- Design requirements for second-generation, recoverable-reusable booster and spacecraft systems, outlined by NASC staff member Dr. Eugene B. Konecci at American Astronautical Society meeting in Wayne, N.J., included: new propulsion capability; materials, structures, and subsystems suitable for reuse; and system designed for "full use of the human factor capabilities in the control loop, . . . minimum (limited) ground support participation, and for easy maintenance and repair." (Text)
- President Charles de Gaulle, speaking at Moscow State Univ., suggested that France and the U.S.S.R. should work toward a "new alliance" for scientific and other intellectual pursuits. He said that "culture, science, and progress justify national ambitions in our epoch in place of former dreams of conquest and domination." (Grose, *NYT,* 6/23/66, 1)
- Generator-burner assembly of the LORHO pilot facility was successfully run at full power at Arnold Engineering Development Center, marking

first application of magnetohydrodynamic principles in achieving high-velocity flow in a ground environmental test facility. (AEDC)

June 22: ComSatCorp denied ITT World Communications, Inc.'s accusation of "unjustly and unreasonably" discriminating against ITT in quoting charges to Defense Communications Agency (DCA) for 30 satellite circuits between Hawaii and Japan, the Philippines and Thailand: "The rate quoted to the DCA for each of the . . . circuits is, in fact, higher than the rate quoted to the carriers, as it takes into account the additional service costs by Comsat to carry out a responsibility to provide end-to-end service to DCA. . . . ITT Worldcom may have been confused by advance knowledge as to the Corporation's planned tariff for satellite channels to be furnished to authorized common carriers, in which the $48,000 charge to the carriers would be revised downward to $45,000 [and] erroneously assumed that this lower charge, which would apply only to authorized common carriers, had been quoted by Comsat to the DCA." (ComSatCorp Release)

June 23: NASA's PAGEOS I (Pageos A Passive Geodetic Earth-Orbiting Satellite) was launched with Thrust-Augmented Thor-Agena D booster from WTR into near-circular polar orbit: 2,645-mi. (4,258-km.) apogee, 2,603-mi. (4,191-km.) perigee, 182-min. period, and 87° inclination. PAGEOS I was an aluminum-coated Mylar balloon of the ECHO I type. Balloon was folded and packaged inside spherical canister ejected into space as launch vehicle reached orbital velocity. Canister then separated in half by explosive device, and folded satellite automatically inflated to 100-ft.-dia. sphere. PAGEOS I carried no instruments. By reflecting sunlight, satellite would provide orbiting point source of light to be photographed over projected five-year period to determine size and shape of earth to a degree never before possible; in orbit, satellite should be as bright as star Polaris. Simultaneous observations from 41 portable camera stations around world would be used to construct three-dimensional geodetic reference system. Resulting satellite triangulation network would make it possible to obtain distance between two surface points on earth 3,000 mi. apart to an accuracy of 32 ft. Project was managed by LaRC. (NASA Release 66-150; ESSA Release 66-32; NASA Proj. Off.; LaRC *Researcher*, 7/1/66, 1, 6; *Wash. Post*, 6/24/66, A5)

- NRX-A5 nuclear reactor—operated at full power June 8 at Jackass Flats, Nev.—was restarted and operated an additional 14½ min. at design power of 1,100 mw (55,000 lbs. thrust). Tested by Aerojet-General and Westinghouse Corp. under joint NASA-AEC program, reactor consumed 218,000 gal. liquid hydrogen. (AEC Release J-160)

- NASA would negotiate six-month study contracts totaling $200,000 with General Electric Co. and RCA to determine "feasibility of a satellite capable of broadcasting directly to conventional home FM radio and/or short wave radios." Contracts would be managed by NASA OSSA. (NASA Release 66-161)

- NASA and Vocational Rehabilitation Administration signed agreement "to accelerate the application of new technology to the problems of the disabled" by making results of aerospace research available for application to rehabilitation "on a continuing basis." This would involve evaluation of NASA technology by VRA scientists; adaptive engineering to apply that technology to needs of disabled persons; demonstration of resulting devices, systems, and procedures; and commercial introduction of new devices and services. (NASA Release 66-160)

June 23: House Committee on Science and Astronautics reported favorably H.R. 14832 incorporating changes and improvements in organization and operation of NSF recommended in Subcommittee on Science, Research, and Development report, "The National Science Foundation—Its Present and Future." (House Rpt. 1650)
- FCC ruled that users of satellite communications must operate through common carriers instead of dealing directly with ComSatCorp to prevent "serious adverse effects upon the well-being of the commercial telecommunications industry and the general public it serves." Ruling, which resolved controversy between ComSatCorp and common carriers, would protect carriers from reduced traffic volume resulting from direct user-ComSatCorp relations. (FCC; *WSJ*, 6/24/66, 6)

June 24: NASA awarded Univac $30-million, five-year contract to provide "new computing system" for MSFC beginning in 1967. Computer Sciences Corp. was selected to provide support services for MSFC Computation Lab. under $5.5-million, cost-plus-award-fee contract with provisions for four one-year extensions. Services would include computer operation, maintenance, and programing. (NASA Releases 66-164, 165)
- NASA had awarded $22-million renewal contract to TWA for continued support services at KSC. Two contracting methods would be followed: cost-plus-award-fee for supply operations and general maintenance; and fixed-price for remaining services. (NASA Release 66-166)
- NASA named L/Col. Robert A. Rushworth (USAF) to receive Exceptional Service Medal for his "outstanding contributions" to U.S. aeronautical research programs "both as a pilot and as an engineer" on X-15 rocket aircraft. Col. Rushworth, who would leave X-15 program in summer 1966 to attend Armed Forces Staff College, would be presented medal at NASA awards ceremony in October. (NASA Release 66-159)
- USAF space launch crews were being familiarized with new low-cost Burner II upper-stage booster developed by Boeing to inject payloads into orbit and then orient the payload accurately. Small, guided, solid-fuel stage, scheduled for operation in late 1966, would "economically bridge the payload gap between the DOD-NASA Scout launch vehicle and the Delta and Agena upper stages." (Boeing Release)
- Potential impact of aeronautical technology on social patterns of the nation was discussed by Dr. Raymond L. Bisplinghoff, chairman of MIT's aeronautics and astronautics department, at meeting of Aviation/Space Writers' Assn. in New York City. In only the 10th year of the space age, he said, the world was ". . . at the beginning of a new surge in science and technology stimulated by their interactions with each other and with social needs, nourished by the resources and needs of space exploration." (Text; *AF/SD*, 8/66)

June 25: France's President Charles de Gaulle watched launch of COSMOS CXXII unmanned satellite from Baikonur cosmodrome, Kazakhstan, into 625-km. (388-mi.) altitude circular orbit with 65° inclination and 97.1-min. period. Tass said all instruments aboard satellite were functioning normally. De Gaulle was first Westerner to witness a Soviet launching and to visit Baikonur space center. (Tass, 6/25/66; Tanner, *NYT*, 6/26/66, 1; *Wash. Post*, 6/26/66, A20; AP, Wash. *Sun. Star*, 6/26/66, A8)
- NASA Administrator James E. Webb told National Conference of Lieutenant Governors in Cleveland that new satellite weather reporting system

meant "you say to every other nation the United States is developing this power technique of space, not to get power over you, but to develop power together with you over the limitations of nature." (AP, *Miami News*, 6/26/66)

June 25: First full-disc photograph of earth, taken from U.S.S.R.'s third MOLNIYA I, had been received from Soviet Novosti Press Agency by Donald Zahner of St. Louis, publisher of *Review of Popular Astronomy*. Picture, transmitted May 30 and authenticated by U.S. Government officials and by Dr. Gerard P. Kuiper, director of Univ. of Arizona's Lunar and Planetary Laboratory, would appear in July-August issue of magazine. Zahner said Soviet officials had probably sent photograph out of gratitude for article published by magazine in spring 1966 on 100th anniversary of Pulkovo Observatory. (*Tech. Wk.*, 7/11/66, 12; AP, *Balt. Sun*, 6/26/66)

June 26: New National Oceanography Assn. (NOA) "to raise the level of public awareness of what impact oceans will have on future generations" became operative. In first press release, NOA said it would promote "opportunities—for national advancement, for profit, for pleasure, and for meeting basic human needs—which lie just across the threshold of the ocean." It said U.S. lagged behind U.S.S.R., Japan, and other nations in ocean exploration and that a "high priority, fullscale national oceanography program" was needed to catch up. (NOA Release)

- Siberian-born scientist Mikhail K. Yangel, elected to Communist Party's Central Committee in April and to Supreme Soviet June 12, had been tentatively identified as new head of U.S.S.R.'s space program. Several other scientists known or believed to be associated with space program had been selected to Supreme Soviet for first time; however, since Yangel was only one also to occupy seat in Central Committee, this was taken as indication he occupied top position. (Shabad, *NYT*, 6/26/66, 3)

June 27: U.S.S.R. would continue intensive exploration of space in 1966-70 period, Soviet Academy of Sciences president Mstislav V. Keldysh reported to general meeting of Academy. He listed these general programs: manned space flights; exploration of physical conditions on moon and nearest planets; and development of use of satellites and rockets for long-distance communications systems. (Tass, 6/28/66)

- Rep. Hale Boggs (D-La.) told House: ". . . no businessman would invest tens of thousands of dollars in a locomotive and then allow it to rust in the yard for lack of a $5 part. Neither can we invest tens of billions in a space program and leave it to falter for the lack of funds." Calling the present budget "austere," Boggs warned that in future budgets we could not put important elements of our capability into mothballs. "We must use it or see it rust." (*CR*, 6/27/66, 13707-08)

- NASA announced conversion of contract with Douglas Aircraft Co. for development of Saturn launch vehicle's S-IVB stage from cost-plus-fixed-fee to cost-plus-incentive-fee. Under revised contract, company's fee would be increased or decreased depending on "attainment of the incentive for cost, schedule and performance." Estimated cost was $700 million, plus fee. (MSFC Release 66-141)

- Lockheed Aircraft Corp. displayed 273-ft., full-scale mockup of its 1,800-mph revised supersonic transport model to be submitted to FAA in Sst competition with Boeing Co. Sept. 6. Made primarily of titanium and scheduled for first flight by 1970, the 266-passenger Lockheed 2000 would have double-delta wing to provide "outstanding flight handling characteristics at all speeds" and to permit cruising at 7,000-ft. altitude, and "weather vision" nose which could move down 15° from supersonic cruise position to give pilot visibility during takeoff, landing, and subsonic flight. Boeing design—revealed June 15—had featured variable wing swept forward for takeoffs, landings, and subsonic flight and back for supersonic cruise. (Lockheed Release)

June 27: June 8 loss of XB-70 experimental aircraft would delay NASA-DOD Sst research program for two or three months, Hal Taylor reported in *Technology Week* following interview with Charles Harper, Director, NASA Office of Aeronautical Research, OART. (*Tech. Wk.*, 6/27/66, 18)

- NASA-sponsored space science course—including lectures, demonstrations, and experiments—began at (Washington) D.C. Teachers College to help local elementary school teachers understand space-age fundamentals and interpret them to their pupils. (*Wash. Eve. Star*, 6/26/66, D9)

- Physicists at Columbia Univ., under the direction of Dr. Paolo Franzini, and at State Univ. of New York at Stony Brook, directed by Dr. Juliet Lee-Franzini, published experimental evidence that would disprove theory of charge conjunction invariance—theory that oppositely charged particles behave symmetrically. Photographic analysis of decay of eta meson, a neutral nuclear particle, into three pions—one positively charged, one negatively charged, and one neutral—indicated that positive pion traveled away from decomposed eta meson at a greater velocity than negative pion. Results, based on 1,441 photographs, Dr. Franzini said, had "far-reaching consequences regarding antiparticles" and provided means of identifying "whether a particle or body anywhere in the universe is positively or negatively charged." Experiment —conducted in Columbia's Nevis Laboratory, Brookhaven National Laboratory, and Stony Brook—was reported in *Physical Review Letters*. (*Phys. Rev. Lett.*, 6/27/66, 1224; AIP Release, 6/26/66)

- Question of "garbage problem" in space was raised during recent House Appropriations Committee hearings on Spacetrack, which maintained surveillance on over 1,800 orbiting objects, Drew Pearson reported in *Washington Post*. Assistant Secretary of the Air Force Alexander H. Flax replied there was no danger of collision "because of the tremendous amount of volume that these 1,800 objects are disposed in. It is like the problem of stepping on a flea in Grand Central Station. It could happen, but it is highly unlikely." (Pearson, *Wash. Post*, 6/27/66)

- AFSC had awarded Bell Aerospace Corp. a one-year study contract to develop Dual-purpose Maneuvering Unit (Dmu) that could be worn by astronaut or operated by remote radio and TV signals. Dmu would combine "best features" of Astronaut Maneuvering Unit (Amu)— which Astronaut Eugene Cernan was unable to test during GEMINI IX-A flight—and Remote Maneuvering Unit (Rmu)—which had been tested in laboratory. (AFSC Release 116.66)

- American Astronautical Society had granted following awards for achievements in 1965: W. Raymond Lovelace Award to Dr. Jeanette Ridlon Piccard "for past experiments in ballooning and continued contributions to space progress"; AAS Space Flight Award (posthumously) to Dr. Hugh L. Dryden "for outstanding technical and administrative leadership"; Melbourne W. Boynton Award to Dr. Charles A. Berry "for managing successfully medical program for astronauts"; AAS Flight Achievement

Award to Astronauts Walter M. Schirra, Jr., Thomas P. Stafford, Frank Borman, and James A. Lovell "for dual outstanding achievements for first successful in-plane rendezvous (GEMINI VI) and sustained orbit operations (GEMINI VII)"; Victor A. Prather Award to Richard Johnston "for outstanding contributions to design and technology of spacesuits for astronauts." (*Av. Wk.*, 6/27/66, 19)

June 27: Pointing out that "space triumphs" had made "secrecy and the hoarding of scientific data less worthwhile," editorial in the Baltimore *Sun* concluded: ". . . it is not surprising that, despite the relationship between military defense and some space activity, Russia and France seem about to agree to cooperate in space research—and the United States and Russia likewise." (Balt. *Sun*, 6/27/66)

- Several small solid motors had been test-fired during "last three months" at Hercules, Inc.'s Bacchus, Utah, facilities under classified program; highest impulse recorded in Free World was reported to have been reached. (*Tech. Wk.*, 6/27/66, 3)

June 28: Analysis of data from aborted March 16 GEMINI VIII mission had indicated three of 10 experiments were successfully conducted and a fourth might later yield data, NASA announced. Bioassays of body fluids had provided data on flight crew response to spaceflight; frog-egg growth experiment had indicated fertilized frog eggs divide normally even under conditions of weightlessness; nuclear emulsion experiment had functioned correctly and had completed 17% of its planned schedule when flight was terminated. Fourth experiment was micrometeoroid collection device aboard GATV; exposed surfaces were expected to yield useful data if they could be retrieved from GATV during Gemini X mission scheduled for no earlier than July 18. Six experiments had had to be canceled when GEMINI VIII mission was terminated after electrical short circuit in spacecraft caused continuous firing of roll thruster. (NASA Release 66-163)

- Nike-Cajun meteorological sounding rocket launched from Churchill Research Range carried 19-grenade payload in GSFC experiment coordinated with launch from Point Barrow, Alaska, to study atmospheric parameters of wind, temperature, pressure, and density as summer progressed toward time of maximum noctilucent cloud sightings. Rocket and instrumentation performed satisfactorily. (NASA Rpt. SRL)

- Pratt & Whitney Div. announced it had successfully tested second version of engine being built for U.S. supersonic transport program. Engine developed more than 600,000 lbs. thrust at West Palm Beach Facility. (*WSJ*, 6/28/66, 9)

- U.S.S.R. was encountering difficulties in development of its supersonic jet aircraft so that "on the present showing it would seem that the Concorde will not only take to the air before the Tu-144 but will be flying commercially earlier, too," experts at Radio Liberty reported. (DJNS, Wash. *Eve. Star*, 6/28/66)

June 28-29: NASA's SURVEYOR I, which had softlanded on moon June 2, did not respond to daily 3½-hr. attempt by JPL to reactivate its instruments after two-week lunar night. JPL spokesman said: "It may or may not be dead. Variations in the moon's temperature might have caused Surveyor's radio frequency to change." Beginning June 30, JPL's Deep Space Facilities, Goldstone, Calif., would try nightly to contact spacecraft using preplanned sequence of commands. If response were received, SURVEYOR I's television camera would be commanded to take one picture of spacecraft's footpad to verify camera was still operational and preparations would be made to resume full program of transmission of lunar pictures. (AP, Wash. *Eve. Star*, 6/29/66, A23; UPI, *Wash. Post*, 6/29/66, A9; UPI, Phil. *Eve. Bull.*, 6/30/66)

June 29: First nuclear power generator in space marked fifth anniversary. Developed by AEC to supplement solar power on USN's 175-lb. experimental navigational satellite TRANSIT IV-A, 5-lb. Snap-3 generator had traveled 724,000,000 mi. (over 25,000 times around earth). TRANSIT IV-A, launched June 29, 1961, from Cape Canaveral, became oldest operating U.S. satellite in May 1964. (AEC Release J-159)

- Senate passed and cleared for President's signature H.R. 6125, which would convert National Air Museum to National Air and Space Museum with NASA representation on advisory board. Bill authorized construction of museum building. (*CR*, 6/29/66, 14041-2)
- Senate-House conference committee approved $17.5 billion DOD FY 1967 authorization bill—$377.3 million less than House version and $310.7 million more than Senate version. Approved budget was $553.8 million more than DOD request. Bill included $137.9 million increase in R&D funds over DOD budget request but House proposal to add $80 million to MOL program was cut to $50-million increase. R&D funds also included $14.4 million for Condor medium-range guided missile; $11.8 million for Amsa (Advanced Manned Strategic Aircraft) follow-on bomber; and $130.5 million for nuclear guided missile frigate authorized in FY 1965. (*Av.Wk.*, 7/4/66, 19; *NYT*, 6/30/66, 13)
- S-IB stage for sixth flight Uprated Saturn I booster (S-IB-6), powered by eight Rocketdyne H-1 engines, was successfully captive-fired at MSFC for 145 sec., producing 1.6 million lbs. thrust. (MSFC Release 66-124)
- Hughes Aircraft Corp. received $280-million contract from NATO Air Defense Ground Environment Project to build integrated early-warning and weapon-control system—consisting of radar, data-handling, and communications equipment—extending from Norway to Turkey. (*WSJ*, 6/30/66, 4; Hughes)

June 30: On the floor of the House, Rep. Chet Holifield (D-Calif.) discussed Committee changes made in Dept. of Transportation bill, H.R. 13200. A "clean bill," H.R. 15963, was introduced giving proposed Secretary of department responsibility for aviation and making Federal Aviation Administration (new name for FAA) a division of Dept. of Transportation. Civil Aeronautics Board aircraft accident investigation unit would also be under Secretary of Transportation. (*CR*, 6/30/66, 14272-5)

- Soviet Foreign Minister Andrey Gromyko and French Foreign Minister Maurice Couve de Murville signed agreement in Moscow providing basic framework for space cooperation during next ten years. Major objectives of compact—which initially scheduled several common experiments, including an astronomy test to be conducted from Archangel, U.S.S.R., and Kerguélen, French islands in the Indian Ocean —were missions with Imp-type satellites and Lunar Orbiter spacecraft. (*SBD*, 6/28/66, 344)
- NASA selected Northrop Corp. for final negotiations on $5-million contract for design, development, fabrication, and testing of "engineering mechanics subsystems for Mariner Mars '69 spacecraft." Contract would be managed by JPL. (NASA Release 66-170)
- Sens. Carl T. Curtis (R-Neb.) and Jennings Randolph (D-W.Va.) urged that appropriations not be sought for construction of National Air and

Space Museum—authorized by H.R. 6125 passed June 29—until the war in Vietnam ended or its expenditures decreased. (*CR*, 6/30/66, 14112-14)

June 30: NASA might not in future be able to order aircraft carriers and large ships for recovery of astronauts, *Chicago Tribune* sources reported. USN felt vessels could be put to better use to support or relieve forces in Vietnam. USAF had proposed to NASA use of new long-range helicopters. (*Chic. Trib.*, 7/1/66)

- NASA tracking station at East Grand Forks, Minn., ceased operation after five years of space tracking. Advances in tracking technology and orbit determination had made site no longer essential for support of scientific satellites; some equipment would be sent to other stations. (NASA Release 66-117)
- USN's Pacific Missile Range reported 2,608 live missile firings in record total of 13,155 operations during FY 1966. (*SBD*, 7/19/66, 93)

During June: Lurain as seen by SURVEYOR I Science Evaluation Analysis Team after preliminary analysis of data was a dark, relatively smooth, gently rolling plain "the consistency of a freshly turned field," studded with craters ranging in diameter from one inch to several hundreds of feet and littered with fragmental debris ranging in size from few hundredths of an inch to more than three feet. Dr. Eugene Shoemaker, U.S. Geological Survey, said terrestrial scenery most closely resembling the lurain as seen by SURVEYOR I would be Sedan crater at AEC's Nevada Test site—large crater covered with small craters and "secondary ejecta" resulting from explosion of subsurface nuclear device. Lurain appeared to be composed of soil-like substance, or a fine sand, with rougher material through it. It appeared to be about three feet deep with harder surface beneath. Surface had dynamic bearing strength of 6-10 psi. Leonard Jaffe, NASA Project Surveyor scientist, noted that results of SURVEYOR I could be summarized by list of previous theories on nature of lurain that had been disproved: surface was not composed of hard rock; it was not composed of one piece of very porous rock; it was not covered by a layer of loose dust either thick or thin; it was a finely granulated material with particles that cohered to each other but did not seem particularly attracted to spacecraft; it was heavy enough for man to walk and work on and for a properly designed spacecraft to land on; and it reflected accurate radar signal from surface, not from subsurface substance. Jaffe noted there were a few hazards to manned spacecraft, mostly from large rocks and possibility of sinkage, but emphasized data so far indicated surface was adequate for landing Apollo Lem.

There appeared to be two different types of rocks near SURVEYOR I: one with pitted, spongy surface that appeared to be caused by molten rock that had cooled and from which gas had escaped; the other harder, smoother surfaced, and smaller grained.

SURVEYOR I photographs were considered superior to U.S.S.R.'s LUNA IX results because (1) camera was higher so that horizon was 10 times farther away; (2) camera resolution was greater (down to 1/50th of an inch); camera worked longer and took more photos under different light angles. SURVEYOR I's pictures of its own footpad had yielded data on surface and subsurface composition. (Clark, *NYT*, 6/8/66, 1; 6/17/66, 15; Hines, Wash. *Eve. Star*, 6/17/66, A4; Sullivan, *NYT*, 6/19/66, 14F; *SBD*, 6/20/66, 297)

During June: NASA Deputy Administrator Robert C. Seamans, Jr., writing in *Astronautics & Aeronautics*, outlined choices in "potential manned-spaceflight missions of the next generation" and argued for taking up the preliminary options now. He selected what he called "three natural, potential targets" for major extensions of manned space flight. First of these he saw as "a spacefaring research and operations center in earth orbit," described as a "large, permanent, manned space facility" that would "make possible whole new ranges of experiments, investigations, and operations." Second choice was "the moon itself... following up the initial exploratory landings with a permanent lunar base." Third objective could be "manned surface exploration of Mars" within context of "extension of the lunar-landing experience to the near planets." Citing both advantages and difficulties of the latter, he said: "Such an effort would represent a greater national commitment of far longer duration than our commitment in 1961 to a manned lunar landing, and would generate proportionately increased industrial, university, and government activities. Significant advances in virtually every field of space-related technology would be required. The increase in scientific knowledge from direct, firsthand observation and experimentation on another planet, however, has not yet been measured against the direct cost of its achievement, or against the more subtle cost of longterm dedication to a given course of action, with its inherent loss of national flexibility in allocation of the limiting resources—technical and scientific manpower." (*A&A*, 6/66, 30-33)

- Measurement of angular size of x-ray source Sco X-1 [see March 8, May 23, June 17-18] was reported by Herbert Gursky, Riccardo Giacconi, Paul Gorenstein, and John R. Waters of American Science and Engineering, Inc. (ASE) and MIT's Minoru Oda, Hale Bradt, Gordon Garmire, and B. V. Sreekantan in *Astrophysical Journal*. Accompanying article by ASE's Oscar P. Manley suggested source was a protostar; Giacconi group said source, if observable, would appear in visible light as starlike object. (*Astrophys. J.*, 6/66, 1249, 1253)

- AIP Center for History and Philosophy of Physics received grant from Ford Foundation "for a long range planning study to determine the future level and direction of the Center's growing activities in the history, philosophy and sociology of twentieth century physics." (AIP Center, *Newsletter*, 6/66)

- Purposes of Saturn/Apollo Applications program, M/G David M. Jones (USAF), NASA Deputy Associate Administrator for Manned Space Flight, said in an interview in *Data* magazine, were to serve as transition program from Apollo to next space goal and to accomplish "significant scientific experiments."

  Apollo Program Director M/G Samuel C. Phillips (USAF) said in an interview in the same issue that "barring some major setback, some major failure in a static test on the ground or major failure on a flight, it is reasonable to expect that the lunar landing... will be accomplished before the end of 1969." (*Data*, 6/66, 27, 32)

- Univ. of Toronto Chancellor O. M. Solandt was named chairman of newly created Science Council of Canada; Univ. of Montreal Rector Roger Gaudry was named Vice Chairman. (NAS-NRC-NAE *News Report*, 6-7/66,8)

- More than 70 tethered balloon ascensions to 1,500-ft. altitude for vehicle drop tests had been conducted by AFCRL at Holloman AFB over past three years to test Surveyor spacecraft's retrofire descent and impact capabilities, OAR *Research Review* reported. (OAR *Res. Rev.*, 6/66, 17)

During June: Dr. Dorothy Martin Simon, Avco Corp. vice president and director of Corporate Research, received 1966 Achievement Award of Society of Women Engineers. She was cited for "significant contributions to engineering administration and space engineering, especially in the fields of combustion and ablative coatings." (*Tech. Wk.*, 6/27/66, 46)

- Star RU Camelopardalis, that four years ago varied its light output rhythmically every 22 days between visual magnitudes 8.2 and 9.1, had become constant in brightness at magnitude of 8.5—too faint to be seen with naked eye. Cause of change, reported in *Sky and Telescope*, was unknown. (Sci. Serv., *NYT*, 7/4/66, 32)
- Ten-month survey was reported by Cal Tech radioastronomy research fellow Dr. John D. Wyndham. He found that at least a fourth of the radio sources reported beyond this galaxy belonged in the quasar category. If a fourth of the thousands of radio sources beyond the Milky Way are quasars, the latter must be very abundant, the report said. (Henry, *Wash. Eve. Star*, 6/29/66, A26)
- U.S. aerospace exports in 1965 totaled $1,474,000,000—up 21.6 per cent from 1964—Aerospace Industries Assn. reported. (*WSJ*, 6/27/66, 5)

July 1966

July 1: NASA successfully launched EXPLORER XXXIII (Imp-D) Anchored Interplanetary Monitoring Platform (Aimp) from ETR using Thrust-Augmented Delta booster with FW-4S 3rd stage to study interplanetary radiation and magnetic fields. Because tracking data indicated at 3:30 GET spacecraft was in excess-energy orbit, alternate mission plan was put into effect at 4:30 GET: retrorocket was fired, injecting EXPLORER XXXIII into elliptical earth orbit with 278,990-mi. (449,174-km.) apogee; 18,975-mi. (30,550-km.) perigee; 13.5-day period; 28.9° inclination. Planned parameters: 4,000-mi. (6,440-km.) apolune; 800-mi. (1,288-km.) perilune; 10-hr. period; 175° inclination. NASA later revealed booster's 2nd stage had produced 70-80 fps over-velocity which would have caused spacecraft to reach moon's vicinity too early to achieve lunar capture. Despite EXPLORER XXXIII's failure to achieve lunar orbit, all active experiments were operative and high scientific yield was expected.

Fourth of seven Interplanetary Explorers planned by NASA and first to attempt lunar orbit, 206-lb. EXPLORER XXXIII carried six scientific experiments and one engineering experiment—a solar cell damage study. Primary mission objectives were to study at lunar distances the earth's magnetic tail and magnetosphere in interplanetary space twice a month by means of lunar anchored spacecraft; and to measure interplanetary magnetic fields, solar plasma, and energetic particles in cislunar space. Program was managed by GSFC under overall direction of NASA Office of Space Science and Applications (OSSA). (NASA Proj. Off.; NASA Release 66-162; UPI, *NYT*, 7/2/66, 5; Strothman, *Wash. Post*, 7/2/66, A2)

- L/Col. Robert A. Rushworth flew X-15 No. 2 to 1,023 mph (mach 1.70) and 44,800-ft. altitude to evaluate handling qualities with full external fuel tanks. (X-15 Proj. Off.)
- U.S. National Academy of Sciences and Czechoslovak Academy of Sciences announced initiation of three-year exchange program in which scientists of each country would visit the other country from one month to one year to lecture, exchange professional views, and conduct research. Similar exchanges with science academies in Yugoslavia, Poland, and Romania had gone into effect this year, NAS reported, and an earlier exchange agreement with Soviet Academy of Sciences remained effective. (NAS Release)
- *Administrative History of NASA, 1958-63* (SP-4101), by Dr. Robert L. Rosholt and with a foreword by NASA Administrator James E. Webb, was published by NASA. Volume—first in history series covering major programs and other aspects of NASA operations—would be available through Superintendent of Documents, GPO. (NASA Release 66-158)

July 1: GEMINI IV Astronaut Edward H. White II, first U.S. astronaut to walk in space, received the General Thomas D. White Space Trophy from Secretary of the Air Force Dr. Harold Brown at National Geographic Society ceremony in Washington, D.C. Established in 1961 by Dr. Thomas W. McKnew, chairman of the National Geographic's Board of Trustees, trophy was awarded annually to the USAF officer or unit that made the foremost contribution to U.S. progress in aerospace.

NASA Administrator James E. Webb said at the ceremony: "I think . . . that no factor has been more important in the advances made in aviation and in space than the continued and dedicated driving leadership of the officers of the Air Force and the personal willingness to take the risks involved of these outstanding test pilots and combat pilots and those who have done the intervening jobs from transportation to helicopter work. We have a great Air Force." (Text; *NYT*, 7/12/66)

- Defense Communications Agency (DCA) said it would begin "immediate negotiations" with ComSatCorp for contract to provide new military message-carrying links to Far East as part of continuing buildup of activity in Vietnam. (Denniston, Wash. *Eve. Star*, 7/2/66, 2)

July 2: Collection of more than 1,500 tektites—dark, glassy objects suspected to be of lunar origin—had been presented to Geology Dept. of Pomona College, Claremont, Calif., by Dr. Harvey H. Nininger, a 1917 alumnus. Dr. Nininger had gathered most of the tektites in South Vietnam. (AP, *NYT*, 7/3/66, 24)

- France exploded an experimental atomic bomb from tower on Mururoa Atoll in the Pacific. Estimated yield was less than one megaton. (*Wash. Post*, 7/3/66, A1)

July 3: Clotaire Wood, former chief of Vehicle Technology Flight Experiments in NASA's OART, became NASA European Representative, Office of International Affairs. He replaced Gilbert W. Ousley, who would leave Paris office and return to GSFC Aug. 1. (NASA Ann.)

- John A. Edwards, Director of Gemini Flight Operations, became Deputy Director, Gemini Program, in NASA Office of Manned Space Flight. He replaced LeRoy E. Day, who became Director of Apollo Test. (NASA Release 66-176)

July 4: U.S.S.R. was entering second phase of space medicine research, seeking the "why" of physiological changes, V. V. Parin, a Soviet physician-scientist with a major role in selecting and training cosmonauts, told *Technology Week* in a Moscow interview. In phase one "we merely made objective observations of animals . . . we were interested in just keeping them alive, in generating the atmosphere, maintaining thermal control, humidity, pressure, and the rest," Parin said. "Now we want to know the intimate details of animal adaptation" to space, because lower animals could be more intensively studied in space flights than man. Parin cited four areas in which improved medical technology could aid the study of physiology: measuring calcium loss in astronauts; analyzing process of water loss in astronauts; measuring muscle tone; and devising better telemetry. Parin said he was satisfied with the interchange of space-medical data between U.S.S.R. and U.S. (Beller, *Tech. Wk.*, 7/4/66, 12, 13)

- NASA might have attempted to send Surveyor spacecraft to the moon in December 1965 but was effectively discouraged by severely critical report on the program published by House Committee on Science and Astronautics' Subcommittee on Space Science and Applications, *Avia-*

tion Week contended. NASA had wanted to rush Surveyor into launch that would have landed during lunar night rather than preferred daylight opportunity, to ensure that U.S. achieved first soft-landing before U.S.S.R. (*Av. Wk.*, 7/4/66, 15)

July 4: Soviet tests of carrier rockets begun April 24 "to develop new space systems" had been completed, Tass announced. Conducted in the Pacific in two areas west and northwest of Midway Island, tests were to have ended July 31; early completion was not explained. (AP, *NYT*, 7/5/66, 28)

- NAA's Rocketdyne Div. reported single J-2 rocket engine had operated continuously for 635.9 sec.—more than two minutes longer than burn time required for an Apollo lunar launch. Test firing was one of a series of 104 during which engine had accumulated a total of 12,120 sec.—more than three times engine's specified operating life. (*Tech. Wk.*, 7/4/66, 4)

July 5: NASA's Apollo/Saturn AS-203 mission was successfully launched from Complex 37 down ETR to obtain flight information on 2nd stage (S-IVB) and instrument unit (Iu), which reflected Saturn V configuration as nearly as possible. Two-stage Uprated Saturn I (Saturn IB) launch vehicle boosted unmanned payload composed of S-IVB stage, Iu, and nose cone into 117-mi.-altitude circular orbit (88.21-min. period, 31.94° inclination). S-IVB engine burned once in earth's atmosphere and was shut down. Engine's capability to restart after coast, as would be necessary during the Apollo mission, was demonstrated. No reignition was planned on this mission. Photos of liquid hydrogen fuel behavior were transmitted to four ground stations by one operable TV camera; attempts to repair malfunctioning second camera before launch had failed. During fourth orbit, while performing pressure differential test of S-IVB stage common bulkhead, internal pressure in the stage built up "well in excess of design values," and the stage fragmented.

Initial evaluation of flight data indicated that all mission objectives were attained. Success of mission—second of three preliminary missions preceding first manned Apollo flight—was critical in meeting lunar exploration timetable. 58,500-lb. S-IVB, heaviest satellite ever orbited by the U.S., was developed by MSFC and launched by KSC under direction of NASA Hq. OMSF. (NASA Proj. Off.; NASA Release 66-157; Wilford, *NYT*, 7/6/66, 1; UPI, *Wash. Post*, 7/7/66)

- Reliable sources said France would build her own "Saros" comsat system to avoid U.S. domination in the field, UPI reported. Decision stemmed from realization that if France did not act soon, the Anglo-French supersonic transport Concorde would have to be flight-controlled by U.S. satellite when it entered service within the next decade. (UPI, *Wash. Post*, 7/6/66)

- FAA awarded contracts totaling $1.95 million to Philco Corp. and IBM Corp., for first two installations of nationwide semiautomatic air traffic control system at FAA's National Aviation Facilities Experimental Center (NAFEC), Atlantic City, N.J. (FAA Release 66-62)

July 6: PROTON III unmanned scientific space station was launched by U.S.S.R. into earth orbit with 630-km. (391-mi.) apogee; 190-km. (118-mi.) perigee; 92.5-min. period; and 63.5° inclination. Instruments —including "special scientific apparatus" for continuing complex investigations of cosmic rays—were functioning normally. (Tass, *Krasnaya Zvezda*, 7/8/66, 1, USS-T Trans.)

July 6: NASA had accepted two miniature TV cameras built by Teledyne Systems Co. under $350,000 contract for experimental use in space research. Designated "Microeye," camera measured 1½ in. by 4½ in., weighed less than 1½ lbs., and was powered by small rechargeable battery. One camera, which included low-power transmitter and could telecast distances up to 100 ft., was delivered to NASA Hq. OART to evaluate possible applications in biotechnology and human research work. Second camera, connected by wire to a monitor screen, was delivered to MSFC to evaluate potential use with more powerful transmitter for monitoring functioning of launch vehicle subsystems. Feasibility of mounting camera on spacecraft and focusing on astronauts to record optical responses to spacecraft's rotations would be studied by U.S. Naval Aerospace Medical Institute, Pensacola. (NASA Release 66-171; MSFC Release 66-144)

July 6-21: NASA's SURVEYOR I, resting on the moon, revived and responded to signals from Tidbinbilla, Australia, tracking station after inactivity during two weeks of lunar night and one week of morning sunlight. Responding to command, spacecraft angled its solar panel toward sun to receive more operating power and began transmitting engineering data. Within a few hours after first response, it was declared "a fully operational spacecraft"; its battery was recharging; its transmitters, receivers, and command system were working; and machinery that moved its solar energy collector and high-gain antenna had been exercised. Performance amazed JPL officials who doubted spacecraft could survive the $-260°$ cold of lunar night.

Spacecraft continued to perform satisfactorily for two days until rapid temperature rise in battery, apparently caused by short circuit, threatened to end spacecraft's life. Temperature was rising at 3° per hour, and reached 141°F. Near-fatal temperature subsided by July 10 to 115°F and SURVEYOR I later transmitted 257 photos—including pictures of a shattered glass mirror on metal box containing its battery. Box was insulated to maintain temperature between 40°C and 125°C; segments of mirrored glass helped radiate away heat produced when instruments were operating. JPL officials believed shattering was caused by combined stresses of heat at lunar noontime and cold during lunar night rather than by impact of meteorites or heat from inside compartment.

Unexpectedly long life of SURVEYOR I ended July 14 after spacecraft transmitted 9 out of 10 photos commanded by Johannesburg tracking station to search for traces of lunar atmosphere that might scatter light from ring of gases erupting from the sun. JPL officially terminated mission July 21; spacecraft's solar panels would be checked during next period of advantageous sunlight direction, around Aug. 2, to determine if SURVEYOR I were transmitting again. Other checks would be made later to assure that spacecraft's transmissions would not interfere with Surveyor B, scheduled for launch in fall 1966. (Miles, *Wash. Post*, 7/7/66, A4; *NYT*, 7/9/66, 1; *Wash. Post*, 7/11/66; Clark, *NYT*, 7/7/66, 1; 7/13/66, 12; 7/15/66, 11; *Av. Wk.*, 7/25/66, 30)

July 7: NASA Nike-Apache sounding rocket launched from Thumba Equatorial Rocket Launching Station carried proton magnetometer and Langmuir probe to 104-mi. (167-km.) altitude in NASA-Indian National Commission for Space Research (INCOSPAR) experiment to determine intensity of ionospheric electric current system near magnetic equator. Payload had been developed by Physical Research Laboratory in

Ahmedabad, India; NASA provided launcher, ground telemetry system, and rocket. (NASA Rpt. SRL)

July 7: NASA established new program offices at MSC and MSFC to "handle the increasing level of activity involving Apollo Applications." Offices—headed by George M. Low at MSC and Leland Belew at MSFC—would manage "activities concerned with projects using Apollo hardware for purposes in addition to the manned lunar landing." To provide "program management and direction" for MSFC's consolidated research and technology program and to coordinate activities with NASA Hq., MSFC established new Experiments Office headed by Dr. William G. Johnson. (NASA Release 66-172; MSFC Release 66-151)

- NASA and ComSatCorp had signed agreement for satellite communications services to support Apollo program. ComSatCorp would provide voice/data channels and teletype channels in synchronous satellites to be positioned over Pacific and Atlantic oceans by NASA. Draft tariff filed by ComSatCorp with FCC estimated that cost of services would be $8.95 million annually. (NASA Release 66-178; ComSatCorp Release)
- NASA awarded Zia Corp., Las Cruces, N. Mex., a one-year, $5-million, cost-plus-award-fee contract to provide maintenance, operation, and miscellaneous services at WSMR. (NASA Release 66-173)
- 1966 AIAA fellows were selected in recognition of prolonged, significant achievements in aerospace technology: Dr. Robert Jastrow, Director of NASA's Goddard Institute for Space Studies; Christopher C. Kraft, Jr., Director of Flight Operations, MSC; Prof. Frank E. Marble, Cal Tech's Guggenheim Aeronautical Laboratory; Dr. George E. Mueller, NASA Associate Administrator for Manned Space Flight; Francis M. Rogallo, LaRC; Prof. Ascher H. Shapiro, MIT; Dean E. Wooldridge, director of TRW, Inc.; and Robert B. Young, vice president of Aerojet-General Corp. Awards would be presented Dec. 2 at AIAA's annual meeting in Boston. (*NYT*, 7/8/66, 12)
- Pratt & Whitney Div., United Aircraft Corp., had signed agreement with U.K.'s Bristol Siddeley Engines, Ltd., and France's SNECMA to develop jet engine for twin-engine air-bus being designed to transport 250 to 300 passengers on "medium-range" flights—as from London to Rome—by 1970 or 1972. (*WSJ*, 7/7/66, 32)

July 8: U.S.S.R. launched COSMOS CXXIII carrying scientific instruments for "further investigation of outer space" into earth orbit with 529-km. (329-mi.) apogee; 263-km. (163-mi.) perigee; 92.2-min. period; and 48.8° inclination. Equipment was functioning normally. (Tass, *Izvestia*, 7/9/66, 4, USS-T Trans.)

- "Surveyor I—A Preliminary Report," covering first five days of successful SURVEYOR I lunar mission, had been compiled and published by NASA and was available from Clearinghouse for Federal Scientific and Technical Information. In foreword, Dr. Homer E. Newell, NASA Associate Administrator for Space Science and Applications, emphasized that the publication "does not attempt to present detailed measurements and analyses of the immense amounts of scientific data that the spacecraft has telemetered to Earth; this task will take months and even years." (NASA Release 66-177)
- ESRO launched British Skylark sounding rocket from Salto di Quirra Range, to study photochemical processes in comets under meteorological research program which began in 1964 and would include 400 sounding rocket launches by 1972. (Reuters, *Wash. Post*, 7/13/66, A21)

July 8: Mars polar caps are white with dry ice, not frozen water in the form of frost, wrote Dr. Robert B. Leighton and Dr. Bruce C. Murray of Cal Tech in *Science*. Their argument was based on MARINER IV's July 14, 1965, flyby which indicated that planet's atmosphere is very thin and largely composed of carbon dioxide which, in frozen form, is dry ice. Theory that white on Mars polar caps is ordinary frost was based on observations of its reflectivity in the infrared and on manner in which it affected orientation of light waves. Authors believed there might be frozen water on Mars, but suggested it is all underground in the form of permafrost. (*Science*, 7/8/66, 136-44)

- International Assn. of Machinists (IAM) struck five major airlines including TWA, which provided base support services at KSC. Some work on Project Apollo was curtailed, but NASA officials said essential operations were continuing. (UPI, *NYT*, 7/9/66)
- Vice president of U.S.S.R. Academy of Sciences Boris Konstantinov discussed mission of Proton spacecraft series in Tass interview: "Research already carried out with the help of the Proton stations gives reason to expect that the growing weight of such laboratories will allow scientists to delve deeper into the processes occurring within the universe." He revealed that PROTON III, launched July 6, would attempt to experimentally detect "quarks," fundamental particles believed to appear during collision of cosmic ray particles with atoms of interstellar environment. Spacecraft carried "new, complicated instrumentation which is more sensitive than the old equipment." (Tass, 6/8/66)
- Oklahoma City welcomed home a favorite son, GEMINI IX Astronaut Thomas P. Stafford, and conferred honorary state citizenship on his crewmate Eugene A. Cernan. Both astronauts were inducted into the National Cowboy Hall of Fame, dedicated to the memory of Western frontiersmen. (AP, *NYT*, 7/9/66, 29)
- Aeronautical exhibition tracing aviation's history from supercharger to turbojet opened in new Hall of Aeronautical Propulsion at Smithsonian Institution, Washington, D.C. (*Wash. Post*, 7/9/66, B2)
- ITT had protested to U.S. Comptroller General that DOD decision to purchase communications services directly from ComSatCorp was contrary to recent FCC decision requiring all Federal agencies to buy services from international carriers. (*WSJ*, 7/8/66)

July 8-10: During plenary meeting in Paris, ELDO members agreed to (1) expand the present rocket development program from three-stage Europa 1 to five-stage ELDO Asp capable of launching 200-kg. (440-lb.) comsats by 1971 or 1972; (2) limit ELDO spending to $331 million and cut U.K.'s contribution from 39% to 27%; and (3) launch their comsats from base in French Guiana. U.K.—which had threatened to withdraw—announced decision to remain in ELDO. (AP, *Kansas City Star*, 7/10/66; *Tech. Wk.*, 7/18/66, 20)

July 9: Sun had blinded GEMINI IX Astronauts Thomas P. Stafford and Eugene A. Cernan for more than two minutes of the critical liftoff period June 3, preventing them from reading control gauges or dials, KSC officials disclosed. As a result, NASA was considering fitting outside of spacecraft windows with tinted coverings on launchings when sun might be a serious problem. Coverings would be sprung free of windows when spacecraft had achieved orbit. Cernan's suggestion of sunglasses would probably be rejected, NASA said, because it would be impossible to wear them under helmets. (*NYT*, 7/10/66, 45)

July 9: ComSatCorp would conduct communications test in which U.S. and European computers would "talk with each other" for a month in fall 1966 via EARLY BIRD I comsat, reported *New York Times*. Test could lead to daily exchange by satellite of business, scientific, and technical information, and to establishment of U.N. "voice of peace" agency—a world bank of medical, technical, and educational knowledge—proposed at November 1965 White House conference on international cooperation. (Clark, *NYT*, 7/10/66, 43)

- William A. Hyman, internationally known air and space lawyer and author, died in New York City. (AP, *Wash. Post*, 7/11/66)

July 10: William R. Berry reached 19,000-ft. altitude before landing his hot-air balloon "City of San Francisco" near Antioch, Calif., after 65-min., 30-mi. flight from Livermore, Calif. If endorsed by National Aeronautic Assn., altitude would break 15,600-ft. world record for hot-air balloons set by Don Piccard in 1965. (AP, *Wash. Post*, 7/11/66, A8)

- L/C Vladimir Komarov, pilot-cosmonaut on U.S.S.R. VOSKHOD I space flight Oct. 12-13, 1964, told second Japan-U.S.S.R. youth festival at Lake Yamanaka, Yamanashi Prefecture, Japan, that U.S.S.R. soon would softland an unmanned spacecraft on moon and return it to earth, then make similar flight with dog. Commenting on U.S. rendezvous and docking achievements, Komarov said: "The United States experiment was a very interesting one. But I do not believe that in the series of plans to conquer space the Soviet Union is behind the United States." (Tass, 7/14/66; *SBD*, 7/25/66, 122)

July 11: NASA Nike-Apache sounding rocket launched from Churchill Research Range reached 126-mi. (202-km.) altitude in NASA-Univ. of Michigan experiment to measure ambient neutral and ion composition and density as function of altitude and compare data with those obtained on February 1965 flights. Rocket and instrumentation performed satisfactorily. (NASA Rpt. SRL)

July 12: NASA Javelin sounding rocket launched from NASA Wallops Station reached 631-mi. (1,015-km.) altitude in Univ. of Pittsburgh-Syracuse Univ. experiment to observe ionization levels of exospheric helium. Rocket and instrumentation performance was satisfactory, with 18-min. telemetry signal. Unanticipated high-energy electron particles were measured, and data were under study to determine their source. (NASA Rpt. SRL)

- U.S. and U.S.S.R. introduced their draft space law treaties at opening session of U.N. Committee on Peaceful Uses of Outer Space's Legal Subcommittee, meeting in Geneva to discuss peaceful cooperation in exploring the moon and other celestial bodies. U.S. Ambassador Arthur J. Goldberg told Subcommittee that although drafts were different in scope, they could be reconciled if there were good will and common purpose to reach agreement. Soviet representative Platon Morozov charged U.S. should not speak of peace in outer space while continuing its "shameful aggressive war on the peaceful people of Vietnam." (Reuters, *NYT*, 7/12/66; *Wash. Post*, 7/13/66, A23)

- X-15 No. 1 was flown to 3,652 mph (mach 5.34) and 130,000-ft. altitude by Maj. William J. Knight (USAF) to check out electrical loads, nonglare glass, stick kicker, and shade window. (X-15 Proj. Off.)

- M2-F2 lifting-body vehicle, piloted by NASA test pilot Milton O. Thompson, was successfully air-launched from B-52 bomber at 45,000-ft. altitude and maneuvered into 200-mph landing four minutes later in first free flight at Edwards AFB. Thompson, Chief Project Pilot for FRC's Lifting Body Program, noted that "With today's test we have just about completed our whole test program . . . the next area to be explored is whether such a craft could survive reentry into the Earth's atmosphere." The 2½-ton wingless vehicle was being studied "to establish the technological base" for design of future spacecraft and to ferry men and equipment between earth and satellites. (FRC Release 14-66; AP, *NYT*, 7/13/66, 1; *Tech. Wk.*, 7/18/66, 17)

July 12: USAF launched unidentified satellite with Atlas-Agena D booster from WTR. (*U.S. Aeron. & Space Act., 1966*, 153)

- NASA had tentatively selected six experiments to be carried on two Mariner Mars spacecraft scheduled for launch by Atlas-Centaur boosters between early February and mid-April 1969: two television cameras, infrared spectrometer, infrared radiometer, ultraviolet spectrometer, celestial mechanics experiment, and S-band occultation experiment. Experiments were selected to extend knowledge of Martian atmosphere and visible features of Martian terrain and to gather additional data to continue planning for landing instrumented capsules on the planet. JPL had project management responsibility for Mariner Mars 1969 missions. (NASA Release 66-174)

- NASA had selected Garrett Corp., Los Angeles, Calif., for negotiation of $15-million contract for final design, development, construction, and testing of small research ramjet engines under Hypersonic Research Engine Project directed by NASA Hq. OART. Engines—which were to be compatible for mounting beneath aft fuselage of X-15 No. 2—would be useful for hypersonic transport aircraft, boosters, and for spacecraft flying within earth's atmosphere. (NASA Release 66-182)

- Paramount objective of NASA's Apollo program was not to land man on the moon before U.S.S.R. but to make "U.S. first in space by the end of this decade, and to make this pre-eminence unmistakably clear to all the world," NASA Deputy Administrator Dr. Robert C. Seamans, Jr., told IEEE Aerospace Systems Conference in Seattle, Wash. Apollo program was a means of acquiring the ability to operate in space on a variety of missions, Dr. Seamans explained: ". . . we are building much more than a rocket and a spaceship. We are developing, in government, in industry, in our universities, one of the most remarkable teams that has ever been assembled. There are more than 400,000 men and women, 20,000 industrial companies, and more than 150 universities actively engaged in the NASA program. More than 90 percent of NASA's five-billion-dollar-a-year budget goes to contractors." He warned against failing to look beyond the near-term goal of manned lunar exploration and said: "Programs of the future must be determined within the total context of national need and the availability of resources. The Nation's interest in space projects, and the level of support accorded them must in the long run be related to tangible benefits that can be derived from the emerging technical and operational capabilities." (Text)

- U.S. had reaped many unexpected benefits from NASA's manned spaceflight program, NASA Deputy Administrator Dr. Robert C. Seamans, Jr., told Federal Executive Board in Seattle, Wash. Specifically, NASA was (1) "making a definite effort to stimulate the development of technology generally . . . [and had] set up what may well be a unique, formal endeavor to transfer technology which has been developed in the space program to more general usage in the industrial community . . . (2)

striving to promote a rising educational level among the people . . . (3) developing industrial skills, management techniques, and a high degree of quality control that are of immense benefit to the Nation in such mundane things as washing machines, television sets, transistor radios, and other household items . . . (4) helping to assure the country of a hard core of engineers and scientists, technicians and managers, laboratories and industrial facilities, which have responded to the challenge of working on important projects in aeronautics and astronautics. . . ." (Text)

July 12: Five leading European aerospace companies announced formation of consortium to seek European space contracts. Called the European Satellite Team, group's major goal was $18-million design and construction contract for ESRO's 800-lb. TD-1 and TD-2 research satellites scheduled for launch by NASA in 1969 and 1970. Consortium was composed of Elliott-Automation, U.K.; Compagnie Francaise Thomson-Houston, France; Fokker, Netherlands; Allmanna Svenska Elektriska, A.B., Sweden; and Fabbrica Italiana Apparecchi Radio, Italy. General Electric Co.'s Missile and Space Div. would serve as consultant. (Wilford, *NYT*, 7/13/66; *Chic. Trib.*, 7/13/66)

July 12-13: NASA launched five Nike-Apache sounding rockets from NASA Wallops Station between 9:00 p.m. and 5:00 a.m. EDT. Each rocket was programed to eject a vapor cloud of reddish or bluish color as its altitude increased from 50 to 125 mi. Experiments were to measure wind velocities and directions at various altitudes; motion of the trails was photographed from five camera sites within a 100-mi. radius of Wallops. Launches were conducted for GCA Corp., under contract to GSFC. (NASA Release 66-181; Wallops Release 66-38)

July 13: Vice President Hubert Humphrey told group of young scientists visiting Washington, D.C., under National Youth Science Camp program to devote some of their talents to solving problems on earth as well as in space. "If we can put a Surveyor on the moon to take pictures, why can't we get some one in from La Guardia Airport to downtown New York City before he collects his old age pension . . . we must try to harness our divergent and separated resources and our energies to solve the problems of our earthly environment." The Vice President forecast scientific break-throughs for the year 2000, including the elimination of bacterial and viral diseases, the correction of hereditary defects, landing of men on Mars, farming of the ocean, manufacture of synthetic proteins, control of weather on regional basis, and creation of life in the laboratory. (Text)

- NASA would negotiate $1.25-million contract with Texas Instruments, Inc., for design, fabrication, and testing of telemetry subsystem for 1969 Mariner missions to Mars. Motorola, Inc., was selected for design, fabrication, and testing of command subsystem for Mariner spacecraft under $2-million contract. Both contracts would be managed by JPL. (NASA Release 66-186)
- NASA had assigned MSFC project management of Apollo Telescope Mount (Atm) for Apollo Applications (Aa) missions. A combination of accurate, high-resolution, solar-oriented telescopes, Atm would be attached to spacecraft so that astronauts could manually adjust it to select and focus upon a specific area on the sun. (NASA Release 66-185)
- More than 1,500 building tradesmen staged one-day strike at KSC to protest NASA's method of awarding construction contracts. Work on three

key Apollo projects was delayed by walkout, which had been called by Brevard (County) Building and Construction Trades Council, but preparations for July 18 Gemini X mission continued uninterrupted. (UPI, NYT, 7/14/66, 21)

July 13: Nation's most modern ship for oceanographic research, USC&GS *Oceanographer*, was commissioned at Washington [D.C.] Navy Yard in commemoration of Environmental Science Services Administration's (ESSA) first anniversary. President Johnson, speaking at the ceremony, called for international cooperation in oceanographic research and invited 11 nations, including U.S.S.R., to participate in first round-the-world expedition of the new research ship. (Text; ESSA Release 66-41)

July 14: Nike-Apache sounding rocket launched from NASA Wallops Station carried 52-lb. variable frequency impedance probe to 121-mi. (195-km.) altitude in cooperative West German-U.S. experiment to measure electron density in the ionosphere. NASA furnished two radio propagation experiments and sounding rocket; West Germany furnished probe. There was no exchange of funds between cooperating agencies—NASA and West German Federal Ministry for Scientific Research. Results would be made available to world scientific community. (NASA Release 66-188; Wallops Release 66-40)

- USAF's OV1-VIIIT satellite was launched with Atlas D booster from Vandenberg AFB into near-circular retrograde orbit: apogee, 627 mi. (1,009 km.); perigee, 619 mi. (996 km.); period, 105 min.; inclination, 144°. Balloon-shaped satellite—made of soft aluminum wire mesh covered with a plastic "skin" which disintegrated after inflation—would act as a passive comsat to reflect signals transmitted to it between earth stations, determine whether wire would be rigid enough to maintain spherical shape, and test theory that "open-mesh" structure would reduce drag and solar pressure in space environment. (AFSC Release 140.66)

- COSMOS CXXIV was launched into earth orbit by U.S.S.R. for continued space research. Orbital parameters: apogee, 303 km. (188 mi.); perigee, 208 km. (129 mi.); period, 89.4 min.; inclination, 51.8°. Instruments were functioning normally. (Tass, 7/14/66)

- NIMBUS II meteorological satellite, launched by NASA May 15, completed primary test objective of two months (800 orbits) continuous operation; all subsystems were operating according to plan. The 912-lb., 10-ft.-tall satellite had traveled more than 20 million mi., taken more than 150,000 pictures, and received more than 23,000 commands from ground controllers. NIMBUS II was now being readied for extensive infrared photo coverage of hurricane breeding areas in Atlantic Ocean. (NASA Proj. Off.; NASA Release 66-187; *Marshall Star*, 7/27/66, 3)

- First anniversary of first close-up photos of Mars taken by NASA's MARINER IV from distance of 134 million mi. Spacecraft was launched Nov. 28, 1964. (*NYT*, 7/26/66, 49)

- All participants in the Gemini X mission rehearsed their parts in preparation for July 18 flight. Astronauts John Young and Michael Collins went over flight plan from liftoff to splashdown and conducted review of all spacecraft systems. (Wilford, *NYT*, 7/15/66, 11)

- New program to disseminate information to industry was initiated by NASA and AEC's Argonne National Laboratory. Purpose was to identify technological innovations resulting from Argonne's research and development projects; prepare brief, business-oriented summaries of innovations; and distribute them to industry. Summaries—to be known as

AEC-NASA Tech Briefs—would contain information not customarily reported in scientific literature such as descriptions of devices, processes, and techniques developed to meet specific or engineering needs at Argonne and would be combined with similar information currently published by NASA. (NASA Release 66-183)

July 14: Dr. Robert C. Seamans, Jr., NASA Deputy Administrator, told World Affairs Council in Los Angeles U.S. civil space program "affords perhaps the most effective global projection of the American personality and the American achievement today . . .

"The elements of this favorable projection are the openness with which the program is conducted, the direct benefits to others through space communications and weather systems, the willingness to share both research and results in projects of mutual interest, the energetic forward thrust toward technological and managerial pre-eminence, and . . . the evidence of high national purpose." (Text)

- If MSFC Director Dr. Wernher von Braun should obtain assignment to develop an orbital payload capability between Uprated Saturn I's 35,000-40,000 lbs. and Saturn V's 250,000-275,000 lbs., "the art of rocketry will take a step backward, a promising branch of space technology will suffer another reversal, and the American taxpayer will be several hundred million dollars poorer," wrote William Hines in the Washington *Evening Star*. Hines contended that both von Braun's suggestions—increasing the thrust of the Uprated Saturn I at cost of $1.5 billion or reducing the capability of the Saturn V at cost of $3.25 billion—would be "fantastically expensive." He suggested that, "assuming that there actually is need for an 80,000-100,000-pound orbital capability, this requirement could be filled more cheaply, and sooner, by pushing development of the solid [propellant] rocket. . . ." (Hines, Wash. *Eve. Star*, 7/14/66, A14)

July 15: NASA Aerobee 150 launched from WSMR carried GSFC-instrumented payload to 90-mi. (145-km.) altitude in experiment to measure spectral irradiance of stars Alpha Lyrae, Lambda Scorpii, Zeta Ophiuchi, and Delta Scorpii. Flight marked first successful operation above earth's atmosphere of Star Tracking Rocket Attitude Positioning (Strap) system. The 330-lb. telescope device telemetered almost 80 sec. of ultraviolet data and was recovered undamaged about 55 mi. from launch site for reuse in fall 1966. Data would give experimenters information on star surface temperatures, gravitational field intensity, and relative ages. (NASA Rpt. SRL; GSFC Release G-16-66)

- 50th anniversary of founding of Boeing Co. (EH)
- Possible explanation of astronaut's inability to see stars on earth's daylight side was presented by physics professors Edward P. Ney and W. F. Hugh in *Science*: a "spacecraft corona"—cloud of particles formed by waste materials shed from spacecraft—might produce glow that would hide stars dimmer than fourth magnitude. Corona had been discovered by former Astronaut John H. Glenn, Jr., who reported during his Feb. 20, 1962, orbital mission in FRIENDSHIP 7 that he was being accompanied by a cloud of "fireflies." (*Science*, 7/15/66, 297-9)
- Analysis of available evidence concerning possibility of an advanced form of life on Mars by Univ. of Maryland astronomy professor Dr. Ernst J. Opik appeared in *Science*. Dr. Opik said the most heavily eroded craters visible in photos taken by MARINER IV during the July 14, 1965, flyby dated from the earliest period of the planet's history, implying that

Mars had never had a dense, moist, earth-like atmosphere for any extended period and would therefore be a poor prospect for life. Dr. Opik argued strongly for at least a primitive form of vegetation, citing in particular a region of 1.2 million sq. km. that had darkened rapidly from 1946 to 1954. He likened this to the change a distant observer would have seen on earth as a consequence of the virgin lands development program in U.S.S.R. Vegetation was the best explanation for such changes; however, if it were not vegetation, "it must be something specifically Martian." The planet, he added, "may become a source of great surprises in the future." (*Science*, 7/15/66, 255-65)

July 16: Circular insignia of Project Apollo astronauts was displayed for first time at KSC. Capital letter "A" was in center with its crossbar formed by part of a curving track joining earth (at right) and moon (at left) of the "A." Centered on the crossbar were three twinkling stars—symbolic of the three-astronaut crew of each Apollo mission. Earth symbol was dominated by North American continent; moon was represented in third quarter phase with dark part dominated by a man's profile. Moon, earth, and letter "A" were displayed against dark blue background studded with stars—representing outer space. Entire design was ringed by a circle with "Apollo" at top and "NASA" at bottom. (Wash. *Eve. Star*, 7/16/66, A1)

July 16-17: Five vapor cloud experiments were launched from NASA Wallops Station on Nike-Apache sounding rockets between 8:56 p.m. and 5:08 a.m. EDT. Vapor trails were ejected at altitude ranges of 30 to 130 mi. (48 to 209 km.) to measure wind velocities and directions. Data were obtained by photographing motion of trails from five camera sites within 100-mi. radius of Wallops. Launches were conducted for GCA Corp. under contract to GSFC. (Wallops Release 66-41)

July 17: IAM strike at KSC ended when an agreement was reached between Local 773 of IAM and TWA, sending 1,100 machinists back to work. Labor-management observers cited the local agreement as a "first" for KSC. (KSC Historian)

- Commenting on the tiny U.S. flag packed into SURVEYOR I which softlanded on the moon June 2, Sydney Harris said in the *San Francisco Examiner & Chronicle*: "If we begin to export nationalism to the moon, we are defeating the very purpose of science and exploration." (Harris, *S.F. Examiner & Chron.*, 7/17/66)

July 18: NASA Aerobee 150 launched from WSMR carried two far-ultraviolet image converters to 83-mi. (134-km.) altitude in NRL experiment to obtain stellar spectra and photographs; rocket and instrumentation did not perform as expected, and no useful data were obtained. (NASA Rpt. SRL)

- X-15 No. 3 was flown by NASA test pilot William H. Dana to 3,218 mph (mach 4.71) and 96,100-ft. altitude to check out MH inertial system, cockpit display, horizontal tail loads, and stick kicker. (X-15 Proj. Off.)

- System of two interconnected balloons 815-ft. long designed to test parachute deceleration system of NASA's Voyager spacecraft—scheduled to land on Mars in 1973—was lofted by USAF from Holloman AFB. Helium in small upper balloon inflated main balloon—largest ever launched—to 26-million-cu.-ft. capacity as both rose to 130,000-ft. altitude where atmosphere was believed to be as thin as that on Mars. 1,800-lb. simulated Voyager spacecraft payload was jettisoned by radio command seven hours after launch, successfully parachuted to earth, and was later recovered near Phoenix, Ariz. Main balloon rose to higher altitude and ruptured as expected. Parachute program was managed by LaRC; Voyager program by JPL. (Sullivan, *NYT*, 7/19/66; *Av. Wk.*, 7/25/66, 34)

July 18: Probability statistics for astronaut reactions to Apollo lunar mission radiation based on NASA-sponsored study at AEC's Oak Ridge National Institute of 2,100 persons exposed to radiation for medical reasons: basing estimates on a "better than 50% reaction of those exposed to each dose," it appeared that a 100-Rad dose would produce loss of appetite; 150 Rads, nausea and vomiting; and 225 Rads, diarrhea. (*Tech. Wk.*, 7/18/66, 4)

- House passed H.R. 14838 revising operation of National Science Foundation (NSF) to (1) give new emphasis to NSF's basic missions; (2) strengthen and increase functions of National Science Board; (3) unify and strengthen operational authority of director; and (4) modify and streamline NSF's organization and structure. (*CR*, 7/18/66, 15180-4)

- John F. Stearns, former chief of National Referral Center at Library of Congress, became Director of NASA's Science and Technology Information Div., Office of Technology Utilization. He succeeded Melvin S. Day, who had been appointed Deputy Assistant Administrator for Technology Utilization June 5. (NASA Ann.)

- USAF's Manned Orbiting Laboratory (Mol) was "moving quietly but quickly toward its first test launch, now scheduled for Oct. 28 from Cape Kennedy," *Aviation Week* reported. Operational Mol would consist of modified Gemini spacecraft connected to a "can-shaped" laboratory. USAF astronauts, while orbiting in space, would pass through door in spacecraft's heatshield to reach laboratory. October test would show whether configuration would endanger astronauts in spacecraft during reentry. (*Av. Wk.*, 7/18/66, 25)

- Delays at major U.S. airports totaling 20 million minutes cost aircraft operators $63.6 million in 1965, FAA reported in a staff study. One third of total delays was caused by air traffic controls, and the remainder by airport limitations such as weather conditions, ground congestion, and construction. (FAA Release 66-68)

July 18-21: GEMINI X rendezvous and docking mission—eighth manned flight in Gemini series—touched on all aspects of mission objectives for which two-man spacecraft was designed. Mission began at ETR's Complex 14 with launch by Atlas booster of Gemini Agena Target Vehicle (Gatv), followed 100 min. later by launch from Complex 19 of Titan II booster orbiting GEMINI X spacecraft with Astronauts John W. Young as command pilot and Michael Collins as pilot. GATV X was placed in near-circular orbit with 187-mi. (301-km.) apogee; 180-mi. (290-km.) perigee; and 90.5-min. period. The 8,248-lb. GEMINI X was launched into elliptical orbit with initial apogee of 167 mi. (268 km.) and 99-mi. (159-km.) perigee.

Maneuvers were executed as planned, with docking of GEMINI X and GATV X occurring at 5:58 GET. Only about 380 lbs. maneuvering fuel of 940 lbs. at liftoff were left, whereas a 680-lb. remainder had been expected. An out-of-plane error had necessitated a major correction, with excessive fuel consumption. Mated spacecraft, using GATV X's primary propulsion system, attained 476-mi. (766-km.) record altitude for manned flight. Burn of GATV X's secondary propulsion system at 22:38 GET brought docked configuration into same orbit as GATV VIII—launched March 16 for GEMINI VIII mission—preparatory to later rendezvous.

Standup Eva began at 23:27 GET with Collins photographing stellar ultraviolet radiation through open hatch. Young ordered termination of Eva when both he and Collins experienced severe watering and irritation of eyes from an unidentified source. Separation of GEMINI X-GATV X—which had remained docked 38 min. 42 sec.—occurred at 44:40 GET.

Series of maneuvers brought GEMINI X within 50 ft. of GATV VIII. Umbilical Eva began at 48:08 GET: Collins removed fairing from GATV VIII and retrieved micrometeoroid detection experiment—completing first productive work ever accomplished on Eva mission. Collins lost Eva camera; photographic coverage was limited to pictures taken by Young from spacecraft. Eva was terminated after 28 min. to conserve maneuvering fuel; 50 lbs. had been used, as against 35 lbs. estimated consumption for planned 55 min.

Third hatch opening of mission occurred when astronauts, cramped by unstowed equipment, jettisoned 12 items.

Reentry July 21 after 70 hrs. 47 min. in space and 43 revolutions was normal. At 5:07 p.m. EDT, GEMINI X splashed down in Atlantic Ocean 544 mi. east of Cape Kennedy—less than three miles off target and five miles from recovery ship U.S.S. *Guadalcanal*. Within minutes, USN pararescue team dropped from helicopters had attached flotation collar to spacecraft, and Young and Collins were transported by helicopter to recovery ship.

GEMINI X had accomplished (1) two rendezvous, one involving longest docking yet achieved, the other a close approach to a passive satellite from which an object was retrieved; (2) two Eva exercises; (3) successful maneuvering by astronaut from one spacecraft to another and back; and (4) use of powered, fueled satellite to provide primary and secondary propulsion for docked spacecraft. (NASA Release 66-179; NASA Proj. Off.; Wilford, *NYT*, 7/19/66, 1; Hines, *Wash. Eve. Star*, 7/19/66, A1, A3; 7/20/66, A1, A6; Wilford, *NYT*, 7/20/66, 1; Reistrup, *Wash. Post*, 7/22/66, A1, A4; *Av. Wk.*, 7/25/66, 26-30)

July 19: House-Senate Conference Committee on NASA authorization reached agreement on a $5,000,419,000 FY 1967 bill. Compromise was $11.581 million below original NASA request; $13.5 million above House version; and $7.5 million below Senate version of bill. The $18 million for 1967 Mariner Venus flyby mission that House Science and Astronautics Committee's Space Science and Applications Subcommittee had sought to cancel was restored. Three main additions to NASA's request were $13 million for Voyager program; $4 million for 260-in. solid-propellant motor development; and $2 million for Snap-8 nuclear electric generator program. Conference report (H.R. 14324) would be debated in House and Senate. (*CR*, 7/20/66, 15551-4)

- NASA had selected Philco Corp. for negotiation of $35-million, incentive-fee contract for "continued systems engineering and operational support" of Mission Control Center at MSC. (NASA Release 66-190)
- M2-F2 lifting body vehicle was successfully air-launched from B-52 aircraft at 45,000-ft. altitude in second glide flight at Edwards AFB. Purposes of flight, piloted by NASA research pilot Milton O. Thompson, were determination of lateral stability control, longitudinal trim, vehicle performance, and landing characteristics, and checkout of onboard systems. (NASA Proj. Off.)
- NASA Marshall Space Flight Center awarded Bendix Corp. and Boeing Co. six-month, $350,000 contracts for parallel studies to "develop specifica-

July 21: GEMINI X spacecraft, its heatshield forward, aboard recovery ship U.S.S. *Guadalcanal*.

tions and drawings" for vehicles being proposed to transport astronauts on lunar surface. (MSFC Release 66-157)

July 20: NASA Nike-Apache sounding rocket was launched from Churchill Research Range in engineering test preceding series of five GSFC launches to measure intensity and energy spectra of low-energy proton, helium nuclei, and heavier nuclei during Polar Cap Absorption event. Rocket performance was near predicted; however, nose cone did not retract because of override switch heating problem, and data were poor in quality. (NASA Rpt. SRL)

July 20: U.S.S.R. launched COSMOS CXXV carrying scientific instruments for continued study of outer space into circular orbit with 250-km. (155-mi.) altitude, 89.5-min. period, and 65° inclination. Instruments were functioning normally. (Tass, 7/20/66)

- 28-member Legal Subcommittee of the U.N. Committee on the Peaceful Uses of Outer Space began article-by-article examination of drafts by U.S. and U.S.S.R. for proposed treaty governing exploration of space. U.S. Ambassador to U.N. Arthur J. Goldberg announced that U.S. had agreed to extend negotiations beyond proposal for treaty to ban military use of moon and other celestial bodies to cover "space itself along the lines proposed by the Soviet Union." Agreed text would be presented to U.N. General Assembly in New York in September. (*NYT*, 7/22/66, 11)

- NASA announced appointment of Maj. Michael J. Adams (USAF), pilot in USAF's Manned Orbiting Laboratory (Mol) program, to X-15 research program, replacing Capt. Joe H. Engle (USAF), who had transferred to MSC April 4 to become an astronaut. Following extensive training program, Major Adams would become 12th man to fly the rocket-powered aircraft. (FRC Release 15-66)

- ARC's new Hypervelocity Free Flight Facility would be fully operational by the end of July, NASA announced. "Light-gas guns" would fire free-flying spacecraft models into three wind tunnels. Muzzle velocities up to 20,000 mph against an air stream flowing up to 10,000 mph in the opposite direction would enable models to achieve test speed of 30,000 mph, simulating atmosphere reentry flight from moon or planets. (NASA Release 66-189; ARC Release 66-10)

- Flight plan for Lunar Orbiter I, scheduled for launch between Aug. 9 and Aug. 13, had been changed to shift trajectory for mission so photographic spacecraft would pass over SURVEYOR I twice. Stereographic photos, with resolution down to three feet, could thus be obtained. First seven areas selected for Lunar Orbiter photography remained unchanged and included examples of all major types of lurain to permit assessment of suitability for Apollo manned landing missions. (NASA Release 66-193)

- France exploded an atomic bomb, air-dropped over her test site at Mururoa Atoll in the Pacific, almost directly under flight path of GEMINI X one hour before the spacecraft passed. NASA, advised of the impending test, had warned Astronauts John W. Young and Michael Collins to "keep your heads down" and "not to look at the earth while flying over the atoll." (*WSJ*, 7/22/66, 1; Sullivan, *NYT*, 7/20/66, 22)

- House passed $58.6 billion DOD FY 1967 appropriation bill (H.R. 15941) —almost $1 billion more than the Administration had requested and largest since World War II—by 393 to 1 vote. Negative vote was cast by Rep. George E. Brown, Jr. (D-Calif.), who was registering opposition to Vietnam war. (*CR*, 7/20/66, 15516-51)

- Achievements in space since first flight in Gemini series March 1965 had been "phenomenal," stated *Washington Post* editorial: ". . . the people of this country can feel pride in the officials of . . . [NASA] who have deliberately but confidently advanced the space program without catering to pressures for spectacular stunts. The total achievement, in and of itself, is spectacular beyond the wildest hopes of a few years ago." (*Wash. Post*, 7/20/66, A20)

July 21: U.S. and U.S.S.R. agreed in discussions at Geneva to treaty article on exploration of space barring any state from claiming sovereignty over space, including the moon and planets. Article was also approved by 28-nation U.N. Legal Subcommittee drawing up treaty from draft accords submitted by U.S. and U.S.S.R. Subcommittee also accepted article binding states to conduct space exploration in accordance with international law and in the interest of international peace. (Reuters, *NYT*, 7/22/66, 8)

- Maj. William J. Knight (USAF) flew X-15 No. 2 to 3,614 mph (mach 5.12) and 192,300-ft. altitude for pilot altitude buildup, star tracker experiments, and base drag studies. (X-15 Proj. Off.)
- House adopted joint House-Senate conference report (H.R. 14324) on 1967 NASA authorization by voice vote. Authorization was $5,000,419,-000. (*CR*, 7/21/66, 15887-96)
- MSFC awarded Emerson Electric Co. a 10-wk., $34,990 study contract to determine feasibility of recovering small metallic coupons from wings of PEGASUS III meteoroid-detection satellite, orbited by NASA July 30, 1965. Using sample wing with attached coupons, Emerson would analyze possible approaches an astronaut could make to satellite and types of equipment he would need to perform coupon-retrieving mission. (MSFC Release 66-162)
- Application of space age technology to traffic control was near, Chairman of the Board of the Ford Motor Co. Henry Ford II told the Young Men's Business Club of Greater New Orleans meeting in New York. He said plans would soon be presented to the Government for a nationwide traffic control system based on earth survey satellites for aerial reconnaissance. They would be linked by computers to urban traffic control centers and "to the stoplight in the corner and even the car radio. . . . As fantastic as it may seem, we believe such a system will be technically feasible and economically sound." Technical capacity for development of system had been gained from experience of the Philco Corp., a Ford subsidiary, in designing and operating MSC's Flight Control Center. Philco's communication system was being used on Gemini missions. (*NYT*, 7/22/66, 27)
- NSF announced that, under $5.4-million contract awarded Univ. of California's Scripps Institute of Oceanography, scientists would bring up core samples from 1,000 to 3,000 ft. below Atlantic and Pacific ocean floors. Two-year drilling program, to begin in 1967, was outgrowth of Project Mohole. (*Wash. Post*, 8/22/66, A7)
- The *Will Rogers*, 41st and last of Nation's Polaris ballistic missile submarines, was christened in Groton, Conn., ceremony by Mrs. Hubert H. Humphrey. Launching culminated 8½-yr. construction program of nuclear-powered submarines capable of indefinite submergence and equipped with 16 2,500-mi.-range missiles with nuclear warheads. (Baldwin, *NYT*, 7/21/66, 5)

July 22: Senate approved by voice vote $5,000,419,000 1967 NASA authorization recommended in joint House-Senate conference report (H.R. 14324). Bill was then sent to White House for President Johnson's signature. (*CR*, 7/22/66, 15981-4; *Wash. Post*, 7/23/66, A6)

- GEMINI X Astronauts John W. Young and Michael Collins flew from recovery ship U.S.S. *Guadalcanal* to KSC to begin 10 days of debriefing on July 18-21 mission. (Strothman, *Wash. Post*, 7/23/66, A3)

July 22: First move of 402-ft.-high, 9-million-lb. mobile service structure by crawler-transporter from parking area at KSC to Pad 39-A. (KSC Historian; *Av. Wk.*, 8/1/66, 27)
* NASA responded to 47 questions on post-Apollo programs submitted by Space Science and Space Technology panels of President's Science Advisory Committee; bibliography of advanced study contract reports was appended to 107-page document. Information would be used in preparation of PSAC report on post-Apollo goals. (Text)
* Implications of the GEMINI X mission received editorial comment.

    Watertown [N.Y.] Daily Times: "The mind and the body of man have demonstrated once again what brilliance and discipline can accomplish. They have also demonstrated once again that fear born out of ignorance is overcome by those who will prepare themselves through training and perseverance." (*Watertown Daily Times*, 7/22/66)

    Los Angeles Times: GEMINI X flight moved U.S. "a few steps closer to the goal of landing a man on the moon before 1970," and it showed the time had come to look "beyond the Apollo moon landing, and to map new goals for manned spaceflight." (*L.A. Times*, 7/22/66)

    New York Times: "Useful work" accomplished by Michael Collins in retrieving micrometeoroid experiment from GATV VIII portended the day when "men working in space will put huge structures together, joining vehicles and parts that were rocketed separately and even at widely separated intervals." Use of GATV X's propulsion system to power docked GEMINI X-GATV X "opens breathtaking new possibilities in both the economics and technology of tomorrow's space travel." (*NYT*, 7/22/66, 28M)

July 23: U.S.S.R. objected to a U.S. proposal to allow military equipment to be used on the moon or other celestial bodies for peaceful purposes, Platon Morozov, deputy acting permanent U.S.S.R. U.N. representative, told the Legal Subcommittee of the U.N. Committee on the Peaceful Uses of Outer Space meeting in Geneva to draft a treaty governing outer space exploration. He said such a provision would create a loophole for violations; he was prepared, however, to accept with a few minor modifications other provisions of U.S. draft article. (Reuters, *Wash. Post*, 7/23/66)
* Locating GATV VIII during Gemini X flight was called "a major feat of three-dimensional navigation" in a *Hartford Times* editorial, which said "use of Agena 8's fuel also brings closer the day when manned laboratories permanently parked in orbit around the earth will be in use for purposes both experimental and immediately useful." (*Hartford Times*, 7/23/66)

July 23-27: Three-axis stabilization for NASA's OGO III spacecraft was incapacitated after 46 days of flawless operation by failure of a magnetic amplifier which shorted out the attitude control system power inverter. Backup spin stabilization system was activated July 27, and reorientation affected only six experiments. Primary scientific objectives were expected to be achieved, and spacecraft's spin axis would be reoriented in late December to increase power input. (NASA Proj. Off.)

July 24: Launch crews began erecting Titan II booster on Pad 19 at ETR in preparation for September launch of Gemini XI mission. (AP, *Houston Post*, 7/24/66)
* U.S. Astronaut Alan Shepard and Soviet Cosmonaut Boris Yegorov had same positive response to *Parade* editor Jess Gorkin's suggestion that

an American and a Russian fly together in the same spacecraft. Shepard, contacted at MSC, said it would be of "enormous value." Yegorov indicated that U.S.S.R. might already be preparing for joint space exploration, saying: "If the Americans are really keen on the idea of joint flights, they had better start learning Russian. All of us are studying English already." (Anderson, *Parade*, 7/24/66, 4-5)

July 24: Harris poll indicated that Americans, in case of national emergency, favored first curtailing space and aid-to-cities programs and last curtailing health-assistance and college-education aid programs. (*Parade*, 7/24/66)

July 25: Boosted Arcas 2 rocket was successfully launched for first time from WSMR. Equipped with new Marc 42A1 1st stage, rocket carried 16-lb. payload to 90-mi. (145-km.) altitude. (*Tech. Wk.*, 8/1/66, 10)

- SURVEYOR I bounced two to three inches upward on first impact with lurain June 2, JPL announced. According to data returned by 620-lb. spacecraft, signals generated by strain gauges indicated the three shock-absorber legs had all landed simultaneously on a horizontal surface. Eight-inch-diameter circular footpads rebounded clear of surface before settling on moon. Spacecraft exerted pressure of about 0.5 psi on lurain after it came to rest. Pressure of footpads at impact was about 8 psi. (AP, *Wash. Post*, 7/25/66, A3; *Wash. Eve. Star*, 7/25/66, A12; AP, *NYT*, 7/25/66, C3; *Tech. Wk.*, 8/1/66, 9)

- GEMINI X Astronauts John W. Young and Michael Collins flew from KSC to MSC for reunion with families and continued debriefing. News conference had been set for Aug. 1. (AP, *Wash. Eve. Star*, 7/25/66, A12)

- Editorial on five years of "outstanding service" by NASA Administrator James E. Webb appeared in *Technology Week*: ". . . it is time to state . . . our belief that Mr. James E. Webb has done an outstanding job as NASA Administrator. He has presided over an explosive growth of the agency which would have overwhelmed a less capable administrator; he has met and solved some exceedingly thorny management problems; he has dealt effectively with some of the world's most temperamental scientists; he has successfully steered the space agency's budget through the perils of both Administration and Congressional financing; he has given the space program a nationwide base of facilities and political strength; and, most important of all, he has forged a team which is writing an unparalleled record of technical accomplishment. . . ." (Coughlin, *Tech. Wk.*, 7/25/66, 152)

- Ludlow King, head of Ludlow King Associates and former executive of Owens-Corning Fiberglas Corp., had been appointed as a consultant to NASA. He would study means of making relationship between NASA and industry more effective. (NASA Ann.)

- MSFC had selected Boeing Co. and Westinghouse Electric Corp. for parallel feasibility studies of reflector satellite to provide illumination over land masses at night, *Technology Week* reported. Studies, which would be worth about $120,000 each for 90-day effort, would be made for NASA and DOD. (MSFC PAO; *Tech. Wk.*, 7/25/66, 16)

- General Electric Co. announced "successful initial running" of 600,000-lb.-thrust GE4 turbojet engine developed under FAA contract. Engine "was started and accelerated to idle speed on July 18, and run up to 100 per cent speed on July 20, nine days ahead of the contract testing date. . . ." GE was competing with Pratt & Whitney Div., United Aircraft Corp., for Government contract to build engine for SST. (GE Release)

July 25: Tribute was paid in *Aviation Week* to "some of the men whose efforts played a significant role in bringing the Surveyor program through its technical and managerial morass" to success: W. Eugene Giberson; Robert Garbarini; Benjamin Milwitzky; Rep. Joseph E. Karth (D-Minn.); Fred Adler; Leo Stoolman; James D. Cloud; John Bozajian; Richard Cheng; Robert Roney; Ralph Colbert; Theodore F. Gautschi; Edward Pfund; Marshall Johnson; Richard Gunter; Robert J. Parks; Fielding Hedges; Thomas Lund; Donald Zimmet; Richard Davis; R. L. Roderick; Richard Iverson; R. E. Sears; and George Kerster. (*Av. Wk.*, 7/25/66, 21)

- "Businesslike international negotiations" in Geneva to draft treaty ensuring peaceful exploration of space received comment in *New York Times*: "... Space law will help confine national rivalries within the edges of the atmosphere, beyond which all men must work together in the gigantic task of exploration, a task to which there is no conceivable end." (*NYT*, 7/25/66, 26)

- General aviation (nonairline) pilots in 1965 flew record 2.6 billion miles in 16.7 million hours with 95,442 aircraft, while achieving lowest fatal accident rate in history, according to FAA's "Selected General Aviation Statistics." The 8% increase during 1965—largest annual increase since 1954—included 382 million more miles and one million more hours than in 1964, with "business flying" continuing to be busiest segment of general aviation operations. (FAA Release 66-72)

- Boeing Co.'s board of directors approved building 747 jet aircraft capable of carrying 350-490 passengers at speeds up to 600 mph. (UPI, *Wash. Post*, 7/26/66, A12)

- About 80 U.S. companies were experimenting with crystal fibers to produce strong lightweight materials for aerospace use, the *Wall Street Journal* reported. Laboratory tests had shown that tiny crystal fibers added to metals and plastics produced rigid, almost unbreakable plastics and featherweight metals that remained hard under temperature and pressure extremes. (Martin, *WSJ*, 7/25/66, 1)

- U.S. domestic airlines flew 25 per cent more revenue passenger miles during first six months of 1966 than during comparable 1965 period, according to Air Transport Assn. report. Nearly 30.2 billion revenue passenger miles were flown during first half of 1966 by scheduled trunk, local service, and helicopter carriers compared with 24.1 billion revenue passenger miles during first half of 1965. (*Av. Wk.*, 7/25/66, 40)

During week of July 25: Patent was granted to Wolfgang G. Offik, senior staff engineer for Chrysler Corp., for escape system to rescue workers from booster-launch gantry in case of explosion or fire. Equipment included series of rescue cabins, each suspended in its own cable at a different level in front of gantry. In emergency, workers and astronauts still outside spacecraft would enter cabins and drop in them into underground shelter, protecting them from flames on gantry and on ground. (Jones, *NYT*, 7/30/66, 29)

July 26: U.S. and U.S.S.R. agreed in principle that countries were internationally liable for damage caused to other states by objects they launched into outer space. U.S. accepted, with minor changes in wording, article on liability in Soviet treaty on space law under discussion in Geneva by Legal Subcommittee of U.N. Committee on Peaceful Uses of Outer Space. (*Wash. Post*, 7/26/66, A12)

July 26: At Pentagon ceremony, Chief of Naval Operations Adm. David L. McDonald presented astronaut wings to Astronaut Eugene A. Cernan (LCdr., USN) who set a world record for length of "walk in space" during June 3-6 GEMINI IX-A mission. McDonald suggested to NASA officials that "you get your own Navy" for astronaut recovery operations. USN felt smaller ships not needed in the Vietnam war could be used as effectively for this task as the aircraft carrier and dozen ships assigned to NASA.

Speaking later at NASA Hq., Cernan said that what a man needed in space to do useful work was a "three-point restraint system"—three points of contact with the spacecraft for stabilization. (Norris, *Wash. Post*, 7/27/66, A1, A6)

- Negotiations were underway between LRC and Martin Marietta Corp. on a $150,000 study contract to investigate feasibility of obtaining data on Venusian atmosphere using "a balloon or other inflatable buoyant device or devices floating in the planet's clouds." (NASA Release 66-196)
- MSFC awarded Univ. of Michigan two-month, $29,930 study contract to develop 20-channel "infrared sensing technique to survey the earth's natural resources." Operating on orbiting satellite, system would survey possible mineral deposits, analyze soil for growth potential, diagnose "health" of crops, and predict bountiful fishing areas. (MSFC Release 66-167)
- Hundreds of people along Eastern seaboard reported sighting several Ufo's. Robert A. Bennett, FAA watch supervisor at Fulton County (Ga.) Airport Tower, sighted three or four objects which slowly changed colors from red to green to blue. (UPI, *Wash. Post*, 7/27/66, A6)

July 27: Eleven scientists formerly associated with the International Geophysical Year (IGY) and International Quiet Sun Year (IQSY) urged permanent cooperation in international studies of earth in letter to the *London Times*: "These worldwide cooperative scientific projects have shown conclusively that however many and serious are the political problems that trouble the human race, it is possible for all the nations of the world to work closely together in great enterprises for the common good." Signers included U.S. representatives on IGY Committee Dr. Lloyd V. Berkner, former chairman of the NAS Space Science Board, and Dr. Homer E. Newell, NASA Associate Director for Space Science and Applications, and U.S.S.R. representatives Dr. Vladimir V. Belossov and N. V. Pushkov. (*NYT*, 7/27/66, 12)

- USAF pilots Col. Robert L. Stephens and L/Col. Walter F. Daniel received official certificates from President Johnson for establishing three absolute and six jet-class world records May 1, 1965, in YF-12A aircraft. Col. Stephens had set new speed record of more than 2,070 mph and new altitude record of 80,259 ft. Col. Daniel had set new speed records for straight course of over 2,000 mph and on closed course of 1,688 mph. Records were formerly held by U.S.S.R. (*Pres. Doc.*, 8/1/66, 1000)
- Sydney (Australia) Univ. physicist Stuart Thomas Butler said that U.S., U.K., and U.S.S.R. were studying the possibility of using a "nuclear rocket" to prevent the asteroid Icarus from colliding with earth, UPI reported. Icarus would pass within four million miles of earth in June 1968. Smithsonian Astrophysical Observatory, world clearinghouse for all astrophysical information, said it had no knowledge the three nations were studying the possibility. Two authorities on asteroids,

Dr. Paul Herget, Univ. of Cincinnati, and Dr. Gerald Clemence, Yale Univ., said there was not "one chance in a billion" Icarus would strike earth. (UPI, *Chic. Trib.*, 7/28/66)

July 27: Westinghouse Electric Corp.'s Defense and Space Center had developed a 7¼-lb. TV camera that could be carried inside Apollo spacecraft and by astronauts as they explored the lurain. Prototype had been delivered to NASA and a commitment obtained for live TV coverage. (Gibbons, *WSJ*, 7/27/66, 1)

July 28: U.S.S.R. launched COSMOS CXXVI to "continue exploration of outer space" into earth orbit with 359-km. (223-mi.) apogee; 212-km. (132-mi.) perigee; 51.8° inclination; and 90-min. period. Equipment was functioning normally. (Tass, 7/28/66)

- X-15 No. 1 was flown by NASA test pilot John B. McKay to 3,682 mph (mach 5.19) and 241,800-ft. altitude. Mission objectives included micrometeoroid collection and evaluation of pace transducer, horizon scanner, nonglare glass, and stick kicker. (X-15 Proj. Off.)

- NASA in conjunction with the Brazilian Space Commission and Canadian Research Council would launch 10 Nike-Cajun sounding rockets with 80-lb. acoustic grenade payloads from Natal, Brazil; Churchill Research Range; Point Barrow, Alaska; and NASA Wallops Station. Four rockets would be launched from Churchill and two from each of the other sites. Experiments would begin whenever noctilucent clouds appeared over Churchill area and would be conducted over 24-hr. period. Project would be coordinated by GSFC. (NASA Release 66-199; Wallops Release 66-42)

- ComSatCorp announced in financial statement that revenues from operating EARLY BIRD I comsat had totaled $2,107,000 for first half of 1966 while income from temporary cash investments was $4,472,000. Cash and temporary cash investments totaled $186,534,000 on June 30. (ComSatCorp)

- Expansion of behavioral and social research to keep pace with scientific and technical advance was urged by Dr. Eugene B. Konecci, NASC staff member, at NDEA Institute for Advanced Study in Industrial Arts Education at Univ. of Maryland. "One of the biggest problems confronting us," he said, "is that we tend to train our young people for today's world rather than prepare them for the new developments of the future. Aerospace systems approach can help to neutralize this problem by outlining the parameters and making a model to help predict the future requirements.... Fear of innovation is based generally on lack of knowledge, so we must educate our entire population to our rapidly changing world." (Text; *AF/SD*, 10/66)

July 29: Saturn V 3rd stage (S-IVB) was static-fired for full flight duration at Douglas Aircraft Co.'s Sacramento, Calif., test facility, MSFC announced. During captive firing—simulating operation of propulsion system during normal 3rd stage flight in a lunar mission—stage burned 150 sec., shut down for 1½-hr. simulated coast period, then reignited and operated 290 sec. Stage was powered by Rocketdyne J-2 hydrogen-oxygen engine that would develop 200,000-lbs. thrust in flight. (MSFC Release 66-173)

- First launch of USAF's Titan III-B-Agena launch vehicle from Vandenberg AFB placed an unidentified satellite into polar orbit. (SAC TWX; *Av. Wk.*, 8/8/66, 29)

July 29: NASA pilots Robert Champine and Ron Gerdes were among 15 pilots participating in "extensive evaluation" of USA's XV-5A V/Stol aircraft at Ft. Eustis, Va. Pilots' observations and evaluations were being compiled and studied by USA "for application on future XV-5A tests" and V/Stol aircraft in general. Test program would continue through September 1966. (*Langley Researcher,* 7/29/66, 4)

- DOD announced that USA's OV-1 Mohawk surveillance aircraft had claimed five new world aviation records: (1) time to climb 3,000 meters (9,842 ft.)—3 min. 46 sec.; (2) time to climb 6,000 meters (19,685 ft.) —9 min. 9 sec.; (3) sustained horizontal flight altitude—32,000 ft.; (4) nonstop straight-line distance—2,422 mi. at 255-mph average speed; and (5) average speed over 100-km. (63-mi.) closed-circuit course at 5,000-ft. altitude—292 mph. Records claimed had been filed with Fédération Aéronautique Internationale under new category for land-based turboprop aircraft weighing 13,227 to 17,636 lbs. (DOD Release 643-66)
- NASA-sponsored space exhibit opened in Arts and Industries Building of the Smithsonian Institution, Washington, D.C.; exhibit would begin U.S. tour Sept. 5. (EH)

July 30: Dr. C. V. Raman, 1930 Nobel Prize winner in physics, said in convocation address at Indian Institute of Technology in Madras that it was "nothing but sheer raving lunacy" to spend millions of dollars "to shoot men into space and make them walk there." Dr. Raman said it was "mere pretence" to say these exploits were intended to find out what was happening on the moon. "It is militarism, very thinly disguised." Dr. Raman said he thought spaceflight was the "most sinister aspect" of the progress of science in the last 60 yrs. (*NYT,* 8/1/66, 42)

During July: Teófilo M. Tabanera, president of Argentina's Comisión Nacional de Investigaciones Espaciales (CNIE), reviewed accomplishments and current plans of space program in the *Air University Review.* Since creation of CNIE in 1960, Argentina had organized meetings to promote interest in space research and conducted experiments in aeronomy, ionospheric studies, cosmic radiation, and meteorology, launching balloons and rockets from Chamical Rocket Range. Current plans included: (1) continuation of Centaure sounding rocket launches for wind studies, using "luminous-trails" technique; (2) cooperation in inter-American experimental meteorological sounding rocket research network (EXAMETNET) under agreement with NASA to launch boosted Darts and Arcas rockets; and (3) launchings of sounding rockets to study cosmic radiation and ionospheric phenomena. (*Air Univ. Review,* 7-8/66)

- Quasars, once believed rare, might account for 28 per cent of the radio sources in space, Dr. John D. Wyndham, research fellow at Cal Tech, reported on the basis of 10-mo. research project. Dr. Wyndham had located all 328 sources listed in Third Cambridge Catalog of Radio Objects, then identified many of them with visible light sources, increasing the number of identifications to more than 150. Identification of radio noise source with a visible light source was first step in determining its distance and confirming that it was a quasar, he explained. (AP, *NYT,* 7/6/66, 35)
- Six years of exploratory work at JPL and NASA centers on practical problems of landing on Mars was summarized by R. P. Thompson of JPL in *Astronautics & Aeronautics.* He concluded that: (1) design of first

lander should be simple and conservative, and capable of thorough testing so that no single failure mode would cause "catastrophic failure of mission"; (2) independent backup should be provided for every critical event in landing sequence with "selective" redundance to achieve high reliability; and (3) capsule should be instrumented to allow diagnosis of failures. (*A&A*, 7/66, 66-73)

During July: David H. Stoddard, M.D., Director of NASA Occupational Medical Division, cited in *Hospital Topics* two "ideal" physiological monitors developed in space medicine—biotelemetry, which permitted immediate observation of data, although subject was aware he was being monitored; and electrocardiocording, which did not allow immediate observation but "permits the subject to go where he pleases, do what he pleases and to almost forget that he is being monitored." (*Hospital Topics*, 7/66, 39)

- Czechoslovakian radio correspondent Jan Petranek reported: "Usually well-informed Moscow circles have stated that in the next few years the Soviet Union will have a spaceship weighing 100 tons at its disposal." (*Wash. Post*, 7/25/66, A2)
- Patent for system with direction finder that would assure ground stations continuous and efficient communication with orbiting satellites was granted electronics engineers Henry P. Hutchingson, Applied Research Lab., Sylvania Electronic Systems, and Dr. Paul R. Arendt, Army Electronic Command Labs., Fort Monmouth, N.J. Equipment could also correct "Doppler frequency effect" caused by relative movement of satellite and station. (Jones, *NYT*, 7/23/66, 29)
- 60 per cent of the Nation's commercial airlift capability was immobilized by the International Assn. of Machinists walkout, which began July 8. (*Av. Wk.*, 7/25/66, 41)

August 1966

August 1: In MSC ceremony, NASA Deputy Administrator Dr. Robert C. Seamans, Jr., presented GEMINI X Astronauts John W. Young and Michael Collins NASA Exceptional Service Medals for "outstanding contribution to space flight and engineering," and the astronauts reviewed their July 18-21 mission for the press. Pilot Collins revealed that maneuvering difficulties during his walk in space had forced him to approach GATV VIII three times before successfully retrieving a micrometeoroid experiment. He felt the "basic problem" in EVA was that "without some sort of handholds or restraining devices, a large percentage of the astronaut's time is going to be devoted to torquing his body around until it is in the proper position to do some useful work." Command Pilot Young said mission's fuel shortage had been caused by large out-of-plane error, not human error as previously believed: "We had to use a brute force method of rendezvous [with GATV X]. That takes quite a lot of fuel." (Transcript)

- NASA had signed $339 million supplemental agreement with Chrysler Corp.'s Space Div. converting contract for production of Uprated Saturn I (Saturn IB) 1st stages (S-IB) from cost-plus-fixed-fee to cost-plus-incentive-fee. Under contract managed by MSFC, Chrysler would manufacture, assemble, and test 12 stages and provide support services through February 1969. (NASA Release 66-201)
- NASA selected Warrier Constructors, Inc., for final negotiations of a $3.5-million, cost-plus-incentive-fee contract to complete construction and equip a lunar receiving laboratory at MSC by end of 1967. A central complex where samples of lunar surface material collected by Apollo astronauts could be received, examined, and processed, laboratory would also be equipped to quarantine spacecraft and crew after flight to moon. (NASA Release 66-200)
- USAF fired Minuteman II ICBM equipped with Mark 12 reentry vehicle from silo at ETR to target site about 5,000 mi. downrange. Test was 15th straight success for Minuteman II. (AF News Bureau, *NYT*, 8/2/66, 2)
- In unpublished letter to *Science*, Dr. J. Allen Hynek, head of Northwestern Univ.'s Dept. of Astronomy and USAF consultant on Ufo's since 1948, criticized the "American scientific establishment" for failure to investigate persistent Ufo phenomena [see Oct. 21]. (Lewis, *Wash. Post*, 8/29/66, A12)
- Radio Corp. of America was being awarded a $4,450,000 USAF definitive contract to supersede previously awarded letter contract for production of communication and electronic components for unspecified space satellites. Contract would be managed by ASFC's Space Systems Div. (DOD Release 652-66)
- N. A. Zakharov, chief of Soviet Civil Aircraft Research Institute, an-

nounced that U.S.S.R. would construct 220-passenger, 560-mph Tu-154 subsonic airliner to replace the Tu-104, Il-18, and An-10. New Yak-40 short-haul aircraft, designed to carry 24 passengers up to 375-mi. distances at 275 mph, was already under construction and would replace the Li-2 and An-2. (*Pravda,* 8/1/66, 6)

August 1: Team of 33 U.S. technical experts from Lockheed Aircraft Corp. arrived in Noervich, W. Germany, to reorganize maintenance on F-104G Starfighter jet aircraft and correct "technical difficulties" which had caused crashes of 61 Starfighters and deaths of 36 Luftwaffe pilots since 1961. (Reuters, *Wash. Post,* 8/2/66, C1)

- In reply to March 2 invitation by FCC to submit comments on whether or not private entities should be allowed to establish special-purpose communications satellite systems for domestic use, 18 organizations filed responses.

  Ford Foundation proposed that consideration be given to formation of a nonprofit nationwide television system—operating under Broadcasters Nonprofit Satellite Service (BNS)—which would carry an extensive schedule of educational programs financed by transmission of commercial TV programs. Foundation contended FCC had power to authorize such a satellite system and that the act creating ComSatCorp had anticipated additional systems to meet unique national needs.

  ComSatCorp filed brief saying there was no legal basis for authorizing private organizations to establish special-purpose communications satellite systems for domestic use and that it would be in the public interest to assign ComSatCorp a monopoly on all such systems. Opinion was accompanied by full technical proposal for multi-use, high-capacity, four-satellite system for domestic distribution of TV, voice, and data signals that would be operational by 1970 at estimated cost of between $110 and $126 million.

  AT&T argued it would be against "expressed national policy" to authorize satellite systems outside the framework of commercial communications carriers and urged FCC "to conclude in this finding that there can be no authorization of private satellite systems."

  Carnegie Commission on Educational Television said it was encouraging a study of a comsat system that would be specifically designed for noncommercial television. While conceding that national satellites might be advantageous in some cases, Carnegie Commission warned that such a system presented problems of "educational monopoly and centralized control of instructional and other material." FCC was requested to take no definitive action until Carnegie's forthcoming report on educational television was published in late 1966 or early 1967.

  Briefs of broadcasting industry cited urgency for domestic satellite service to cut costs of distributing television and radio programs. This cost reduction for the three networks—CBS, ABC, and NBC—was estimated at approximately $50 million a year. NBC said a determination as to how this satellite service should be obtained could not be made in the abstract but "would depend on ... a comparison of the respective merits of specific proposals."

  National Assn. of Manufacturers told FCC that manufacturing community should have full freedom of choice either to lease channels from ComSatCorp directly or establish private systems. (*CR,* 8/2/66, 17131-3; Gould, *NYT,* 8/2/66, 1, 18; *Av. Wk.,* 8/8/66, 27-8)

August 1: All transatlantic flights out of major European ports of embarkation were sold out, and 2,100 tourists were stranded in Europe as a result of IAM strike against five major U.S. airlines which began July 8. (*Av. Wk.*, 8/8/66, 42)

August 2: NASA's Lunar Orbiter A spacecraft was mated to its Atlas-Agena D booster in preparation for scheduled Aug. 9 launch from ETR. (*Tech. Wk.*, 8/8/66, 3)

- NASA was negotiating with RCA for $13-million cost-plus-award-fee renewal of previous contract to operate and maintain three data acquisition facilities. Renewal called for engineering and operations services for NASA's unmanned Space Tracking and Data Acquisition Network (STADAN), operation of weather satellite control centers at GSFC, and operation and maintenance of tracking stations at Rosman, N. C., and Fairbanks, Alaska. (NASA Release 66-202)
- Dr. T. Keith Glennan, first NASA Administrator (1958-1961), was elected a director of Air Products and Chemicals, Inc., producer of industrial gases and chemicals. Dr. Glennan was currently president of Associated Universities which operated Brookhaven National Laboratory for AEC and National Radio Astronomy Observatory under NSF sponsorship. (*NYT*, 8/3/66, 50)

August 3: X-15 No. 2 was flown to 3,443 mph (mach 4.85) and 249,000-ft. altitude by Maj. William J. Knight (USAF) on star-tracking mission. (X-15 Proj. Off.)

- Two tape recorders in NASA's NIMBUS II meteorological satellite had failed but spacecraft continued to transmit daytime and nighttime cloud-cover photos. Loss of first recorder, used to store data from Medium Resolution Infrared Radiometer (Mrir), would prevent scientists from receiving information on earth's heat balance, water vapor, and temperatures in the atmosphere. Failure of second recorder, used in measuring satellite's engineering performance, would force NIMBUS II to transmit information each time it passed a ground station instead of storing it for more convenient playback. Since its launch by NASA May 15, NIMBUS II had met all mission objectives and transmitted more than 200,000 weather photos on a global scale. (NASA Release 66-203)
- First manned Apollo mission might be launched in late 1966, predicted Astronaut Virgil Grissom during news conference held by Apollo crew and backup crew at North American Aviation, Inc.'s Downey, Calif., facility. Mission was officially scheduled for 1967. (Transcript)
- Senate Committee on Appropriations reported FY 1967 Independent Offices Appropriation bill (H.R. 14921), which included $4,991,600,000 appropriation for NASA: $4,246,600,000 for R&D; $95,000,000 for construction of facilities; and $650,000,000 for administrative operations. (NASA LAR V/126)
- Tokyo Univ.'s Institute of Space and Aviation announced successful launch of Japan's first television-equipped sounding rocket: 1.4-ton rocket carried 22-lb. TV camera to 200-mi. (322-km.) altitude. (UPI, *Wash. Post*, 8/4/66)
- NASA modified existing cost-plus-fixed-fee contracts with North American Aviation, Inc., and IBM Corp. for Apollo program services with new agreements totaling $252.6 million: under $145.6 million cost-plus-incentive-fee contract extended through December 1968, NAA would provide 52 additional J-2 engines for launch vehicles as well as support services; IBM would receive $107 million under multiple-incentive con-

tract, extended through February 1970, for design, development, implementation, maintenance, and operation of MSC's Real Time Computer Complex (RTCC). (NASA Release 66-205; MSFC Release 66-176)

August 3: Dr. Leonard Roberts, a specialist on reentry heat shielding at LaRC, was named Director of the Mission Analysis Div., NASA Office of Advanced Research and Technology, effective Sept. 1. He would succeed Clarence A. Syvertson, who would become Assistant Director for Astronautics at ARC. (NASA Release 66-198)

- U.S. Court of Customs and Patent Appeals rejected a challenge to the validity of a key patent in laser development granted to Dr. Charles H. Townes, former provost of MIT, and Dr. Arthur L. Shawlow, Stanford Univ., in 1958. Plaintiff R. Gordon Gould, a Columbia Univ. graduate student when Dr. Townes was doing research there, argued that he had initiated a critical portion of the laser design—the mirror principle—in November 1957. The patent for his idea, granted after the Townes-Shawlow patent, was held by a subsidiary of Control Data Corp. (*NYT*, 8/6/66, 10)

August 4: Legal Subcommittee of U.N. Committee on Peaceful Uses of Outer Space reached accord on nine substantive articles for proposed treaty on space law before adjournment after three weeks of negotiation in Geneva. Agreement provided that (1) planets should be freely accessible to all nations; (2) no nation should appropriate any part of them; and (3) man's conduct in space should be reserved to peaceful purposes and ruled by concepts of international law and U.N. Charter. Two key provisions proposed by U.S. on reporting of outer space activities and on open access to stations, equipment, and spacecraft on planets were rejected by Soviet delegate Platon D. Morozov, but U.S. Ambassador Arthur Goldberg told the press he hoped outstanding issues could be resolved when Subcommittee reconvened in New York in September. (AP, Wash. *Eve. Star*, 8/5/66, C4; *Wash. Post*, 8/5/66, A16)

- X-15 No. 3 was flown by NASA test pilot William H. Dana to 3,682 mph (mach 5.34) and 132,700-ft. altitude in his third flight. Primary purpose was pilot checkout. (X-15 Proj. Off.; UPI, *Chic. Trib.*, 8/5/66)

- USAF launched OV3-III satellite with Scout booster from WTR to measure charged particle hazards to space payloads. Orbital parameters: apogee, 2,781 mi. (4,475 km.); perigee, 223 mi. (358 km.); period, 136.9 min.; inclination, 81°. (*U.S. Aeron. & Space Act., 1966*, 154)

- NASA assigned LRC responsibility for development of space vehicle design criteria in the area of chemical propulsion. Howard W. Douglass, former director of LRC's Flox Project Office, was appointed Chief of Design Criteria Office and Assistant Chief of Chemical Rocket Div. (LRC Release 66-40)

- President Johnson appointed BOB Director Charles L. Schultze chairman of committee to oversee final review of recommendations made by Nov. 28-Dec. 1, 1965, White House Conference on International Cooperation. Schultze would be assisted by Conference's executive director, Raymond D. Nasher, and White House special assistants Walt W. Rostow and Joseph A. Califano, Jr. (*Pres. Doc.*, 8/8/66, 1025)

- American Federation of Government Employees (AFGE) at KSC had charged NASA with preferential hiring of retired military officers for choice civilian jobs—a practice which was "demoralizing" career officials and damaging the space program. AFGE's charge was supported by House Civil Service Manpower Subcommittee, which had been investigating the situation for several months. (Young, *Wash. Eve. Star*, 8/4/66, A2)

August 5: President Johnson signed H.R. 14324 authorizing $5,000,419,000 NASA appropriation for FY 1967, and then warned that he might have to cut back expenditures on space and other programs if prices and wages "rise in an inflationary way." He made no direct mention of current steel price hike or demands of International Assn. of Machinists in airline strike, but added: "If we are to continue our space effort and continue to make the magnificent progress represented by our past achievements, we can do so only if business and labor leaders will make their contribution by responsible pricing and bargaining decisions." (*Pres. Doc.*, 8/8/66, 1026)

- MSFC awarded North American Aviation, Inc., a $23,438,532 modification to an existing contract for "additional work in building and testing the Saturn V launch vehicle's second stage [S-II]." Work would include redesign of stage's umbilical system and additional testing of stage's hardware under simulated flight conditions. (MSFC Release 66-181)

- German scientists Drs. Otto Hahn, Lise Meitner, and Fritz Strassman were named joint recipients of AEC's 1966 Enrico Fermi Award for "their combined and individual efforts in discovering nuclear fission, and for their extensive experimental studies which led to this vital discovery." 1966 winners were first foreigners to receive award; Dr. Meitner was first woman recipient. (*NYT*, 8/6/66, 4)

August 6: GEMINI VI Astronauts Walter M. Schirra, Jr., and Thomas P. Stafford and GEMINI VII Astronauts James A. Lovell, Jr., and Frank Borman were named joint winners of the 1966 Harmon International Aviator's Trophy for achieving the first rendezvous in space by two separately launched, maneuverable spacecraft Dec. 15, 1965. Award, established by late pioneer aviator and balloonist Col. Clifford B. Harmon (USAF) "for exceptional feats of pilotry," would be conferred in fall 1966. (*NYT*, 8/7/66, 4; *Wash. Sun. Star*, 8/7/66, A18)

- Third anniversary of limited test-ban treaty, ratified by 112 nations, barring nuclear tests in the atmosphere, under water, and in outer space. (Marder, *Wash. Post*, 8/7/66, 1)

August 6-23: 25th anniversary of first U.S. manned rocket-assisted aircraft flights made by Homer A. Boushey, then Capt., USA Air Corps, in Ercoupe monoplane specially fitted with jet-assisted takeoff (Jato) units "to determine if military aircraft could use rockets to get into the air faster from short runways with heavy payloads." During Jato test series, conducted by USA Air Corps from March Field, Calif., under direction of the late Dr. Theodore von Kármán, Boushey made: two takeoffs with Ercoupe on rocket power alone; 11 takeoffs combining rocket and aircraft power; and four flights in which rockets were fired while airplane was at altitude. On Aug. 6, he first flew on rocket power; on Aug. 12, he made first rocket-assisted takeoff; and on Aug. 23 he made two takeoffs on rocket power alone. Three rockets under each of Ercoupe's wings produced total thrust of 168 lb. for 12 sec. Von Kármán's group formed company now known as Aerojet-General Corp. to manufacture Jato units. (*Tech. Wk.*, 8/8/66, 12)

August 7: New "self-healing" solar cell believed at least 50 times more radiation resistant than conventional power supply cells had been developed by RCA for GSFC. Unlike previous cells which were made of silicon and protected by transparent shields of quartz or sapphire, new cells contained small lithium additive which sealed gaps produced by bombardment of high-energy radioactive particles and reduced weight of shielding by up to 90%. GSFC physicist Milton Schach described the new cell as a "major accomplishment" and predicted its use would permit flights of over 1,000 days through center of Van Allen radiation belts. Applications were also foreseen for communications satellites, high-altitude military reconnaissance spacecraft, and missions to Jupiter, surrounded by high-energy electron radiation belt. (Wilford, *NYT*, 8/7/66, 40)

August 7: NASA Nike-Apache sounding rocket was launched from Churchill Research Range in GSFC-Univ. of Michigan pitot-static tube experiment to observe the fine atmosphere temperature structure at noctilucent-cloud altitude. Rocket and instrumentation performed satisfactorily. (NASA Rpt. SRL)

- Two series of coordinated NASA Nike-Cajun sounding rockets were launched from NASA Wallops Station, Churchill Research Range, Point Barrow, Alaska, and Natal, Brazil, in GSFC experiment to gather data—by means of exploding grenade technique—on atmospheric wind, temperature, pressure, and density as summer maximum of noctilucent cloud sightings approached. Rockets and instrumentation performed satisfactorily, except one rocket launched from NASA Wallops Station did not achieve predicted altitude. (NASA Rpt. SRL)

- First new comet in 1966 was discovered by Stephen Kilston, recent Harvard Univ. graduate, at Lick Observatory, San Jose, Calif., and confirmed Aug. 9 by Central Telegram Bureau, Smithsonian Astrophysical Observatory. Named for its discoverer, Comet Kilston had a magnitude of plus 10.6 and could be seen in the constellation Hercules, apparently moving southeast. (AP, *NYT*, 8/10/66, 5; AP, *Wash. Post*, 8/10/66, A5; AP, Wash. *Eve. Star*, 8/10/66, B4)

- First microscope that could either superimpose or rapidly alternate images was being used at Cal Tech to detect changes in position or size of metal crystals that might cause structural failure. Microscope was assembled by David S. Wood, Thad Vreeland, Jr., and David P. Pope. (Sci. Serv., *NYT*, 8/7/66, 39)

- Proposal for a permanent international program of science cooperation made by 11 leaders of the 1957-58 International Geophysical Year program and 1964-65 International Year of the Quiet Sun program in letter to London *Times* [see July 27] was praised by the *New York Times*: "The initiative taken by the eleven scientists represents a promising new attempt to cut through . . . [political] limits on cooperation and thus promote more effective research on behalf of all mankind." (*NYT*, 8/7/66, 8E)

- Hughes Aircraft Co. engineer's smuggling of a small U.S. flag on NASA's SURVEYOR I spacecraft which softlanded on the moon June 2 was a breach of discipline and a violation of international space policy, wrote Joshua Lederberg in the *Washington Post*. ". . . NASA policy requires 'minimum contamination' so that the moon can remain a useful preserve for the detection of any life that might be transported by, say, meteorites. This calls for 'clean room' standards during the assembly of spacecraft components.

  "A bootlegged flag which could not have been subjected to such control for decontamination is a clear violation of a policy to which this country's honor has been attached and calls for strenuous disciplinary measures to insure the integrity and credibility of purpose of our future exploratory missions." (Lederberg, *Wash. Post,* 8/7/66)

August 7: French-Soviet June 30 agreement on space cooperation was arousing speculation about Soviet participation in French Guiana launch site, reported Richard Lewis in *Chicago Sun Times.* Use of large Soviet launch vehicles—which were also missiles—at French Guiana range would "open a missile corridor to the soft underbelly of the United States early warning system," and "undoubtedly generate serious tension in the Western world and in Europe," Lewis suggested. But French restriction that all satellites launched from site use French boosters and crew "would make the Guiana port useless to the Russians for space research, since the only space launcher the French have developed is the Diamant" whose payload capability was equivalent to that of the Scout—smallest booster in U.S. inventory. (Lewis, *Chic. Sun Times,* 8/7/66, 11)

August 8: COSMOS CXXVII was launched into earth orbit by U.S.S.R. to continue exploration of outer space. Orbital parameters: apogee, 279 km. (173 mi.); perigee, 204 km. (127 mi.); period, 89.2 min.; inclination, 51.9°. Instruments were functioning normally. (Tass, 8/8/66; *Krasnaya Zvezda,* 8/10/66, 1, USS-T Trans.)

- Senate began consideration of FY 1967 Independent Offices Appropriation bill (H.R. 14921), which included $4,991,600,000 NASA appropriation. Two amendments proposed by Sen. William Proxmire (D-Wis.) to reduce NASA's appropriation—one by $998,320,000, the other by $150,000,000—were defeated. (NASA LAR V/128)

- NASA had selected Bendix Corp., Federal Electric Corp., and Philco Corp. for competitive negotiations of a five-year, $60-million contract for operation and maintenance services for Deep Space Network. Contract would be managed by JPL. (NASA Release 66-210)

- NASA Pasadena Office (NaPO), formerly the NASA Resident Office-JPL, was established "as a component field activity" of the Office of Space Science and Applications (OSSA), NASA Hq. In addition to negotiating, executing, and administering NASA contracts with Cal Tech for operation of JPL, new NaPO would provide procurement, contract administration, patent and technology utilization, and related services in support of OSSA and other NASA organizational elements. (NASA Hq. Bull.)

- Lack of major manned spaceflight goal beyond Apollo Applications (Aa) program was causing NASA to delay Saturn booster development for at least two years, reported George Alexander in *Aviation Week.* Major Saturn contractors, facing end of current production contracts, wondered which derivatives would be chosen for development and when development and production would begin. Although current inventory of Uprated Saturn I (Saturn IB) and Saturn V vehicles was sufficient for programs through Aa, future missions might require a "super booster" with 280,000-lb. payload capacity or an intermediate vehicle with 40,000-lb. payload capacity. (Alexander, *Av. Wk.,* 8/8/66, 59-78)

- Atlas F advanced ballistic reentry system ICBM fired from Vandenberg AFB apparently exploded and fell into the Pacific. USAF was investigating causes of the failure. (UPI, *NYT,* 8/9/66)

- FAA was proposing extensive changes in regulations for determining crashworthiness of transport aircraft and for meeting passenger evacuation standards. Proposed changes included: (1) possibility of evacuation of aircraft in 90 sec.—instead of 120 sec.; (2) increase in number and size of emergency exits and more uniform distribution of exits throughout passenger cabin; (3) improvement of emergency lighting systems; (4) extensive preflight briefing of passengers; (5) increase in number of flight attendants per passenger; and (6) changes in design of landing gear, fuel lines, and electrical cables for greater safety. (*Av. Wk.*, 8/8/66, 34)

August 8: New method of averting aircraft crashes on landing or takeoff was demonstrated at London's Royal Aircraft Establishment when a jet fighter traveling 80 mph down runway stopped short in three seconds on 400-ft.-long, 7-ft.-deep gravel bed. Establishment spokesman estimated that pilots of large passenger aircraft could stop in 10 sec. on long strip of ordinary gravel. (*NYT*, 8/9/66, 62)

August 9: NASA should submit to Congress by Dec. 1 its recommendations on possible major national space objectives, including costs and benefits of specific missions, concluded 439-page staff report prepared for House Science and Astronautics Committee's NASA Oversight Subcommittee. Subcommittee would hold hearings "at an early date" to define future space goals. Study—based on testimony of NASA center directors, industrial managers, scientists, and other interested parties—recommended greater interagency coordination in planning and urged NASA to "aggressively pursue possible mechanisms for more comprehensive program of international cooperation . . ." Future space programs should be evaluated in relation to national economy and include "increased emphasis on minimum cost logistics vehicles within the state of the art to assure that economical logistics and rescue systems are available in the early 1970s and that similar emphasis be placed on new generation of spacecraft design capable of ground recovery at lower cost." (Text)

- USAF launched unidentified satellite with Thor-Agena D booster from WTR. (*U.S. Aeron. & Space Act., 1966*, 154)
- Computer technique developed by JPL scientist Dr. Robert Nathan and applied to 12 photos of the lunar surface transmitted by NASA's SURVEYOR I spacecraft revealed details as small as 1/50 of an inch and, in some cases, apparently doubled observable characteristics. Enhancement process had been used successfully to intensify photos transmitted by NASA's three Ranger spacecraft and MARINER IV Mars probe. (NASA Release 66-206; JPL Release)
- NASA awarded Dow Chemical Co. a $4-million, cost-plus-fixed-fee extension to previous contract for KSC support services through June 30, 1967. (NASA Release 66-212)

August 10: NASA's LUNAR ORBITER I (Lunar Orbiter A) unmanned spacecraft was successfully launched by Atlas-Agena D booster from ETR in first U.S. attempt to orbit the moon and photograph possible landing sites for Apollo astronauts [see Aug. 14-31].

Agena 2nd stage fired to boost 850-lb. spacecraft into 100-mi. (161-km.) altitude parking orbit, reignited after 28-min. coast period, injecting spacecraft into 90-hr. translunar trajectory, and separated. On schedule, LUNAR ORBITER I deployed its four solar panels and two antennas and locked its five solar sensors on the sun. Only difficulty occurred when spacecraft's startracker failed to lock on Canopus and flight engineers were forced to lock it on the moon—a weaker navigational reference point. Assistant Project Manager James S. Martin, Jr., of LaRC was "still quite confident" about achieving lunar orbit, but noted it might be more difficult to obtain desired precision and exact positions required for optimum photos of lunar surface. At approximately 00:28 GMT Aug. 11 planned midcourse maneuver was successfully executed, and JPL scientists predicted spacecraft would miss original aiming point by only 50 mi.

Primary objectives of NASA's LUNAR ORBITER I mission, first in series of five, were (1) to place three-axis stabilized spacecraft into lunar orbit; and (2) obtain high-resolution photos of various types of lunar surface to assess their suitability as landing sites for Apollo and Surveyor spacecraft and improve knowledge of the moon. Photos would cover 3,000-mi. strip along moon's equator, concentrating on nine potential landing areas. Spacecraft would also attempt to photograph SURVEYOR I landing sites; provide precision trajectory information; and monitor meteoroids and radiation intensity in lunar environment. Lunar Orbiter program was managed by LaRC under direction of NASA's Office of Space Science and Applications. Tracking and communications were the responsibility of JPL-operated Deep Space Network. (NASA Release 66-195; Wilford, *NYT*, 8/8/66, 3; 8/11/66, 1, 12; 8/12/66, 51; O'Toole, *Wash. Post*, 8/11/66, A1; 8/12/66, A1; 8/13/66, A1; 8/14/66, A1, A6; AP, *Wash. Eve. Star*, 8/11/66, A3)

August 10: Senate passed FY 1967 Independent Offices Appropriation bill (H.R. 14921) 82–2 with NASA appropriation at $4,991,600,000 as reported by Committee on Appropriations. (NASA LAR V/130)
- Rice Univ. scientists successfully launched 500-ft.-long helium-filled balloon from National Center for Atmospheric Research balloon flight station, Palestine, Tex., to obtain electronic data on radioactivity within the Crab Nebula. When balloon reached 132,000-ft. altitude, 900-lb. electronic gamma-ray telescope began recording data. Payload was dropped by parachute on electronic signal from ground and recovered near Midland, Tex. Study was third in series financed under $166,500 USAF grant and was termed "highly successful." (*Houston Post*, 8/11/66)

August 11: NASA test pilot John B. McKay flew X-15 No. 1 to 3,511 mph (mach 5.06) and 251,000-ft. altitude to collect micrometeorites. As secondary mission, pilot performed maneuvers to check horizon scanner, electrical loads, and wing-pod flutter. (X-15 Proj. Off.)
- MSFC would negotiate with Chrysler Corp. and Douglas Aircraft Co. for procurement at $5-10 million of long lead time items for additional Uprated Saturn I (Saturn IB) launch vehicles. Interruption of production capability after completion of presently approved 12 vehicles would thus be avoided. Chrysler was building the S-IB 1st stage; Douglas was manufacturing S-IVB 2nd stage. (MSFC Release 66-183)
- Horace Sheely of the National Survey of Historic Sites, Dept. of the Interior, inspected Pakachoag Hill, Auburn, Mass., site of first successful liquid-fuel rocket flight made by Dr. Robert H. Goddard. Accompanying Sheely were members of the Auburn Rotary and Lions Clubs joint committee to preserve the Goddard site. (Joint Comm. Ltr.)

August 12: X-15 No. 2 was flown by Maj. William Knight (USAF) to 3,473 mph (mach 4.90) and 231,000-ft. altitude to conduct startracker experiment and base drag study and test alternate pitot static system. (X-15 Proj. Off.; AP, *NYT*, 8/14/66, 8)
- Comprehensive report on June 8 collision of XB-70 and F-104 aircraft near Barstow, Calif., compiled by two investigative boards—one to determine causes, the other to examine command responsibility—was transmitted

to Secretary of Defense Robert S. McNamara by Secretary of the Air Force Harold Brown. Inadvertent movement by F-104 into position "from which recovery was virtually impossible" was cited as "most probable cause" of collision; failure of Maj. Carl Cross (USAF) to eject from XB-70 before crashing was attributed to an escape system malfunction. Among report's recommendations were: improvement of AFFTC's operational procedures; modification of XB-70's escape system; and establishment of standards for formation flights. In accompanying memorandum, Secretary Brown said that Col. Joseph F. Cotton (USAF), XB-70 test director; Col. James G. Scott (USAF), Edwards AFB information officer; and John S. McCollom, Director of Research Vehicles for Aeronautical Systems Div., had been reprimanded and Col. Albert M. Cate (USAF) relieved of duties as Deputy for Systems Tests at AFFTC. (CR, 8/15/66, 18505-10; UPI, NYT, 8/16/66, 15; Maffre, Wash. Post, 8/16/66, 6)

August 12: New absorption bands, probably caused in part by reduced gases in the Martian atmosphere, had been found in near-infrared spectrum of Mars by Fourier spectroscopy, Lewis D. Kaplan, JPL, and Janine and Pierre Connes, Observatoire de Meudon and Centre National de la Recherche Scientifique, France, reported in *Science.* The authors believed the presence of such constituents might have "important implications for the possibility and nature of life on Mars." (*Science*, 8/12/66, 739-40)

- NASA had selected Honeywell, Inc., for negotiation of $5-million contract to design, fabricate, and test attitude-control and scan-control subsystems for two Mariner spacecraft to be launched in 1969. (NASA Release 66-216)
- Boeing Co. awarded subcontracts totaling $1.5 billion to Avco Corp., Martin Co., North American Aviation, Inc., Northrop Corp., Fairchild Hiller Corp., and LTV, Inc., for work on Sst prototype. (Boeing Release S-8928)
- Dr. T. L. K. Smull resumed his duties as Director of the Office of Grants and Research Contracts, NASA announced. Since Dec. 1, 1965, he had been on leave from that Office, serving as Special Assistant to the Administrator on NASA-university relationships. (NASA Ann., 8/12/66)
- M2-F2 lifting body vehicle, with NASA research pilot Milton O. Thompson as pilot, was successfully air-launched from B-52 aircraft at 45,000-ft. altitude in third glide flight from Edwards AFB. Purposes of flight were determining effect of increasing mach number and minimum damper requirements, and testing of longitudinal and lateral stability and control. (NASA Proj. Off.)

August 13: Five-man NASA board and House Science and Astronautics Committee's NASA Oversight Subcommittee had begun separate inquiries into failures of OGO I, II, III, OAO I, and NIMBUS I. Questions under investigation included: (1) Is the expenditure on observatory satellites wise? (2) Are the projects being carried out on an efficient basis? and (3) Is the taxpayer's interest being protected?

GSFC Director Dr. John F. Clark told AP he believed there had been "technical goofs but no management problems" associated with the failures. "Scientifically, we have not had a failure in OGO. Something less than 100 percent but more than 75 percent of OGO's experiment return has been outstandingly successful. OGO 1 and OGO 2—although classified as spacecraft failures—gave us more data than all other satellites combined up to that time." (*NYT*, 8/14/66, 1, 70; AP, Wash. *Eve. Star*, 8/14/66, A23)

August 13: Recent Trendex poll taken for Thiokol Chemical Corp. indicated that 71.4 per cent of Cincinnati residents were in favor of the Nation's lunar program. (Haggerty, *J/Armed Forces*, 8/13/66, 8)

- Research by NASA and Air Transport Assn. indicated that wet-pavement skidding by aircraft could be substantially reduced by cutting $\frac{1}{8}$- to $\frac{1}{4}$-in.-deep grooves one to two inches apart in runways. Test runways would be equipped with grooves at cost of $70,000 by early 1967 for further experimentation. (Hudson, *NYT*, 8/14/66, 90; UPI, *Chic. Trib.*, 8/15/66)

August 13-14: Alternate system kept U.S. tracking station at Robledo de Chavela, Spain, in operation when 500-600 ft. of cable were burned by forest fire raging nearby. Station, which played a critical role in NASA's Lunar Orbiter program, was expected to receive photos of the lunar surface in several days. (NASA Release 66-219)

August 14-30: NASA's photographic spacecraft LUNAR ORBITER I became first U.S. spacecraft to enter lunar orbit—only 15 mi. off target—after 92-hr., 236,319-mi. flight from ETR. Orbital parameters: apolune, 1,152 mi. (1,854 km.); perilune, 119 mi. (192 km.); period, 3 hrs. 37 min.; inclination, 12°. JPL reported drop in spacecraft's apolune Aug. 17 to 1,150 mi. (1,851 km.), and rise in perilune to 126 mi. (202.8 km.); moon's gravitational field was considered cause of orbital variations. After taking engineering photos transmitted to JPL Aug. 18-20, spacecraft was injected into close-in orbit by retrothrust maneuver initiated Aug. 21 for medium- and high-resolution photography of lurain. Tracking experts said 36-mi. (58-km.) perilune was "within a city block" of where it should be.

Spacecraft's cameras had exposed 211 dual frames of film to photograph nine potential Apollo and Surveyor landing sites during mission's photo acquisition phase, announced complete on Aug. 30. Readout was expected to be completed by Sept. 15. Medium-resolution camera returned good images; however, except for an excellent photo of moon's far side, all exposures made with the high-resolution camera were disappointing. Camera's malfunction was attributed Aug. 25 to spurious signal generated by spacecraft's film transfer motor at 31-mi. perilune. Signal was tripping high-resolution camera shutter and causing film smear; LUNAR ORBITER I's perilune was lowered to 23 mi. (37 km.) to eliminate signal. Unprogramed photograph of earth, commanded Aug. 23 and received at NASA's tracking station at Robledo de Chavela, Spain, Aug. 25, showed appearance of earth's terminator—line dividing sunlit and shadowed portion of planet—from distance of about 240,000 mi. (386,400 km.). Spacecraft also photographed SURVEYOR I landing site. (NASA Proj. Off.; NASA Releases 66-228, 66-230, 66-233; Wilford, *NYT*, 8/15/66, 1, 23; 8/16/66, 15; 8/19/66, 1; 8/20/66, 8; 8/21/66, 1; 8/22/66, 8; AP, *NYT*, 8/24/66, 15; *NYT*, 8/25/66, A5; 8/26/66, 13; AP, *NYT*, 9/1/66, 2; O'Toole, *Wash. Post*, 8/15/66, A1, A13; 8/19/66, A1; 8/20/66, A1; 8/21/66, A1; AP, *Balt. Sun*, 8/18/66; *Av. Wk.*, 8/29/66, 18)

August 14: Coordinated NASA Nike-Cajun sounding rocket series was launched from NASA Wallops Station, Churchill Research Range, Point Barrow, Alaska, and Natal, Brazil, in GSFC experiment to study atmospheric parameters of wind, temperature, pressure, and density as

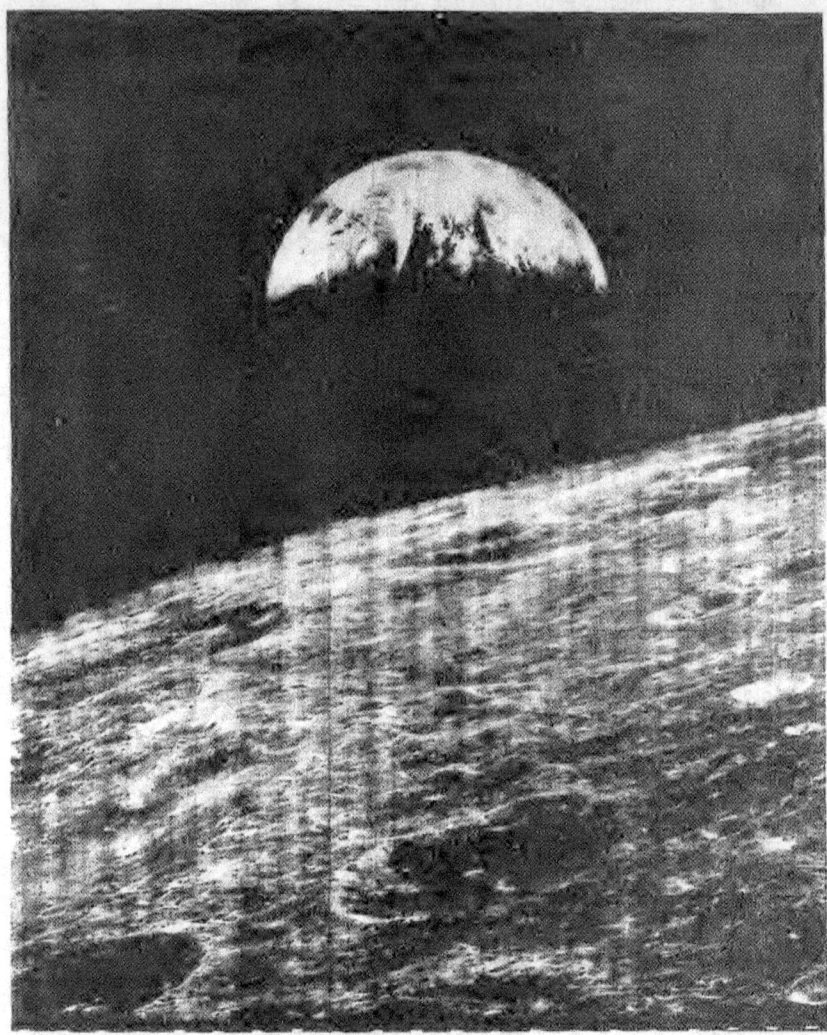

August 25: First picture of earth from the vicinity of the moon, taken by NASA's LUNAR ORBITER I.

summer maximum of noctilucent cloud sightings approached. Rocket and instrumentation performances were satisfactory. (NASA Rpt. SRL)

August 14: The critical voices raised during Aug. 8-9 Senate debate on the NASA portion of the Independent Offices Appropriation bill (H.R. 14921) indicated the U.S. space program was in "political trouble," concluded *New York Times* editorial. "Fundamental questions are being asked about whether this country should have gotten involved in the moon race in the first place, and invidious comparisons are being drawn about the dividends from money spent in space and those that might have been realized had those same funds been used to meet urgent human needs....

"There are important and useful things to be done in a post-Apollo space program, but it is doubtful in the extreme that the nation should go on indefinitely paying five billion dollars or more annually for such a program...." (*NYT*, 8/14/66, E10)

August 14: Capt. Chester M. Lee (USN, Ret.), Operations Planning Chief in Mission Operations, NASA Office of Manned Space Flight, was appointed Assistant Mission Director, Apollo. (NASA Ann., 9/15/66)

- Successful test firing of supersonic Chaparral, USA's newest air defense guided missile system, was conducted at WSMR. One of two weapon systems selected by USA for new air defense battalions, Chaparral met all test objectives when it intercepted Firebee target missile in continuing series of development firings. (AF News Bureau, *NYT*, 8/14/66, 16)
- Hurricane seeding plans for Project Stormfury—joint USN-ESSA hurricane control study program—were announced. Between Aug. 1 and Oct. 15, 17-plane force would be on alert in Puerto Rico to seed storms with silver iodide and then monitor temperature, atmospheric pressure, and structure of cloud. Project Stormfury was begun in 1961: experiments had been performed on hurricanes Esther in 1961 and Beulah in 1963, and on tropical cumulus clouds in 1963 and 1965. (DOD Release 654-66)
- Comment in the Washington *Sunday Star* called U.S. Sst program a "dubious" venture moved by extraordinary pressures: "This is a project the taxpayer should view with more than a little interest, if not apprehension, for its cost is coming out of his pocketbook and the price tag is escalating. Industry may repay the cost some day, but no one is sure just when. The SST, in fact, has become a symbol of national prestige. If we do not produce such a plane, American leadership in the aircraft industry will sustain a loss of face from which it could not recover. Perhaps it is impossible to back away from the situation. But the American public is at least entitled to know what kind of mare's nest this rivalry has got us into, and the hazards that must be faced." (*Wash. Sun. Star*, 8/14/66, C1)

August 15: NASA Javelin sounding rocket with new 4th stage ignition system was launched from Wallops Station to 434-mi. (698-km.) altitude. Instrumented payload measured positive ion composition of ionosphere, electron density, and temperature; and mapped distribution of Lyman-alpha radiation to determine amount of atomic hydrogen in earth's atmosphere. (NASA Rpt. SRL)

- NASA Nike-Apache sounding rocket launched from WSMR reached 100-mi. (161-km.) altitude in Dudley Observatory (Albany, N.Y.) micrometeorite particle collection experiment at a time of 33-year maximum for Leonid meteor shower. Rocket and instrumentation performed satisfactorily; since radar signals were lost after 265 sec., intensive search was necessary for payload recovery. Chief objective was not accomplished because of range scheduling, and launch was used for control shot. (NASA Rpt. SRL)
- Chrysler Corp. had announced it was building 10 more Redstone rockets for Sparta—joint U.S.-U.K.-Australian reentry measurements program—*Technology Week* reported. TRW Systems had prime contract for program; LTV Aerospace Corp. was building two-stage solid fuel velocity package to propel reentry vehicles back into earth's atmosphere at speeds simulating ICBM reentry. (*Tech. Wk.*, 8/15/66, 12)

August 15: George A. Lemke, Director of Apollo Reliability and Quality, NASA Office of Manned Space Flight, died at age 54. A veteran of more than 26 years in aircraft and space industry, Mr. Lemke had formerly been NASA Apollo Project Resident Manager at North American Aviation, Inc. (NASA Hq. Bull.)

- Program Evaluation Review Technique (Pert)—multi-level scheduling program used to spot potential problems before they developed—had made it possible during 1966 to meet every "major milestone" in NASA's Saturn-Apollo site activation program for KSC's Complex 39, *Technology Week* reported. Earlier, only about 2/3 of the target dates were being met. Using computer programed with data categorized into three levels of detail on individual contractor's needs, abilities, and timing, Pert set up work schedules based on key milestones as periodic progress reminders. Frequent computer checks of programed work provided officials with definite reading of status of remaining tasks. (*Tech. Wk.*, 8/15/66, 40)
- B/G Paul T. Cooper, Vice Commander of AFSC's Space Systems Div. (SSD), would replace M/G Ben I. Funk, retiring Sept. 1, as Commander of SSD, AFSC Commander Gen. Bernard A. Schriever announced. (AFSC Release 169.66)

August 16: NASA successfully launched 550-lb. Project Scanner instrument package from Wallops Station to 380-mi. (612-km.) altitude on 13.5-min. suborbital flight: two-channel radiometers measured infrared energy emitted from earth's horizon by carbon dioxide and water vapor; star-mapper telescope provided attitude orientation data. Managed by LaRC, Project Scanner was advanced research program to obtain data for designing and developing improved horizon-scanning instrumentation for space missions. (Wallops Release 66-43)

- USAF launched two unidentified satellites with Atlas-Agena D booster from WTR. (*U.S. Aeron. & Space Act., 1966*, 154)
- AFSC Commander Gen. Bernard A. Schriever received National Aviation Club's (NAC) 1966 Award of Achievement for his "significant contributions to the creation of the Air Force strategic missile force." The award would be presented Aug. 16 in Washington, D.C. (*Tech. Wk.*, 8/8/66, 13)
- More than 1,000 sheet metal workers out of 1,300 staged strike at KSC to protest NASA's permitting nonunion workers to install plumbing and sheet metal on Saturn V mobile service tower. Confined to construction jobs, strike was not expected to seriously affect major schedules unless prolonged. (AP, *NYT*, 8/17/66, 18)
- North American Aviation, Inc., received a one-year, $48,229 contract from MSFC to "investigate the best methods and hardware for performing manned Mars and Venus flybys with maximum use of Apollo/Saturn systems . . . as a stepping stone to manned landing missions." (MSFC Release 66-187)
- General Dynamics Corp. was being issued a $5,723,878 increment to a previously awarded (USAF) contract for work on design and development of an unspecified standard launch vehicle. Contract would be managed by AFSC's Space Systems Div. (DOD Release 701-66)
- Senate Judiciary Committee reported S. 1609, Federal Inventions Act of 1966. Bill would "establish a uniform national policy concerning property rights in inventions resulting from the expenditure of public funds for experimental, developmental, or research work," superseding present law governing NASA patent policy. (NASA LAR V/145; Text)

August 16: In *Houston Post* Edward W. O'Brien cited need for definition of future national space effort, pointing out that "the mighty industry-government team that has been mobilized for manned flight is tapering off." Employment, at the peak, had been 400,000; by the end of 1967, total would be 200,000; by 1968, only 100,000. What the U.S. must soon determine is "whether this unique assembly of brainpower and mechanical skill should be held together, or whether it should be applied elsewhere to better public advantage." (*Houston Post*, 8/16/66)

- In scheduling manned space flights, NASA Associate Administrator for Manned Space Flight Dr. George E. Mueller told the National Space Club in Washington, D.C., crew safety was the principal consideration, followed by accomplishment of program goals. He pointed out the "open-ended" aspect of Apollo mission scheduling and expressed confidence that lunar landing would be achieved before the end of the decade based on present program posture. During question and answer session, Dr. Mueller said that while NASA and its contractors had studied possible methods of rescuing astronauts in space and would continue to do so, the cost, complexities, and uncertainties made it appear to date that the money and effort would be better spent in making flights safer in the first place. He added: "We don't exactly have instant rescue for people flying around in airliners and this is a large segment of the population." (Text; Clark, *NYT*, 8/17/66, 16)

August 17: NASA's PIONEER VII (Pioneer B), second in new Pioneer series, was successfully launched from ETR's Launch Complex 17A by Thrust-Augmented Improved Delta booster into orbit around the sun where it would chart magnetic fields emanating from the sun; measure the quantity, energy levels, and direction of the solar wind; and differentiate between solar and galactic cosmic rays. The 140-lb., drum-shaped satellite had perihelion of 1.01 astronomical units (au), or 93.93 million mi. (151 million km.); aphelion of 1.125 au, or 104.625 million mi. (168.45 million km.); period of 402.95 days; inclination of 0.0946°. All systems were reported turned on and operating satisfactorily. Two days following launch, PIONEER VII successfully completed maneuver which locked its high gain antenna on earth and stabilized its orbital path. First of five flights in current Pioneer program—managed by ARC—was PIONEER VI, launched Dec. 16, 1965. Overall program objective was to provide continuing measurements over the solar cycle at widely separated points in interplanetary space. (NASA Release 66-211; NASA Proj. Off.; AP, *Wash. Post*, 8/18/66, L1; Wilford, *NYT*, 8/18/66, 15; UPI, *Wash. Post*, 8/20/66, A9)

- Aerobee 350 sounding rocket was launched by NASA from Wallops Station on second development flight of series. The 53-ft.-long rocket developed 60,000 lbs. thrust and reached 222-mi. (357-km.) altitude before impacting in Atlantic Ocean some 106 mi. from launch site. Preliminary examination of data indicated rocket's performance was close to predicted. Direction of sounding rocket program was under OSSA. (NASA Release 66-221; Wallops Release 66-44)

- U.S.S.R. announced that COSMOS CXXII, launched June 25, was collecting meteorological data which would be sent to other nations in accordance with 1962 agreement with U.S. Satellite carried "instruments for taking TV pictures of the clouds, cameras to photograph clouds by infrared rays on the day and night sides of the earth, and instruments to measure radiation in the earth's atmosphere." (Tass, 8/17/66)

August 17: Dr. Carl Schreiber, Procurement Specialist with NASA Hq. Office of Industry Affairs since Aug. 2, 1964, died in Washington, D.C. Dr. Schreiber, NACA employee since 1939, had been Assistant Procurement and Supply Officer from 1958-64. (NASA Hq. Bull.)

- Soviet Air Marshal V. A. Agaltsov claimed in Tass interview that the U.S.S.R. could fire long-range nuclear missiles from aircraft. He said this meant Soviet bombers could be beyond limits of enemy's antiaircraft defense zone and accurately fire missiles from distance of several hundred miles. (AP, Wash. *Eve. Star*, 8/17/66, 18)
- Martin Co. had been awarded a three-month, $75,000 contract to design and develop a flight-qualified, lightweight, hand-held lunar core driller for use on Apollo missions, MSC announced. Similar contract, also for $75,000, had been awarded Northrop Space Laboratories. (*Chic. Trib.*, 8/18/66)

August 17-18: Senate Commerce Committee's Subcommittee on Communications held hearings concerning Ford Foundation's Aug. 1 proposal to FCC to create a nonprofit domestic satellite system which would carry "an extensive schedule" of educational television programs financed by transmission of commercial TV programs.

Ford Foundation president McGeorge Bundy said that although the use of profits from a commercial business to finance a noncommercial undertaking might be novel to television, creation of a special carrier service in the public interest was not without precedent. He cited the Government's granting of special mailing privileges to newspapers and magazines. Bundy disclosed the Ford Foundation was planning to arrange "a series of meetings in various parts of the country with potential users of a noncommercial satellite system."

Fred W. Friendly, TV consultant to Bundy, described what he said was the underlying irony of modern TV: commercial networks had the money to do qualitative public service but lacked sufficient time on the air, while educational TV had the time but lacked the money.

ComSatCorp chairman James McCormack said the heart of the Ford proposal called for the "funneling of savings" that might be enjoyed by commercial communications carriers into support of a cultural undertaking. While acknowledging the value of educational TV, McCormack said he felt only Congress could rule whether one segment of a diversified society could be compelled to support another segment without benefit of specific legislation to that end.

AT&T vice president Richard Hough, ITT executive Bertram B. Tower, and Western Union Telegraph Co. vice president Earl D. Hilburn challenged the proposal saying that any such communications system should adhere to the common carrier principle of serving all users rather than any particular group such as broadcasters. While these witnesses supported the objective, they questioned the feasibility of the plan on economic and technical grounds. Tower suggested that consideration be given to permitting special rates for educational TV services through existing common carriers. Hilburn said that rate adjustments in favor of educational broadcasters might be authorized if some form of subsidy were determined to be in the national interest.

Sen. John O. Pastore (D-R.I.), who had called the hearings, said they were a "prelude" to action by FCC and noted that Congress almost certainly would want to explore the implications of the Ford proposal as it related to a private corporation. Hearings were completed Aug. 23. (ComSatCorp Releases; Gould, *NYT*, 8/18/66, 59; *Wash. Post*, 8/18/66, A12; AP, *Wash. Post*, 8/19/66, A2)

August 18: U.S.S.R. transmitted to U.S. for first time information obtained from its only known meteorological satellite, COSMOS CXXII, launched June 25. Previously, U.S.S.R. had relayed only conventional observations from land stations, ships, and balloons. Direct telecommunications channel between Moscow and Washington, D.C., had been established after March 1963 signing of bilateral agreement for exchange of meteorological satellite data under June 1962 space cooperation accord. (Bird, *NYT*, 8/20/66, 1; AP, *Wash. Post*, 8/21/66, A6)

- USAF launched unidentified satellite with Scout booster from WTR. (*U.S. Aeron. & Space Act., 1966*, 155)
- NASA test pilot William H. Dana flew X-15 No. 3 to 3,545 mph (mach 5.20) and 178,000-ft. altitude in flight test for altitude buildup. As secondary mission, pilot performed maneuvers to check out boundary-layer noise, horizontal tail loads, and heat transfer panels. (X-15 Proj. Off.)
- U.N. Ambassador Arthur J. Goldberg told Senate Committee on Foreign Relations he was "encouraged" about chances for reaching agreement with U.S.S.R. on treaty to ensure peaceful uses of outer space but that "several key issues"—including U.S.S.R.'s position on reporting its activities in space—still posed obstacles after 24 days of negotiations in Geneva. Goldberg said that despite this and other problems, he believed "the basis for resolving the outstanding issues has already been agreed upon." (AP, *Wash. Post*, 8/18/66, A1)
- Joint Senate-House Comference Committee on Independent Offices Appropriations reported out H.R. 14921 which included a $4,968,000,000 NASA FY 1967 appropriation: R&D, $4,245,000,000; construction of facilities, $83,000,000; administrative operations, $640,000,000. (NASA LAR V/136)
- National Labor Relations Board settled AFL-CIO sheet metal workers strike that began at KSC Aug. 16 to protest use of nonunion labor to install sheet metal and plumbing on Saturn V mobile service tower. Over 1,000 of 1,300 construction workers had honored picket lines. (AP, Phil. *Eve. Bull.*, 8/18/66)
- Helicopter-like atmosphere-entry-and-descent system employing unpowered rotor was undergoing wind tunnel testing at ARC as means of returning to earth from space. System could land on any solid, level surface and return vehicles of wide range of shapes from orbit; most of the lift would be supplied by auto-rotor and not by aerodynamic shape of vehicle. In orbit, blades for rotor system would be folded and stowed. On approaching earth, blades would be released and passage through atmosphere would start them turning. (NASA Release 66-217)
- Government of Pakistan announced plans to use comsats to relay telephone communications between East and West Pakistan. Two satellite tracking stations would be built—one in each province—and would be operative by 1968. (Reuters, *NYT*, 8/18/66, 38)
- Colonization of ocean floor would be more useful to man than colonization of the moon, said B. G. Anderson, senior research psychologist for General Dynamics Corp.'s Electric Boat Div. at joint AIAA-USN Marine Systems Conference in Los Angeles. The food and minerals available to man in and under the oceans were more important natural resources—

and more easily retrieved—than any likely to be found on the moon or elsewhere in the solar system. (L.A. Times, *Wash. Post*, 8/18/66, H4)

August 19: USAF launched USA's SECOR VII geodetic satellite, ERS-XV satellite, and an unidentified satellite from Vandenberg AFB, using Atlas-Agena D booster. (*U.S. Aeron. & Space Act., 1966*, 155)

- NASA selected McDonnell Aircraft Corp. for negotiation of $9-million, fixed-price contract to provide an airlock for an experiment in which astronauts would enter empty hydrogen tank of a spent Uprated Saturn I (Saturn IB) 2nd stage (S-IVB). Airlock would be stacked on space vehicle between Saturn and Apollo units using lunar module mounts. In orbit, command and service modules would separate and dock with airlock unit, and the crew would activate systems to pressurize spent hydrogen tank for habitation; hatch in airlock would permit astronauts to egress into space without depressurization of tank or spacecraft. First mission in MSFC-managed program would be carried out "no earlier than 1968." (NASA Release 66-223; MSFC Release 66-189)
- Fuel line leading to Saturn V launching pad at KSC ruptured, spilling more than 800,000 gallons of liquid oxygen. Spokesman said it was not yet known whether line break would delay qualification of pad to receive flight version Saturn V during September 1966 for launching early in 1967. (AP, *Wash. Post*, 8/20/66, A7)
- About 600 construction workers staged strike at KSC—one day after end of three-day sheet metal workers' strike—in dispute with McGregor-Werner Co., which provided administrative services to KSC. (AP, *NYT*, 8/20/66, 26)
- International Assn. of Machinists (IAM) voted 17,727 to 8,235 to ratify three-year contract negotiated Aug. 15, ending 43-day strike that had halted five major U.S. airlines. (*F on F*, 8/18-24/66, 317)

August 22: NASA had canceled hardware development and fabrication of Apollo experiments pallet which was designed to hold modular experiments and to fit into a section of the Apollo service module. Four firms selected in November 1965 for four-month Phase 1A design studies had been advised of decision: Lockheed Missiles & Space Co.; Martin Co.; McDonnell Aircraft Corp.; and Northrop Space Laboratories. (NASA Release 66-224)

- Twenty-two non-U.S. airlines—14 that had reserved delivery positions for Sst and eight that might be potential customers—had been invited to conduct independent evaluations of the competing U.S. Sst designs, *Aviation Week* reported. (*Av. Wk.*, 8/22/66, 26)
- Many Ufo's might be corona discharge—balls of ionized air—originating along high-tension electric power lines, suggested Phillip J. Klass in *Aviation Week*. After comparing corona discharge and reported behavior of Ufo's, Klass found that they were similar in everything but size. Discrepancy in size might be from illusionary nature of light, fright of the observers, or lack of familiar objects for comparison. Klass emphasized that his theory could not explain all such sightings, but urged USAF to continue investigating corona discharge theory. (Klass, *Av. Wk.*, 8/22/66, 48-60)
- The study of future national space objectives by the House Committee on Science and Astronautics' Subcommittee on NASA Oversight was termed a "timely and thoughtful" document in *Technology Week* editorial by William J. Coughlin. Noting that its major contribution "is to focus attention on the urgent need for a major decision concerning

the nation's space program," Coughlin added: "Not only does Congress have a role to play in such decisions, but the Teague subcommittee is moving to fill a vacuum in the decision-making process. The Administration has shown itself reluctant to bring forth decisions in this area. NASA itself has backed away from defining the next space objective. The subcommittee is performing a valuable national service in attempting to bring the matter to a head." (*Tech. Wk.*, 8/22/66, 50)

August 22: In editorial on current public debate over need for space rescue system, Robert Hotz noted in *Aviation Week* that NASA had "spent considerable time, money and ingenuity in designing safety into its manned spaceflight hardware" and in "developing sound operational procedures" that minimize the opportunity for disaster. He added that hearings conducted by Rep. Olin Teague (D-Tex.), chairman of the Manned Spaceflight Subcommittee of the House Committee on Science and Astronautics, and a RAND Corp. study commissioned by NASA should provide "much useful data on which to base a technically sound and economically feasible program that can provide the maximum space safety possible without jeopardizing the basic goals of the missions." (*Av. Wk.*, 8/22/66, 11)

- USAF School of Aerospace Medicine had concluded 15-day test to evaluate effects of increased carbon dioxide on four airmen in spacecraft cabin simulator. One of a series of such experiments, the latest took the level of experience up to the 4% CO_2 mark, compared with normal ground level conditions of 0.033% CO_2. Aim was to assess problems astronauts might have to face with a partial failure of their life support system. (*Tech. Wk.*, 8/22/66, 4)

August 23: AEC Chairman Dr. Glenn T. Seaborg discussed space role of nuclear energy in keynote address at Western Electronic Show and Convention in Los Angeles. In addition to advantages of nuclear propulsion, Dr. Seaborg said Snap (Systems for Nuclear Auxiliary Power) systems offered: (1) independence of sunlight, permitting use in shadow of a planet or moon and at great distances from the sun; (2) reduced atmospheric drag and propellant weight for maintaining low orbits of manned orbiting stations; and (3) larger quantities of electric power. He suggested nuclear power systems assembled on the moon could not only provide heating, cooling, and electric power, but could extract water and minerals and help produce synthetic food. (Text)

- General Electric Co. announced that GE4 turbojet engine had reached full power without augmentor in excess of 40,000 lbs. thrust. GE was competing with Pratt & Whitney Div. of United Aircraft Corp. for Sst engine contract. (GE Release 66-51)

August 24: U.S.S.R. launched 3,608-lb. LUNA XI space station toward moon for "further testing of systems of an artificial moon sputnik and scientific explorations of near lunar space." Preliminary data indicated that spacecraft was traveling close to calculated trajectory and equipment was functioning normally. (Tass, 8/24/66)

- Senate unanimously approved House-Senate conference report on FY 1967 Independent Offices Appropriations (H.R. 14921). Funds for NASA were: R&D, $4,245,000,000; construction of facilities, $83,000,000; and administrative operations, $640,000,000. (*CR*, 8/24/66, 19598-600)

- NASA extended for one year a contract with Documentation, Inc., to operate its College Park, Md., Scientific and Technical Information Facility —containing world's largest collection of aerospace literature—during

FY 1967 at cost of $5,150,000. Monitored by NASA Hq. Scientific and Technical Information Div., contractor would also operate Selective Dissemination of Information Program, using computer techniques to notify NASA scientists and engineers individually of new developments of direct interest to their work. (NASA Release 66-227)

August 24: NASA Nike-Apache sounding rocket launched from Churchill Research Range reached 98 mi. (157 km.) in NASA-Univ. of New Hampshire experiment to provide data on neutron intensity at different latitudes, solar x-ray fluxes, Lyman-alpha radiation, and ionospheric electron densities. Rocket performed satisfactorily and instrumentation performance was considered excellent. (NASA Rpt. SRL)

- Two NASA Nike-Apache sounding rockets launched 90 min. apart from NASA Wallops Station reached 102-mi. (164-km.) and 105-mi. (169-km.) altitudes in GCA Corp.-Univ. of Illinois experiment to measure recombination coefficients of major E-region constituents and investigate importance of metallic ions in lower ionosphere at three solar zenith angles. Both rockets performed satisfactorily, and instrumentation on first flight—except for ion-mass spectrometer—obtained good data. Telemetry deteriorated at lift off on second flight, and transmitter failed at about 145 sec. Some data might be extracted through special treatment of telemetry record. Third flight in series was postponed to allow investigation of ion-mass spectrometer difficulty. (NASA Rpt. SRL)

- M2-F2 lifting body vehicle was air-launched from B-52 aircraft at 45,000-ft. altitude in fourth glide flight at Edwards AFB with NASA research pilot Milton O. Thompson as pilot. Purpose of flight was determination of control damper requirements, lift-drag ratio, elevon response and flap effectiveness, and longitudinal stability and control. (NASA Proj. Off.)

- House debated H.R. 935 on establishment of a Dept. of Transportation. Rep. Benjamin J. Rosenthal (D-N.Y.) proposed that it include an Office of Aircraft Noise Control and Abatement to assume duties of noise abatement research groups presently operative in several Government agencies, including NASA. (*CR*, 8/24/66, 19509-33)

August 25: NASA's Apollo/Saturn AS-202 mission was successfully launched from ETR's Complex 34 at 1:16 p.m. EDT: 56,000-lb. unmanned Apollo spacecraft (011) was boosted into suborbital flight by Uprated Saturn I launch vehicle generating 1,600,000 lbs. thrust in second flight test of major spacecraft systems; second performance check of command module (Cm) heatshielding; and third flight test of Uprated Saturn I. Liftoff and powered flight were as programed. After spacecraft separation, 21,500-lb.-thrust service module (Sm) propulsion engine burned 3 min. 35 sec. to boost spacecraft to 706-mi. (1,128.6-km.) altitude. Sm's engines ignited three more times to test rapid restart capabilities, with last burn separating Sm. Cm reentered earth's atmosphere at more than 19,900 mph. Maximum temperature of spacecraft's surface was calculated to be about 2,700°F; temperature inside Cm was 70°F. Main parachutes deployed at 23,850-ft. altitude, lowering Cm to splashdown in Pacific Ocean some 500 mi. southeast of Wake Island—200 mi. from target—at 2:49 p.m. EDT. Recovery was by aircraft carrier U.S.S. *Hornet*. Apollo heatshield well withstood high heat-load test, and the spacecraft was in "stable condition 1."

Officials said that during flight, minor problem developed in unit which was to cool drinking water and various electrical components. They did not consider problem serious.

NASA Associate Administrator for Manned Space Flight Dr. George E. Mueller told postflight press conference that "the results of today's flight—once examined—will provide us with the information necessary" to make a final decision whether to commit the next Apollo flight to a manned mission. (NASA Release 66-213; NASA Proj. Off.; Hines, Wash. *Eve. Star*, 8/26/66, A4; AP, *Wash. Post*, 8/26/66, A3; Wilford, *NYT*, 8/26/66, 1)

August 25: X-15 No. 1 was flown to 3,511 mph (mach 5.00) and 257,500-ft. by NASA test pilot John B. McKay to conduct series of high-altitude scientific experiments: micrometeorites and extraterrestrial dust were collected in special container in wing-tip pod; intensity and spectral distribution of daytime sky conditions measured; and horizon scanner, electrical loads, and wing-pod flutter checked out. (X-15 Proj. Off.)

- Rep. Joseph E. Karth (D-Minn.), chairman of the Space Sciences Subcommittee of the House Science and Astronautics Committee, said at a luncheon of the Aviation/Space Writers' Assn. in Washington, D.C., that the war in Vietnam and the needs of "the Great Society in general" would keep the space budget from expanding soon. He said the budget of "more than $6 billion" that NASA planned to ask Congress for in January 1967 "is financially not in the cards for the near future." (Text)

- Lawrence Levy, founder and president of Allied Research Assoc., Inc., and former Defense Adviser to U.S. Ambassador to NATO, was sworn in as a consultant to NASA Administrator James E. Webb on "cooperation with western Europe and future space programs. . . ." (NASA Release 66-231)

- NASA selected Martin Co. for negotiation of $3-million, incentive-fee contract to build 11 experimental spacecraft equipped with parachute payloads "to investigate parachute designs and techniques for landing instrumented capsules on Mars." Four would be launched by high-altitude balloon systems and seven by Honest John-Nike rockets under LaRC's Planetary Reentry Parachute Program. (NASA Release 66-229)

- A phased-out Bomarc missile, serving as target for USAF and USN missiles, exploded shortly after it was fired from Vandenberg AFB. (UPI, *Wash. Post*, 8/27/66, A3)

- Gov. Edmund Brown of California told aerospace executives meeting in Los Angeles of plans to create state-level Office of Science and Technology to provide liaison between Government and industry in using aerospace skills to solve civilian problems. Five state contracts had been let for preliminary study of the problems, using the aerospace "systems engineering" approach. Brown said state's studies "not only demonstrated that the systems analysis concept would work" but also that "in some cases it was the only concept that would cut through the red tape and the customs arising from generations of solving social problems in the same old ways." (Sederberg, *L.A. Times*, 8/26/66; *CR*, 8/29/66, 20129-30)

- U.S.S.R. began rocket test series in the Pacific Ocean. (Tass, 9/6/66)

August 26: NASA officials revealed at Hq. news briefing that data accumulated thus far from LUNAR ORBITER I indicated the moon's shape departed from that of a perfect sphere with a bulge of about ½ mi. at its north pole, a depression of about ⅛ mi. around the Northern Hemisphere, a ⅛-mi. bulge around the Southern Hemisphere, and a depression of about ½ mi. at the South Pole. Conclusions were based on assumption moon's density was uniform. (AP, Wash. *Eve. Star*, 8/27/66, A3; *Av. Wk.*, 8/29/66, 18)

August 26: USAF Titan III-C carrying eight Initial Defense Communications Satellite Program (Idcsp) repeaters destroyed itself some 80 sec. after launch from ETR's Launch Complex 41. Preliminary observations indicated breakup of payload fairing at about 85,000-ft. altitude; trouble-detecting system sensed erratic behavior of rocket and triggered device that destroyed it. Available data showed "no abnormality in any of the vehicle's other systems." Detailed study of telemetry and film record would be necessary before cause of fairing malfunction could be determined. Launch was attempt to duplicate June 16 mission of identical Titan III-C that placed seven Idcsps and one gravity-gradient satellite into random, near-synchronous equatorial orbits as part of worldwide military communications system which would eventually include 23 comsats. (Wash. *Eve. Star*, 8/26/66, A1; UPI, *NYT*, 8/27/66, 14; AP, *Wash. Post*, 8/27/66, A3)

- Four sounding rockets—an Aerobee 150A, two Nike-Tomahawks, and a Nike-Apache—were launched by NASA from Wallops Station at predetermined intervals between 2:13 p.m. EDT and 3:11 p.m. EDT in conjunction with passage of EXPLORER XXXII satellite. Objective of series was to correlate measurements of properties, characteristics, and conditions of upper atmosphere obtained by rocket-borne experiments with similar measurements made by EXPLORER XXXII, second U.S. Atmosphere Explorer, launched May 25. Project was conducted by GSFC under overall direction of OSSA. (Wallops Release 66-45; NASA Rpt. SRL)

- President Johnson spoke at AEC's National Reactor Testing Station in Arco, Ida. (where world's first electricity from nuclear power had been produced), on his hopes for compromise agreements preventing spread of nuclear weapons and ensuring peaceful uses of space. He urged recognition "that at the heart of our concern in the years ahead must be our relationship with the Soviet Union. . . .

    "I believe that the Soviets share a genuine desire to enlarge the area of agreement. This summer we have been negotiating . . . a treaty that would limit future activity on celestial bodies to peaceful purposes. This treaty would, for all time, ban weapons of mass destruction, not only on celestial bodies, but also in orbit around the earth. . . ."

    He announced that treaty negotiations would resume Sept. 12, and continued:

    ". . . Peace will not dramatically appear from a single agreement or a single utterance or a single meeting.

    "It will be advanced by one small, perhaps imperceptible, gain after another, in which neither the pride nor the prestige of any large power is deemed more important than the fate of the world." (Text, *Pres. Doc.*, 9/5/66, 1160-64)

August 27: NASA Nike-Apache sounding rocket launched from NASA Wallops Station reached 94-mi. (151-km.) altitude in GSFC-Univ. of Michigan pitot-static probe experiment to measure atmospheric density, pressure, and temperature. Experiment, designed to observe diurnal variations in fine structure of region from 9–68 mi. (15–110 km.), was launched in conjunction with thermosphere probe experiments launched Aug. 26. Rocket and instrumentation performed satisfactorily. (NASA Rpt. SRL)

August 28: COSMOS CXXVIII was launched by U.S.S.R. into earth orbit with 364-km. (226-mi.) apogee; 212-km. (132-mi.) perigee; 90-min. period; and 65° inclination. Equipment, carried to "continue space investigations," was functioning normally. (*Pravda*, 8/28/66, 3)

- NASA Nike-Tomahawk sounding rocket launched from NASA Wallops Station reached 187-mi. (301-km.) altitude in GSFC-Univ. of Michigan experiment coordinated with Aug. 26 and 27 flights. Simultaneous measurement of N_2 and electron density and temperature, measurements of atmospheric ion and neutral composition, and comparative data from two mass spectrometers and other experiments were obtained; rocket and instrumentation performed satisfactorily. (NASA Rpt. SRL)
- Damage to a 900,000-gallon stainless-steel storage tank for Saturn V booster's liquid oxygen fuel would delay booster's first flight—scheduled for first quarter of 1967—by at least 45 days, NASA announced. Damage had occurred during Aug. 19 first-stage tanking test when pipeline had ruptured spilling 800,000 gallons of liquid oxygen; vacuum thus created inside tank had caused depression in tank's 2½-in.-thick dome. Laboratory tests would determine extent of resulting structural weakness. (*NYT*, 8/29/66, 8)
- U.S.S.R.'s official silence on fate of LUNA XI artificial moon satellite launched Aug. 24 prompted speculation that spacecraft had switched off its beam in orbit, overshot the moon, or crashed. (*Wash. Post*, 8/28/66, A10; Wash. *Sun. Star*, 8/28/66, A1)
- Cal Tech scientists Eric E. Becklin and James A. Westphal reported findings from 19-day telescopic study of comet Ikeya-Seki that began Oct. 14, 1965: temperature of comet was "entirely dependent" on the sun, varying from 700°F to 1,200°F in direct proportion to distance from sun; nucleus and tail were same temperature; and comet was composed of "lots of metallic material" rather than ice and dust as previously suggested. (*Wash. Post*, 8/29/66, A9; Bird, *NYT*, 8/29/66, 11)
- Commenting on President Johnson's Aug. 26 appeal for agreement on nuclear nonproliferation and peaceful uses of space, *New York Times* editorialized that issues of reporting space activities to U.N., and reciprocity and timing of visits to space vehicles and installations "would all appear to be soluble in time given a genuine desire for a treaty on both sides.... But the unanswered question is whether an East-West breakthrough on space can in fact come at all while the Vietnam impasse continues." (*NYT*, 8/28/66, E6)

August 29: Two NASA Boosted Arcas sounding rockets were launched one hour apart from WSMR in GSFC experiment to provide simultaneous measurements of D-region ions and electrons. Flights marked first launches of high-velocity Boosted Arcas rockets from rail launcher. Radar coverage was not provided, but rockets appeared to perform as predicted. Instrumentation functioned normally; however, parachute systems—designed to gather data for Polar Cap Absorption Program—were deployed too high and "streamered" entering dense atmosphere. (NASA Rpt. SRL)

- NASA Aerobee 150 sounding rocket launched from WSMR carried GSFC-instrumented payload to 128-mi. (206-km.) altitude to obtain dayglow measurements. Rocket and instrumentation performed well. (NASA Rpt. SRL)

August 29: Pictures from U.S.S.R.'s LUNA XI spacecraft that apparently entered lunar orbit were received and tape recorded by Jodrell Bank Experimental Station. Although unable to immediately decipher pictures, Station officials said "... they appear to be similar to those of Luna 9 and the American Lunar Orbiter transmissions." U.S.S.R. had made no official statement since Aug. 24 launch. (UPI, *NYT*, 8/29/66, 11; UPI, *Chic. Trib.*, 8/30/66; UPI, *Phil. Eve. Bull.*, 8/30/66)

- Plans for manned solar astronomical mission to orbit earth in Apollo spacecraft during period of maximum solar activity beginning in 1968 were announced by NASA. Mission objectives would be to acquire high-resolution measurements and observations of structure and behavior of sun above earth's atmosphere and to test man's capabilities for conducting astronomical observations in space. MSFC would have project and experiment development responsibility. (NASA Release 66-232)

- NASA Deputy Administrator Dr. Robert C. Seamans, Jr., told *Space Business Daily* that NASA had presented several proposed FY 1968 budgets to BOB, including one for "just over $6 billion" and another for $5.5 billion. He said that in the $5.5-billion budget about $500-$600 million would be for Apollo Applications (Aa); the $6-billion budget would include about $1 billion for Aa. Emphasizing that major funds for Aa must be included in the FY 1968 budget if the billions of dollars spent on Apollo technology were not to be wasted, he indicated that a budget of about $5 billion would end plans for an adequate Aa program to follow Project Apollo. He noted that under a $5-billion budget, there would be a greater percentage cut back in contractor operations as opposed to inhouse NASA efforts, but that with a $5.5-billion budget the current ratio would remain about the same. Seamans confirmed that the first manned Apollo lunar landing could come as early as 1968, but said it is "most likely" to occur in 1969. (*SBD*, 8/29/66, 317-8)

- ComSatCorp confirmed it was developing a plan which would have all commercial users of a domestic satellite communications system—telephone and telegraph companies as well as commercial television networks—underwrite educational TV. Funds would come from a portion of anticipated savings to communications users after a domestic satellite system was established. Details of plan would be submitted to Carnegie Commission on Educational Television—a private group making study on needs of educational TV. Carnegie study—to be released in late 1966 or early 1967—was expected to be relied on when Congress considered renewal of Educational Facilities Act—due to expire in 1967—which concerned educational TV. (*WSJ*, 8/30/66, 10; AP, *Wash. Post*, 8/30/66, D8)

- AT&T would sell its Andover, Me., ground station to ComSatCorp for $4,981,000, pending FCC approval. Station, which transmitted and received signals from EARLY BIRD I for transatlantic telephone, television, and data communications, had been leased by ComSatCorp since January 1965. (ComSatCorp Release)

- Death of Nick Piantanida, 33-yr.-old parachutist in a coma since oxygen supply failed at 57,000-ft. altitude during May 1 attempt to break world's freefall record. (*NYT*, 8/30/66, 41; *Wash. Post*, 8/31/66, A6)

- Aerial Coast Patrol of U.S. Coast Guard established 50 years ago.

August 30: X-15 No. 2 was flown to 3,614 mph (mach 5) and 102,200-ft. altitude by Maj. William J. Knight (USAF) to conduct heat tests. Pilot conducted base-drag studies and checked out stability and control, ablative materials, and wing tip accelerometer. It was 170th flight of the X-15 and 7th during August. (FRC Release 16-66; X-15 Proj. Off.)

August 30: Tass announced that LUNA XI had become second U.S.S.R. satellite to circle moon when it entered lunar orbit Aug. 27 with 1,200-km. (745-mi.) apolune; 160-km. (99-mi.) perilune; 2 hr. 58-min. period; and 270° inclination. In first official statement since Aug. 24 launch, agency reported that midcourse maneuver had been successfully executed Aug. 26 and that onboard equipment was functioning normally. LUNA X began orbiting moon April 3—first spacecraft to do so—but did not transmit photos to earth. (Tass, *Pravda,* 8/30/66, 1; AP, *Wash. Post,* 8/25/66, A3)

- First of series of high-altitude experiments to investigate parachute designs and techniques that might be incorporated into Voyager spacecraft scheduled for unmanned soft landing on Mars in 1973 was conducted by NASA from Walker AFB. The 15-ft.-diameter, 1,600-lb. disc-shaped flight unit containing packaged test parachute, instruments to record loads and parachute deployment, and ring of 12 small rockets for acceleration was carried by balloon to 120,000- to 125,000-ft. altitude then released on ground command. Acceleration rockets apparently ignited as planned, propelling unit upward at about 850 mph into arching trajectory. Parachute deployed; system descended in planned recovery area at WSMR. Test was apparent success; extensive performance analyses were being made. Balloon system had been tested July 18. Parachute project was managed by LaRC and coordinated with JPL, which managed Voyager program. (NASA Release 66-225; AP, Balt. *Sun,* 8/31/66)
- House passed H.R. 15963, bill to establish Cabinet-level Dept. of Transportation. Amendments proposed by Reps. Benjamin J. Rosenthal (D-N.Y.) and John W. Wydler (R-N.Y.) that would have established an Office of Aircraft Noise Control and Abatement in the Department were defeated. (*CR,* 8/30/66, 20342-84)
- *New York Times* editorial: "The Soviet-American race to the moon is heating up. Even as the United States-built satellite, Lunar Orbiter, finished photographing the moon, Moscow's huge Luna 11 also went into lunar orbit. The indicated success of the second unmanned flight of the Apollo moonship—another step toward putting the first Americans on the moon—undoubtedly will speed initial tests of a Soviet lunar vehicle." (*NYT,* 8/30/66)

August 31: U.S.S.R.'s LUNA XI that entered orbit around the moon Aug. 27 was described as containing "an entire orchestra of rocket engines." Vladimir Orlov, writing in *Pravda,* said the spacecraft contained in addition to its main power system a braking engine installation and "four directing engines with comparatively small thrust." (AP, *NYT,* 9/1/66, 2)

- Columbia Univ. professor of business administration Dr. Leonard R. Sayles was sworn in as consultant to NASA Administrator on organization policy and managerial practices and procedures. (NASA Release 66-235)
- Photographic proof that two clouds of "cosmic rubble" were orbiting the earth in the same path as the moon was announced by Lockheed Missiles & Space Co. engineer J. Wesley Simpson. The "rubble" might have come from all over the universe and could offer clues to the earth's origin, solar winds, and the possibility of life in outer space.

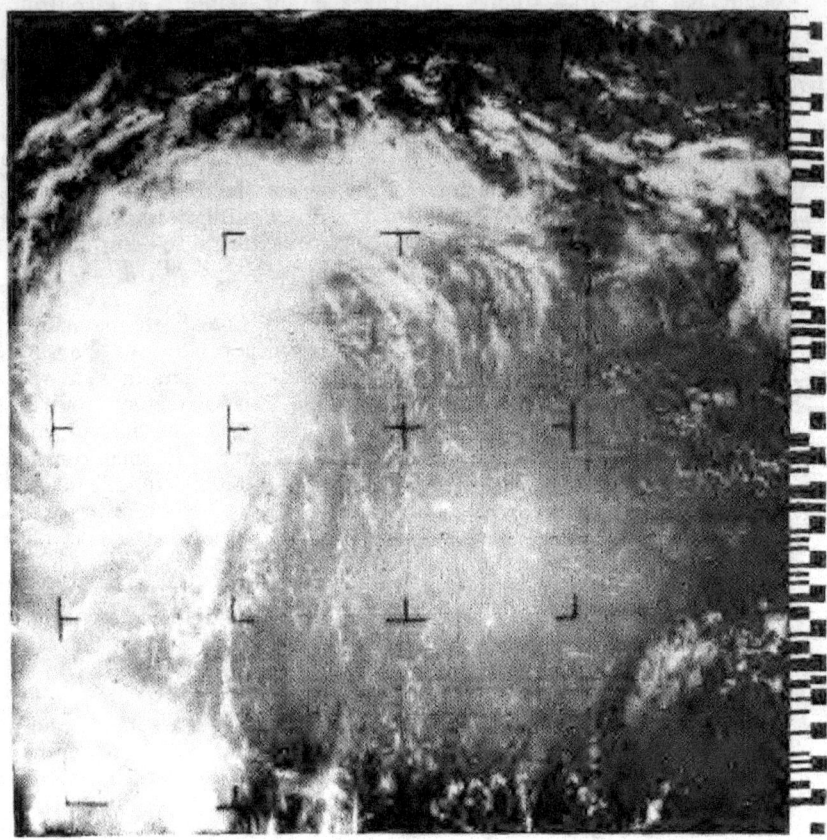

August 30: Hurricane Faith (white mass at left) as photographed from 700-mi. altitude by NASA's NIMBUS II satellite.

Although photographic images were too faint to be reproduced on paper, they were visible on negatives and when projected on a smooth white surface. (UPI, *NYT*, 9/1/66, 24)

During August: Dr. Raymond L. Bisplinghoff, head of MIT's Dept. of Aeronautics and Astronautics and Special Consultant to NASA Administrator, wrote in *Air Force and Space Digest* that there appeared to be widespread lack of public understanding that aeronautical and space developments might have a profound effect on transformation of society: "It is not so much the possession of the new technologies ... that counts. Of more importance are the quickened spirits, the sharpened intellects, and the developed resources that can permeate the whole fabric of a society." (*AF/SD*, 8/66, 86-7)

- In letter to House Committee on Science and Astronautics' Subcommittee on NASA Oversight, Dr. H. H. Hess, chairman of NAS Space Science Board, said that the national space program had become too big to make it practical to focus on any single future goal: "To pick one and virtually abandon the remainder would give a far smaller return than a balanced program aimed at several major objectives, even though the advantage of strong focus is being given up." Three possible major goals suggested by Hess were: (1) unmanned exploration of the planets, primarily Mars and Venus, "with a subsidiary effort" on Jupiter; (2) continued manned lunar effort; and (3) earth-orbiting astronomy platform with optical, X-ray, gamma-ray, and radio telescopes "with ultimately manned maintenance." (*Av. Wk.*, 8/22/66, 95)

During August: First space disaster was among predictions for 1967 in new edition of *Old Moore's Almanack*, a volume published in U.K. for 270 years. (UPI, *NYT*, 8/22/66, 37)

- Role of Venus in NASA's future space sciences program was reviewed in report by OSSA to House Science and Astronautics Committee. Report suggested that NASA might increase interest in Venus with combinations of Mars-Venus flights and single missions to Venus using Mariner, Voyager, and advanced Automated Biological Laboratory spacecraft to investigate "the possibility of life forms. . . . From almost every aspect . . . Venus is an important, puzzling, and paradoxical planet, meriting continued examination by space flight missions." (Text; *Tech. Wk.*, 9/12/66, 18-9)

- U.S. Sst program had been assigned "Dx" priority—highest manufacturing priority usually reserved for key defense and space programs—by President Johnson in April without public announcement, the press reported. The *New York Times* questioned secrecy and priority implied by President's decision: "It was an error to keep the decision secret." (Clark, *NYT*, 8/7/66, 1; *WSJ*, 8/9/66, 1; *NYT*, 8/8/66, C26)

- Summarizing two years of negotiations between France and U.S.S.R. on cooperative satellite communications program in *Astronautics & Aeronautics*, Andrew G. Haley said questions of (1) access of French scientists and technicians to Soviet launch sites, and (2) size of orbit for French satellite to be launched with Soviet vehicle were still unsettled. (Haley, *A&A*, 8/66, 15-6)

- Sharing of research with U.S. industry under NASA's technology utilization program was commended in *Petroleum Today* article by Clay Hornick. December 1965 Conference on Selected Technology for the Petroleum Industry at Lewis Research Center, Hornick said, had encouraged utilization of space research discoveries in fields of combustion, lubrication, and nuclear resonance. (Hornick, *Petroleum Today*, Summer 1966, 16-19)

- U.S. Army's aviation role would increase, not diminish, as result of recent USA-USAF Chiefs of Staff agreement on tactical airlift capabilities, USA Director of Aviation B/G Robert R. Williams told *Data*. ". . . I assure you that the Army will continue to exploit the inherent capabilities of aircraft to support the conduct of prompt and sustained combat operations on land. Just as we will continue to receive aviation support from the other services, so will we continue to provide organic aviation responsive to the direct control and planning of the Army commander. Further, we will develop and acquire aircraft and aircraft command and control systems which are uniquely capable of fulfilling . . . the requirements of immediate availability for sustained operation in a field environment. . . ." (*Data*, 8/66, 9-10)

- Comment on LUNAR ORBITER I mission:

    New York Times: "Unfortunately, the decision made a half decade ago was to give priority to the Apollo program for landing a man on the moon... Lunar Orbiter's achievements remind us that there is no pressing necessity for putting manned lunar flight first. If President Johnson should decide to slow down the Apollo project by cutting its budget appreciably, scientific investigation of the moon with instrument satellites could continue and be richly productive.

"In a few years pictures of the earth's orb as seen from distant space will become commonplace. Tomorrow's schoolchildren will find it hard to believe there was a time when no human eyes had seen the full crescent arc of earth's globe from the perspective of the moon or beyond."

Philadelphia *Evening Bulletin*: "Looking rather like a crumbly piece of Roquefort, there it was—Earth, as seen from the Moon by a camera aboard Lunar Orbiter 1. The picture was a black-and-white, but no matter. Since the Moon has no appreciable atmosphere, Earth will never take on the bright orange and yellow hues of a harvest moon. And since there are no forests, no lakes, no oceans, no snowcovered fields on the Moon, Earth will never be framed, for lunar lovers, in the sylvan or aquatic settings that lend so much enchantment and mystery to earthly views of the Moon."

Sam Shenton, leader and secretary of the International Flat Earth Society, claimed that the circular body shown in LUNAR ORBITER I's photos Aug. 25 was not really earth, but "one of the non-luminous bodies between us and the moon." He admitted that photos of the earth had been "a great shock" to the 24-member society. (*NYT*, 8/19/66, 29; 8/27/66; *Phil. Eve. Bull.*, 8/27/66; Reuters, *Wash. Post*, 8/29/66, A3)

During August-September: USAF Chief of Staff Gen. J. P. McConnell discussed military future of manned aircraft in *NATO's Fifteen Nations*. Aeronautical technology in fields of propulsion and composite materials—and less radical improvements in performance, navigation aids, fire control systems, and conventional ordnance—indicated that "by 1975 we will have increased by major increments the speed, range, altitude, payload and over-all versatility of manned aircraft for military roles." Noting that X-15 could fly faster and higher than V-2 ballistic missile, General McConnell said he believed "that well in advance of the estimates that most people would make, Air Force pilots will be flying genuine aerospace craft that can take off under their own power, operate in space and return to a controlled landing." (*NATO's Fifteen Nations*, 8-9/66)

During Summer: First four sessions of Legal Subcommittee of the U.N. Committee on the Peaceful Uses of Outer Space were summarized in an article by NASA Deputy General Counsel Paul G. Dembling and Daniel M. Arons, Attorney-Adviser, NASA Office of General Counsel, in *Journal of Air Law and Commerce*. (*J/Air Law and Commerce*, Summer/66, 329-86)

September 1966

September 1: Lewis Research Center awarded Aerojet-General Corp. a $10-million contract for fabrication and static test firing of 260-in.-dia. solid propellant rocket motor. An advanced version of two motors of the same diameter previously tested by Aerojet, new motor would use propellant with higher burning rate and was expected to develop 5.25 million lbs. peak thrust when test-fired for 80 sec. in June 1967. Earlier versions had produced 3.6 million lbs. thrust during 114-sec. tests. (NASA Release 66-236; LRC Release 66-52)

- Explosion of Saturn V booster's 2nd stage (S-II-F) at MTF May 28 was caused by inadequate test procedures and faulty hydrogen line fitting, NASA investigative board, chaired by KSC Director Dr. Kurt H. Debus, concluded. Board reported that first shift workmen had failed to attach pressure-sensing line to stage's hydrogen tank or to inform second shift workmen of disconnect, with result that test conductor's gauge registered zero reading when stage was actually under pressurization with gaseous helium. Tank ruptured at 23 psig—far below 38-psig design pressure—because of poor hydrogen line fitting. Board recommended "tighter control" over MTF test procedures. (MSFC Release 66-196)

- NASA Nike-Tomahawk launched from Churchill Research Range reached 150-mi. (242-km.) altitude in first of a series of four NASA-Univ. of California, Berkeley, experiments to obtain proton and electron measurements over energy range of 1–300 kev. Rocket and instrumentation performed satisfactorily [see Sept. 6, 16]. (NASA Rpt. SRL)

- NASA appointed Paul A. Barron Deputy Director of Procurement and Gordon H. Tyler, formerly GSFC procurement officer, to succeed Barron as Assistant Director of Procurement for Policy and Review. (NASA Release 66-234)

- Gen. James Ferguson, Deputy Chief of Staff/R&D, Hq. USAF, assumed command of AFSC, replacing Gen. Bernard A. Schriever, who retired Aug. 31. (AFSC Release 184.66)

- USAF would make electronic portraits of satellite shapes to develop catalog of radar images of possible space vehicle shapes, AFSC announced. Measurements on some 24 satellite models would be taken by AFSC Electronic Systems Div. at Holloman AFB in 2,000 hrs. of tests extending through December. (AFSC Release 156.66)

- L/C Harry W. Peterson (USAF, Ret.), former chief of Contract Division, Contract Administration Services, Hq. USAF Defense Supply Agency, was appointed Director, Contract Administration Division of NASA Procurement Office. (NASA Ann., 10/7/66)

September 2: Tape recorder for Advanced Vidicon Camera System on NASA's NIMBUS II meteorological satellite had failed, but camera could still transmit cloud-cover photos when within range of North American receiving stations. Launched from WTR, NIMBUS II had successfully passed its final test objective of two months of continuous operation July 15. Automatic Picture Transmission (Apt) camera system and High Resolution Infrared Radiation (Hrir) system were still operating. (NASA Release 66-243)

September 2: Data from onboard cameras and other recovered instruments had confirmed success of NASA's first high-altitude experiment to investigate parachute designs and techniques that might be used to land instrumented capsules on Mars. Launched aboard a balloon Aug. 30 from Walker AFB, experiment had "provided a solid data base with which we can compare and correlate the results of later flight tests involving parachutes and flight units of different sizes and shapes," Project Manager John C. McFall reported. (NASA Release 66-241)

- Eight Government agencies, headed by FAA, formed Inter-Agency Bird Hazard Committee to exchange and consolidate data in an effort to develop methods for reducing danger of collisions between birds and aircraft: Dept. of the Interior; USN; HEW; CAB; NASA; USA; and USAF. Since 1961, FAA had spent more than $500,000 for research on bird habitats, migration, and preventive and corrective methods to reduce collisions—which had averaged 300 per year. (FAA Release 66-84)
- NASA Langley Research Center had invited four aircraft manufacturers— Boeing Co., Lockheed Aircraft Corp., Douglas Aircraft Co., and General Dynamics Corp.—to submit by Sept. 30 proposals for a research program to reduce noise generated by commercial jet transports. (NASA Release 66-242)
- New Eastman Kodak film processing system for fast reaction, high-quality aerial reconnaissance photography that simultaneously developed a negative and a high-resolution positive transparency in 30 sec.; had been demonstrated by USAF at Wright-Patterson AFB. (AFSC Release 142.66)
- H.S. Becker, former North American Aviation, Inc., executive, had been appointed Deputy Director of MSFC's Advanced Systems Office. (MSFC Release 66-199)
- North American Aviation, Inc., was awarded a $1,550,000 initial increment to a $5,671,000 USAF contract for development work on vertical takeoff and landing (Vtol) aircraft. Contract would be managed by AFSC's Systems Engineering Group. (DOD Release 760-66)
- M2-F2 lifting body vehicle, piloted by NASA research pilot Milton O. Thompson, was launched from B-52 bomber at 45,000-ft. altitude in fifth glide flight at Edwards AFB. Purposes of flight were evaluation of a 360° overhead approach using visual cues only and determination of control damper-off handling qualities. (NASA Proj. Off.)

September 2-3: Three NASA Nike-Apache sounding rockets were launched from Churchill Research Range in GSFC experiment to measure intensity and energy spectra of low-energy protons, helium nuclei, and heavier nuclei present during a Polar Cap Absorption event. Rockets and instrumentation—except for solid-state detectors on first flight— performed satisfactorily. (NASA Rpt. SRL)

September 3: Reports by Volunteer Flight Officers Network—new tracking system in which 22,000 commercial pilots on 51 airlines searched the skies for reentering space hardware—had proved valuable to Smithsonian Institution's Astrophysical Observatory and USAF in improving computer methods and clearing records. Network was founded by amateur astronomer Herbert E. Roth, director of Denver Moonwatch team. (Clark, *NYT*, 9/4/66, 68)

September 4: U.S.S.R.'s 12-ton PROTON III unmanned spacecraft—launched into earth orbit July 6 to study cosmic rays and atomic particles—was still functioning normally and transmitting data to earth regularly, *Trud* reported. (AP, *NYT*, 9/6/66, 74)

- Invention of reusable turbo-rocket engine, which could launch spacecraft for 1/10 current U.S. costs, was announced by Rolls Royce at British Assn. for the Advancement of Science meeting in Nottingham, U.K. Engine had been designed in titanium, but compressor blades could be made of hifil—new fiber-reinforced plastic five times as stiff as titanium, more reliable, lighter, and less expensive. Partly air-breathing turbo-rocket—cross between jet engine and rocket—would be fueled on kerosene and liquid oxygen. Battery of 48 turborockets—each weighing 1,000 lbs. and capable of generating 22,000 lbs. thrust—could boost Apollo spacecraft to 25,000-ft. altitude in fraction of a second and then be landed by remote control. Reported cost of launch would be $2.8 million compared to $28 million with U.S. booster. (*Houston Chron.*, 9/5/66)

- Soviet cosmonauts had not been troubled by perspiration, breathing difficulties, or high heartbeat rates as was Astronaut Richard F. Gordon, Jr., during GEMINI XI mission, Soviet physician-cosmonaut Dr. Boris Yegorov said at press conference in Geneva. Probable reason was that cosmonauts worked in "completely normal" atmospheric conditions and breathed ordinary air, while astronauts used artificial air mixture containing more oxygen than normal, he said. Yegorov, one of three crew members in Oct. 12, 1964, VOSKHOD I mission, was attending congress on space biology. (Reuters, *Wash. Post*, 9/22/66, A3)

September 5: Administration officials had no indication that U.S.S.R. was man-rating a new booster for manned lunar landing program, *Technology Week* reported. They also doubted reports that U.S.S.R. would attempt landing mission in October 1967—fiftieth anniversary of Bolshevik Revolution—but anticipated another Soviet manned earth-orbital flight in 1966. (*Tech. Wk.*, 9/5/66, 3)

- Gen. Bernard Schriever, who retired as AFSC Commander Aug. 31, was saluted by *Technology Week*: "The nation owes an immense debt to Gen. Schriever, for he and a handful of other far-sighted men almost literally dragged this country into the missile era at a time when there were few who believed that the intercontinental missile was a feasible weapon.

    "Moreover, once the decision had been made it was to Gen. Schriever that the task fell of putting together the organization to tackle such a formidable assignment.

    "The success story that followed is well-known in the industry and the nation...." (Coughlin, *Tech. Wk.*, 9/5/66, 50)

- Future of consortium agreements between U.K. and European nations was still uncertain despite success of European Launcher Development Organization (ELDO) and progress of Concorde supersonic transport program, L. L. Doty reported in *Aviation Week*. "... dissensions over the distribution of work loads, allocation of procurement, financial sharing and miscalculations of projected costs are threatening present consortiums, and possibly discouraging formation of others."

    In addition, nationalism "which is accelerating the drive for independence in aerospace activities . . . tend[s] to draw governments away from the consortium principle. . . ." (Doty, *Av. Wk.*, 9/5/66, 29)

September 5: Dr. Clifford A. Spohn (Col., USAF, Ret.) had been appointed director of operations at ESSA's National Satellite Center, succeeding Arthur W. Johnson, who had been appointed deputy director of the facility. (ESSA Release 66-8)

- Prevailing theory that nuclear particles are symmetrical throughout nature, challenged two months ago by Dr. Paolo Franzini, Columbia Univ., and his wife Dr. Juliet Lee-Franzini, State Univ. of New York [see June 27], was reinforced by results of experiment conducted in Geneva by European Center for Nuclear Research (CERN) scientists and reported at Rochester Conference at Univ. of California, Berkeley. Validation of Franzini discovery, which had stirred wide interest in scientific circles, would force physicists to derive new or more sophisticated explanations for basic nature of matter. CERN experiment, based on analysis of 10,665 photographs of eta-meson decay into three pions, found no significant difference in energy levels of positive and negative pions. Franzini had observed greater energy in positive pion than negative. Franzini and CERN groups planned further experiments, and similar test would be conducted by Columbia Univ. physicist Leon Lederman. (*NYT*, 9/6/66, 30)

- "Nobody has yet found any space goals that will stir the Europeans the way manned space flight and the race toward a lunar landing have generated tremendous technical impetus in the U.S. and U.S.S.R.," Robert Hotz reported in *Aviation Week*. "If new formulae can be found to fit changing European requirements, the prospects for applying the spearhead technology of the aerospace industry will never be brighter. For the United States, this will require a more realistic government policy in permitting more first-line technology to be exported to our allies. For Europe, it will require a more realistic appraisal of how its technical resources can be applied to compete successfully in the world market." (Hotz, *Av.Wk.*, 9/5/66, 21)

September 6: President Johnson signed FY 1967 Independent Offices Appropriation Bill (H.R. 14921), which included $4,968,000,000 NASA appropriation—below $5 billion for first time since FY 1963. (NASA LAR V/144)

- Boeing Co., Lockheed Aircraft Co., General Electric Co., and Pratt & Whitney Div., United Aircraft Corp., submitted final designs for 1,800-mph Sst to FAA. Evaluation team of Government and airline experts was expected to submit engine and airframe choices to FAA Administrator William F. McKee by Nov. 1. Final selection would be announced by President Johnson in late 1966. (AP, *Wash. Post*, 9/7/66, D7; Clark, *NYT*, 9/7/66, 26)

- Rebutting Aug. 14 Washington *Sunday Star* editorial which called the U.S. Sst program "a dubious venture moved by extraordinary pressures," Sen. Warren G. Magnuson (D-Wash.) told Senate that program was "moving forward on a logical basis" to make Sst available for commercial use in 1974 at total cost of about $4.5 billion—25% of which would be paid by manufacturers. Senator Magnuson said that planners were "well aware" of potential hazards of excess radiation, internal heat, lack of maneuverability, and structural stress suggested by *Star* and assured the Senate that aircraft had been designed "to preclude catastrophic conditions...." (*CR*, 9/6/66, 20909-10; *Tech. Wk.*, 9/5/66, 17)

September 6: Two NASA Nike-Tomahawks launched about six hours apart from Churchill Research Range reached 145-mi. (238-km.) and 150-mi. (243-km.) altitudes. They were second and third in series of four NASA-Univ. of California, Berkeley, experiments to obtain proton and electron measurements. Both rockets and instrumentation performed satisfactorily; first rocket's nose cone did not eject until after apogee, but good data were received [see Sept. 1, 16]. (NASA Rpt. SRL)

- Soviet rocket tests begun Aug. 25 in the Pacific had been completed six weeks ahead of schedule, Tass announced. (Tass, 9/6/66; UPI, *NYT*, 9/7/66)

- NASA awarded facilities grants totaling $3.2 million to Univ. of Wisconsin and Univ. of Washington to provide "urgently needed" quarters and specialized facilities for space-related research and training. (NASA Release 66-244)

- NASA awarded George Washington Univ. a three-year, $750,000 grant "to undertake a multidisciplinary program of policy studies in science, technology, and public administration." Program was expected to contribute to analysis of various policy questions related to development of the space program and utilization of knowledge derived from it. (NASA Release 66-240)

- A 1967 world conference on space to be held in Geneva, Paris, or Vienna was proposed by Soviet professor Anatoli Blagonravov to Legal Subcommittee of U.N. Committee on Peaceful Uses of Outer Space meeting in New York. U.S. had favored 1968 meeting in New York. (*NYT*, 9/7/66)

- Proposal to produce first worldwide live telecast in June 1967 was outlined at Geneva meeting of European Broadcasting Union. ABC, CBS, and NBC news executives expressed interest in proposal but would not commit themselves until details of program, under supervision of British Broadcasting Corp., were resolved. Live television transmission around the world would be possible after ComSatCorp's planned launching of two communications satellites in fall 1966—one over the Atlantic, the other over the Pacific. (Adams, *NYT*, 9/7/66)

- 1966's second new comet—first with a tail—had been discovered by Dr. Roberto Barbon, Italian astronomer visiting Mt. Wilson and Mt. Palomar Observatories, and confirmed by Smithsonian Astrophysical Observatory. Named for its discoverer, ninth magnitude object was in constellation Cetus. (Sci. Serv., *NYT*, 9/6/66, 46)

September 7: Thunder and lightning brought down NASA meteorological balloon near Austin, Tex. The 10-million-cu.-ft. balloon, launched from Minneapolis Aug. 24 by Litton Systems, had failed to deflate automatically after its payload had been detached and was slowly drifting toward the Gulf of Mexico. USAF had been prepared to shoot it down because it "was a menace to jet aircraft." (UPI, Phil. *Eve. Bull.*, 9/7/66)

- New mineral named "djerfisherite"—a copper-iron sulfide—had been found in meteorites. Since it does not exist naturally on earth and could be formed only at high temperatures, it was speculated that djerfisherite might have originated in the sun's outermost layer. Report of finding was released through American Assn. for the Advancement of Science by Dr. Louis H. Fuchs. (Henry, Wash. *Eve. Star*, 9/7/66, D2)

September 8: NASA test pilot John B. McKay made successful emergency landing of X-15 No. 1 at Smith's Ranch, Nev., after malfunction developed in aircraft's fuel system. There was no damage to aircraft and no injury to pilot. Primary purpose of flight was photometer measurements of sharpness of earth's horizon for use in Apollo program. (X-15 Proj. Off.; UPI, *Wash. Post*, 9/9/66, B10)

- Rep. Ogden Reid (R-N.Y.) introduced H.R. 17604 to amend National Aeronautics and Space Act to provide a NASA R&D program for aircraft noise abatement. Program would: attempt to develop workable measuring system for correlating intensity and quality of aircraft noise with effects on persons on the ground; develop quieter aircraft through research and development; compile knowledge on methods and devices for aircraft noise abatement; coordinate research relating to aircraft noise abatement; and determine cause and develop solutions for other "related nuisances." (NASA LAR V/146)

- Dr. Winston E. Kock, first director of ERC, had resigned to return to Bendix Corp., effective Oct. 1. James C. Elms, NASA Deputy Associate Administrator for Manned Space Flight, would replace him. (NASA Release 66-247)

- U.S. should divert funds from NASA's Apollo program and from Federal highway projects to improve urban housing and transportation systems, Sen. Eugene J. McCarthy (D-Minn.) suggested to International Conference of Social Work in Washington, D.C. (Honsa, *Wash. Post*, 9/9/66, A14)

- Development of a "magnetic hammer" at MSFC had saved the Government some $2,130,000 in 1965 and netted its inventors $2,500 each—largest joint invention award ever made at MSFC. A portable device which created an intense magnetic field to move metal uniformly without marring the surface, hammer had been used to replace mechanical stretching of Saturn V bulkhead gore segments and restore previously rejected segments. Inventors were Robert J. Schwinghamer, MSFC, and Leslie D. Foster, formerly of MSFC and currently with Comprehensive Designers, Inc. (MSFC Release 66-206)

- FCC's Oct. 1 deadline for filing comments on whether or not private entities should be allowed to establish special-purpose communications satellite systems for domestic use was extended to Nov. 3. Initially FCC had requested that comments and proposals be submitted by Aug. 1. Complexity of proposals submitted by ComSatCorp and the Ford Foundation on that date and resulting questions and discussion had necessitated two extensions of the deadline. (Adams, *NYT*, 9/9/66, 77)

- In response to allegations that F-111 aircraft was unacceptable as an all-purpose military aircraft, DOD told Sen. Stuart Symington (D-Mo.) that: (1) production of F-111A for USAF was on schedule with first deliveries planned in 1967; (2) performance of the aircraft would be close to 1962 evaluations; (3) aircraft would be superior to any other tactical weapons system in the world in "critical" factors of range, payload, speed, and versatility; (4) cost variations were "typical of large scale development programs"; and (5) goal of one basic aircraft for two services was being met. (*CR*, 9/8/66, 21207)

September 9: Validity of satellite measurements of micrometeoroid fluxes in which piezoelectric microphones had been used as detectors was challenged in *Science* by Carl Nilsson, Australian scientist working at GSFC. Although these measurements had given rise to theory that earth was surrounded by dust cloud, analysis of 100 hrs. of data from NASA's OGO II satellite revealed: (1) there had been no detectable signals from sensors that could have been caused by micrometeoroid impact; and (2) the microphone systems had been emitting noise. Nilsson concluded that in past satellite measurements, microphone noises had not been caused by micrometeoroid impact but by reaction of microphone crystals to temperature changes. (Nilsson, *Science*, 10/9/66, 1242-6)

September 9: House of Commons' Committee of Public Accounts said development costs for Anglo-French supersonic airliner Concorde would total $1.4 billion—compared to original $476 million estimated cost—before its completion in 1973, according to the *New York Times*. The "massive increase" in cost was attributed to changes in design, increases in wages, and "underestimation in the earlier estimates." (*NYT*, 9/9/66, 76)

September 10-16: NASA Administrator James E. Webb made unpublicized trip to West Germany to discuss three new space agreements; cooperative projects would involve a German-built advanced solar exploration spacecraft to be launched by NASA, a joint aeronomy satellite to study solar radiation, and participation in Apollo Telescope Mount for observation of sun's structure from above earth's atmosphere. (*Tech. Wk.*, 10/3/66, 18)

September 10: Chemical reactions in gases produced by a "Q-switched" ruby laser suggested a new method for studying very high temperature reactions, reported Westinghouse Research Laboratories scientists L. M. Epstein and K. H. Sun in a letter to *Nature*. Studies establishing dependence of yield on intensity and configuration of focal spot were recommended to prove usefulness of method as an absolute dosimeter for laser shots. (Epstein and Sun, *Nature*, 9/10/66, 1173-4)

September 11: Andrew G. Haley, early advocate of outer space rule of law and author of the "metalaw" concept to govern the conduct of terrestrial beings in outer space, died in Washington, D.C. At the time of his death, Haley was general counsel to International Astronautical Federation (IAF), counsel and fellow of AIAA, IAF observer to U.N., and academician of International Institute of Space Law. He authored the book *Space Law and Government* in 1964. (*Wash. Post*, 9/12/66, B7; *Wash. Eve. Star*, 9/12/66, B5)

- USAF had launched reconnaissance satellite over U.S.S.R. that could photograph Soviet military bases and return film packages on command from ground control, Dick Lyons reported in *Washington Post*. Key components were high-powered camera that could photograph objects several feet in diameter from 100-mi. altitude; cluster of film packages—cassettes—that could be dropped from satellite at random and picked up by recovery aircraft; and an Agena vehicle which could be maneuvered from ground. A new version of Samos spacecraft, satellite could remain in orbit for weeks or months—until its supply of film packages was exhausted—and might also be able to transmit TV pictures to U.S. ground stations. (Lyons, *Wash. Post*, 9/12/66)

September 12: Balloonist Tracy Barnes landed at Rio Grande, N.J., completing first hot-air balloon flight across the continental U.S. Barnes left San Diego April 10. (*Wash. Post*, 9/12/66, D11)

- Shift of management of NASA's Apollo Telescope Mount mission from GSFC to MSFC was evidence of NASA's effort "toward inhouse development of major space projects" to "offset threatened loss of personnel" at NASA manned space flight centers as Apollo program passed its personnel peak, *Aviation Week* reported. (*Av. Wk.*, 9/12/66, 80)

September 12: U.S.S.R. was developing booster with 7.5- to 10-million-lbs. thrust—enough to send a man to the moon or past Mars, unnamed sources reported. Evidence presumably had been gathered by U.S. reconnaissance satellites. (Clark, *NYT*, 9/13/66, 28; Simons, *Wash. Post*, 9/11/66, A1)

September 12-15: Three-day GEMINI XI mission—ninth manned flight in NASA's Gemini series—began with launch of Gemini Agena Target Vehicle (GATV) by Atlas booster from ETR's Complex 14, followed 97 min. later by launch of GEMINI XI by Titan II booster from Complex 19. Astronauts were Charles Conrad, Jr., command pilot, and Richard F. Gordon, Jr., pilot. GATV entered near-circular, 185-mi.-altitude orbit; GEMINI XI entered elliptical orbit with 175-mi. (280-km.) apogee; 100-mi. (161-km.) perigee; 90-min. period; and 33° inclination. GEMINI XI docked with GATV 94 min. after liftoff to become first manned spacecraft to achieve first-revolution rendezvous and docking—mission's primary objective. Other accomplishments included extravehicular activity (Eva), maneuvering of docked GEMINI XI-GATV configuration to record 851-mi. altitude, station-keeping using tether to link two spacecraft, computer-controlled reentry, and completion of several experiments.

After initial docking, crew executed four practice docks and spent sleep period attached to GATV. Gordon began planned 107-min. Eva on schedule at 23:58 GET, but ten minutes later, after setting up movie camera, retrieving micrometeoroid experiment, and attaching Agena's 100-ft. tether to GEMINI XI's docking bar, had to rest astride GATV. When blinded in his right eye by perspiration, Gordon was ordered by Conrad at 24:02 GET to cancel power-tool evaluation experiment and return to cabin.

At 40:58 GET, following crew's second sleep period, mated spacecraft used GATV's primary propulsion system (pps) to attain orbit with 851-mi. apogee—record altitude for manned space flight. After two revolutions, second pps burn restored docked configuration to approximately 185-mi.-altitude circular orbit. At 46:58 GET Gordon opened hatch to begin 2 hr. 8 min. standup Eva during which several photographic experiments were conducted. Tether dynamics exercise followed: GEMINI XI undocked from GATV and moved to end of 100-ft. tether which was supposed to hold two spacecraft on vertical line pointing earthward, proving that two objects could fly in formation for long duration without using maneuvering fuel. When vehicles began to drift, Conrad initiated slow rotation with appropriate spacecraft maneuvers, creating slight centrifugal force which kept tether taut and spacecraft at controlled distance. After separation from Agena during sleep period, Conrad executed successful re-rendezvous—added to flight plan because of GEMINI XI's favorable fuel supply.

Reentry Sept. 15 in 45th revolution after 71 hrs. 17 min. of flight was first in U.S. space program to be guided automatically by computer which fed commands directly to thrusters. Splashdown of GEMINI XI in the Atlantic 700 mi. from Cape Kennedy—within two miles of target—and recovery operations by crew of U.S.S. *Guam* were carried on live television via EARLY BIRD I comsat. USN pararescue team, dropped from helicopters, attached flotation collar to spacecraft. Astronauts were flown by helicopter to recovery ship. (NASA Proj. Off.; NASA Release 66-226; Hines, Wash. *Eve. Star*, 9/13/66, A1, A6; 9/14/66, A1, A14; 9/15/66, A1, A6; Reistrup, *Wash. Post*, 9/13/66, A1, A3; 9/14/66, A1, A3; 9/15/66, A1, A6; 9/16/66, A1, A5; Wilford, *NYT*, 9/13/66, 1, 28; 9/14/66, 1, 32; 9/15/66, 1, 26; 9/16/66, 1, 24)

September 12-16: Joint Royal Aeronautical Society (RAeS) Centenary Congress and Fifth Congress of the International Council of the Aeronautical Sciences (ICAS) held in London.

Prince Philip, Duke of Edinburgh and honorary RAES president, said in a centenary address that ". . . making the right selection in research programmes and development projects has become crucial to the whole of aeronautical evolution. . . ." Risks involved in the total investment necessary for success, he said, had led European nations to "the obvious answer"—international cooperation: "If it is important to see that aviation is fitted into the national economy in such a way as to enable it to function most efficiently, it is even more important to work out an international structure which will enhance and encourage the chances of success rather than frustrate and defeat them." (Hersey, *A&A*, 11/66, 126-30; Text, *A&A*, 12/66, 32-4)

September 13: U.N. Ambassador Arthur Goldberg, attempting to clear obstacles to early completion of space law treaty, announced two U.S. compromises to Legal Subcommittee of U.N. Committee on Peaceful Uses of Outer Space meeting in New York: (1) U.S. would not require that space stations and vehicles be "open at all times" for inspection, but would allow that visits be made on "a basis of reciprocity" with "reasonable advance notice" to host government; (2) U.S. would not demand submission of comprehensive reports on space exploration to U.N. members and scientific community, but would have reports submitted "to the extent feasible and practicable." U.S.S.R. demand that nations granting tracking facilities to any space power provide same facilities on identical terms to other space powers was modified by Soviet delegate Platon D. Morozov to include provision that expenses incurred in rendering tracking aid would be reimbursed. (Teltsch, *NYT*, 9/14/66, 32)

- Rep. George P. Miller (D-Calif.), chairman of House Science and Astronautics Committee, praised NASA's GEMINI XI mission on the floor of the House: ". . . the Gemini 11 crew successfully accomplished the first rendezvous and docking in space within one revolution, by far the quickest of any rendezvous attempted to date. It is significant to note at this point in the mission that the . . . flight is following the planned schedule of events closer than any other Gemini flight to date. . . ." (*CR*, 9/13/66, 21495)

- U.S.S.R. pilot Aleksey Anosov had established new world's distance record for helicopter with 1,509-km. (938-mi.) flight in MI-1 helicopter from Sumi in the Ukraine to the Orenburg region in the Urals. Previous record of 900 mi. (1,449 km.) had been set by Capt. Chester R. Radcliffe, Jr. (USAF), July 5, 1962. (*Wash. Post*, 9/14/66, B2)

September 13-14: First Logistics Management Symposium attended by some 400 industrialists and government officials met at MSFC to discuss logistics support of activities in the free world and in space. In the keynote address, NASA Associate Administrator for Manned Space Flight Dr. George E. Mueller stressed that effective logistics management was essential to the success of NASA's Apollo program. "For the first time in a NASA manned space flight program, the greater part of logistics is not being provided by the Department of Defense. . . .

"However, the need for increasing the emphasis on logistics management for Apollo is great. . . .

"The Apollo Saturn space vehicle involves 20,000 contractors and subcontractors and has more than 900,000 individual parts. The Saturn V first stage holds 56 tank cars of propellants. The second and third stages of Saturn V transported by water during the Apollo program will spend a total of 700 days at sea. Apollo program transportation by all modes will require coordination with nine Government agencies. The launch windows for the Apollo lunar mission are relatively small, malfunctions on the pad must be kept to a minimum while corrective maintenance must be extremely fast and reliable. All of these elements make the Apollo logistics program both complicated and costly. . . ." (MSFC Release 66-202; Text)

September 13-15: Series of five NASA Nike-Apache sounding rockets was launched from Churchill Research Range in GCA Corp. experiment to compare, by means of Langmuir probe and trimethyl-aluminum (TMA) cloud techniques, electron density profile with wind profile. A NASA aircraft obtained supporting photographs of vapor clouds. All rockets and instrumentation performed satisfactorily; TMA did not eject during second flight in series, but useful data were obtained from Langmuir probe. (NASA Rpt. SRL)

September 14: X-15 No. 3 was flown to 3,580 mph (mach 5.11) and 254,200-ft. altitude by NASA pilot William H. Dana for altitude buildup, micrometeoroid collection, measurements of total solar flux and broadband light distribution by JPL spectroradiometer, and radiometer measurements of ultraviolet exhaust plume characteristics. (X-15 Proj. Off.; FRC Release; JPL Release 408)

- LUNA XI, U.S.S.R.'s second lunar orbiting satellite, launched Aug. 24, was continuing to orbit moon and had completed 1,410 revolutions, Tass announced. All equipment was functioning normally. (Tass, 9/14/66)

- U.S.S.R. astronomers had scanned the Milky Way in six-centimeter wavelength for first time, using Pulkovo Observatory's huge radiotelescope and parabolic antenna. Observations did not confirm current theory that radio emanations were primarily linked with radio glow of "hot gas" (ionized hydrogen) in our galaxy, but showed that "even in these wave-lengths a considerable role is played by the emanation of fast electrons moving in magnetic fields at speeds close to that of light." Simultaneously, Pulkovo astronomers conducted detailed study of 80 other clouds of ionized hydrogen to "help determine more exactly the physical conditions which exist in the most interesting parts of the Milky Way and their relationship with the continuous radio emanation of our galaxy as a whole." (Tass, 9/14/66)

September 15: USAF launched unidentified satellite from Vandenberg AFB with Thor-Burner II booster. (*Tech. Wk.*, 9/26/66, 15; *U.S. Aeron. & Space Act., 1966*, 156)

- Recent developments in space technology were discussed by President Johnson and President of Republic of the Philippines Ferdinand E. Marcos in White House discussions. President Marcos expressed desire to encourage greater training of Philippine scientists and engineers in the peaceful application of space technology, and President Johnson offered appropriate fellowships for study in U.S. institutions. (*Pres. Doc.*, 9/19/66, 1296)

September 16: U.S. and U.S.S.R. reached impasse over question of access to tracking facilities that would be part of U.N. treaty on space exploration. U.S. Ambassador Arthur J. Goldberg blamed U.S.S.R. for halting agreement on treaty, voicing U.S. contention that tracking facilities must be negotiated by bilateral agreements and not guaranteed by treaty. Compulsory access to tracking facilities, Goldberg said, would constitute sovereignty violation. Soviet Ambassador Platon D. Morozov defended proposal as means to correct a discriminatory situation in which U.S. had certain advantages because of its alliances with other nations. (Teltsch, *NYT*, 9/17/66; Sanger, *Wash. Post*, 9/17/66, A4)

- NASA Nike-Apache sounding rocket launched from NASA Wallops Station reached 103-mi. (166-km.) altitude. NASA-Univ. of Maryland experiment was designed to (1) evaluate pulse and thermal equalization probes as research tools, (2) investigate electron energy distribution in normal daytime ionosphere, and (3) investigate use of wing-slope techniques with Langmuir probes. Rocket and instrumentation performed satisfactorily. (NASA Rpt. SRL)
- NASA Nike-Tomahawk launched from Churchill Research Range reached 151-mi. (243-km.) altitude in last of four NASA-Univ. of California, Berkeley, experiments [see Sept. 1, 6]. Rocket and instrumentation performed satisfactorily, and measurements of flux, energy spectrum, pitch angle distribution, and time variations of mirroring and precipitating charged particles were obtained. (NASA Rpt. SRL)
- USAF launched two unidentified satellites with Atlas-Agena D booster from WTR. (*U.S. Aeron. & Space Act., 1966*, 156)
- GEMINI XI Astronauts Charles Conrad, Jr., and Richard F. Gordon, Jr., arrived at KSC from recovery ship U.S.S. *Guam* to undergo physical examination and to begin debriefing. (Wilford, *NYT*, 9/17/66, 11)
- A 40-ft. strip of 16-mm. film "of excellent quality" shot by Astronauts Richard F. Gordon, Jr., and Charles Conrad, Jr., during Sept. 12-15 GEMINI XI mission was released at MSC. One sequence showed Gordon straddling Agena to tie tether from GATV to GEMINI XI. Movie revealed Conrad had described scene accurately when he had radioed to earth: "He's riding it like a cowboy."

  In Washington, D.C., White House announced Gordon would be promoted from USN lieutenant commander to full commander. (AP, *Wash. Post*, 9/17/66, F1; AP, Wash. *Eve. Star*, 9/17/66, A5; UPI, *NYT*, 9/18/66, 1, 82)
- M2-F2 lifting body vehicle, piloted by NASA test pilot Bruce Peterson, was air-launched from B-52 aircraft at 45,000-ft. altitude in sixth glide flight at Edwards AFB. Purpose of flight was checkout of new pilot. (NASA Proj. Off.)
- President Johnson announced his intention to nominate Dr. Werner A. Baum, Vice President for Scientific Affairs of New York Univ., as Deputy Administrator of ESSA to succeed V/Adm. H. Arnold Karo who would retire Jan. 1, 1967. (*Pres. Doc.*, 9/19/66, 1305)
- LRC Director Dr. Abe Silverstein discussed LRC studies on use of liquid methane as an aircraft fuel at the 5th Congress of the International Council of the Aeronautical Sciences (ICAS) meeting in London. Since liquid methane had higher heating value, greater cooling capacity, and lower price per pound than present jet fuels, its use in commercial

supersonic transport might increase Sst payload by 30 per cent and similarly reduce direct operating cost. Dr. Silverstein pointed out, however, that much more would have to be learned about both flight and ground equipment before liquid methane could be seriously considered as an aircraft fuel. (LRC Release 66-56)

September 16: New York Times editorial: "Space historians will have to invent new superlatives to describe the contributions of Gemini 11 and astronauts Conrad and Gordon to man's entry into the cosmos. From the link-up with the Agena in the very first orbit to the incredibly accurate automatic return to earth, the three-day flight demonstrated a new level of capability far surpassing that shown in any previous flight by any nation." (*NYT*, 9/16/66, 34M)

- Soviet newspaper *Trud* said that although Sept. 12-15 GEMINI XI mission had set new altitude record, it otherwise only repeated "already achieved results." *Trud* added that some of these results were "worse" than before, citing the lesser length of time spent in Eva by Astronaut Richard F. Gordon, Jr. (AP, Wash. *Eve. Star*, 9/16/66, A2)

- Air Force Assn. honored four officers for exceptional achievement at its second annual fall meeting in Washington, D.C.: Col. Robert A. Berman, Wright-Patterson AFB, Logistics Executive Management Award; Col. Spencer S. Hunn, Hanscom Field, Mass., AFSC Distinguished Award for Management; Col. Gregory C. Frese, Jr., Washington, D.C., AFSC Meritorious Award for Support Management; and L/Col. Martin H. Brewer, Vandenberg AFB, AFSC Meritorious Award for Program Management. (*J/Armed Forces*, 9/10/66, 25)

September 17: U.S.S.R. launched five- to six-ton spacecraft, possibly from Cosmos series, which exploded into 51 pieces detectable by radar. GSFC *Satellite Situation Report* would later reveal that nine pieces of debris remained in orbit on Oct. 15, the rest having reentered beginning Sept. 20. Launch was first unannounced U.S.S.R. space flight since Jan. 4, 1963. Evert Clark said in the *New York Times* that launch marked first U.S.S.R. 49° inclination launch from Tyura Tam and that Kapustin Yar had been used for all previous 49° launches. (GSFC *SSR*, 10/15/66, 22-5; Clark, *NYT*, 11/1/66, 19; Wilson, *Wash. Post*, 11/1/66, 3)

- NASA reported that smudgy spacecraft windows and exposure problems had compromised quality of most of movie film taken by Astronauts Charles Conrad, Jr., and Richard F. Gordon, Jr., during Sept. 12-15 GEMINI XI mission. Still pictures were all "of good quality" and showed earth's curvature more clearly than on any other manmade photographs. (AP, Wash. *Sun. Star*, 9/18/66, A8)

September 18: Records set by GEMINI XI mission that had made "space history" were detailed in *Chicago American* editorial: "The greatest speed ever attained by man—18,129 mph—as well as the slowest yet flown by anyone in orbit . . . 15,402 mph; the greatest altitude ever reached, 850 miles; the first 'direct ascent' rendezvous with an orbiting satellite; the first multiple docking in space . . .; the first rendezvous using on-board equipment, with no help from the ground." (*Chic. Am.*, 9/18/66)

- Optimism about man's ability to perform useful work in space was expressed by Dr. Charles Berry, MSC Director of Flight Research and Operations, in telephone interview with *Washington Post* writer J. V. Reistrup. Astronaut Richard F. Gordon, Jr., had demonstrated problems involved when he had left GEMINI XI spacecraft Sept. 13 to perform seemingly simple tasks and had exerted enough energy to raise his heart rate to peak 180 beats a minute. Berry said there was a "tremendous amount of effort involved in getting into position for performing basic tasks"; once "positioning" was solved, jobs could be done. The problem was to set up each activity to require minimum effort. (Reistrup, *Wash. Post,* 9/18/66, A4)

September 18: USAF Aerospace Medical Div. chief scientist Dr. Hubertus Strughold stated in telephone interview with Washington *Sunday Star*'s William Hines that NASA had not given adequate attention to keeping astronaut's vision unimpaired during Eva. He said that in normal earth gravity, man could orient himself by means of sight sense (eyes); balance sense (otoliths of inner ear); and pressure sense (Meissner corpuscles concentrated primarily in palms of hands and soles of feet). In space there was no gravity, so otoliths and Meissner corpuscles could not function. When vision was also lost or impaired, orientation was impossible. Strughold pointed out that vision impairment had been a factor in curtailment of three of four Eva missions to date. He said advances in training and in spacesuit design were urgently needed. USAF had been "working on the problem" but had not been asked by NASA for advice or assistance. (Hines, Wash. *Sun. Star,* 9/18/66, A7)

September 19: NASA Aerobee 150 launched from WSMR reached 102-mi. (165-km.) altitude in GSFC-Princeton Univ. Observatory experiment to study ultraviolet radiation from the star Epsilon Canis Majoris. Although spectrograph recording was limited by rocket's low altitude, good spectral resolution was obtained and wavelength limit was extended shortward to lines never before observed. (NASA Rpt. SRL)

- Proposal by Senate Committee on Aeronautical and Space Sciences to establish nonmilitary global navigation system had been rejected by interdepartmental committee after two-year study. Leonard Jaffe, Director of NASA's Communication and Navigation Programs Div., and chairman of the committee, said in interview with *Aviation Week* that further research and development were required. "Low-keyed" study effort planned by NASA would include award of more feasibility study contracts to industry and test of navigation equipment on Applications Technology Satellite (Ats) mission, during which voice transmissions would be made to aircraft. Report of committee, which included representatives from NASA, FAA, DOD, and Dept. of Interior, would be submitted to Congress in several weeks. (*Av. Wk.,* 9/12/66, 38; *Tech. Wk.,* 9/19/66, 3)

- Employment in the aerospace industry would reach 1,349,000 in December 1966—an increase of 11 per cent over December 1965—Aerospace Industries Assn. estimated in a semiannual survey. Estimate was based on reports from 267 plants and facilities of 59 companies which represented nearly 80 per cent of entire industry. (AIA Release 66-55)

- Gen. Bernard A. Schriever (USAF, Ret.), former AFSC commander, became chairman of Aerojet-General Corp.'s newly established advanced planning advisory board. He would serve as senior consultant. (Wash. *Eve. Star,* 9/20/66, A16)

- The general-aviation (nonairline) fleet would "be 80 per cent larger and 90 per cent busier" in 1975 than in 1964, FAA estimated in a staff study and forecast. Study, which would be used to plan aviation facilities and services during next decade, predicted that 160,000 general-aviation aircraft would fly 30 million hrs. in 1975, compared to 88,742 aircraft and 15.7 million hrs. in 1964. (FAA Release 66-87)

September 19: In telegram to companies which had submitted Sst airframe and engine designs Sept. 6—Boeing Co., Lockheed Aircraft Corp., General Electric Co., and Pratt & Whitney Div., United Aircraft Corp.—FAA said it would no longer approve "the public release of any information on the supersonic transport program which could be construed as being intended to influence anyone's judgment," relative to evaluation of Sst designs, until "conditions permit." Selection of contractors was expected in late 1966. (Clark, *NYT*, 9/24/66, 50)

- FAA Administrator William F. McKee told the Economic Club of Detroit that by bringing closer together the industrialized and developing nations, supersonic aircraft would make possible "a better way of life for all.... The social meaning of this greater ability to communicate is difficult to imagine." Calling air travel one of the Nation's fastest-growing businesses, General McKee said that rate of increase had averaged 14 per cent over recent years and had jumped to 25 per cent for the first half of 1966. (*NYT*, 9/20/66, 76)
- Dr. W. C. J. Garrard, George K. Williams, and William W. Williams, Lockheed-Georgia Co. engineers, had been named recipients of Wright Brothers Medal by Society of Automotive Engineers for group report on development of soft-field and rough-field landing gear. (*Av. Wk.*, 9/19/66, 25)

September 20-22: NASA's 2,204-lb. SURVEYOR II (Surveyor B) was successfully launched from ETR's Complex 36 at 8:32 a.m. EDT. After excellent injection by Atlas-Centaur (AC-7) launch vehicle on trajectory toward moon's Sinus Medii, spacecraft successfully accomplished all required sequences up to midcourse maneuver. Initiated at 1:00 a.m. EDT Sept. 21, the 9.8-sec. thrust phase was unsuccessful because one of three vernier engines failed to ignite, which caused spacecraft to tumble at about 1 rps. Thirty-nine additional attempts were made to start all three engines. In all cases, the same two engines ignited; the third did not.

JPL engineers, uncertain of cause of malfunction, speculated it might be result of stuck valve or insufficient power to operate engine's triggering mechanism. They said unless it was corrected, SURVEYOR II could not softland since all three vernier engines were needed to stabilize spacecraft during decelerating descent to lunar surface.

On Sept. 22, when it became apparent the mission could not be completed, engineering experiments were performed to obtain as much data as possible on spacecraft performance. All communication with spacecraft was lost 30 sec. after main retro ignition at 5:34 a.m. EDT. Probable cause was mechanical failure within telecommunications subsystem due to centrifugal forces generated by high tumbling rate. SURVEYOR II impacted in area of moon southeast of crater Copernicus about 11:18 p.m. EDT.

Spacecraft was second in series of seven flights to prove out design, develop technology of lunar soft-landing, and provide basic scientific and engineering data in support of Project Apollo; it contained survey television system and instrumentation to measure lunar surface bearing strength, temperatures, and radar reflectivity. SURVEYOR I, identically equipped, was launched May 30, successfully softlanded on moon June 2 in region of Oceanus Procellarum, and transmitted 10,338 pictures to earth. Surveyor program was under direction of OSSA's Lunar and Planetary Div.; project management was assigned to JPL; Atlas-Centaur booster was managed by LRC; prime contractor for spacecraft development and design was Hughes Aircraft Co. (NASA Release 66-248; NASA Proj. Off.; AP, *Wash. Post*, 9/21/66, A1; Wilford, *NYT*, 9/22/66, 2; *NYT*, 9/23/66, 25)

September 20: During White House meeting with President Johnson, NASA Administrator James E. Webb discussed current status of national space program and showed President two previously unreleased color photographs—one of southern Texas, the other of Baja California—taken from 180-mi. altitude during Sept. 12-15 GEMINI XI mission. Following meeting, Webb announced that Astronauts Richard F. Gordon, Jr., and Neil A. Armstrong would leave Oct. 7 for three-week tour of 10 Latin American nations: Venezuela, Ecuador, Colombia, Peru, Bolivia, Brazil, Paraguay, Uruguay, Argentina, and Chile. (*Wash. Eve. Star*, 9/21/66, A4)

- Project Eros (Earth Resources Observation Satellite)—program to gather facts about earth's natural resources from earth-orbiting satellites with remote sensing observation instruments—was announced by Secretary of the Interior Stewart L. Udall: "Facts on distribution of needed minerals, our water supplies and the extent of water pollution, agricultural crops and forests and human habitations, can be used for regional and continental long-range planning." Udall named Dr. William T. Pecora, Director of U.S. Geological Survey, to head program. (Beckman, *Chic. Trib.*, 9/21/66; *CR*, 9/27/66, 23171-2)

- Simulated Aerobee 150 was launched from WSMR to 2-mi. (3.2-km.) altitude in GSFC flight test of proposed new Aerobee booster. Instrumentation performed satisfactorily, and booster performance was excellent. (NASA Rpt. SRL)

- USAF launched unidentified satellite with Thor-Agena D booster from WTR. (*U.S. Aeron. & Space Act., 1966*, 156)

- Rep. George P. Miller (D-Calif.), chairman of the House Committee on Science and Astronautics, told the National Space Club in Washington, D.C., that the national space program provided the best possible workshop for gathering of vital science and technology necessary for other important goals on earth: "In other words, from space research and development have come new methods of thinking, new techniques of management, new marriages of scientific disciplines, and a new illumination on possible interrelationships of human problems to scientific knowledge." During question and answer period, Rep. Miller said there would be a "very definite mission" for the 260-in.-dia. solid propellant motor in the U.S. space program, although that mission could not presently be strictly defined. He said the Nation would have to develop space rescue capability and one apparently promising way to do this was through the use of solids. Solids, he pointed out, were storable and could be readied and launched on short notice. (Text; *M/S Daily*, 9/21/66, 22)

- M2-F2 lifting body vehicle, piloted by L/C Donald Sorlie (USAF), was air-launched from B-52 aircraft at 45,000-ft. altitude in seventh glide flight at Edwards AFB; purpose of flight was checkout of new pilot. (NASA Proj. Off.)

September 21: Juan T. Trippe, chairman and chief executive officer of Pan American World Airways, in remarks made when he received National Defense Transportation Assn. Award in Dallas, urged development of both Boeing and Lockheed supersonic transport prototypes: "Only further competition between these two manufacturers, during the three years required to build and flight-test the prototypes, can insure that the American supersonic transport ultimately selected will have the maximum attainable performance." (Text; *Av. Wk.*, 9/26/66, 21)

September 21: NASA awarded Honeywell, Inc., a $4.2-million, fixed-price contract to provide six computer complexes to support Apollo command-module and lunar-module simulators at MSC and KSC by March 1967. (NASA Release 66-254)

- Dr. Robert H. Cannon, Jr., professor of aeronautics and astronautics at Stanford Univ., succeeded Dr. Robert G. Loewy as USAF's chief scientist. Appointment was for one year. (DOD Release 791-66; *M/S Daily*, 9/20/66, 21)

September 22: Dr. Otis E. Lancaster, professor of engineering education at Pennsylvania State Univ., was sworn in as a consultant to NASA Administrator. He would advise on the subjects of NASA-university relationships, methods for strengthening engineering education programs, and ways to provide stronger engineering/administrative ties. (NASA Release 66-256)

- In proclamation designating Oct. 12 Columbus Day, President Johnson said Columbus' conquest of the Atlantic, "the 'outer space' of the fifteenth century," was as meaningful to Americans of the space age "as it was to our forefathers who pushed across the vast expanses of this continent." All our frontiers were not conquered, for "new shores of promise await those who, like Columbus, push on undaunted by the failures of the past or fear of the uncharted future." (*Pres. Doc.*, 9/26/66, 1340)

- In address to U.N. General Assembly in New York, U.S. Ambassador Arthur J. Goldberg said U.S. would provide "tracking coverage from United States territory" for Soviet rocket launchings if a "mutually beneficial agreement" could be made. Stressing that tracking facilities were a matter for "bilateral negotiation and agreement," he said the treaty to govern activities in outer space was "too urgent and too important to be delayed. . . . It is all the more urgent because of man's rapid strides toward landing on the moon." Impasse had been reached in discussions in Geneva because U.S.S.R. had insisted pact guarantee that states granting tracking facilities to any country provide same facilities to other countries. (Text; *NYT*, 9/23/66, 12)

- M2-F2 lifting body vehicle, piloted by NASA test pilot Bruce Peterson, was air-launched from B-52 aircraft at 45,000-ft. altitude in eighth glide flight from Edwards AFB. Purposes of flight were pilot checkout and tests of longitudinal and lateral stability and control with dampers on and off at slightly higher speeds. (NASA Proj. Off.)

September 23: NASA Lewis Research Center had awarded General Dynamics' Convair Div. a $15,565,331, fixed-price contract to fabricate and assemble five additional Centaur stages for future space missions. (NASA Release 66-253)

- S-IC-3 booster for first Saturn V flight was shipped from NASA's Michoud Assembly Facility aboard *Poseidon* to MSFC for static-testing scheduled for late October. (NASA Release 66-218)

- Practical dividends from investment in space program were cited by MSFC Director Dr. Wernher von Braun in an address at Midwest Space Exposition, Jackson, Mich. In addition to "tangible returns" from advances in navigation, meteorological, and communications satellites, he noted uses of space technology for surveying earth's resources and suggested ultimate benefits, as yet unpredictable, would "dwarf" those he had named. (Text)

September 24: Four-stage Javelin sounding rocket launched from NASA Wallops Station ejected multicolored barium clouds over eastern U.S. at 310-mi. (499-km.) and 570-mi. (917-km.) altitudes. Cooperative NASA-German Ministry of Scientific Research (BMWF) experiment measured electric fields and wind motion in upper atmosphere by photographing and tracking ionized clouds; secondary objective was observation of interaction between cloud and solar wind. Launch was followed by reported sightings of "brilliant" Ufo's. (NASA Release 66-245; Wallops Releases 66-46, 66-48; AP, *Wash. Post*, 9/25/66, A23; AP, *Wash. Sun. Star*, 9/25/66, A3; NASA Rpt. SRL)

- Lockheed Aircraft Corp., Boeing Co., General Electric Co., and Pratt & Whitney Div., United Aircraft Corp., charged that FAA was attempting a form of censorship by restricting advertisements and release of information on proposed Sst designs. FAA replied that the temporary restriction had been imposed Sept. 19 to keep the public from getting only fragments of overall Sst story and to relieve burden of "clearing" new advertisements for publication while designs were being evaluated by the Government. (Clark, *NYT*, 9/24/66, 50)

- 10-nation ESRO's new rocket-launching base opened at Sekkujokki, Sweden —inside the Arctic Circle. ESRO would launch some 400 British Skylark and French Centaure rockets from $7.6-million base during next eight years to study aurora borealis. (Reuters, *NYT*, 9/25/66, 42)

- Evidence for volcanic origin of lunar craters was presented in a letter to *Nature* by Univ. of Western Australia geologist G. J. H. McCall. Eleventh frame of photographs by NASA's MARINER IV spacecraft revealed scalloping in walls of Martian craters comparable to scalloping in terrestrial calderas—such as Krakatau, Indonesia, Aso, Japan, and Crater Lake, Ore. McCall had also detected scalloping in lunar craters, which he believed "closely resemble terrestrial volcano-tectonic craters." Taking "pronounced scalloping ... to be a valid indication of volcano-tectonic subsidence," McCall questioned whether lunar craters could be explained by hypervelocity impact. (McCall, *Nature*, 9/24/66, 1384-5)

- U.S. offer at U.N. Sept. 22 to provide U.S.S.R. with tracking facilities on U.S. territory for Soviet space program received editorial comment in *New York Times*: "If all the Russians are after is improved tracking facilities for their own space program, the ... offer ... should be most agreeable to Moscow. Conclusion of a Soviet-American treaty for this purpose would very likely help smooth the way for similar bilateral agreements between the Soviet Union and other non-Communist countries.... Ambassador Goldberg's suggestion points the way to satisfying legitimate Soviet needs without forced infringement on the national sovereignty of others." (*NYT*, 9/24/66, 22)

September 24-25: Society of Experimental Test Pilots' conference in Beverly Hills, Calif. Astronaut John W. Young said space crews needed "another set of arms" before they could do useful work in orbit; he suggested rigid arm-like clamps built into or attachable to spacesuit. Discussing early fears ... that zero g (weightlessness) might prove a physical problem, Young commented: "There is simply not that difference between zero-G and one-G [earth's gravity]. Men will die of old age before they are bothered by zero-G." Referring to present landing mode, Young said Gemini data had helped substantiate belief that land landings

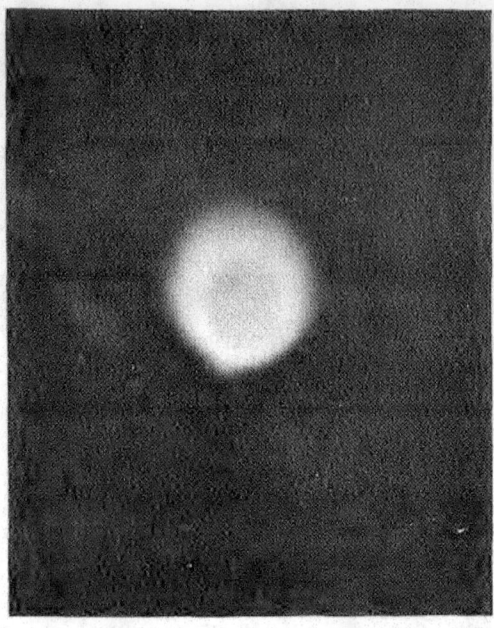

September 24: Javelin rocket launched from Wallops Station releases barium cloud visible over much of eastern United States.

were feasible and that "development of a land-landing capability must be vigorously pursued."

NASA test pilot Milton O. Thompson was awarded SETP's 1966 Iven C. Kincheloe Award, presented annually to the outstanding experimental test pilot of the year, "in recognition of outstanding professional accomplishment in the flight testing" of M2-F2 wingless lifting-body vehicle. (NASA Release 66-252; FRC Release 18-66; Wash. *Eve. Star,* 9/24/66, A2)

September 25: Nike-Tomahawk sounding rocket launched from NASA Wallops Station ejected mixture of barium and copper oxide at about 125-mi. (201-km.) altitude during ascent and at 160-mi. (258-km.) altitude during both ascent and descent. Three distinct clouds, visible for hundreds of miles, were photographed and tracked in second NASA-German Ministry of Scientific Research cooperative launch to measure electric fields and wind motions in upper atmosphere [see Sept. 24]. (NASA Release 66-248; Wallops Releases 66-46, 66-49; NASA Rpt. SRL)

- Indications were increasing that President Johnson would accede to pressure from White House fiscal advisers to cut $1 billion from Project Apollo budget as part of effort to reduce proposed FY 1967 Government spending by $4 billion, Leslie Carpenter reported in the Washington *Sunday Star.* It was contended that reduction would postpone manned lunar landing by only two years, putting it between 1970-72 rather than 1968-70. NASA Administrator James E. Webb, opposing the move, argued it would raise the overall cost and permit the U.S.S.R. to beat the U.S. to the moon. (Carpenter, Wash. *Sun. Star,* 9/25/66, A21)

- Data transmitted from NASA's PIONEER VII spacecraft (launched from ETR Aug. 17) to Johannesburg DSN station indicated spacecraft had penetrated earth's magnetic tail at 3.25 million mi. from earth, NASA Ames Research Center announced. Spacecraft measurements of solar wind flux suggested PIONEER VII had emerged from and repenetrated tail several times. PIONEER VI (launched Dec. 16, 1965) had observed that solar wind does not flow from the sun in straight lines but blows magnetic tail from side to side; this "flapping" was reason spacecraft entered tail several times. (ARC PAO, 9/26/66)

September 26: Astronaut Richard F. Gordon, Jr., pilot of Sept. 12-15 GEMINI XI mission, said at MSC mission summary news conference that 80 per cent of his energy during EVA was spent trying to hold still and in keeping his arms and hands in front of him while connecting spacecraft to Agena with tether: ". . . a little simple task that I had done many times in training to the tune of about 30 seconds lasted about 30 minutes." Gordon recommended development of restraint system to keep astronauts in position to use both hands while outside spacecraft. He said stiff spacesuit worn on mission hampered his movements; his arms and legs tended to float upward because of spacesuit's configuration. (MSC *Roundup,* 9/30/66, 1-2; Waldron, *NYT,* 9/27/66, 14; AP, Wash. *Eve. Star,* 9/27/66, A4)

- Japan's first attempt to orbit satellite failed when Lambda 4S-1 rocket went off course after ignition of 4th stage. The 8.48-ton, 55-ft. booster was launched from Uchinoura Range, Kyushu Island, and carried 57.2-lb. instrument package designed to radio back scientific information on the ionosphere. Scientists from the Univ. of Tokyo had scheduled second launch attempt—using 39-ton Mu rocket—for December. (*NYT,* 9/27/66, 16; *Tech. Wk.,* 10/3/66, 13)

September 26: U.S.S.R. began carrier rocket tests in the Pacific, to last until Oct. 25. Tass said center of area with 74-km. (46-mi.) radius had coordinates of 37°14′ north latitude and 172°49′ east latitude. Ships and aircraft were requested not to enter this area daily between noon and midnight local time. (Tass, *Pravda*, 9/25/66, 2)

- NASA and NAS-NRC jointly announced that a "limited number" of appointments as scientist-astronauts was being made available. To qualify, a scientist must be a U.S. citizen on or before March 15, 1967, be no taller than six feet, be born after Aug. 1, 1930, and have a doctorate in the natural sciences, medicine, or engineering. Applicants would also be required to meet physical qualifications for pilot crew members. After the Academy had done initial screening, NASA would make final selection. Deadline for applications was midnight, Jan. 8, 1967. Applicants would be informed of their status by March 15, 1967. (NASA Release 66-255; NAS-NRC Release)

- NASA Administrator James E. Webb expressed appreciation to USAF, on its 19th anniversary, in a letter to Air Force Assn. President Jess Larson: "The technology developed by the Air Force for its needs in the period before 1958 has served as the foundation for the manned and unmanned missions which NASA is carrying out. . . .

    "An. . .important aspect of the growing aerospace strength of the United States is the close cooperation between the Air Force and NASA in planning for future projects. . . ." (Text; *AF/SD*, 11/66, 85)

- Comparing earth to a "manned spaceship hurtling through the infinity of space," Vice President Hubert H. Humphrey said in address at San Fernando State College that exploration of space would bring home the fact that "earth is a mere speck in the universe." He said it would seem increasingly absurd we had not better organized our living together on this planet: "Our experience in space can be a powerful stimulus to all of us, wherever we live, to move towards a world of law, a world without war. And may we find also in the space effort that 'moral equivalent of war' for which philosophers have been searching over many centuries. Like war, it stretches our human capabilities to the utmost. It demands all that we possess of ingenuity, determination, persistence, and intelligence—and, on the part of our gallant astronauts, the highest degree of courage and resourcefulness in the face of danger. Space activities —even competition in space—can be a substitute for aggression, a bridge for mutual understanding and the identification of common interests with other nations, and a major tool of arms control and disarmament." (Text)

- Columbia Univ. oceanographic research ship *Vema* was fixing position off the coast of Iceland by means of three USN Transit navigational satellites. This marked first civilian use of space navigation systems for oceanographic research. Primary function of satellites, launched in July 1964, was to help guide USN ships and Polaris submarines. (Wilford, *NYT*, 9/26/66, 41)

- Pan American World Airways' fan-jet Falcon claimed new speed record for business aircraft when it flew from St. John's, Newfoundland, to Lisbon, Portugal, covering 2,388 mi. in 4 hrs. 38 min. 28 sec. North American Aviation's Sabreliner had set previous record of 4 hrs. 45 min. 59.4 sec. on Oct. 26, 1963. Both flights had been piloted by George Eremea of Petit Jean Air Service. (*NYT*, 9/29/66, 86)

September 26: U.S.S.R.'s "period of inactivity in manned space flight" received editorial comment in *Kansas City Star*: "About all the West can do is wait until something happens. Meanwhile, the U.S. program moves ahead with significant achievements. But while guessing can be fun, a word of caution: Let's not make the mistake of counting the Russians out of the space race." (*Kansas City Star*, 9/26/66)

September 27: U.S.S.R.'s LUNA XI—which entered lunar orbit Aug. 27—was studying gamma rays and x-rays emitted from lunar surface "to determine its chemical composition more exactly," Tass reported. (Tass, 9/27/66; AP, Wash. *Eve. Star*, 9/27/66, A1)

- Space negotiations played important role in visit of West German Chancellor Ludwig Erhard to U.S. Following discussions with President Johnson, joint statement was released from White House: "The President and the Chancellor discussed possibilities for increased cooperation in technology and science and in particular in the field of space research. The Chancellor expressed his satisfaction that effective steps toward increased cooperation in space research have been initiated since his last meeting with the President in December 1965. The President and the Chancellor welcomed the decision to expand the present cooperative satellite program reached as a result of the recent discussions in Bonn between NASA Administrator Webb and Minister of Science Stoltenberg.

  "The President and the Chancellor agreed that scientific cooperation should be pressed forward for the mutual benefit of both countries and the advancement of human knowledge, preserving opportunities for additional nations to participate and contribute."

  President Johnson and Chancellor Erhard, accompanied by NASA Administrator James E. Webb and Secretary of Defense Robert S. McNamara, later flew from Washington, D.C., to Cape Kennedy for tour of KSC. During speech in Vehicle Assembly Building, President Johnson said he was discussing "with the Chancellor, as well as other leaders, my hope that our scientists can join in joint endeavors to reap the full benefits of space research." He said he had authorized Administrator Webb "to discuss whether solar physicists from Europe may wish to be associated with the American solar physicists" in program for Apollo Telescope Mount (Atm) to be flown "on an Apollo flight in order to study the sun at the height of the solar cycle."

  Appealing to the U.S.S.R. for early conclusion of a treaty on outer space exploration, Johnson said: "I earnestly hope that the Soviet Union—whose space achievements have been very great—will feel as we feel in America: that the rapid evolution of space technology makes ... a treaty between us governing the use of space a most urgent matter ... as we explore the vastness of space, and as we dream of new horizons, we work, too, for the manmade controls that will keep these efforts at the service of man and at the service of peace." He noted that the U.S. sought and would continue to seek cooperation in space with U.S.S.R.: "We have an agreement to exchange certain kinds of space data. . . . We will soon publish jointly American and Soviet material on space biology and medicine." The President also mentioned that the U.S. was "on schedule in our plan and our determination to put men on the moon before 1970." (*Pres. Doc.*, 10/3/66, 1365-6, 1370; Wash. *Eve. Star*, 9/28/66, C1)

September 27: West Germany's Chancellor Ludwig Erhard and President Johnson are briefed by NASA officials at Kennedy Space Center.

September 27: President Johnson, in 10th anniversary edition of Russian-language *America Illustrated,* published by USIA for distribution in U.S.S.R., appealed to Soviet people for new era of friendship and understanding between the two countries. Johnson said "constructive steps" toward that goal would be "progress in the field of disarmament and in greater cooperative efforts . . . in space exploration, medical research, and communications." He added: "We've got to get into the habit of peaceful cooperation. The test ban treaty was a significant step. There have been others since 1963. We have agreed not to put bombs in orbit, we are working together on . . . desalination, weather information, exchanges of scientists, artists, and yes, magazines." (*Pres. Doc.*, 10/3/66, 1360-3)

- In awards ceremony at NASA Hq., Administrator James E. Webb presented Charles Conrad, Jr., command pilot of Sept. 12-15 GEMINI XI mission, gold ribbon attachment to NASA's Exceptional Service Medal awarded to him after August 1965 GEMINI V spaceflight; pilot Richard F. Gordon, Jr., received Exceptional Service Medal.

  Webb announced that during Sept. 10-16 visit to West Germany for preliminary negotiations of space agreement, he had been authorized to offer room on manned and unmanned spacecraft for valid scientific experiments submitted by West Germans. Offer applied also to scientists of other nations. (AP, *NYT*, 9/28/66, 21; *Wash. Post*, 9/28/66, A4; *Tech. Wk.*, 10/3/66, 18)

- NASA, to "allow a better time and motion evaluation of the workloads and stresses" experienced by an astronaut during extravehicular activity, had canceled plans to test USAF Astronaut Maneuvering Unit (Amu) on Gemini XII mission. NASA and DOD would study plans to reschedule experiment on later manned spaceflight. (NASA Release 66-258)

September 27: Prominent details of the moon's far side photographed by ZOND III in July 1965 had been named by Soviet astronomers for great scientists, Tass reported. Among them were the Korolev thalassoid and the craters of Confucius, Spinoza, Mendel, Zhukovskiy, and Galois. Information relayed by ZOND III would be included in second part of *Atlas of the Other Side of the Moon*, currently under preparation. First part was published in October 1960. Prof. Yuri Lipski, in charge of preparing the atlas, said one of the photos taken by LUNAR ORBITER I that had been "kindly supplied by American researchers," had revealed approximate $2/3$ overlap of details in photo taken by ZOND III while remainder of photo coincided in part with photos taken by LUNA III. (Tass, 9/27/66)

- Dr. Eugene M. Shoemaker, chief adviser on geology for Interior Dept.'s Geological Survey, proposed 10-yr. lunar exploration program following first Apollo astronaut landings late in this decade that "would have nothing to do with exploiting natural resources of the moon." He explained: "We are using our mineral resources at a fantastic rate. What we really need to sustain our economic needs in the years ahead is a . . . theory about how the earth was formed. A good way to develop this theory would be by comparison with another celestial body—and the moon is just such a body." Cost would be some $1.5 billion a year. Proposal was made during interview with William Hines, Washington *Evening Star*. (Hines, Wash. *Eve. Star*, 9/27/66, A9)

September 28: USAF launched an unidentified satellite from Vandenberg AFB using Titan III-B-Agena D launch vehicle. (UPI, *Wash. Daily News*, 9/29/66, 24; *U.S. Aeron. & Space Act., 1966*, 156)

- 700-seat auditorium would be built at National Academy of Sciences and a special award established as a memorial to Dr. Hugh L. Dryden, late NASA Deputy Administrator. Financing would be by Hugh L. Dryden Memorial Fund, established by Dr. Dryden's personal friends and scientific associates shortly after his death in December 1965. Announcement was made jointly by Mrs. Dryden, NASA Administrator James E. Webb, and NAS President Dr. Frederick Seitz.

    Commenting on proposed memorial, Webb said: "To a very large degree NASA and NASA's program owe their existence and success to Hugh Dryden. . . . His wise leadership as Deputy Administrator of NASA and as NASA's first scientist, played a leading role in creating effective working relationships between Government, industry, and the universities of the Nation to meet the tremendous new scientific and engineering tasks of the space age."

    Dr. Seitz said: "It is particularly appropriate that funds raised in the name of Hugh Dryden, who served the Academy for ten years as its Home Secretary, be used to provide the Academy with an auditorium in which scientific and engineering questions of national importance could be discussed and considered. Dryden's remarkable personal gifts as fundamental scientist, engineer, and public administrator were responsible, in great measure, for this nation's preeminence in aviation and its successes in space. To all his activities, he brought ethical and professional standards that were deeply admired by all who knew him." (NAS-NRC Release; NAS-NRC-NAE *News Rpt.*, 1; *Wash. Post*, 9/28/66, B4)

September 28: LaRC awarded 7-mo. study contracts, valued at $390,000 each, to Northrop Corp. and Fairchild-Hiller Corp. to design V/Stol aircraft for research on handling qualities and operation during instrument approach in the terminal area. Aircraft would be used by NASA and USAF to gather flight research information necessary to improve operation of V/Stol jet aircraft in hovering and transition speed range. (NASA Release 66-259)

- U.S. invited NATO allies to use its military comsat system temporarily and to collaborate in joint development of new system specifically for the alliance. U.S. would make time available on its seven-satellite military network launched by USAF Titan III rocket June 17. To adapt system to NATO's needs, it was proposed that new ground station be built in Naples to operate with one under construction in the Netherlands and another operating in West Germany. In early 1968, satellite would be launched from U.S. site into synchronous orbit at 22,300-mi. (35,903-km.) altitude for alliance's exclusive use. It could be built by consortium of member countries. Eventually, comsat system was envisaged whose relays would permit use of ground stations small enough to be carried on a truck which could be used by military units on the move. Cost of plan was said to be some $50 million. Proposal would be on agenda of next meeting of NATO's foreign ministers, set tentatively for Dec. 15-16. (Mooney, *NYT*, 9/29/66, 1; Nossiter, *Wash. Post*, 9/29/66, A22)

- M2-F2 lifting-body vehicle, piloted by L/C Donald Sorlie (USAF), was air-launched from B-52 aircraft at 45,000-ft. altitude in ninth glide flight at Edwards AFB. Purposes of flight were to complete pilot checkout and to extend the M2-F2's flight envelope. (NASA Proj. Off.)

- Soviet physicians, according to Tass, had reported that highlanders or people who had lived for a long time in mountain areas made the best astronaut candidates since they "spend the energy of their organism much more economically." They used less oxygen when doing physical work; had approximately 30 per cent greater resistance to oxygen shortage; and possessed "heightened resistance to infrared, ultraviolet, and cosmic radiations." (Tass, 9/29/66)

September 29: Crew for second manned Apollo mission, scheduled for 1967 as open-ended earth orbital flight up to 14 days, was announced by NASA: Walter M. Schirra, Jr., command pilot; Donn F. Eisele, senior pilot; and Walter Cunningham, pilot. Backup crew: Frank Borman, command pilot; Thomas P. Stafford, senior pilot; and Michael Collins, pilot. It would be first space mission for Eisele and Cunningham. (NASA Release 66-260)

- NASA Marshall Space Flight Center had awarded three new study contracts totaling $400,000 to investigate launch vehicle needs and best methods for sending manned spacecraft on planetary flyby trips: North American Aviation, Inc., received $100,000 to study feasibility of modifying Saturn V 2nd stage (S-II) for use as an orbital injection stage; Douglas Aircraft Co., $100,000 for feasibility of using Saturn V 3rd stage (S-IVB) as part of a planetary vehicle; and TRW Systems, Inc., $200,000 for alternate mission modes for manned Mars and Venus orbital and landing missions. (MSFC Release 66-223)

- Soviet scientists, in report distributed by Tass, said it was now possible to protect spacecraft from radiation with an electromagnetic field in place of heavy metal shielding that had begun to limit possibilities of space travel. "The proposed electrostatic protection is much lighter, with its weight consisting only of the electrodes, insulators, and sources of power." Tass also quoted Soviet scientists as saying that danger of man's becoming overheated in space during prolonged work in an airtight suit would disappear if the suit were ventilated from within. Experiments had shown that man wearing an unventilated spacesuit might develop first symptoms of overheating on fourth day; on fifth day, body temperature might rise and heart rate increase, resulting in considerably decreased ability to work. (Tass, 9/29/66; AP, *Wash. Post*, 9/29/66, A1)

September 29: Boeing Co. unveiled a $2-million, full-scale mockup of 1,800-mph Sst—featuring variable-sweep wings, 350-passenger capacity, and a movable nose—built in Government-sponsored competition for Sst production contract. Specifications were submitted to FAA Sept. 6, for evaluation. (Boeing Photos; Clark, *NYT*, 9/30/66, 45; Wash. *Eve. Star*, 9/30/66, B9)

- AIAA-sponsored tour of European aerospace companies began in Paris with briefing arranged by Eurospace, the Council of European Industrial Groups. AIAA President Raymond L. Bisplinghoff led the combined tour, which then broke into three groups for visits to facilities in France, U.K., Belgium, the Netherlands, West Germany, and Italy.

  In a welcoming address in Paris, Eurospace President Jean Delorme said his group would (1) urge establishment of a "European Space Authority" modeled after ESRO, (2) encourage greater national investment in cooperative R&D progress, and (3) point out discrepancy between Europe's high economic potential and low investment in aerospace R&D. (Newbauer and Hersey, *A&A*, 1/67, 46-56)

September 30: NASA's Blossom Point, Md., tracking station, prototype station for Project Vanguard, was closed "because of technical progress in space communications." Blossom Point Station, which had tracked every unmanned earth-orbiting NASA satellite since EXPLORER I was launched Jan. 31, 1958, would become part of STADAN Network Test and Training Facility at GSFC. Facility, to be operational by Jan. 1, 1967, would test equipment, modes, and procedures and train personnel for STADAN stations. (NASA Release 66-117)

- Three earth-orbiting objects not identified "with any launching or country of origin" were reported by GSFC *Satellite Situation Report*. Two had inclination of 85° and the third, 35°. (GSFC *SSR*, 9/30/66; Clark, *NYT*, 11/1/66, 19)

September 30-October 2: Series of Nike-Cajun and one Nike-Apache sounding rocket were launched from NASA Wallops Station and Natal, Brazil, in GSFC experiment to obtain temperature, wind, pressure, and density data at intervals throughout 24-hr. periods near fall equinox. Using exploding grenade technique, experiment investigated thermally driven diurnal tide theory of upper atmosphere. Wind motion was studied by photographically tracking smoke puffs from upper grenade using Apache rocket motor. Rockets and instrumentation performed satisfactorily. (NASA Rpt. SRL)

During September: Vice President Hubert Humphrey, in an interview with *Technology Week*, predicted that U.S. would land men on the moon in 1969—probably before U.S.S.R. He said the Johnson Administration had definite plans for "very active post-Apollo schedule," but refused to "put a time schedule on what comes next." Post-Apollo activities would include exploring the moon, increasing our knowledge of the planets, expanding earth-orbiting capability, both manned and unmanned, and developing maneuverable, reusable spacecraft. The Vice President foresaw no possibility of merging USAF's Manned Orbiting Laboratory (Mol) program with any of NASA's manned projects. (*Tech. Wk.*, 9/5/66, 12, 13)

During September: U.S. and U.S.S.R. were exchanging daily six to eight weather satellite photographs with nephanalyses. First U.S. transmittal began Sept. 6 with pictures from ESSA I meteorological satellite, launched Feb. 3. On Sept. 11, U.S.S.R. began sending pictures from COSMOS CXXII, launched June 25. Regions photographed—primarily Atlantic Ocean and Western Europe—had been matched so operations of satellite systems could be compared. Joint effort to improve transmission quality had been proposed since transmission seemed to degrade photographs: U.S. scientists found Soviet photos "poor" in quality; Soviet scientists described U.S. photos as "usable."

Receipt of photos from COSMOS CXXII had enabled U.S. to analyze satellite's instrumentation: COSMOS CXXII carried pair of 48° TV cameras canted 16.5° to each side of orbital plane to provide coverage of about 690 mi. of earth. Cameras overlapped field of coverage by 1°; frames transmitted so far had 10 per cent overlap. (*Av. Wk.*, 9/26/66, 26-7; Schmeck, *NYT*, 9/27/66, 15)

- Scientific and technological knowledge must be directed to the "making and keeping of peace," Dr. Donald F. Hornig, Director of Office of Science and Technology and science adviser to the President, told *General Electric Forum*.

  "Science will not be the sole contributor toward this end. . . . Its role must be to relieve those pressures that have traditionally been the causes of political, social, and economic instability—pressures which have always led to an upheaval of social order. By helping to alleviate poverty, ignorance and disease, and by improving food supplies and living conditions, science can help to remove the causes of war. . . ." (*GE Forum*, 7-9/66, 8)

- Honors and elections reported by *Aviation Week*: (1) George Galipeau, president of Van Dusen Aircraft Supplies, was elected president of National Pilots Assn. (NPA); (2) Paul Poberezny, president of Experimental Aircraft Assn., was named 1966 NPA Pilot of the Year; and (3) H. M. Horner, United Aircraft Corp.'s chairman and chief executive officer, was named recipient of National Aviation Club's Award for Achievement for his "life-long devotion to the advancement of air power and space power for world commerce and Free World Defense." (*Av. Wk.*, 9/12/66, 23)

- Rep. Joseph E. Karth (D-Minn.), chairman of House Science and Astronautics Committee's Space Sciences Subcommittee, recommended in *Astronautics & Aeronautics* that manned and unmanned earth-orbital missions be given major emphasis in the next few years. He suggested lower priority be assigned manned planetary exploration and establishment of manned lunar bases, but that unmanned Voyager Mars and Venus missions be given more extensive support. (*A&A*, 9/66, 12-14)

- NASA released its 1,000th Tech Brief, one which described development of improved titanium alloy of possible value in field of medicine.

  Potential industrial uses of inorganic fibers developed as reinforcements in materials for spacecraft were discussed in *Non-Glassy Inorganic Fibers and Composites*, issued by NASA Office of Technology Utilization.

Techniques used to monitor astronauts' blood pressure could be adapted to prolong the lives of persons with certain types of heart disease, concluded Westinghouse Research Laboratories engineers William J. Jones and Wyatt C. Simpson in monograph entitled *NASA Contributions to Cardiovascular Monitoring*. Study was conducted for NASA's Office of Technology Utilization. (NASA Releases 66-238, 66-246, 66-250)

During September: USAF had begun evaluation studies of Adam (Air Deflection and Modulation) II—a revolutionary type of aircraft with engines located inside the wings—at Wright-Patterson AFB. A full-span, powered model of the propulsive-wing V/Stol aircraft would be designed, built, and tested by LTV Aerospace Corp. under a 15-mo., $439,000 contract jointly funded by USAF and USA. Wind-tunnel tests would be made at LaRC. Aircraft could be used in both military tactical and commercial passenger versions. (AFSC Release 154.66)

October 1966

October 1: Eighth anniversary of NASA, established by the National Aeronautics and Space Act of 1958.
* Rufus E. Miles, Jr., director of mid-career programs at Princeton Univ.'s Woodrow Wilson School of Public and International Affairs and former HEW Assistant Secretary for Administration, was sworn in as a NASA consultant on management operations. (NASA Release 66-277)

October 2: ESSA III (Tos-A), third meteorological satellite in Tiros Operational Satellite (Tos) series and first to include Advanced Vidicon Camera System (Avcs), was successfully launched for ESSA by NASA from WTR using three-stage Thor-Delta booster on mission to provide daily global cloud-cover photos. Satellite achieved nearly polar, sunsynchronous orbit with 922-mi. (1,486-km.) apogee; 858-mi. (1,381-km.) perigee; 114.5-min. period; and 101° inclination. It was expected to transmit first photos to Gilmore Creek, Alaska, receiving station Oct. 3. An advanced version of the cartwheel configuration, 325-lb. ESSA III carried two Avcs systems to provide cloud coverage and an infrared sensor (Ir) to measure earth's heat balance. Launch was first for Thor-Delta booster from WTR.

ESSA financed, managed, and operated the Tos system; GSFC was responsible for procurement, launch, and initial checkout of spacecraft in orbit. First two satellites in system—ESSA I and ESSA II—were launched Feb. 3 and Feb. 28, respectively. (NASA Proj. Off.; AP, *NYT*, 10/3/66, 11; AP, Wash. *Eve. Star*, 10/3/66, A12)

* NASA ended series begun Sept. 30 of ten Nike-Cajun sounding rockets carrying acoustic grenades launched at six-hour intervals from Natal, Brazil, and NASA Wallops Station to gather meteorological information on atmosphere between 20-mi. and 60-mi. altitude. Grenades were ejected and detonated at programed altitudes to study diurnal variations in winds, temperatures, densities, and pressures. Launchings completed international series of experiments for meteorological research in the upper atmosphere carried out under Nov. 15, 1965, agreement between Brazilian Space Commission (CNAE) and NASA. (NASA Release 66-265; Wallops Release 66-52)
* Zenith Radio Corp. demonstrated experimental systems for displaying television pictures which substituted laser beam for conventional cathode-ray tube. Small nozzle placed on one wall projected television image on opposite wall in same fashion as movie projector. Clarity of images produced by laser beam was only slightly below that of conventional TV sets. (Kotulak, *Chic. Trib.*, 10/3/66)
* NAS Council approved statement urging scientists to "help maintain ICSU's [International Council of Scientific Unions'] nonpolitical traditions by avoiding controversial nonscientific activities while participating in union-sponsored meetings." (NAS-NRC-NAE *News Report*, 10/66, 11)

October 2: President of the Republic of the Philippines Ferdinand Marcos was seeking to postpone Asian summit conference on Vietnam from Oct. 18 to Oct. 24 to allow time for comsat, scheduled for launch over Pacific Oct. 19 by NASA for ComSatCorp, to become operative. Comsat would assure worldwide mass media coverage of event. (Reuters, *Wash. Post,* 10/3/66)

- "The answer to the housing problem lies on the way to the moon," technologist and Southern Illinois Univ. research professor Buckminster Fuller told Wolf Von Eckardt of the *Washington Post.* Much of the complex apparatus used in development of closed ecological systems for long manned space flights could be used in terrestrial housing, Fuller said. He envisioned enclosing the necessary machinery in a "little black box. . . . We are investing $7 billion in it. It weighs 500 pounds. We can probably mass-produce it for $2 a pound.

  "That means we can give all the things they need foremost in a house —water, heat, cooling and waste regeneration—for $1000. And we can give it to them without tying them down to real estate with its wastefully expensive water pipes and sewers . . . with high-pressure sprays and chemical regeneration, a man can quench his thirst and clean himself for a long time with just a glass of water.

  "[It] . . . also solved the pollution problem. . . ." (Von Eckardt, *Wash. Post,* 10/2/66, G11)

October 3: USC&GS ship *Oceanographer,* $9.2-million "floating laboratory," left Jacksonville, Fla., on 11-week, 14,000-mi. expedition to conduct marine scientific studies off eastern coast of South America and participate in observations of total eclipse of the sun Nov. 12. Expedition was being conducted by ESSA. (ESSA Release 66-57)

- Absence of statement by President Johnson and West German Chancellor Erhard after their September meeting in Washington, D.C., on suggested U.S.-Western Europe program of unmanned exploration of Jupiter indicated that the proposal had been "killed for the foreseeable future," *Technology Week* speculated. (*Tech. Wk.,* 10/3/66, 3)

- Anticipating launching of Pacific satellite in late 1966, ComSatCorp filed tariff with FCC for Pacific services. Monthly rates ComSatCorp would charge authorized common carriers would be $2,700 between U.S. mainland and Hawaii; $4,900 from U.S. mainland to Japan; and $3,800 from Hawaii to Japan. Tariff for service to NASA for Project Apollo would be equivalent to level of charges applied to other commercial satellite channels. ComSatCorp also said it planned reduced rates for Atlantic services, to apply when second satellite became operational. (ComSatCorp Release)

- M/G John D. Stevenson (USAF, Ret.), former commander of Joint Task Force Eight of the Defense Atomic Support Agency, was appointed Special Assistant to NASA Associate Administrator for Manned Space Flight Dr. George E. Mueller to assist with general management of manned space flight activities. (NASA Ann.)

- Tonnage of U.S. bombs dropped on Vietnam in 1966 would surpass that dropped by U.S. in the entire Pacific during World War II or during three years of the Korean War, Ted Sell reported in the *World Journal Tribune.* U.S. planned to drop 638,000 tons of bombs in 1966 compared to 502,781 tons dropped by USAF in the Pacific between 1941 and 1945. (Sell, *WJT,* 10/3/66, 4)

October 3-8: MSFC-sponsored space sciences exhibit tracing history of rocketry into the future of space exploration, with models of rockets, rocket engines, and spacecraft, was featured at Alabama State Fair in Birmingham. (MSFC Release 66-224)

October 4: Ninth anniversary of orbiting of SPUTNIK I by U.S.S.R.

- U.S.S.R.'s LUNA XI spacecraft, which entered lunar orbit Aug. 27, stopped transmitting at 2:30 GMT Oct. 1 when its batteries were exhausted, Tass reported. Spacecraft had completed 277 orbits around moon and had sent 137 transmissions of scientific information and trajectory data to monitoring stations on earth. Results would be published later. (Tass, 10/4/66; Tass, *Pravda*, 10/5/66, 1, USS-T Trans.)

- NASA announced plans to launch two ion engines into polar orbit in late 1968 satellite mission to advance development of ion engines as propulsion units for future long-duration space missions. Second part of LRC's Space Electric Rocket Test (Sert) program, mission would evaluate inflight performance of electron-bombardment ion engines for six months or more and analyze effects of thrusters and their electric fields on other satellite components. LRC conducted first successful operation of ion engine in space July 20, 1964, in 50-min. Sert I ballistic flight from NASA Wallops Station. (NASA Release 66-262)

- Dr. René Dubos, microbiologist and pathologist at Rockefeller Univ., was selected to receive Pacific Science Center's 1966 Arches of Science Award for his "outstanding contribution to the public understanding of the meaning of science to contemporary man." (Wash. *Eve. Star*, 10/5/66, A8)

- France exploded last nuclear device in its test series at Mururoa Atoll in the South Pacific. There was no official estimate on size of the explosion. Test series began July 2. (Reuters, *Wash. Post*, 10/5/66, A21; AP, Wash. *Eve. Star*, 10/5/66, A9)

October 4-7: NASA-Lewis inspection held at LRC to commemorate 25th year of operation as aeronautical and space propulsion research laboratory; Lewis-managed launch vehicles and the GEMINI VII spacecraft were on display for congressional, Government, university, and industrial leaders touring facility. (LRC Release 66-60)

October 5: Revised Soviet draft for space law treaty circulated to U.N. members contained three apparent concessions to U.S. position: (1) equal access to all nations' tracking facilities would be arranged by bilateral negotiations; (2) all parties would inform U.N. Secretary General, the public, and international scientific community of nature and results of their activities in space "to the greatest extent feasible and practicable"; and (3) space stations and vehicles of any party on the moon and/or other celestial bodies would be open to visitors "on a basis of reciprocity" with "reasonable advance notice." (Teltsch, *NYT*, 10/6/66, 6; *Wash. Post*, 10/6/66, A20)

- USAF launched USA's SECOR VIII geodetic satellite and unidentified satellite with Atlas-Agena D booster from WTR. (*U.S. Aeron. & Space Act., 1966*, 156)

- M2-F2 lifting-body vehicle, piloted by L/C Donald Sorlie (USAF), was air-launched from B-52 aircraft at 45,000-ft. altitude in 10th glide flight at Edwards AFB; purpose was to explore lateral and longitudinal stability and control characteristics with dampers on and off. (NASA Proj. Off.)

October 5: NASA Astronaut Michael Collins presented U.S. flag carried on July 18-21 GEMINI X mission to St. Albans School in Washington, D.C., as "a token of my appreciation for getting me started on this road." Collins and three other alumni were honored by St. Albans at a banquet and reception. (Adams, Wash. *Eve. Star*, 10/6/66, B1)

October 6: In his first X-15 flight test, Maj. Michael J. Adams (USAF) flew X-15 No. 1 to 2,045 mph (mach 3.0) and 75,400-ft. altitude before making successful emergency landing at Cuddeback Dry Lake, Calif., when fuel tank unported 90 sec. after launch. There was no aircraft damage or pilot injury. (X-15 Proj. Off.)

- Press briefing was held at LaRC on results of NASA's LUNAR ORBITER I mission, launched Aug. 10 from ETR to orbit the moon and photograph possible landing sites for Apollo astronauts.

  The more than 200 photos transmitted by the spacecraft revealed a rugged lunar landscape covered with thousands of craters. Ridges and hills were more gently sloped near moon's center and became progressively smoother in the west. Presence of large rocks, or "blocks," proved that surface was firm enough to support manned spacecraft. Of the nine sites photographed, the Ocean of Storms where SURVEYOR I was resting appeared safest for manned landing. Classified as a dark mare—flat, waterless plain with relatively few craters—it had 20 per cent fewer craters than any of the other sites.

  U.S. Geological Survey scientists Dr. Harold Masursky and Dr. Lawrence Rowan suggested that the photos showed evidence of volcanic activity. Citing a small cone-shaped hill on line of a geologic rift near SURVEYOR I and a large crater on the back of the moon, apparently filled with congealed lava, Dr. Masursky said: "The moon is not a cold, dead, lifeless planet. It looks close to being as dynamic as the earth." Dr. Rowan noted that major faults on the lunar surface seemed to run in two general directions—northeast-southwest and southeast-northwest —and concluded that much of the volcanic activity probably occurred along these lines.

  LaRC scientist William Michael said nonphotographic data on the satellite's orbit indicated moon might turn flat side toward earth, not bulgy side as generally believed. He said displacement of moon's center of mass toward earth could explain both a flat face and moon's slight wobble, which had been accounted for previously by the "bulge" theory.

  Micrometeoroid detection experiment had not recorded any "hits" during satellite's eight weeks in lunar orbit in contrast to earth-orbiting satellites which often registered a hit every two weeks.

  Project Manager Clifford H. Nelson announced that Lunar Orbiter B—second of five satellites in series—would be launched between Nov. 6 and 11 to photograph area slightly north of that scanned by LUNAR ORBITER I. (Clark, *NYT*, 10/7/66, 5; Wash. *Eve. Star*, 10/7/66, A3; O'Toole, *Wash. Post*, 10/7/66, A1)

- NASA Javelin sounding rocket launched from NASA Wallops Station reached 404-mi. (650-km.) altitude in GSFC-Canadian Defence Research Board Telecommunications Establishment experiment to measure ion and electron density and temperature. Although rocket did not perform as expected, all instrumentation functioned normally and "excellent" data were received. (NASA Rpt. SRL)

- Apollo Command Module (CM) hardware could be recovered and refurbished for use on extended earth-orbital missions and long stays on the lunar surface, North American Aviation, Inc., Vice President Dr. John McCarthy suggested at AIAA-AAS Forum discussing "After Apollo—What Next?" in Washington, D.C. "Such refurbished hardware could be available within 18 months at no interference to the lunar landing program and would permit the gathering of vital data and experience for the more difficult task of exploring the planets," Dr. McCarthy said. He recommended an integrated space program which would include developing one advanced earth reentry module, single basic laboratory module, and modular launch vehicle system; he emphasized the need to determine the Nation's 10-yr. space objectives "as soon as possible." (Text)

October 6: U.S.S.R. might use data from published accounts of NASA's Gemini space flights to "reap the benefits of our efforts" and "leapfrog" the U.S. into the next phase of the moon race, Astronaut Michael Collins told the Yale Club of Washington, D.C. As examples of data that would assist Soviet scientists, Collins noted that Gemini flights had demonstrated the feasibility of rendezvous and docking missions in space, as well as man's capability to withstand long-duration space flights. (Wash. *Eve. Star*, 10/7/66, A7)

- Because "procedural difficulties" had delayed approval for construction of new high-capacity ComSatCorp earth station at Moorefield, W. Va., ComSatCorp asked FCC to authorize construction of second large antenna at Andover, Me., "to meet pressing demands for greater satellite communications capacity." ComSatCorp stressed that even if the Andover application were approved, it would shortly update the Moorefield application and "continue to seek approval of it," because a second East Coast station site would be required for flexibility, diversity, and survivability. (ComSatCorp Release)

- Clark Univ.'s Robert Hutchings Goddard Library announced plans to bury a time capsule containing 100 "early Space Age" artifacts, including tape recording of former Astronaut John Glenn's communications with ground control during his Feb. 20, 1962, MA-6 flight in FRIENDSHIP 7. Capsule would be set to be opened in 500 yrs. (Hines, Wash. *Eve. Star*, 10/6/66, A4)

- Lockheed-Georgia Co. engineers Dr. W. C. J. Garrard, George K. Williams, and William W. Williams received Society of Automotive Engineers' (SAE) Wright Brothers Medal for their group report on development of soft-field and rough-field landing gear. Presentation was made at SAE's National Aeronautic and Space Engineering Meeting in Los Angeles. (Lockheed *Southern Star*, 9/15/66, 1)

October 6-7: Conference on status of spectroscopic observations of Martian atmosphere was held at NASA's request by NAS Space Science Board in Washington. Estimate of atmospheric pressure would determine spacecraft design for Mars landing. If pressure were no lower than 15 millibars, parachute landing would be possible, but full retrothrust system would be necessary at 5 millibars. Recommendations of the conference would be made available "in the near future." (NAS-NRC-NAE *News Report*, 10/66, 4)

October 7: NASA's Eighth Annual Honor Awards Ceremony was held in Washington, D.C.

NASA Administrator James E. Webb accepted from Mr. and Mrs. C. Thomas Clagett, Jr., for placement at KSC a replica of the bronze Kennedy bust sculpted by Felix de Weldon for the John F. Kennedy

Memorial Library; and from the American Society of Civil Engineers, the 1966 Outstanding Civil Engineering Achievement Award for KSC's Launch Complex 39. Mr. Webb presented NASA's Distinguished Public Service Medal to Dr. Lloyd V. Berkner, chairman, Graduate Research Center of the Southwest's Board of Trustees, who emphasized the need for careful and immediate post-Apollo planning in his keynote address: "We now have a powerful machinery for the command of space—trained specialists, mighty rockets, a technological dexterity that is unsurpassed. But to keep this capability, we must have new strategic objectives. And that there are many worthwhile and compelling objectives there can be no doubt. . . .

"I would assert that it is now imperative that more advanced space strategies be adopted—and quickly—if our space program is to remain efficient and effective. And I would suggest that whatever this further strategy, it employ and advance our full capability in science, men, mechanisms, and skill. . . ."

Recipients of Distinguished Service Medal: Dr. Hugh L. Dryden, former Deputy Administrator (posthumously); Dr. T. Keith Glennan, first NASA Administrator; and Gen. Bernard A. Schriever (USAF, Ret.), former AFSC Commander. Recipients of Exceptional Scientific Achievement Medal: Richard F. Arenstorf, MSFC; Helmut J. Horn, MSFC; Norman F. Ness, GSFC; and George F. Pezdirtz, LaRC. Outstanding Leadership Medal: John F. Clark, GSFC; Edgar M. Cortright, Hq.; Robert L. Krieger, Wallops; George J. Vecchietti, Hq.; Harold B. Finger, Hq.; Harry H. Gorman, MSFC; Edmund F. O'Connor, MSFC; Eberhard F. M. Rees, MSFC; and Herman K. Weidner, MSFC. Exceptional Service Medal: M. Helen Davies, ARC; Herbert A. Wilson, LaRC; Robert A. Rushworth, USAF; Roll D. Ginter, Hq.; David S. Gabriel, LRC; Edmund R. Jonash, LRC; J. Cary Nettles, LRC; Wilfred E. Scull, GSFC; Harry Press, GSFC; Leland F. Belew, MSFC; Lee B. James, MSFC; William G. Johnson, MSFC; and Peter A. Minderman, KSC. Group Achievement Award: JPL; LRC; LaRC (2); and Hq. Space Nuclear Propulsion Office. Public Service Award: Grant L. Hansen, General Dynamics Corp. (Program; Text *AF/SD*, 1/67/54-5)

October 7: NASA announced selection of Bendix Field Engineering Co. to provide maintenance and operations support of the Deep Space Network under a three-year, $37-million, cost-plus-award-fee contract to be managed by JPL. (NASA Release 66-268)

- Willson H. Hunter, Assistant to the Director of LRC, had been appointed NASA's Senior Scientific Representative to Australia, replacing Ray W. Hooker, who had been named Special Assistant for Industrial Affairs at LaRC. (NASA Release 66-267; LRC Release 66-64)

- General Dynamics Corp. received eight-month, $275,000 MSFC contract to investigate concepts of large space structures having possible post-Apollo use. Three structures which could fly as Apollo Applications (AA) payloads on Saturn launch vehicles would be selected, evaluated, and validated for immediate useful application in space. (MSFC Release 66-245)

- USAF awarded Univ. of Colorado a 15-mo., $300,000 research contract to "conduct independent investigations into unidentified flying object (Ufo) reports." Study would be headed by Dr. Edward U. Condon, a professor of physics at the university. (DOD Release 847-66)

- Arthur G. Wimer, Jr., assistant deputy chief of staff for science and technology in the Air Research and Development Command, was named chief scientist for AFSC, replacing Dr. Bernhard H. Goethert who had resigned to become director of Univ. of Tennessee Space Institute. (AFSC Release 206.66)

October 8: NASA's SURVEYOR I spacecraft, which softlanded on the moon's Ocean of Storms June 2, responded to signals from Johannesburg tracking station almost three months after it had been turned off following successful completion of its mission. Batteries had been recharged during latter part of moon's 14-day period of sunlight. JPL scientists said an attempt might be made to obtain additional photos of the lunar surface from the spacecraft before sun set about noon EDT Oct. 9. (Wash. *Sun. Star,* 10/9/66, A4)

- Executive Committee of International Council of Scientific Unions meeting in Monaco approved statement underlining importance of agreement that scientific meetings "shall not be disturbed by political statements or by any activities of a political nature." (NAS-NRC-NAE *News Report,* 10/66, 11)

- Small-scale lunar roughness resulted from impact of small meteorites, not from internal volcanic effects, according to J. A. Bastin, Queen Mary College, London, in a letter to *Nature.* He advanced arguments that (1) roughness lacks horizontal directionality associated with vulcanism, and (2) from known distribution of meteorites by size, taken with estimates of age of lunar surface, "it can be shown that the surface should be virtually covered by craters of millimetre and centimetre size. . . ." (Bastin, *Nature,* 10/8/66, 171-3)

October 9: Astronaut M. Scott Carpenter (Cdr., USN), speaking at national convention of Naval Reserve Officers Assn. in Denver, described his astronaut-aquanaut occupation as a liaison between space and undersea exploration. He said there were plans to train astronauts under water to better withstand the tiring work of maneuvering outside spacecraft—a major problem of recent space flights. Models of spacecraft would be used; astronauts would wear spacesuits and would work without swim fins. Carpenter, who had volunteered in May 1964 for USN's Project Sea Lab, said aquanauts could work in water until they became tired or hungry. (*Denver Post,* 10/9/66)

October 10: NASA Administrator James E. Webb told press conference in New York that U.S.S.R. had "a better chance now than two years ago" to land a man on the moon before the U.S. Basing his view on the Soviet lead in payload-orbiting capability, Webb said: "They have been launching and recovering 10,000-pound spacecraft for four years. You cannot work in space without large boosters." He suggested that the 10-mo. period of Soviet inactivity in manned space flight would end "very soon." (Loory, *NYT,* 10/11/66)

- Gemini Astronauts Richard F. Gordon, Jr., and Neil A. Armstrong were received in Bogota by Colombian President Carlos Lleras Restrepo during their three-week tour of Latin America. (UPI, *Wash. Post,* 10/11/66, B2)

- LaRC awarded Douglas Aircraft Co. a one-year, $145,204 contract for management services and facilities related to Design Criteria Program for space vehicle structures. Program—with total contract estimated at $500,000—was established by NASA to acquaint contractors and program managers with recommended design-problem solutions that had evolved from experience in space technology and alert them to possible pitfalls. (NASA Release 66-264)

October 10: Johnson Administration could not increase NASA's FY 1968 budget because of Vietnam war, but it could make a policy decision on a post-Apollo program, wrote Robert Hotz, in *Aviation Week*.

"What it will take to get this effort going is some political courage to restate the necessity of proceeding beyond Apollo, a decision to chart a specific course from the available alternatives and some modest Fiscal 1968 funding to retain the technical resources that will be required for the future—particularly engineering and scientific manpower. This combination of activity has already been deferred one year beyond its critical moment by the Johnson Administration. Delay for another fiscal year could impose a braking effect from which eventual acceleration would be much more costly and much less effective." (Hotz, *Av. Wk.*, 10/10/66, 21)

- Dr. James A. Hootman, executive secretary of NASA's Inventions and Contributions Board, retired after 30 years Government service to become a professor of physics at Pensacola (Fla.) Junior College. (NASA Ann.)

October 10-15: XVIIth International Astronautical Federation Congress was held in Madrid. Opening ceremony, attended by Prince Juan Carlos, son of pretender to Spanish throne; Princess Sophia, his wife; and Queen Mother Fredericka of Greece, included speeches by IAF President Dr. William H. Pickering; International Academy of Astronautics President Dr. Charles Draper; International Institute of Space Law President Dr. I. Pépin; Spanish Astronautical Assn. President Pedro Huarte-Mendicoa y Larraga; and Spanish Air Minister Gen. José Lacalle Larraga. (GE *Congress Reporter*, 10/11/66, 1)

Dr. Hilliard W. Paige, vice president of General Electric Co., reported first known satellite collision: in April 1965 two experimental gravity-gradient-stabilized satellites launched pickaback by Naval Research Lab. March 9, 1965, locked booms in their 1,756th revolution. After collision, satellites separated and quickly were stabilized by gravity-gradient systems. They were still in orbit. Paige said NASA was currently reviewing GE study to equip a Saturn V 3rd stage with three-axis gravity-gradient system to orient vehicle as manned space laboratory. (GE *Congress Reporter*, 10/11/66, 1, 2)

Jack L. Bromberg and T. J. Gordon, Douglas Aircraft Co., Inc., Missile and Space Systems Div., said the asteroids between Mars and Jupiter could be investigated by utilizing basic Apollo/Saturn launch vehicles, launch facilities, tracking system, and manufacturing and engineering expertise. Bromberg said a three-stage Saturn V could propel a 7,000-lb. unmanned research payload into the asteroid belt and a four-stage Saturn V could propel a 22,000-lb. payload to same target. (GE *Congress Reporter*, 10/11/66, 3)

Dr. Luis Tapis Salinas, president of IX Colloquium on Law of Outer Space, urged changing present laws to cover situations that could arise in outer space. (GE *Congress Reporter*, 10/13/66, 1)

Richard Johnston, Chief of MSC Crew Systems Div., said that to work effectively in space, a future astronaut would require a more streamlined spacesuit, a better self-propulsion system, and an anchoring device such as "stable work platform which will permit his energies to be used in accomplishing the task and not in overcoming body instability." He said that during extravehicular activity on NASA's Gemini missions nearly 80 per cent of the astronauts' energy had been expended to keep still and avoid tumbling. Johnston also recommended further investigation of metabolic expenditures required to perform work in space: twice Gemini astronauts had been forced to return to their spacecraft, overheated and exhausted, after struggling with relatively simple tasks. (Wilford, *NYT*, 10/12/66, 24)

Dr. George E. Mueller, NASA Associate Administrator for Manned Space Flight, presented a comprehensive review of NASA's Gemini program and Dr. Charles A. Berry, Director of Medical Research and Operations at MSC, gave a detailed biomedical report on the Gemini missions. Observations on the effects of spaceflight on man's system had revealed significant changes involving only cardiovascular, hematopoietic, and musculoskeletal systems, but even these changes appeared to be adaptive in nature and were no cause for concern, Dr. Berry said. Results of a study on red blood cell loss during spaceflight were inconclusive, Dr. Berry reported; hyperoxia, lack of inert diluent gas, relative immobility of crew, dietary factors, and weightlessness were being examined as possible causes. (*Tech. Wk.*, 10/17/66, 16)

Soviet scientists V. E. Belai, P. V. Vassilyev, and G. D. Glod reported "significant changes in the effect of various pharmaceutical preparations" under spaceflight conditions and emphasized that new medicines adapted and tested for space were necessary. Report indicated U.S.S.R. used drugs in manned spaceflight both for "therapeutic purposes or to increase the resistance of the human organism to unfavorable effects." Drugs mentioned included narcotics, to relax cosmonauts; stimulants; cardiovascular preparations; and antiradiation medications. (*Tech. Wk.*, 10/24/66, 13)

Soviet scientists A. A. Gurjian, A. V. Yeremin, and V. I. Stepanzov said the dynamics of men working in free space was one of the most important problems connected with manned spaceflight. They had developed a series of arm and leg movements for cosmonauts to maneuver into working position, which included raising one hand over the head and rotating arm in cone-like fashion to cause an opposite body reaction and enable cosmonaut to face about. (*Av. Wk.*, 10/17/66, 30-31)

Phil Bono, Douglas Aircraft Co., Inc., Missile and Space Systems Div., encouraged by recovery, virtually undamaged, of forward section of GEMINI V-Titan II booster from the Atlantic in September 1965 and of an intact Atlas-Centaur insulation panel in November 1965, proposed a "near-term technique for the land recovery of an earth-orbital stage" to be applied specifically to the Uprated Saturn I's 2nd stage (S-IVB). (*Av. Wk.*, 10/24/66, 32)

H. B. Bjurstedt, Karolinska Institute of Sweden, suggested that moon's low gravity field might lead to an accompanying loss of normal g-tolerance. He acknowledged that this was an uncertain extrapolation and that physiological effects of such exposure were as yet unknown. (*Tech. Wk.*, 10/24/66, 13-14)

Soviet delegate Vassily V. Parin narrated new film showing postflight activities of space dogs Veterok and Ugolyek, launched Feb. 22, 1966, in COSMOS CX. Dogs, who lost up to one third of their preflight body weight during 22-day mission, appeared very weak and dazed when they emerged from capsule, but film sequence taken four months later showed them completely recovered. Parin said flight had studied neurological regulators of cardiovascular system. (*Tech. Wk.*, 10/24/66, 13)

Dr. Luigi Napolitano, Univ. of Naples, was elected new president of IAF. (*Tech. Wk.*, 10/24/66, 13)

October 11: NASA Aerobee 150 sounding rocket launched from WSMR reached 89-mi. (144-km.) altitude in NASA-American Science and Engineering, Inc., experiment to make a high-resolution survey of celestial x-ray sources. Rocket and instrumentation performed satisfactorily. (NASA Rpt. SRL)

- NASC Executive Secretary Dr. Edward C. Welsh, speaking before Metropolitan Washington Board of Trade's Science Industry Committee, replied to criticism that the national space program is highly expensive and wasteful: "The national space program is the largest concerted effort undertaken by any nation to advance the frontiers of human knowledge. As such, it is a seedbed of invention, a spur to our productivity, a source of insurance for our national security, a stimulus to learning, and a world-wide ambassador for peace. Because of it, our chances of improving medical research and finding a cure for cancer or heart disease are greater, not less. Because of it, our chances of improving our educational system and solving a vast range of social problems are greater, not less.

  "The issue is really not that of substituting space progress for progress in some other worthy field, because the space program contributes importantly to advances in practically all other lines of endeavor, and it stimulates the national economy at the same time. . . ." (Text)

- Lockheed Missiles and Space Co. engineer Dede Sonnichsen claimed six new world records when he flew an AX-4 hot air balloon to 21,000-ft. altitude from Tracy, Calif., and landed it 15 mi. away. Of 10 hot air balloon classifications, Sonnichsen's claims were to altitude records in AX-4, 5, 6, 7, and 8 categories and to establishment of record for 15-mi. distance. (*L.A. Times*, 10/12/66)

- International cooperation on patents procedures and mechanized information retrieval was urged at Washington, D. C., meeting of National Association of Manufacturers' Patents Committee. U.S. Commissioner of Patents Edward J. Brenner said that "most of the search or examining work in anywhere from 15% to 80% of the applications . . . could be eliminated" if patent offices exchanged information. (Beller, *Tech. Wk.*, 11/7/66, 20)

October 12: M2-F2 wingless lifting body was launched from B-52 aircraft and flown by Capt. Jerauld R. Gentry (USAF) in 45,000-ft. glide to powerless landing at Edwards AFB; maximum speed was nearly 400 mph. This was 11th test mission for 6,000-lb. research vehicle developed for NASA to study possible design for future spacecraft that could maneuver in atmosphere after reentry from orbit. (NASA Proj. Off.; *L. A. Times*, 10/13/66)

- Dr. Lee Arnold, chairman of New York Univ.'s Dept. of Aeronautics and Astronautics and director of Daniel Guggenheim School of Aeronautics, was sworn in as a consultant to NASA Administrator James E. Webb on technology utilization. (NASA Release 66-271)

- USAF launched two unidentified satellites with Atlas-Agena D booster from WTR. (*U.S. Aeron. & Space Act., 1966*, 157)

- Prof. Leonid I. Sedov, head of Soviet delegation to IAF Congress, told news conference in Madrid that U.S.S.R. had to investigate and solve a "number of difficult problems" before it could attempt a manned lunar mission. Asked if he believed U.S. and U.S.S.R. would ever join together

October 12: M2-F2 lifting body mated to B-52 launch aircraft at NASA Flight Research Center, in preparation for flight tests.

for a space mission, Sedov said it would be economically and scientifically feasible, but "we don't have any plans for joint launchings. The situation is not technical but political and I think the international situation in these days cannot change this." (Wilford, *NYT*, 10/13/66, 12)

October 12: IAF conferees toured the NASA-INTA Robledo de Chavela Space Tracking Station outside Madrid. (GE *Congress Reporter*, 10/14/66, 1)

- NAS Committee on Public Engineering Policy (COPEP) held first meeting in New York City; chairman was Dr. Chauncey Starr, president of Atomics International Division of North American Aviation, Inc., and as of January 1, 1967, UCLA Dean of Engineering. (NAS-NRC-NAE *News Report*, 10/66, 1, 3)
- UCLA announced plans to build its own satellite to study earth's magnetic field and interplanetary space. Feasibility studies financed by $55,000 NASA grant were being conducted by Hughes Aircraft Co. and Philco Corp. (UPI, *NYT*, 10/15/66, 33)

October 13: U. S. and Brazil would cooperate in sounding rocket program to study total solar eclipse in South America Nov. 12. Nineteen sounding rockets would be launched to altitudes as high as 175 mi. (280 km.) from Rio Grande launch site located near path of total eclipse. Program was based on July 1965 agreement between NASA and Brazilian Space Commission (CNAE). (Wallops Release 66-51)

- *Washington Post* commented on report of satellite collision made by Dr. Hilliard Paige at IAF Congress in Madrid: "Belated disclosure that two Air Force space vehicles collided in space in April, 1965, ought to remind the country that part of its space program is being carried on behind a cloak of secrecy as absolute as that which conceals space operations in the Soviet Union. The open policy of NASA, which has done so much to keep the whole nation interested in and excited about space progress, is in sharp contrast to the policy of the Air Force. . . ." [Ed. note: Satellites were USN, not USAF.] (*Wash. Post,* 10/13/66, A22)

October 13: Western Union International, Inc., announced receipt of DOD order for 10 Pacific satellite communications channels, permitting company to file with FCC for rate reductions on all its cable and satellite service in the Pacific area. Order and rate reductions were dependent on FCC's approving Western Union's previously announced request to lease and operate up to 65 satellite channels in the Pacific area. (DJNS, Wash. *Eve. Star,* 10/13/66, A13)

October 13-14: Use of satellite navigation and communications system to guide merchant ships was urged by General Electric Co. engineer Roy E. Anderson at Institute of Navigation meeting at U.S. Merchant Marine Academy, Kings Point, N.Y. Ships could be equipped for $2,000 to automatically obtain position fixes anywhere in the world, Anderson estimated. He believed operational system could be designed and built within two years. (*NYT,* 10/17/66, 60)

October 14: U.S.S.R. launched COSMOS CXXIX into earth orbit with 307-km. (191-mi.) apogee; 202-km. (126-mi.) perigee; 89.4-min. period; and 65° inclination. Equipment "for continuing explorations of outer space" was functioning normally. (Tass, 10/14/66)

- Control of ESSA III meteorological satellite, launched by NASA Oct. 3 from WTR, was turned over to ESSA. (NASA Proj. Off.)
- NASA awarded IBM Corp.'s Federal Systems Div. a $51-million, sole-source contract to provide Saturn instrument unit launch support services at KSC through June 30, 1970. (NASA Release 66-275)
- Sixteenth-century manuscript "On Rockets" by Conrad Haas (circa 1550) received at IAF Congress in Madrid. Concepts of rocket construction included multistage rockets and a design of a house-like upper stage. (Text)
- MSC Director Dr. Robert R. Gilruth accepted the 1966 Daniel and Florence Guggenheim International Astronautics Award at the IAF Congress in Madrid. Dr. Gilruth said in his acceptance speech: "The spaceflight achievements of Mercury and Gemini, now entered in the history of man's exploration of the universe, took origin in the ideas, aspirations, and technical contributions of many persons in many countries. Astronautics has an international birthright, just as its future has great meaning for all of mankind." (*A&A,* 1/67, 66)
- Dr. Oleg G. Gazenko, member of Soviet delegation to IAF Congress, told press conference in Madrid that U.S.S.R. was "doing very serious preparatory work" for another manned mission that would be "a serious new step in space exploration." He disclosed no details but indicated mission might include extravehicular activity. Asked why there had been no Soviet manned spaceflight since March 18, 1965, VOSKHOD II mission, Gazenko explained it reflected Soviet desire to make a major advance with each flight rather than repeat proved techniques. (Wilford, *NYT,* 10/15/66, 11)

October 15: President Johnson signed into law H.R. 15963 creating Cabinet-level Department of Transportation. (*Pres. Doc.,* 10/24/66, 1498)

October 15: Close of IAF Congress was dramatized when Spain launched her first sounding rocket—Carabela 4—from Arencillo base, near Huelva. British-built rocket, which cost approximately $1,700, carried 12-lb. instrumented payload to 81-km. (50-mi.) altitude and then impacted 40 mi. downrange in the Atlantic. Two NASA officials—William Hausman, Assistant Administrator for International Affairs, and Clotaire Wood, NASA European Representative—observed the launch, first of six in a long-range program to study meteorology. (Wilford, *NYT*, 10/16/66, 78)

- Communist sources speculated that U.S.S.R. would launch "a large satellite carrying several men . . . just before the Nov. 7 anniversary of the Bolshevik Revolution," and another, the world's largest satellite and carrying eight men, by the end of 1966, AP reported. (AP, *NYT*, 10/16/66, 28)

October 16: Third year in space began for two experimental Vela nuclear detection satellites launched by DOD Oct. 16, 1963, into near-circular orbit at 52,174-mi. (84,000-km.) altitude. Satellites had demonstrated successful R&D program and had provided effective capability to detect nuclear explosions in space. DOD had overall responsibility for Vela program; ARPA had overall direction with support of AEC. (AFSC Release 205.66)

October 17: NASA had begun study of how to keep two or more small monkeys in earth orbit for up to one year to assess long-term effects of weightlessness on primates and evaluate problems related to humans undertaking extensive work in space. Industrial firms had been asked to submit by Oct. 27 "preliminary design study" for Orbiting Primate Spacecraft capable of being carried into earth orbit during Apollo program. LaRC would manage study. (NASA Release 66-273)

- France's Dragon sounding rocket was successfully launched from Norwegian range on Andoeya Island by Centre National d'Etudes Spatiales (CNES). (*Tech. Wk.*, 10/31/66, 13)

- At dedication of Sherman Fairchild Technology Center, Germantown, Md., plans to use Nimbus B meteorological satellite—scheduled for launch late in 1966—in elephant-tracking experiment were revealed in speech prepared for delivery by MSFC Director Dr. Wernher von Braun and read in his absence by MSFC PAO Director, Bart J. Slattery, Jr. Satellite would home in on 25-lb. portable transponder strapped to back of an elephant. Experiment would test use of satellites to monitor wildlife, learn more about animal migration, and develop means of protecting species threatened with extinction. (*Wash. Eve. Star*, 10/18/66, A14; *Wash. Post*, 10/18/66; *Balt. Sun*, 10/18/66)

- New studies of Martian atmosphere had rekindled hope of finding life on Mars, Dr. Lewis D. Kaplan of JPL reported at Western meeting of American Chemical Society in San Francisco. French scientists Drs. Pierre and Janine Connes of the Observatory of Haut-Provence, using improved spectroscopic equipment designed by Dr. Connes, had found concentration of hydrogen compounds around Mars 1,000 times greater than in earth's atmosphere. Dr. Kaplan, who analyzed observations obtained, said the hydrogen compounds—considered necessary to life—probably included methane derivatives and perhaps methane itself. Methane appears on earth as gaseous hydrocarbon product of decomposed organic matter and is known to be produced by anaerobic bacteria not requiring the oxygen essential to most life on earth. Dr.

Kaplan's work was sponsored by NASA and by France's Centre National d'Études Spatiales (CNES) and the Meudon Observatory. (AP, *NYT*, 10/18/66, 9)

October 17: Boeing Co. announced it had awarded $10 million in subcontracts during planning phase (Phase II-C) of Government-sponsored Sst design competition. Phase II-C was being conducted under 18-mo. Government-industry cost-sharing contract which would end Dec. 31. Boeing and Lockheed Aircraft Corp. had submitted final airframe designs to FAA Sept. 6. (Boeing Release)

- Plan for 500-mph, 80-passenger Vtol air-bus service, developed by MIT's Flight Transportation Laboratory to ease transportation congestion on the East Coast in the 1970's, was announced by the Commerce Dept. (UPI, *NYT*, 10/18/66, 76)

During week of October 17: House Committee on Government Operations, in report based on August hearings by Subcommittee on Military Operations, urged a ten- to twenty-fold expansion of DOD's initial comsat system by increasing number and improving performance of both satellites and ground terminals. "Neither the Communications Satellite Corporation nor the other commercial carriers need fear they will lose the Defense Department as a good customer" as a result of this expansion, the report said. (*Av. Wk.*, 10/24/66, 29)

- NASA's Advanced Manned Missions program office issued Rfps on $300,000, nine-month analysis of economic potential of orbiting manned natural resources laboratory. It was specified that firms interested "must possess strong economic expertise at the national and international levels and must not be involved in the production of space or experimental hardware." Replies were due at NASA Hq. Nov. 21. (*Av. Wk.*, 10/24/66, 28)

- MSC Director of Medical Research and Operations Dr. Charles A. Berry received AIAA's 1966 John Jeffries Award "for outstanding contributions to the advancement of aeronautics through medical research." Presentation was made at AIAA Military Aircraft Systems Meeting in Dallas. (MSC *Roundup*, 10/28/66, 1)

October 18: NASA Administrator James E. Webb, in interview with *Look*, rejected report that U.S.S.R. had lost men in space: "Every bit of information we have shows that they have been as careful with the lives of cosmonauts as we have been with astronauts. As far as I know, they have not lost a single man in flight. They have lost them on the ground, they have lost them in airplanes, just like we have." (*Look*, 10/18/66)

- Research and development in essential space programs would not be seriously affected by demands of the Vietnam conflict, USAF Under Secretary Norman S. Paul told a National Space Club meeting in Washington, D.C. Asked during question and answer period for rough estimate of military manned spaceflight expenditures in the 1970s, Paul said that currently DOD was spending $1.3 billion of its $7-billion R&D budget on space and that it was "a safe guess this amount will go up." Asked if there might be NASA programs competitive with USAF's Mol, Paul said there was "not the slightest possibility." (Text; *M/S Daily*, 10/19/66, 20)

- DOD announced USAF would buy 99 improved F4E Phantom jet aircraft equipped for first time with internally mounted 20-mm. Gatling gun capable of firing 6,000 rounds a minute. Aircraft would give U.S. superiority over Soviet-made MiG-21s used by North Vietnam. (AP, *NYT*, 10/20/66, 27)

October 18: Dr. Jan H. Oort, head of Leiden Observatory, the Netherlands, and former head of International Astronomical Union, received 1966 Vetlesen Award in earth sciences. Administered by Columbia Univ. for the G. Unger Vetlesen Foundation, the prize was awarded for "outstanding achievement in the sciences resulting in a clearer understanding of earth."

NASA Associate Administrator for Space Science and Applications Dr. Homer E. Newell noted in speech at award dinner that "through the space approach . . . the domain of the geosciences" had been strengthened and extended: "The theories, instruments, and skills needed and developed to study the earth can now be applied to investigating the moon and planets at first hand. Comparative studies of the planets and their atmospheres, ionospheres, and magnetospheres, promise increased understanding of our own planet, earth." (Text; Sullivan, *NYT*, 10/19/66, 36)

October 19: NASA Administrator James E. Webb and the late NASA Deputy Administrator Dr. Hugh L. Dryden were awarded the 1966 Robert J. Collier Trophy for "representing all of the Gemini program teams which significantly advance human experience in space flight," at special ceremony in Smithsonian Institution's National Air and Space Museum. Trophy was presented annually by *Look* and National Aeronautic Assn. "for the greatest achievement in aeronautics or astronautics in America, with respect to improving the performance, efficiency, or safety of air or space vehicles."

Vice President Hubert H. Humphrey, presenting the trophy, said: "I can think of no more fitting choice for this distinguished award than Project Gemini and all those who made it such a success." He noted that Gemini manned spacecraft had spent 875 hrs. in space; Gemini was first spacecraft to rendezvous and first to dock with another spacecraft; and four Gemini astronauts had walked in space for total 5 hrs. 52 min.

Administrator Webb, accepting trophy, said: "Gemini has broadened our vision, validated our concepts, proved out our technology, and has provided a much sounder structure of knowledge as to the importance of not accepting a position of second best in space." Attributing Gemini program's success to "the application of our scheduling principles," Webb noted "milestones": beginning of unmanned flights in 1963; first manned flight in 1964; rendezvous, docking, and first extravehicular activity in 1965; first use of Agena for orbit change and first closed-loop guidance in 1966; completion of program in 1967. "We missed the first two milestones. From then on, with the exception of docking, we have accomplished each milestone ahead of or on schedule and we do expect to complete the Gemini program ahead of schedule next month." (Text; NASA Ann.; NAA *News*, 10/16/66; *Wash. Post*, 10/20/66, B7)

- Wilhelm Forster Observatory, West Berlin, reported U.S.S.R. had launched a satellite at 7:20 a.m. EDT from space center in Baikonur. Spokesman said it had not been determined whether spacecraft was manned. NASC Executive Secretary Dr. Edward C. Welsh said report appeared to be in error and had apparently stemmed from misinterpretation of Soviet transmissions associated with tracking of previously launched Cosmos spacecraft. He added that if the U.S.S.R. did attempt launching, nothing got into orbit. U.S.S.R. made no comment. (AP, Balt. *Sun*, 10/20/66)
- U.S. did not have women astronauts because "we have not yet been able to find an applicant sufficiently qualified in both scientific and engineering

background, who also possesses the required test pilot experience," said NASA Associate Administrator for Manned Space Flight Dr. George E. Mueller at dedication of Cedar Crest College's science center in Allentown, Pa. He said NASA has always had the same requirements for both men and women; "there had never been any discrimination with regard to sex or on any other basis." (Text)

October 19: MSFC awarded Martin Co. a one-year, $100,000 contract for design feasibility study of 2,000-w nuclear power system consisting of four Snap-29 radioisotope generators—each delivering 500 w—for possible use on long duration post-Apollo missions. Study would be performed at Martin's Baltimore plant. (MSFC Release 66-251)

- Study of Ufo's had been "relegated to the bottom of the barrel" by USAF, Dr. James C. McDonald, Univ. of Arizona professor, charged at Washington, D.C., American Meteorological Society meeting. Attacking USAF's Project Bluebook as "extremely superficial and of a low level of technical competence," he urged that Ufo responsibility be taken away from USAF and put "into scientific hands." The hypothesis that Ufo's might be "extraterrestrial probes" could not be ruled out, he said. (Adams, *Wash. Eve. Star*, 10/20/66, C14)

October 20: U.S.S.R.'s launch of two satellites was observed by leaders of eight Communist nations visiting U.S.S.R. for summit talks on Vietnam and Communist China.

COSMOS CXXX, for continued space research, entered earth orbit with 340-km. (211-mi.) apogee; 211-km. (131-mi.) perigee; 89.9-min. period; and 65° inclination. Equipment was functioning normally.

Fourth MOLNIYA I comsat, for "further testing of . . . long range two-way television and telephonic-telegraphic radio communication," had orbital parameters of apogee, 39,700 km. (24,658 mi.); perigee, 485 km. (301 mi.); period, 11 hrs. 53 min.; and inclination, 64.9°. Equipment, including broadcasting system, orientation system, orbital correction devices, and power supply, was functioning normally. (Tass, 10/20/66)

- Sounding rocket launched by Spain from Arencillo base fell to earth within a hundred yards of launching site. Officials blamed failure in guidance system. Launch was part of series begun Oct. 15 to collect meteorological data from region at 30- to 60-mi. (48- to 97-km.) altitude. (*Chic. Trib.*, 10/21/66)

- No video information was on tape recorded by SURVEYOR I on Oct. 9 and transmitted to Deep Space Tracking Station, Johannesburg, South Africa, Oct. 10, JPL reported. Scientists said they had not been too hopeful, and had been surprised that spacecraft had responded at all on Oct. 8. SURVEYOR I, which had landed on moon's Ocean of Storms June 2, entered its fifth lunar night shortly after Oct. 8 telemetry was returned. It had completed its second lunar day operations on July 13. (*Wash. Eve. Star*, 10/20/66; *Tech. Wk.*, 10/17/66, 10)

- Rep. George P. Miller (D-Calif.), chairman of House Committee on Science and Astronautics, told Clear Lake (Tex.) Chamber of Commerce that he foresaw no curtailment of space program because of Vietnam war. If there were curtailment, "it will be aimed at all Government programs." (Amerine, *Houston Chron.*, 10/21/66)

- Dr. Homer E. Newell, NASA Associate Administrator for Space Science and Applications, visited MSFC to review technological and scientific progress on work being conducted by MSFC for OSSA projects, including Apollo Telescope Mount, x-ray astronomy experiment, and several other astronomy experiments. Accompanying Dr. Newell were his deputies, Edgar M. Cortright and Dr. John E. Naugle, and directors of OSSA program offices. (MSFC Release 66-256)

October 21: U.N. World Meteorological Organization—in report on plan for World Weather Watch to be initiated between 1968 and 1970—warned it would be "foolhardy" for man to attempt to modify weather on large scale with current limited knowledge of atmospheric forces. Plan called for use of satellites; a relatively dense network of land and ocean weather stations, manned and unmanned; drifting sea buoys; balloons; and other instruments to probe the world's atmosphere. High-speed telecommunications systems would centralize findings for analysis and relay latest detailed information to meteorologists everywhere. This would enable them to forecast the weather "both with improved accuracy and for a longer time ahead." (*NYT*, 10/22/66, 12)

- In a letter to *Science*, Dr. J. Allen Hynek of Northwestern Univ. said "we suffer, perhaps, from temporal provincialism" in dismissing the Ufo phenomenon with a shrug. He said that when he was scientific consultant to USAF's Project Blue Book—Ufo registry at Wright-Patterson AFB—some of the most coherent reports had come from scientifically trained people. "Hard data" cases had contained frequent allusions to recurrent kinematic, geometric, and luminescent characteristics. "I have begun to feel that there is a tendency in 20th century science to forget that there will be a 21st century science, and indeed a 30th century science, from which vantage points our knowledge of the universe may appear quite different." He had urged USAF "to ask physical and social scientists of stature to make a . . . scholarly study" of Ufo's. Hynek's letter was written before Oct. 17 announcement of $300,000 USAF contract to Univ. of Colorado for Ufo investigation [see Aug. 1]. (*Science*, 10/21/66, 329; *A&A*, 12/66, 4)

- Radar signature analysis (Rsa) for "spatial detective work" was discussed in *Time*. Using Rsa, scientists could reconstruct characteristics of foreign satellite from radar pulse pattern reflected to tracking station: by measuring amplitude of reflected pulses, satellite's size could be calculated; by analyzing variations in pulse amplitude, satellite's shape could be determined; by determining periodicity with which pulse pattern repeated itself, speed of tumbling, rolling, or spinning around spacecraft's axes could be inferred. Computerized Rsa was being developed so that in event of war, warhead radar signatures could be spotted quickly enough to order interception and destruction by defending missiles. (*Time*, 10/21/66)

- USAF OAR had proved capability of new dust separation device which could provide longer engine life for gas-turbine-powered engines, especially helicopters being used in Vietnam. Device was designed to prevent erosion of compressor section in engine and glass formation on engine's hot surfaces by removing dust particles and other foreign objects from air before they could be sucked into engine. Tests had demonstrated average 92 per cent efficiency. (OAR Release)

October 22: LUNA XII was launched by U.S.S.R. on trajectory toward the moon "to study the systems of the artificial satellite and near lunar space," Tass announced. All systems were functioning normally, and spacecraft was traveling close to planned trajectory. (Tass, 10/23/66)

- NASA Aerobee 150 launched from WSMR reached 93-mi. (149-km.) altitude in third ARC-managed Project Luster experiment to collect meteoric debris during peak of Orionid meteor shower. Luster micrometeoroid sampling instrument deployed successfully during flight; debris was vacuum sealed at recovery. Rocket and instrumentation performed satisfactorily. (NASA Rpt. SRL)

October 22: U.S.S.R. was seeking experienced scientists between ages 40 and 50 as cosmonauts, AP reported Soviet scientist Pyotr Yegorov as saying. Yegorov claimed there were no medical reasons men in this age group could not participate in space flights. (AP, Phil. *Sun. Bull.*, 10/23/66)

October 23: France had requested that Astronauts James A. Lovell, Jr., and Edwin E. Aldrin, Jr., photograph sodium vapor trail of Centaure sounding rocket during NASA's Gemini XII mission scheduled for launch Nov. 9. Sounding rocket—to study winds up to 400 mph at high altitudes— would be launched from Hammaguir Range on signal from Gemini Mission Control at MSC. Experiment would be first by foreign government to be part of a U.S. manned space flight. (*Wash. Post,* 10/23/66, A10; AP, Wash. *Sun. Star,* 10/23/66, A7)

- JPL had awarded $200,000 contract to Avco Corp.'s Space Div. for task work in JPL's new $1.5 million Sterilization Assembly Development Laboratory (SADL). Starting in mid-1967, SADL would be sterilization testing headquarters for Voyager spacecraft, scheduled for Mars landing in the 1970s. (JPL Release)

October 24: NASA Director of Program Review William A. Fleming reiterated NASA hopes of landing a man on the moon by mid-1968 at IEEE conference in Milwaukee. Fleming said plans for post-Apollo space activity had not been completed but might include further exploration of moon, possibly with a roving vehicle; drilling to sample the subsurface of the lurain; or establishing manned space stations. Manned space flight to Mars was still at least 20 yrs. away, he said. (*Milwaukee J.,* 10/25/66)

- Miniature data recording system that could be strapped to pilot's leg to record physiological information during aircraft flights in which missiles were fired was being developed at Naval Missile Center, Point Mugu, Calif. NMC electronic engineer and inventor of device Clifford Phipps said: "In testing the whole weapons system, it is important to know how the man reacts, too." (PMR Release 1411-66)

- Fred W. Friendly, former president of CBS News and current television adviser to Ford Foundation, said at National Assn. of Educational Broadcasters convention in Kansas City that the nonprofit comsat proposed by the Foundation to FCC Aug. 1 would provide a "second service appealing to excellence." He said educational broadcasting was now inefficient because it lacked "the interconnections to bring events to people as they happen." (AP, *NYT,* 10/20/66, 78)

- ComSatCorp proposed to FCC a reduced, flat rate for transatlantic TV transmissions, effective with commercial operation of new Atlantic comsat in late 1966. Proposed rate for one-way, black-and-white transmission: $1,000 for first 10 min. and $30 each additional minute per half channel. Current rates varied from $1,800 for first 10 min. and $32 each additional minute during nonpeak hours, to $3,000 for first 10 min. and $48 each additional minute during peak hours. (ComSatCorp Release)

- Roll D. Ginter, NASA Centaur Program Manager, became Director of OART's newly established Space Flight Programs Div. He would be responsible for mission planning, management, and coordination of technology space flight projects and experiments within OART. (NASA Release 66-276)

October 24: Former Astronaut John H. Glenn, Jr., was appointed chairman of the board of Royal Crown Cola International, Ltd. (AP, *Wash. Post,* 10/25/66, C8)

- Dr. Ivan L. Bennett, Jr., Deputy Director of White House's Office of Science and Technology, warned in speech at seminar on Research in the Service of Man in Oklahoma City that unless scientists "educated" the President and "other intelligent, concerned laymen" about the applicability of their work, their fields of research would suffer. Science, he said, could no longer hope to exist "through some mystique, without constraints or scrutiny in terms of material goals, and isolated from the competition for allocation of resources which are finite." (Sullivan, *NYT,* 10/25/66, 78M)
- USAF was awarding Thiokol Chemical Corp. $1,075,507 contract for design, fabrication, and testing of thrust vector control system for 156-in. solid rocket motor. (DOD Release 897-66)
- Proposed further reduction in NASA FY 1967 appropriation because of Vietnam war would amount to "national disaster," Ira C. Eaker wrote in *San Antonio Express:* "Such desertion of our present space program would give the Soviets an advantage they are now striving mightily to achieve. The U.S.S.R. is now spending about twice what the U.S. is spending for space research." (Eaker, *San Antonio Express,* 10/24/66)

October 25: Propellant tank rupture occurred on Apollo spacecraft 017 service module during proof pressure test at 240 psi conducted by NAA at Downey, Calif.; inflight operating pressure would be 180 psi. No test personnel were injured. NASA had established board of inquiry to investigate. (NASA Release 66-285; NAA S&ID *Skywriter,* 10/28/66, 4; AP, *Wash. Eve. Star,* 10/28/66, D18)

- U.K.'s Jodrell Bank Experimental Station reported that U.S.S.R.'s LUNA XII spacecraft had apparently entered orbit around moon and was transmitting telemetry but no photographic signals. U.S.S.R. had made no official statement since Oct. 22 launch. (UPI, *NYT,* 10/26/66, 10; *Wash. Post,* 10/26/66, A16)
- NASA Aerobee 150 and Nike-Cajun sounding rockets launched 35 min. apart from WSMR reached 116-mi. (186-km.) and 74-mi. (119-km.) altitudes in GSFC-ESSA Institute for Telecommunication Sciences and Aeronomy experiment. Aerobee rocket and instrumentation, designed to measure micrometeoroid and cosmic dust impacts and electron densities, performed satisfactorily. Although Nike-Cajun did not reach predicted altitude and two of seven experiments to obtain data on ionospheric D-region did not function properly, satisfactory results were obtained. (NASA Rpt. SRL)
- MSFC had awarded Univ. of Wisconsin a $679,101 contract to develop sensors for galactic x-ray mapping experiment to be flown on an Uprated Saturn I (Saturn IB) launch vehicle in 1968. Sensors would explore x-ray sources other than sun and Crab Nebula. (MSFC Release 66-258)
- Potential for application of biomedical knowledge acquired in space research was discussed by Dr. Richard L. Lesher, NASA Assistant Administrator for Technology Utilization, at the Conference on Biomedical Knowledge, Oklahoma City. He described two experimental efforts conducted by NASA's technology utilization program: (1) an agreement with Vocational Rehabilitation Administration to make available information from aerospace research to solve "problems of restoring the

disabled to productive life," and (2) interdisciplinary biomedical application teams at research centers under NASA contract to define medical problems and identify aerospace technology applicable to their solution. (Text)

October 25: British Research Council had offered to pay one half the $11.2-million cost of constructing a 150-in. telescope in Australia, largest in Southern Hemisphere. Similar offer from Univ. of California was pending before Australian Government. (*NYT*, 10/26/66, 1)

October 25-28: 75th anniversary convocation, "Scientific Progress and Human Values," was held at Cal Tech. Dr. Lee A. DuBridge in welcoming address noted meeting represented "a time for self appraisal.... We at Cal Tech are seriously seeking to take a hard look at ourselves, at the world of science and technology of which we are part, and at the world of human beings who may either benefit or possibly suffer from what we do."

NASA Associate Administrator for Manned Space Flight Dr. George E. Mueller said that "through the creative use of the capabilities that we are building up, space may . . . be used as one of many approaches to alleviate" problems of "famine, disease, over-population, and the need for more and better education" facing developing nations.

Dreams of eventual space travel were "pure fantasy," according to Cal Tech astronautics professor Dr. Jesse L. Greenstein. He urged, however, that a greater portion of national expenditures be devoted to achieving interstellar communications—a task that "may ultimately become the greatest scientific adventure."

Dr. Robert P. Sharp, professor of geology at Cal Tech, noted that although "earth scientists" were playing important role in national space program, it was also important that they "look downward into our own planet. . . . Our understanding of these distant bodies will depend to a good degree upon how well we understand our own plain earth."

Likelihood that revolutionary improvement in communication on earth would transform society was suggested by John R. Pierce, research director of Bell Telephone Labs. Communications Sciences Div. Satellites would greatly expand potential of backward countries and would help build a sense of national identity in underdeveloped areas.

"Technology is moving faster than our ability to assimilate it," scientist and industrialist Dr. Simon Ramo postulated. Dr. Ramo urged development of new class of men called "socio-technologists" who could "effectively link scientific developments with social betterment."

Dr. Murray Gell-Mann, professor of theoretical physics at Cal Tech, said society must give new direction to technology, diverting it from applications that yield higher productive efficiency into areas that yielded greater human satisfaction. A symbol of this sort of change would occur when man no longer wanted to channel resources into "building bigger, noisier aircraft" or when society decided to divert a new highway around a virgin forest rather than build through it. (Texts; Bart, *NYT*, 10/26/66, 23; Duscha, *Wash. Post*, 10/26/66, A9; *NYT*, 10/30/66, 52)

October 26: NASA's Atlas-Centaur 9 (AC-9) launch vehicle successfully carried out first full-thrust restart in space of liquid hydrogen engine during two-burn indirect ascent mission launched from ETR. Flight—eighth and last planned Centaur development test flight—almost exactly followed prelaunch plans: AC-9 Centaur stage burned for 328 sec. and again for 107 sec., first to drive itself into temporary 100-mi. (161-km.) altitude parking orbit and then to inject mass model of Surveyor spacecraft on simulated lunar transfer trajectory. Preliminary tracking data indicated dummy spacecraft was aimed accurately and on course. Centaur project had been managed by LRC. (NASA Release 66-274; NASA Proj. Off.; LRC Release 66-69; AP, *Wash. Post*, 10/27/66, A7; *Av. Wk.*, 10/31/66, 39)

October 26: U.S.S.R.'s LUNA XII was probably less than 100 mi. from moon, reported director of U.K.'s Jodrell Bank Experimental Station Sir Bernard Lovell. He said: "Last night the signals from the probe were steady and indicated that, contrary to LUNA X and XI, it was ejected into orbit without rotation. That means that potentially it was a more successful exercise." (AP, Balt. *Sun*, 10/27/66)

- Four-stage USAF Trailblazer was launched from NASA Wallops Station in second of a series of six Air Force Avionics Laboratory launches to study communications blackout during space vehicle reentry. (*Tech. Wk.*, 11/7/66, 10)

- Cosmonaut Aleksey Leonov was quoted by Tass as saying that he was in favor of strict international law governing cooperation in space. Through agreement, all stations, equipment, and spacecraft sent to moon or the planets could be available for use by representatives of other states, he noted. "However, all this would be justified only if the moon and the other celestial bodies are used by all treaty participants for peaceful purposes." (Tass, 10/26/66)

- NASA awarded North American Aviation, Inc., a $37-million, cost-plus-award-fee contract supplement for launch preparation and checkout of ten Saturn V 2nd stages (S-II). (NASA Release 66-281)

- M2-F2 lifting body vehicle, piloted by Capt. Jerauld R. Gentry (USAF), made 12th glide flight at Edwards AFB after air-launch from B-52 aircraft at 45,000-ft. altitude. Purposes of flight were to obtain data on stability and control at 7° and 11° angles of attack and on upper flap effectiveness. (NASA Proj. Off.)

October 26-29: Second International Congress on Air Technology met at Hot Springs, Ark.

Need to develop a V/Stol aircraft as an "air bus" for faster short-haul transportation of cargo and passengers within large urban areas was rapidly increasing with growth in population, asserted Jack D. Brewer, program manager for V/Stol research, NASA Hq. By 1985, according to prediction, more than 130 million persons—half the U.S. population—would be living in super metropolitan areas at San Francisco-Los Angeles, Buffalo-Chicago, and Boston-New York-Washington, D.C. (NASA Release 66-282)

Cooperation as well as competition would be necessary to meet challenges of future air transportation, John C. Brizendine, Douglas Aircraft Co., Inc., Aircraft Division vice president for engineering, told the Congress. He said progress in solving problems such as noise, limiting effects of weather, congestion of airports and airways, and rapid transportation needs between airports and cities, should be accelerated. ". . . if society is to realize the progress and benefits technology can offer, we as leaders of large segments of society also must display increasing unity of purpose in striving to overcome the major environmental problems common to all of us." (*CR*, 1/23/67, 5729-31)

October 27: NASA launched ComSatCorp's INTELSAT II-A comsat from ETR by three-stage Thrust-Augmented Improved Delta booster into elliptical transfer orbit in preparation for geostationary orbit. Transfer orbit parameters: apogee, 23,382 mi. (37,646 km.); perigee, 186 mi. (299.9 km.); period, 11 hr. 9 min.; inclination, 26.4°. At ninth apogee—on Oct. 30—ComSatCorp fired the apogee motor. Apparently the apogee motor nozzle was blown off shortly after the motor ignited, and INTELSAT II-A entered orbit with 23,330-mi. (37,545-km.) apogee; 2,072-mi. (3,334-km.) perigee; 12 hr. 16 min. period; and 17° inclination rather than planned 22,300-mi.-altitude synchronous orbit over Pacific. ComSatCorp announced that 192-lb. satellite, nicknamed "LANI BIRD," could receive and transmit television and other forms of communications, but that "significance of this potential remains to be assessed."

First satellite in ComSatCorp's two-satellite Intelsat II system to provide transatlantic and transpacific comsat coverage, INTELSAT II-A had been scheduled to extend commercial service to the Pacific; assist in fulfilling Project Apollo communications requirements; and provide capability for live transpacific TV by handling TV, data transmission, or up to 240 voice channels between "first class" ground stations. Intelsat II-B was planned for launch in early 1967. INTELSAT I ("EARLY BIRD") had been orbited April 6, 1965. (ComSatCorp Releases; NASA Proj. Off.)

- No photographic signals had been detected by Western monitors from U.S.S.R.'s LUNA XII, which had apparently entered lunar orbit Oct. 25. It was speculated that spacecraft had been shut down temporarily to conserve battery power. U.S.S.R. had made no official statement since Oct. 22 launch. (UPI, *NYT*, 10/28/66, 21)
- ComSatCorp asked FCC for authority to construct a $6,500,000 high-capacity earth station in Peach Tree Valley, Calif., to communicate with, and perform tracking, telemetry, and command duties for, orbiting satellites. ComSatCorp would withdraw Oct. 6 application for a second antenna at Andover, Me., station pending FCC action on California application. (ComSatCorp Release)
- NASA Deputy Administrator Dr. Robert C. Seamans, Jr., accepting the New York Board of Trade's 20th annual Business Speaks Award on behalf of NASA, noted "the most important lesson" learned in management of the space effort: ". . . when pressing forward and coordinating the efforts of a large group of teams and individuals . . . it is of paramount importance to provide objectives for each and to assure that these objectives are tailored to the mind and capability of the group attacking them. If the task is ever going to be accomplished, it must seem achievable . . . to the man undertaking it. And . . . each segment of the overall task must be accurately tracked and timed to be properly completed and fitted into the appropriate sequence even as the work goes on." (Text)
- Communist China exploded on target over its own territory a nuclear weapon carried by a guided missile. Test was fourth in series of nuclear explosions: first, Oct. 16, 1964; second, May 14, 1965; third, May 9, 1966. (*NYT*, 10/28/66, 1, 18)

October 28: USAF's OV3-II research satellite was launched from Vandenberg AFB by Scout rocket into orbit with 991-mi. (1,595-km.) apogee; 198-mi. (319-km.) perigee; 104-min. period; and 82° inclination. Launch was planned so satellite would be in orbit during Nov. 12 South American solar eclipse to provide data on charged particle variations in extreme upper atmosphere before, during, and after eclipse. Satellite, designed to remain active in space at least one year, would also extend knowledge of electron and ion density structure found in outer radiation belt by NASA's DISCOVERER XVII. (Vandenberg AFB PIO; UPI, Phil. *Sun. Bull.*, 10/30/66)

October 28: First manned Apollo spacecraft (AS-204) would be orbited in first quarter of 1967, NASA announced. Mission would verify spacecraft systems performance and crew operations. NASA had hoped for December launching, but modification of environmental control system (Ecs) would necessitate further ground testing of spacecraft. Prime crew would be Virgil I. Grissom, command pilot; Edward H. White II, senior pilot; and Roger Chaffee, pilot. Backup crew would be James McDivitt, command pilot; David Scott, senior pilot; and Russell Schweikart, pilot. (NASA Release 66-284)

- NASA awarded Boeing Co. a $4.5-million contract modification for design and procurement of Saturn V 1st stages. Hardware included propellant ducts and valves and pressurization switches and gauges. (NASA Release 66-283)
- General Electric Co. successfully tested prototype GE4 turbojet engine at 52,600 lbs. thrust—believed to be record for air-breathing engines of all types. GE4 was developed in FAA competition with Pratt & Whitney Div., United Aircraft Corp., for Sst engine contract. (GE Release 66-51)
- NASA Administrator James E. Webb said in a speech at Mills College, Oakland, Calif., that the "complex interdevelopment of technological power and of the social organization necessary to control, direct, and exploit it . . . is a central element of the space age. . . . The success of our space program is helping to prove to ourselves and to the world that we have the will and the capacity to establish far-reaching, difficult goals and bring them surely to fruition. This kind of proof is a major basis for national prestige and international cooperation." (Text)
- U.S. Army Corps of Engineers awarded $1,071,205 contract for construction of foundations at ERC to Coleman Bros. Corp., Boston. (MSC *Roundup*, 10/28/66, 8; DOD Release 879-66)
- Recent "booming" in Chicago area precipitated editorial comment in *Chicago Daily News*: "We grant that the supersonic age is upon us. But it is also an age of human beings with human rights. One of these is the right not to be buffeted around by sonic booms, and it is evidently going to take eternal vigilance to keep that right from being obliterated in the name of some totally unsupportable need." (*Chic. Daily News*, 10/28/66)
- Plan to shift headquarters of USN's manned spacecraft recovery fleet (Task Force 140) from Norfolk to Houston was under study by DOD, fleet commander R/Adm. Conrad Abhau told AP. Abhau said he and his staff had been absent from Norfolk on recovery operations 40 per cent of the time during the past 12 mos. (AP, *Wash. Post*, 10/29/66, E16)
- FCC had granted ITT World Communications, Inc., permission to provide facilities on U.S.S. *Wasp* for televising recovery of Gemini 12 spacecraft; three-day mission was scheduled for Nov. 9 launch. TV signals would be transmitted by EARLY BIRD I for relay. (Wash. *Eve. Star*, 10/29/66, A16)

October 29: LUNAR ORBITER I impacted on moon's hidden side at 9:30 a.m. EDT on command from earth. The spacecraft was deliberately crashed to obviate possible interference with mission of second Lunar Orbiter (Lunar Orbiter B) scheduled for launch Nov. 6. First U.S. spacecraft to orbit and photograph moon, LUNAR ORBITER I, launched Aug. 10, became first satellite to be deliberately destroyed in space by NASA. (NASA Release 66-287; O'Toole, *Wash. Post*, 10/30/66, A10)

October 29: LUNA XII had photographed lurain with "a special phototelevision device," Tass announced in first official statement relating to spacecraft since Oct. 22 launch. Satellite had entered lunar orbit Oct. 25 with 1,740-km. (1,081-mi.) apolune; 100-km. (62-mi.) perilune; and 3 hr. 25 min. period. First photos transmitted—shown on Moscow television less than an hour after announcement—showed clearly distinguishable details of craters and mountain ranges.

LUNA XII was ninth photographic mission to moon. Five had been from U.S.: RANGERS VII, VIII, and IX, SURVEYOR I, and LUNAR ORBITER I; four from U.S.S.R.: LUNA III, ZOND III, LUNA IX, and LUNA XII. (Tass, 10/29/66; AP, *NYT*, 10/30/66, 43)

- Dr. Serge A. Korff, director of New York Univ.'s cosmic research program, told a youth science seminar at the Explorer's Club in New York that he believed many Ufo sightings had been of giant balloons sent aloft by NYU to test the upper atmosphere: "[They] are 300 to 400 feet across and because they are partly inflated, assume strange shapes as they ascend. At high altitudes they often reflect sunlight and become visible to viewers below in areas where the sun has set or before it has risen." (*NYT*, 10/30/66, 45)

- "Whole towns" on the moon, complete with dome-like buildings linked by tunnels and parks under transparent hoods, might be constructed in 20th century, Soviet architect Nikolay Kolomiyets suggested in *Krasnaya Zvezda*. Kolomiyets said the moon town should provide a safe shelter for its inhabitants and be attractive as well. Best construction material would probably be lunar rocks. "It can be expected that after treatment with alkalis, lunar rocks can be used to build walls and roofs by pouring, just as they are made on earth by pouring concrete." (Tass, 10/29/66)

October 30: U.S.S.R. Antaeus cargo aircraft lifted 88 tons, 103 kg., in flight from Podmoskovnoye airport (Moscow) piloted by I. Y. Davidov, *Krasnaya Zvezda* announced. Previous cargo record had been 53 metric tons, 479 kg., lifted Dec. 16, 1958, by U.S. MATS C-133 Cargomaster piloted by Col. John M. Thompson (USAF). (Tass, 10/30/66)

- Sst would be limited to speed over land of 660 mph to avoid sonic booms, Evert Clark reported in *New York Times*. (Clark, *NYT*, 10/31/66, 1)

October 31: NASA launched four-stage Pacemaker rocket from Wallops Station to test performance of spacecraft heat shield design. First two stages lifted 135-lb. payload to 75,000-ft. altitude. As payload began to fall back to earth, last two stages were fired to drive it downward at about 6,800 mph. Information was obtained by tracking and photography and from the payload itself, which was recovered after parachute descent into Atlantic. (Wallops PAO; UPI, *NYT*, 11/1/66)

- Japan successfully launched four-stage Mu-1 rocket from Uchinoura Space Center, Kyushu Island, to test 1st-stage fuel combustion and booster propulsion. The 70-ft.-long, solid-fueled launch vehicle was launched at 69° inclination and impacted in Pacific after 3-min. 20-sec. flight. Scientists called test "a success." Mu was scheduled to orbit Japan's first satellite by 1968. (AP, *NYT*, 11/1/66)

October 31: Sst design proposal evaluations were submitted to FAA by 235-member Government team and 30 domestic and foreign airlines; FAA Administrator William F. McKee would study evaluations and make recommendation; President Johnson would make final decision by Jan. 1, 1967. (*Wash. Post*, 11/1/66, B7)

- A revolution in 1970's of air transport industry that would "change its scope and character even more than the switch from piston-engine equipment to jets did in the 1960s," was predicted by Robert Hotz in *Aviation Week*. Transformation would be stimulated by the "sharp spur of advanced technology arriving in the midst of changing economic and social problems." Hotz observed that "no other major business in the industrial history of the world has had to cope with such a series of technical revolutions in such a short time span as the airlines." (Hotz, *Av. Wk.*, 10/31/66, 21)
- U.S.S.R. planned to build two additional satellite tracking stations in Cuba and had sited one near Santiago, *Aviation Week* reported. (*Av. Wk.*, 10/31/66, 23)
- USAF had awarded Boeing Co. $142.3-million firm fixed-price incentive contract to develop and produce AGM-69A Short Range Attack Missile (Sram) to be carried by FB-111 bombers and adaptable to late model B-52 bombers. (DOD Release 925-66)
- Franco-Soviet cooperative space agreement was being viewed with skepticism by French scientists who foresaw effort would break down for same lack of Soviet candor that had stymied useful information exchange between U.S. and U.S.S.R., *Aviation Week* reported. In addition, French Communists were protesting to Kremlin that high Soviet sanction of President De Gaulle's space program was weakening Communist political stature in France. (*Av. Wk.*, 10/31/66, 23)
- Cost-sharing dispute between DOD and NASA over use of ETR was reported by *Aviation Week*. DOD wanted NASA to pay share of continuing costs—possibly as much as $100 million annually—instead of only expenses for launches. (*Av. Wk.*, 10/31/66, 25)
- Discussing the manned spaceflight program in press interview at Texas Technical College, physicist and atomic scientist Dr. Edward Teller said: ". . . to get to the moon first is not important, but who stays there and exploits the knowledge to be gained from the moon is more important." Teller, attending symposium on arid and semi-arid land, said U.S. should settle a colony of scientists on moon to determine feasibility of using it as a base for reaching Mars. (AP, *Balt. Sun*, 11/1/66, 8)
- NASA was planning total of 11 lunar landing missions, reported *Technology Week*, citing procurement request issued by MSC. Number of missions was included as guideline for firms asked to submit bids on development of crew recovery quarantine equipment. (*Tech. Wk.*, 10/31/66, 3)
- FAA predicted flying activity at the 303 airports with FAA control towers would break all records in 1966: an estimated 45.1 million landings and takeoffs were foreseen—a 19 per cent increase over 1965. (FAA Release 66-95)

During October: NASA published *Vacuum Technology and Space Simulation* (SP-105), a comprehensive manual available to public through the Superintendent of Documents, GPO. (NASA Release 66-279)

- FAA cautioned airline executives during briefing in Washington, D.C., not to let desire for a "fly-off competition" keep them from making as clear-cut a first choice as possible for the supersonic transport. If they did not, they might get aircraft they did not want or Congressional critics might use indecisiveness to stall or kill program. Possibility that prototypes of both the Boeing and Lockheed airframe designs would be built was virtually ruled out. (*Av. Wk.*, 10/31/66, 25)

During October: Secretary of the Air Force Dr. Harold Brown summarized USAF's position on its future in space in an interview with *Armed Forces Management:* "It is not a matter of desiring to conduct military campaigns in space but rather to preclude any aggressor from using space as an area of operations for launching attacks against the U.S. . . .

"We should not be doing things just to be doing them. Rather, they must have direct relation to establishing military needs.

"Space is not a mission but a place to perform a mission. When a mission can be performed from space, the Air Force will perform it from there. . . ." (*Armed Forces Management*, 10/66, 69)

- NASA made available a list of abstracts of NASA-owned inventions for foreign licensing through the Assistant General Counsel for Patent Matters, NASA Hq. (NASA Release 66-266)

- AFSC awarded 12-mo. study contracts totaling $900,000 to Lockheed-California Co., North American Aviation, Inc., and McDonnell Aircraft Corp. to develop a high-altitude, hypersonic Scramjet-powered cruise vehicle with potential military applications. (AFSC Release 188.66)

- NASA Deputy Administrator Dr. Robert C. Seamans, Jr., said in *Interavia* that in addition to possible "definite and direct" economic and technological benefits, the "greatest impact the continuing exploration of space will have upon man will be upon his own philosophy: that is, his view of himself in relation to the universe as he imagines it." (*Interavia*, 10/66, 1479)

- USAF School of Aerospace Medicine had completed 46-day simulated spaceflight in which bite-size food diet was tested, *Technology Week* reported. Four airmen, divided into two teams, spent several days at ground-level conditions, then entered two high-altitude chambers at simulated 27,000 ft. Atmosphere was 70% oxygen, 30% hydrogen for one team; 70% oxygen, 30% nitrogen for second. Final sixteen days were spent at ground level. Results had not been released. (*Tech. Wk.*, 10/31/66, 4)

November 1966

November 1: X-15 No. 3 was flown by NASA test pilot William H. Dana to 3,750 mph (mach 5.34) and 306,900-ft. altitude in flight test to collect micrometeoroids. As secondary mission, pilot performed maneuvers to check dual channel radiometer, tip-pod accelerometer, and precision attitude. (X-15 Proj. Off.)

- An 8,189-hr. test of cesium electron-bombardment engine performed for LRC by Electro-Optical Systems, Inc., ended when supply of cesium fuel was exhausted. Conducted in space simulation chamber, test was part of NASA's electric engine research program. Studies had shown that such systems could result in increased payloads in unmanned and manned spaceflight missions. (NASA Release 66-306)
- U.S.S.R. successfully launched and recovered aerostat containing 7.6-ton "high altitude automatic astronomic station," to 20-km. (12-mi.) altitude for "exploration of physical processes in the solar photosphere." Equipment, including optical and radio-technical apparatuses, had functioned normally, and results were being processed. (Tass, 11/5/66; AP, Wash. *Sun. Star*, 11/6/66, E26)
- Test-firings of XE nuclear rocket engine—last major step before developing hardware for flight-rated nuclear engine—would begin in latter half of 1968, Milton Klein, Deputy Manager, NASA-AEC Space Nuclear Propulsion Office, told Atomic Industrial Forum in Pittsburgh. Klein also disclosed that his office had begun preliminary design of new facilities with two test positions, each capable of testing a 200,000-lb.-thrust engine or an entire propulsion module under simulated altitude conditions. (Text)
- Gen. Jacob E. Smart (USAF, Ret.), former Deputy Commander-in-Chief of U.S. European Command, was sworn in as a special assistant to NASA Administrator James E. Webb. (NASA Release 66-288)
- George C. White, Jr., chief of NASA Hq. Apollo Spacecraft and Vehicle Test Performance Office, was appointed Director of Apollo Reliability and Quality. (NASA Ann., 11/18/66)
- Survival of 17 chimpanzees without permanent damage after 3½ min. in simulated space environment offered hope that an astronaut outside his spacecraft could be rescued if his spacesuit were to spring a leak. In 18 tests conducted at Holloman AFB during one year, chimps placed in a decompression chamber for time periods that were gradually increased from 5 sec. to 3½ min. suffered only temporary abnormalities. L/Col. C. H. Kratochvil, commander of Holloman's aeromedical research laboratory, said close relationship between chimps and humans made it possible to assume that findings would be applicable to astronauts. (Nelson, *Wash. Post*, 11/2/66, A4)
- Dr. Hugh Odishaw was appointed executive secretary of the NRC Div. of Physical Sciences. Space Science Board, Geophysics Research Board,

and Committee on Polar Research—groups with which Dr. Odishaw had been closely associated—would be shifted into NRC as joint responsibility of Div. of Earth Sciences and Div. of Physical Sciences. They previously had been among 12 NAS groups that functioned outside the formal NRC divisional structure. (NAS-NRC-NAE *News Report*, 11/66, 1)

November 1: Aerobee 150 sounding rocket launched by AFCRL from WSMR reached 150 mi. (241 km.) in experiment to measure extreme ultraviolet solar radiation. (OAR *Res. Rev.*, 1/67, 8)

- ComSatCorp signed agreements to purchase 210-acre site in Montgomery County, Md., for $7- to $10-million Research and Development Center. Construction would begin in 1967. (ComSatCorp Release)

November 2: U.S.S.R. launched spacecraft, possibly from Cosmos series, which exploded into 32 pieces detectable by radar. GSFC *Satellite Situation Report* would later reveal that 12 pieces of debris remained in orbit Nov. 15; 20 had reentered. Spacecraft was second Soviet satellite to explode in six weeks [see Sept. 17]. (GSFC *SSR*, 11/15/66)

- NASA had extended operational lifetimes of its three Pegasus meteoroid detection satellites 12 additional months because of their successful operation. Satellites would continue to provide meteoroid data, but researchers would concentrate on studying durability of spacecraft systems and components. PEGASUS I was launched Feb. 16, 1965; PEGASUS II, May 27, 1965; and PEGASUS III, July 30, 1965. All had been scheduled to operate 18 mos. (MSFC Release 66-264)

- USAF launched two unidentified satellites from Vandenberg AFB with Atlas-Agena D booster. (*Tech. Wk.*, 11/7/66, 10; *U.S. Aeron. & Space Act.*, 1966, 157)

- JPL selected Litton Industries, Inc., Guidance and Control Systems Div., for negotiation of a $3-million contract to design, fabricate, and check out data automatic subsystem (DAS) for 1969 Mariner (unmanned) flights to Mars. JPL had management responsibility to NASA for Mariner/Mars project. (NASA Release 66-289)

- General Precision Equipment Corp. received $8.6-million NASA contract to provide three sets of equipment to modify Apollo simulators at KSC. (*WSJ*, 11/2/66, 3)

- Man must learn to operate effectively and efficiently in space before he can reap the full benefits of space exploration, Dr. George E. Mueller, NASA Associate Administrator for Manned Space Flight, told the Wisconsin State Chamber of Commerce in Milwaukee: "To employ Apollo flight hardware and capabilities beyond the manned lunar landing, and to extend the capability of the Apollo spacecraft would make efficient use of our national investment. Missions are available to make use of the equipment for flights in earth orbit, in lunar orbit, and to the moon's surface. Experiments are being defined to determine the value of such missions. Users have expressed interest in the results of such experiments. But first we must develop the ability to operate in space. . . ." (Text)

- Leningrad Institute of Cytology scientists had designed special device to permit biologists to study effects on microorganisms of "pure" solar radiation. Device, called "Photostate I," consisted primarily of 18-liter chamber in which wide temperature range could be created and observations made under any intensity of solar radiation in the field of visible and ultraviolet light. (Tass, 11/2/66)

November 2: RCA received $2,500,000 fixed-price USN contract for fabrication, testing, and delivery of six navigation satellites. (DOD Release 933-66)

November 3: XB-70 No. 1 experimental supersonic bomber, piloted by L/Col. Joseph Cotton (USAF) and NASA test pilot Fitzhugh Fulton, reached mach 2.1 and 60,000-ft. altitude during two-hour flight to evaluate sonic booms and conduct noise measurements for FAA. Flight was first since June 8 crash of XB-70 No. 2 near Barstow, Calif. (XB-70 Proj. Off.; UPI, *Chic. Trib.*, 11/4/66)

- USAF Titan III-C booster launched from ETR released unmanned Gemini spacecraft on reentry trajectory and inserted canister containing nine experiments and three satellites into high, circular orbit in successful mission marking first flight test of hardware for DOD's Manned Orbiting Laboratory (Mol).

  Power flight of Titan III-C was close to planned parameters. Modified Gemini spacecraft was released at 125-mi. altitude in test to determine whether new heat shield with hatch could withstand reentry temperatures. Traveling 17,500 mph, spacecraft followed 5,500-mi. trajectory, landing only seven miles off target in the Atlantic. Recovery was made by U.S.S. *LaSalle*. Titan III-C's transtage restarted twice to achieve planned 184-mi.-altitude circular orbit and release 38-ft. experimental canister, proving booster's ability to launch long payload. Canister—OV4-III—ejected OV4-IR and OV4-IT comsats and OV1-VI satellite, which achieved separate, circular orbits. Launch, sixth in series of 12 designed to qualify composite Titan III-C for operational service, was characterized by USAF officials as a "major step" in the Mol program. (Wilford, *NYT*, 11/4/66, 1, 10; AP, Wash. *Eve. Star*, 11/3/66, A1; AP, *Wash. Post*, 11/4/66, A1; *Av. Wk.*, 11/14/66, 30; *U.S. Aeron. & Space Act., 1966*, 158)

- Decisions on the direction of post-Apollo activities in space would require a "two-pronged effort by the part of our economy interested in the evolution of space products," suggested Boeing Co. vice president George H. Stoner before AIAA Forum on "After Apollo, What Next?" in Washington, D.C.: ". . . industry must assist NASA and the federal government in formulating succeeding national space goals for the research that will produce fruitful progress toward understanding our universe and toward building a store of knowledge about specialized space techniques. . . . Industry and government agencies other than NASA must explore and continue to find ways to compete successfully in the world market place and in the international technological race on the large scale development projects that characterize our modern society. . . ." Stoner offered several specific suggestions for post-Apollo activities: (1) communications satellites that would take advantage of the natural access to anywhere on earth provided by spacecraft; (2) applications of space-based activities to ballistic missile defense; (3) expansion and refinement of equipment for survey of earth affairs from space; (4) dramatic reduction of launch costs; and (5) continuing exploration of space and its technologies by NASA astronauts and scientists to learn more about our universe. (Text)

- Apparent inactivity in Soviet manned space program might reflect "growing dismay over the tremendously high cost of astronautics where men are involved," wrote William Hines in the Washington *Evening Star*. "The Russians may feel they have better things to do with their money, just as the United States would have if the Kennedy 'moon message' of

1961 had not been hung like an albatross around the national neck as a commitment involving American prestige before the world." Cost of Project Gemini was about $700,000 per man hour. (Hines, Wash. *Eve. Star*, 11/3/66, A12)

November 3: 1966 Nobel Prizes for chemistry and physics were awarded to Drs. Robert S. Mulliken, Univ. of Chicago, and Alfred Kastler, Ecole Normale Supérieure in Paris, respectively. Professor Mulliken was cited for "his functional work concerning chemical bonds and the electronic structure of molecules by the molecular orbital method"; Professor Kastler, for "the discovery and development of optical methods for studying Hertzian resonances in atoms." (AP, Wash. *Eve. Star*, 11/3/66, A1; Wiskari, *NYT*, 11/4/66, 1, 28)

- U.S. Arms Control and Disarmament Agency announced plans for a "field exercise" in Arizona, Nevada, and California, to test methods of identifying underground nuclear explosions. Project, which would begin in late 1966 and continue through May 1967, would develop methods by which international inspection teams might police a ban on underground nuclear tests. (UPI, *Wash. Post*, 11/4/66, A23)

November 4: U.S.S.R. had launched Yantar I ionospheric laboratory equipped with gas plasma ion engine to altitudes of 100 to 400 km. (62 to 249 mi.) in October "to study the outlook for guided flight in the upper layers of the atmosphere," Tass announced. Data on propulsion-system performance in electrically charged ionosphere would be analyzed and results published in magazines of the Soviet Academy of Sciences.

Both NASA and USAF had already conducted several ground and flight tests of electric engines, and another series was scheduled to begin in late 1966. Jerome P. Mullin, of OART's Nuclear Systems and Space Power Div., told the *New York Times* in a telephone interview that electric engines had "demonstrated their practical utility" for maneuvering and stabilizing spacecraft in flight. (Tass, 11/4/66; Grose, *NYT*, 11/5/66, 12; Wilford, *NYT*, 11/5/66, 12)

- Cosmonaut-pilots would head crews of all spacecraft landing on the moon in the future because landing "cannot be fully trusted to machines," Soviet Cosmonaut Pavel Belyayev predicted in *Aviatsiya i Kosmonavtika*. Belyayev, who manually landed VOSKHOD II March 19, 1965, said that two methods of braking were possible: jet engines and use of aerodynamic forces. (Tass, 11/4/66)

- Fourth stage of Mu-1, Japan's newest and most powerful booster, was successfully tested in low-pressure chamber in Noshiro, Univ. of Tokyo scientists reported. Mu was scheduled to orbit Japan's first satellite by 1968. (AP, Balt. *Sun*, 11/8/66)

- U.S. Ambassador-at-Large and newly appointed Ambassador to U.S.S.R. Llewellyn E. Thompson and Soviet Minister for Civil Aviation Yevgeni F. Loginov signed agreement authorizing direct commercial airline flights between New York and Moscow at State Dept. ceremony. Service would begin in spring 1967 with weekly flights by Pan American World Airways and U.S.S.R.'s Aeroflot. (Eder, *NYT*, 11/5/66, 1; Roberts, *Wash. Post*, 11/5/66, A1)

- Radioastronomers were relying increasingly on ancient Oriental records of novae and supernovae because the phenomena could be observed so rarely, Xi Ze-zong and Po Shu-jen, Academia Sinica, Shanghai, reported in *Science*. Only one supernova had appeared within Milky Way during last 360 yrs., and few of the average of 50 novae that appeared annually could be seen with the naked eye. (Xi and Po, *Science*, 11/4/66, 597)

November 5: NASA conducted first rocket-launched test of parachute landing system for Voyager spacecraft at WSMR. Honest John-Nike rocket boosted 200-lb. experimental package to 104,000-ft. altitude, but ground command transmission did not correctly trigger deployment. Successful balloon-launched test in the LaRC-managed parachute experiment series had been completed Aug. 30. Voyager project was managed by JPL. (NASA Proj. Off.)

- Astronauts James A. Lovell, Jr., and Edwin E. Aldrin, Jr., passed their medical examinations for the Gemini XII mission, scheduled for launch from ETR Nov. 9. (*Wash. Post*, 11/6/66, A17)

November 5-6: FAA, Office of Emergency Planning, and New York City officials successfully staged "Metro Air Support '66," an exercise in which more than 200 Vtol aircraft and helicopters took off and landed at piers, parks, and streets with emergency supplies, technicians, and Government officials to demonstrate how aircraft could be used if other transportation were paralyzed during emergency or disaster. (FAA Releases 66-90, 66-96; Nevard, *NYT*, 11/6/66, 1)

November 6: NASA's LUNAR ORBITER II (Lunar Orbiter B) unmanned spacecraft was successfully launched by Atlas-Agena D booster from ETR in second U.S. attempt to orbit the moon and photograph possible landing sites for Apollo astronauts [see Nov. 10-30].

Agena 2nd stage fired to boost 850-lb. spacecraft into 100-mi. (161-km.) altitude parking orbit, reignited after 14-min. coast period, injecting spacecraft on 94-hr., 232,000-mi. translunar trajectory, and separated. On schedule LUNAR ORBITER II deployed its four solar panels and two antennas and locked its five solar sensors on the sun. Only difficulty occurred when spacecraft's star-tracker lost its fix on Canopus, delaying critical midcourse maneuver eight hours. Correction was accomplished at 44:09 GET and JPL predicted spacecraft would come very close to original aiming point.

Primary objectives of NASA's LUNAR ORBITER II mission, second in series of five, were (1) to place three-axis stabilized spacecraft into lunar orbit; and (2) to obtain high-resolution photos on various lunar surface areas to assess their suitability as landing sites for Apollo and Surveyor spacecraft, and to improve our knowledge of the moon. Photos would cover 13 primary target sites, located generally within northern half of the Apollo zone of interest on the moon's front face. It would also monitor micrometeoroids and radiation intensity in lunar environment and refine definition of moon's gravitational field. Lunar Orbiter program was managed by LaRC under direction of NASA's OSSA. Tracking and communications were the responsibility of JPL-operated Deep Space Network. (NASA Proj. Off.; NASA Release 66-286; O'Toole, *Wash. Post*, 11/7/66, A1, A3; Hines, *Wash. Eve. Star*, 11/7/66, A3; UPI, *NYT*, 11/9/66, 77; AP, *Wash. Post*, 11/10/66, A2)

- President Johnson announced appointment of Under Secretary of Commerce Alan Boyd as Secretary of the new Cabinet-level Department of Transportation. (*Pres. Doc.*, 11/14/66, 1620-21)

- GEMINI X spacecraft was viewed by more than 13,000 people when it was displayed for the day at a Tokyo department store. (*Wash. Post*, 11/7/66, A26)

November 7: Clarence C. Gay, Jr., Acting Director of Gemini Test, became Chief, Apollo Spacecraft Test, Apollo Test Directorate at NASA Hq. (NASA Ann., 11/18/66)

November 7: Western Union Telegraph Co. asked FCC's permission to create domestic comsat system to transmit telegrams. Ground stations would be constructed in Oregon, Iowa, Colorado, Louisiana, Alabama, and, possibly, West Virginia. (UPI, *NYT*, 11/8/66, 15)
- Moscow parade celebrating 49th anniversary of the Bolshevik Revolution included rocket which Russians claimed was immune to any antimissile defense system and could be dropped out of orbit from any direction. (AP, Wash. *Eve. Star*, 11/7/66, A1, A6)
- NASA Administrator James E. Webb was presented Gen. Benedict Crowell Gold Medal by American Ordnance Assn.'s Cleveland Post. (*Lewis News*, 11/10/66, 1)

November 8: NASA's SURVEYOR I spacecraft, which softlanded on the moon's Ocean of Storms June 2, responded to JPL signals almost four months after scientists had announced its "death." Spacecraft did not take additional photos of the lunar surface as commanded. (UPI, *Wash. Post*, 11/10/66, A7)
- In test of hydrogen peroxide attitude control thrusters, INTELSAT II-A ("LANI BIRD") was moved into elliptical orbit with 23,306-mi. (37,523-km.) apogee; 2,609-mi. (4,201-km.) perigee; 17° inclination; and 12-hr. 10-min. period. Systems would be used in future attempt to place comsat in "modified orbit with a 12 hour period," to make possible communications with line-of-sight earth stations for a few hours daily. Preliminary investigations indicated that INTELSAT II-A's failure to achieve planned synchronous equatorial orbit Oct. 30 was due to apogee motor malfunction; launch of Intelsat II-B, scheduled for Nov. 23, would be postponed until exact cause of malfunction was determined. (ComSatCorp Release)
- NASA Nike-Tomahawk sounding rocket launched from NASA Wallops Station reached 134-mi. (216-km.) altitude in GSFC experiment to evaluate the method of the double floating probe as a tool for monitoring ionospheric electric fields. Rocket performance was excellent, and data obtained by one of two independent antenna systems was in quantitative agreement with the double floating probe theory. Other antenna system failed to deploy properly. (NASA Rpt. SRL)
- USAF launched an unidentified satellite from Vandenberg AFB with Thor-Agena D booster. (UPI, *NYT*, 11/9/66, 4; *U.S Aeron. & Space Act., 1966*, 158)
- Secretary of Defense Robert S. McNamara was considering USAF proposal for an advanced manned strategic aircraft (Amsa) which would cost $1.5-$2 billion to develop, William Beecher reported in the *New York Times*. Bomber would reportedly be able to carry internally 25 nuclear-tipped short range attack missiles (Sram), plus nuclear bombs; have unrefueled range of 7,000 to 10,000 mi.; and be capable of traveling 1,625 mph—about 2½ times the speed of B-52 aircraft, the mainstay of current bomber force. (Beecher, *NYT*, 11/8/66, 1)
- Concept of "flying saucers" violated physical laws of solar system, Rev. Francis J. Heyden, Georgetown Univ. astronomer, declared in *Washington Post* interview. Ufo reports, none of which could be adequately studied scientifically, were either misinterpretations or hoaxes, he said. (Casey, *Wash. Post*, 11/8/66, B3)
- FAA awarded $65,000 contract to North American Aviation, Inc., to study competing Boeing Co. and Lockheed Aircraft Corp. Sst designs for possible size reduction that would reduce sonic boom. (*WSJ*, 12/5/66, 12)

November 9: NASA launched from Wallops Station an Argentine-designed and -produced single-stage solid-fuel Orion II sounding rocket with payload for upper atmosphere research. Reaching altitude of 51 mi., payload was captured by helicopter as it descended by parachute to 2,500 ft., marking Wallops' first such mid-air recovery. Test was second successful Orion II flight in three-flight series which began Nov. 4; Nov. 7 launch failed. Six Argentine engineers and technicians from Aeronautical and Space Research Institute (IIAE) assisted Wallops personnel with launch operations. Arrangements were made by NASA and Argentine Space Commission (CNIE). (NASA Release 66-291; Wallops Release 66-53)

- USAF launched Arcas solid-fuel rocket from Vandenberg AFB to collect information on atmospheric temperature, pressure, and movement above 98,400-ft. altitude. Data collected would be used to plan glide paths of vehicles reentering earth atmosphere. (AP, *Phil. Inq.*, 11/10/66)
- AFSC announced plans to use seven defense communications satellites, launched from ETR June 16, in series of experiments to gather data for design of improved communications systems between ground stations 5,000 to 6,000 mi. apart. Using specially equipped C-121 communications research aircraft, Air Force Avionics Lab. engineers would test: (1) relay of voice and teletype communications through satellite from either aircraft or ground station; and (2) transmission of airborne meteorological reconnaissance data to the ground via satellite. Signals often would travel 50,000 mi. between terminal points. (AFSC Release 143.66)
- State Dept. disclosed that "a very small amount" of radioactivity had leaked into the atmosphere from AEC's Sept. 12 underground nuclear test, but the incident did not violate limited nuclear test ban treaty because no radioactive debris had circulated outside U.S. Disclosure was in response to a Soviet inquiry. (UPI, *Wash. Post*, 11/10/66, A2)

November 10: NASA Nike-Apache sounding rocket launched from Churchill Research Range reached 120-mi. (193-km.) altitude in flight to evaluate experiments for NASA-West German cooperative satellite program. Premature ejection of payload cylinder doors caused overheating and malfunction of an electron detector, but all other instruments performed satisfactorily. Rocket performance, though slightly in excess of predicted, was within normal tolerance range. (NASA Rpt. SRL)

- XB-70 No. 1 experimental supersonic bomber, piloted by L/Col. Joseph Cotton (USAF) and NASA test pilot Fitzhugh Fulton, reached mach 2.52 and 60,000-ft. altitude in flight from Edwards AFB for the national sonic boom program. (XB-70 Flight log)
- Dirty windows which obscured astronauts' vision moderately to severely in many of NASA's Gemini missions had been caused by volatile vapors produced by silicones used in window seals, Gemini program director Charles W. Mathews told news briefing at MSC. Problem would be corrected by curing silicone seals in a vacuum. (Wash. *Eve. Star*, 11/11/66, A4)
- Groundbreaking ceremonies were held for ERC's $60-million headquarters complex in Cambridge, Mass. (*Boston Her.*, 11/11/66; AIAA *Daily*, 11/29/66, 1)
- Three USAF Athena missiles were successfully launched within three hours from Green River, Utah, to target area at WSMR, in test to determine

how to distinguish armed missile from decoy missile during reentry. (AP, *NYT*, 11/13/66, 23)

November 10: Secretary of Defense Robert S. McNamara, after conferring with President Johnson, told newsmen at the Texas White House that there was "considerable evidence" that U.S.S.R. was building and deploying an antiballistic missile system around its cities, which might require an increase in U.S. offensive capacity. He said the Administration would probably recommend to Congress that U.S. begin production and deployment of Poseidon missile—a large submarine-launched missile with greater power to penetrate sophisticated defense system than Polaris missile. The Secretary reported that no decision had yet been made on deployment of Nike-X antimissile missile, on which $500 million had already been spent for research and development. (Semple, Jr., *NYT*, 11/11/66, 1, 19; Wilson, *Wash. Post*, 11/11/66, A1, A10; Horner, *Wash. Eve. Star*, 11/11/66, A1)

- MSFC had awarded General Electric Co. a $147,884, 12-mo. contract to study design and systems requirements of an electrically propelled space vehicle for manned Mars landing missions and to identify areas of technology in which concentrated research would produce maximum benefit to manned planetary program. (MSFC Release 66-271)
- Najeeb E. Halaby, vice president of Pan American World Airways and former FAA Administrator, suggested that Government build two Sst prototypes—one Boeing and one Lockheed—using same prototype power plant for both before selecting the final design. "There is no substitute for flight testing," he said. "The whole world would benefit in the economics and safety of a flight-tested plane." Airframe designs by Boeing Co. and Lockheed Aircraft Corp. and engine designs by General Electric Co. and Pratt & Whitney Div. had been submitted to FAA Sept. 6 for evaluation. President Johnson would announce final selection in late 1966. (*NYT*, 11/11/66, 73)
- Columbia Univ. announced creation of Institute for the Study of Science in Human Affairs which would seek to clarify present and potential roles of science in society. Funded by $1-million grant from Alfred P. Sloan Foundation, Institute would work through existing university structure, on both graduate and undergraduate levels, to stimulate teaching and research and possibly establish visiting professorships. The *New York Times* later editorialized: "Its goal is to increase public understanding of the basic issues arising from the scientific revolution—and to develop wise men to lead that revolution.

    "... Its beginnings are modest, but the need for it is great and so, too, we believe, is its promise." (Sullivan, *NYT*, 11/11/66, 32; Aarons, *Wash. Post*, 11/13/66, L4; *NYT*, 11/13/66, 10E)
- Sociological repercussions of the space age might be most important to historians, suggested Robert Toth in the *Washington Post:* "In just nine years since the first man-made moon went aloft, the fact that man can escape his Earth has become so accepted Americans are almost blasé about new manned flights.

    "It is too early for anyone to guess how such developments may affect the deeper thoughts of self and one's place in the world and in the universe. But it is not too much to say that, even before man reaches the moon, he has begun to get a new perspective of himself and his place in the universe from the space adventure." (Toth, *Wash. Post*, 11/10/66, P6)

November 10-30: NASA's LUNAR ORBITER II, launched from ETR Nov. 6, became second U.S. spacecraft to enter lunar orbit; five days later it was successfully transferred to final close-in orbit for photography. Orbital parameters: apolune, 1,163 mi. (1,870 km.); perilune, 122 mi. (196 km.); inclination, 12.2°; period, 3 hr. 38 min. Spacecraft recorded two micrometeoroid hits—first ever detected by U.S. spacecraft in moon's vicinity. It performed 205 attitude changes and responded to 2,421 commands.

A total of 211 medium- and high-resolution photos of 13 potential Apollo landing sites were taken during mission's photo acquisition phase. Readout would be completed Dec. 13. Among photos transmitted, described as "consistently high in quality," was a picture of the crater Copernicus taken from an angle inaccessible to earth cameras. Dr. Martin J. Swetnick, NASA Hq., describing the closeup as "one of the great pictures of the century," said: "It provides new information, which certainly will lead to better understanding of the processes and structures on the moon." (NASA Proj. Off.; AP, *Wash. Post*, 11/16/66, A18; Sullivan, *NYT*, 12/1/66, 1, 36)

November 11-15: GEMINI XII rendezvous-docking and Eva mission, last of Gemini series, began with launch of Gemini Agena Target Vehicle (GATV) by Atlas booster from ETR's Complex 14, followed 1 hr. 38 min. later by launch of GEMINI XII with Titan II booster from Complex 19. Astronauts were James A. Lovell, Jr., command pilot, and Edwin A. Aldrin, Jr., pilot. GATV entered near-circular orbit with 188.5-mi. (303.5-km.) apogee and 182.8-mi. (294-km.) perigee. GEMINI XII entered orbit with 167.9-mi. (270-km.) apogee and 100-mi. (161-km.) perigee.

Docking of spacecraft with GATV was in 2.3 revolutions—4:14 GET. After plans to use GATV propulsion system to place GEMINI XII into higher orbit had to be abandoned because of GATV malfunction, GEMINI XII rendezvoused with Nov. 12 total solar eclipse over South America and made motion and still pictures of eclipse through spacecraft windows. Aldrin then began first of two standup Evas: hatch opening was at 19:29:01 GET with closure 2 hrs. 29 min. later. Nov. 13's activities revolved around Aldrin's umbilical Eva: hatch opening was at 42:46 GET with closing 2 hrs. 8 min. later. While on 30-ft. umbilical, Aldrin performed measured work tasks at Agena docking adapter and at work station in spacecraft adapter section. He also attached 100-ft. tether stowed in Agena adapter to Gemini docking bar in preparation for tethered operations. GEMINI XII backed out of Agena docking collar about 47:37 GET and gravity gradient appeared to be established by one revolution later. Tether exercise lasted 4 hrs. 17 min. After failing twice to sight sodium vapor clouds released in upper atmosphere by French sounding rocket launched from Hammaguir Range and performing experiments with a space sextant, crew began experiencing problems with maneuvering thrusters on spacecraft. In second standup Eva, Aldrin jettisoned unused equipment and conducted additional experiments and photography. Hatch-open time was 66:04 GET for 51-min. duration. Total Eva time for GEMINI XII mission was 5 hrs. 28 min. Retrofire took place Nov. 15 at 94:00:01 GET; reentry in 59th revolution after 94 hrs. 34 min. in space was normal. Splashdown in the Atlantic was at 94:34:31 GET. GEMINI XII landed about three miles from recovery ship U.S.S. *Wasp* and about four miles from aiming point. Astronauts were picked up by helicopter within 20 min.

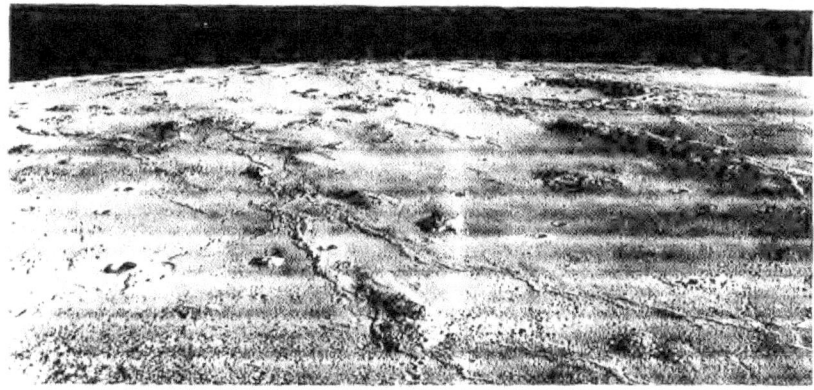

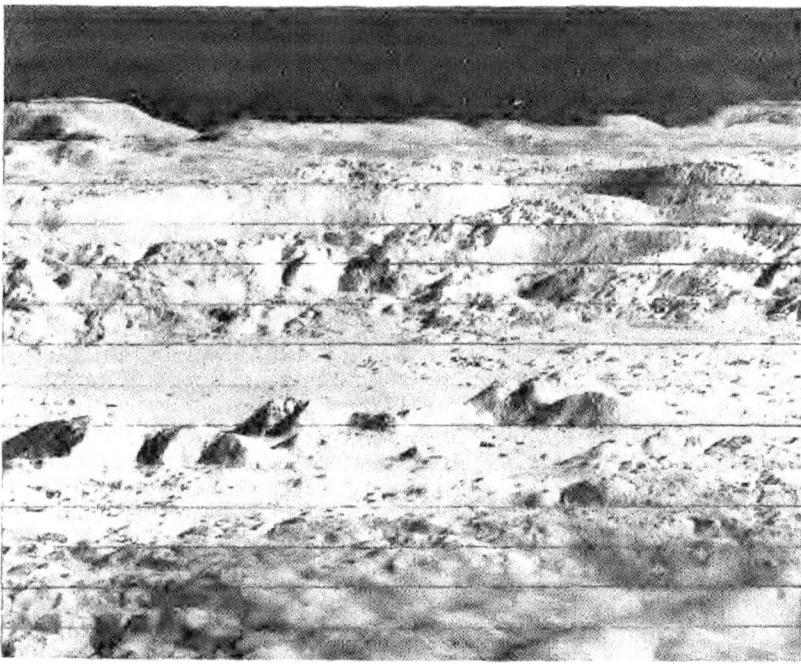

November 10-30: Two historic lunar photographs returned by NASA's LUNAR ORBITER II: one, the first close-up photograph of the crater Copernicus, taken Nov. 23, is an oblique view from 28-mi. altitude; the other, the first detailed view of lunar domes, taken Nov. 25, confirms the moon's long history of volcanic activity. The domes range from 2 to 10 mi. in diameter and from 1,000 to 1,500 ft. high; they are best seen to the south and west of the crater Marius (upper right).

White House announced promotion of Aldrin from USAF major to lieutenant colonel. Lovell had been promoted to USN captain after GEMINI VII mission. A Presidential statement read by Press Secretary Bill Moyers said the flight was "the culmination of a great team effort, stretching back to 1961, and directly involving more than 25,000 people

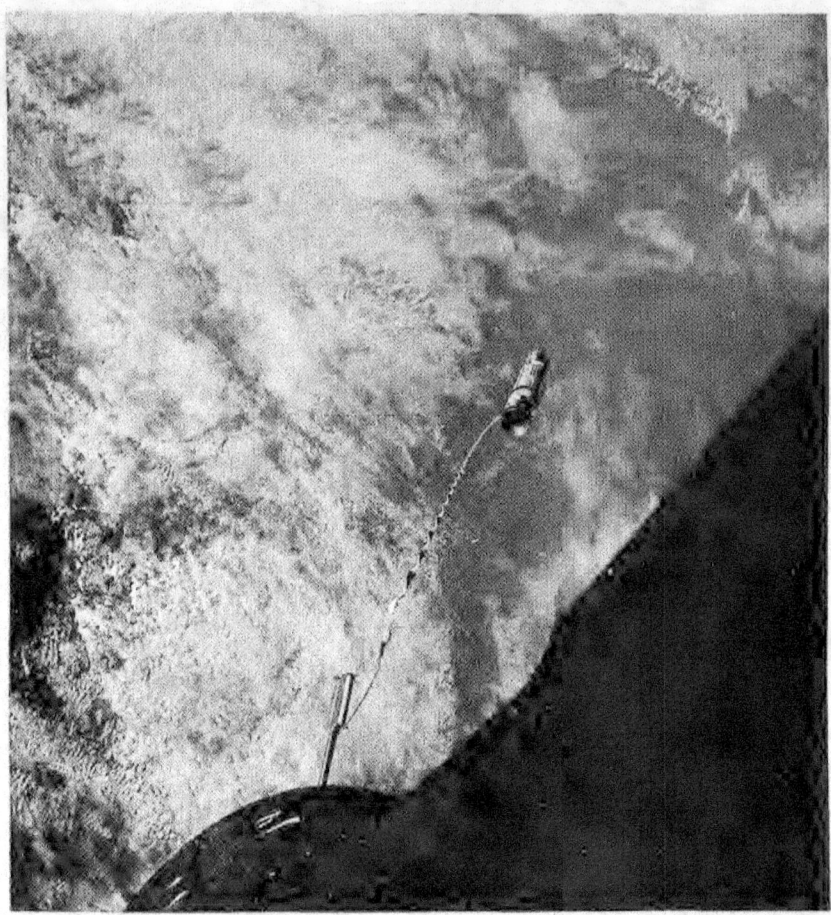

November 11-15: Gemini Agena Target Vehicle is shown at the end of tether securing it to the GEMINI XII spacecraft.

in the National Aeronautics and Space Administration, the Department of Defense, and other Government agencies; in the universities and other research centers; and in American industry." (NASA Proj. Off.; NASA Release 66-272; MSC *Roundup*, 11/25/66, 1, 2; Wilford, *NYT*, 11/12/66, 1, 14; 11/13/66, 1, 34; 11/14/66, 1, 2; 11/15/66, 1, 17; 11/16/66, 1, 30; O'Toole, *Wash. Post*, 11/12/66, A1, A4; 11/13/66, A1, A4; 11/14/66, A1; 11/15/66, A1, A10; *Pres. Doc.*, 11/21/66, 1701-2)

November 12: Total solar eclipse which moved across South America at 2,000 mph for 2 hrs. 42 min. received extraordinary scientific scrutiny: 800 scientists from 12 nations utilized $90 million of equipment. NASA's activities were conducted in cooperation with Argentine and Brazilian scientists as part of an international program. Eclipse formed 55-mi. strip of total blackout beginning in Pacific Ocean west of Peru, crossing southern Peru, Bolivia, northern Argentina, southern tip of Brazil, and extending into South Atlantic. NASA-Ames Convair 990A jet aircraft

"Galileo," a flying research laboratory carrying 26 scientists, successfully rendezvoused with four other instrument-equipped aircraft and raced the eclipse at 575 mph as it moved out over the South Atlantic. Other NASA activities included: five NASA Nike-Apache sounding rockets launched within two hours from Cassino, Brazil, in Univ. of Illinois-GCA Corp. experiment to measure D and E region electron densities and temperatures, absorption profiles 1216 A and 1450 A, and collision frequencies. First rocket ejected trimethyl-aluminum (TMA) cloud for comparison of electron density profile with wind profile; however, no photographs of vapor cloud were obtained. Third rocket carried Brazilian experiment to study extreme ultraviolet. All rockets and instrumentation, except for mass spectrometer on third flight, performed satisfactorily. Specially constructed launch and tracking site was manned and operated by personnel of Wallops Station and the Brazilian Space Commission. Photographs were taken by NIMBUS II meteorological satellite and by Astronauts James Lovell, Jr., and Edwin Aldrin, Jr., orbiting in GEMINI XII spacecraft.

AFCRL launched three Nike-Hydac sounding rockets from a site near Rio Grande, Brazil. Instrumented payloads, which reached 71-mi. (115-km.) altitudes, measured positive-ion mass composition, solar radiation, electron density and temperature, and positive-ion density. Data were telemetered to ground receivers.

Scientists from Argentina, Brazil, Peru, Uruguay, Japan, the Netherlands, U.K., New Zealand, and France participated in observations which would increase knowledge of the solar forces and their effects on earth's atmosphere and provide data for use in space travel, radio communications, and weather prediction. (NASA Release 66-285; Wallops Release 66-55; De Onis, *NYT*, 11/13/66, 35; AP, Wash. *Eve. Star*, 11/13/66, A16; AP, *Wash. Post*, 11/12/66, A16; OAR *Res. Rev.*, 1/67, 8; NASA Rpt. SRL)

November 12: U.S.S.R. launched COSMOS CXXXI for space research into earth orbit with 360-km. (224-mi.) apogee; 205-km. (127-mi.) perigee; 89.8-min. period; and 72.9° inclination. (*U.N. Public Registry*, 1/9/67)

- NASA Aerobee 150 launched from WSMR reached 94-mi. (151-km.) altitude in GSFC experiment to obtain solar x-ray photographs; rocket and instrumentation performed satisfactorily.

  NASA Aerobee 150 launched 25 min. later to 109-mi. (174-km.) altitude in NASA-Naval Research Laboratory experiment photographed outer solar corona with moon near edge of field of view and sunlit particles near the rocket. Flight also obtained heliograms showing inner corona and solar disc and data on solar flux in Lyman-alpha. Rocket and instrumentation, except for a spectroheliograph, performed satisfactorily. (NASA Rpt. SRL)

- Deactivation of Gemini launch pad at ETR's Complex 19 began. Equipment that could be used in other space programs would be salvaged; other components would be reduced to scrap. (AP, Wash. *Sun. Star*, 11/13/66, A6)

- *New York Times* praised success of NASA's LUNAR ORBITER II and GEMINI XII missions: "This week has seen two impressive demonstrations of United States space capabilities. . . .

  "It is no derogation of the important Soviet contributions to conclude that the United States efforts have contributed most of what men now know about the moon and about the possibilities and problems of manned space flight. In part this lead is the result of the long and unexplained halt in the launching of Soviet passenger-carrying rockets." (*NYT*, 11/12/66, 28)

November 12: Congress had "unwittingly" broadly endorsed Sst program by approving obscure phrase in Dept. of Transportation bill (H.R. 15963) empowering Secretary of Transportation to "develop and construct a civil supersonic aircraft," Evert Clark reported in the *New York Times*. Action strengthened position of the Administration and other Sst supporters and lessened chances that congressional critics might abandon the multibillion dollar project. (Clark, *NYT*, 11/13/66, 77)

- Satellite to be launched by U.S.S.R. in near future would carry French instrumentation, French Minister of Research Alain Peyrefitte announced in Caen. France and U.S.S.R. had signed an agreement for joint scientific cooperation June 30. (Reuters, *NYT*, 11/13/66, 45)
- Photos and temperature data gathered by satellites could be used to study ice movements on Canada's lakes and rivers, G. P. Williams, Canadian National Research Council, suggested to Canadian Press. Williams said predictions of the formation and breakup of ice would be valuable to forestry workers who used the waters to float logs, and aviators who landed on northern lakes. (Can. Press, *NYT*, 11/13/66, 46)

November 13: Sunspots as large as 12 times the earth's diameter were detected by Bochum Observatory, West Germany. One group reportedly measured about 97,200 mi. across; smaller sunspots were about 27,000 mi. wide. (Reuters, *NYT*, 11/15/66, 61; Reuters, *Wash. Post*, 11/14/66, A23)

- IBM Federal Systems Div. received $218,000 AFSC contract to investigate means of reducing operating costs and reaction time of future space missions through streamlined guidance and control procedures. (*Wash. Sun. Star*, 11/13/66, D15)

November 14: Team of JPL scientists left Cal Tech on first NASA-sponsored investigation of Antarctica. Scientists would spend 2½ mos. in Wright, Victoria, Taylor, and Beacon Valleys taking samples and conducting tests that might prove valuable in future unmanned exploration of Mars. (JPL Release, 11/13/66)

- NASA selected Martin Co. for negotiations on a $1-million, 18-mo. contract to design, develop, fabricate, and test meteoroid detectors for future NASA experiments to study penetration hazard in space. Contract would be managed by LaRC. (NASA Release 66-293)
- ELDO successfully test-fired three-stage Europa rocket from Woomera Range, Australia. Booster was blown up over Simpson Desert 6 min. 42 sec. after launch. (*Tech. Wk.*, 11/21/66, 8)
- M2-F2 lifting body vehicle, piloted by Capt. Jerauld R. Gentry (USAF), was launched from B-52 aircraft at 45,000-ft. altitude in 13th glide flight at Edwards AFB. Purposes of this unpowered flight were tests of stability and control and determination of vehicle performance characteristics. (NASA Proj. Off.)
- Military should increase its interest in the Nation's technological base rather than rely so heavily on the civilian sector for basic and exploratory research, suggested William Coughlin in *Technology Week*. He noted the Manned Orbiting Laboratory (Mol) program: ". . . . much of the technology on which this military effort is based was developed by the National Aeronautics and Space Administration on programs deliberately tied to the peaceful uses of space. It is a fair question to ask where

a military space program would be today vis-a-vis the Soviet Union if the technological base had not been developed under NASA funding.

"Other areas of military importance have not been as fortunate because they lack a civilian counterpart. Steps must be taken to see that development of a technical base in these areas is not overlooked. It is not easy to justify expenditure of manpower and funds in this somewhat invisible arena. . . ." (Coughlin, *Tech. Wk.*, 11/14/66, 54)

During week of November 14: Increasing application of engineering techniques developed in space program to medical problems was reported by participants at 19th Annual Conference in Medicine and Biology in San Francisco. Among developments reported: application of computer techniques to detection of pathological conditions and to mass diagnosis of heart disease; automation of classical laboratory techniques in hospital, resulting in increased accuracy and speed; use of new engineering devices to obtain information for better diagnoses; application of new engineering to treatment, including refinements on artificial heart and control of photosensitive epilepsy with a real-time computer. (*Tech. Wk.*, 11/21/66, 30)

November 15: NASA's Project Gemini officially ended. From first manned flight in March 1965, Gemini had achieved its goals: demonstration of ability to rendezvous and dock with target vehicle; demonstration of value of manned spacecraft for scientific and technological experimentation; performance of work by astronauts in space; use of powered, fueled satellite to provide primary and secondary propulsion for docked spacecraft; long-duration space flights without ill effect on astronauts; and precision landing of spacecraft. Records set during Gemini program included: (1) longest manned space flight (330 hrs. 35 min.); (2) altitude (851 mi.); and (3) longest total of Eva (5 hrs. 28 min.) in one flight. Total U.S. man hours in space to date: 1,993 hrs. 34 min. MSC Director Dr. Robert R. Gilruth told a news conference: "We have done all the things we had to do as a prelude to Apollo. I believe the Gemini program has been most successful." (NASA Proj. Off.; *Wash. Post*, 11/16/66, A6; *Tech. Wk.*, 11/21/66, 16; *NYT*, 11/20/66, 2E)

- NASA announced that NIMBUS II meteorological satellite had taken nearly 1 million photos since its launch from WTR May 15. "It has been working better than a finely tuned engine," said GSFC project manager Harry Press. "Its performance has exceeded everyone's expectation." (GSFC Release G-19-66)

- NAS Space Science Board chairman Harry H. Hess, Princeton Univ. geology professor, received Geological Society of America's Penrose Medal "for distinguished achievement in the geologic sciences" at society's 79th annual meeting in San Francisco. (NAS-NRC-NAE *News Report*, 11/66, 6)

November 16: GEMINI XII Astronauts James A. Lovell, Jr., and Edwin E. Aldrin, Jr., arrived at KSC to begin debriefing. Addressing some 700 workmen, Lovell said "everyone here has done an outstanding job to get us into space"; Aldrin described the four-day mission as a "treat to me as an individual that I would like to have shared with every person in the world." (Sehlstedt, Balt. *Sun*, 11/17/66)

- NASA released first 140 ft. of more than 1,500 ft. of color movies taken by Astronauts James A. Lovell, Jr., and Edwin E. Aldrin, Jr., during their Nov. 11-15 space flight. Photography was of high quality. (AP, Wash. *Eve. Star*, 11/17/66, A4)

November 16: Eighth Uprated Saturn I booster successfully underwent short-duration—35-sec.—static firing at MSFC. (MSFC Release 66-288)
- Site at Pakachoag Hill outside of Auburn, Mass., where Dr. Robert H. Goddard launched world's first liquid-fuel rocket flight on March 16, 1926, was approved as a national landmark by Secretary of the Interior Stewart Udall. (Dept. of Interior; *SBD,* 11/23/66, 119)
- AFCRL launched Aerobee 150 sounding rocket from WSMR to 117-mi. (188-km.) altitude to collect micrometeorite particles during Leonid meteor shower. (OAR *Res. Rev.,* 1/67, 8)
- ComSatCorp requested FCC permission to construct $6.5-million, high-capacity earth station at Barrio Monte Llano, Puerto Rico, to handle all types of communications in the Caribbean area via EARLY BIRD I and INTELSAT II-A comsats. April 25 application for station at St. Croix, Virgin Islands, was withdrawn because military communications-electronics equipment there "might cause harmful interference. . . ." (ComSatCorp Release)
- Sst development might be delayed because of current pressures to limit Federal spending; growing apprehension that Sst would cause excessive property damage and noise from sonic booms; serious safety and technical problems; and Sst evaluators' indecisiveness, Fred L. Zimmerman reported in the *Wall Street Journal.* (Zimmerman, *WSJ,* 11/16/66, 23)
- Possibility that U.S. astronauts might land on moon in 1968 rather than 1969 "should serve to hasten agreement on international treaty aimed at making outer space exclusively a dimension of man's peaceful activity," postulated editorial in *New York Times.* "Hard decisions" about post-Apollo research should be made soon, especially since there were "far more possibilities for fruitful activities in space than even the United States can afford to carry on simultaneously." (*NYT,* 11/16/66, 42M)
- Daniel J. Haughton, president of Lockheed Aircraft Corp., suggested at American Petroleum Institute meeting in New York that aerospace and petroleum industries combine their technical strengths for joint exploration of earth's continental shelves. He said that within 35 yrs. the world's population would have doubled and a large segment of this population would be hungry: "The oceans represent the last great source for feeding them and providing the chemicals and minerals and water to meet all the future needs of man." (AP, Wash. *Eve. Star,* 11/16/66, C12)

November 16-18: National Conference on the Management of Aerospace Programs was held in Columbia, Mo.

NASA Apollo Program Director M/G Samuel C. Phillips (USAF), in report on "Management Scheme for Apollo," concluded: "There is no single approach to program management that will guarantee success. The development of and experience with various management systems, techniques and tools in the past several years have contributed extensively to a more scientific approach. The 'art' element remains, however. The integrated implementation of those concepts and their associated procedures and disciplines is sometimes confronted with considerable resistance within an organization. This has merit, for it forces management to thoroughly screen new concepts as to validity, feasibility and salability. Once a new management system is approved, however, it deserves and requires the personal support of the program director through line channels to stimulate its overall implementation and acceptance at the grass roots level. ..." (Text)

DeMarquis D. Wyatt, NASA Assistant Administrator for Programming, said that NASA's cost model had "significantly facilitated management decision making in complex on-going programs by providing an overnight computational capability of great capacity. The computer program is capable of handling the interrelated cost influences of up to 60 different hardware configurations at an unlimited number of plants and an unlimited number of production lines within each plant. It can ... summarize program costs for as many as 24 fiscal years and can subdivide and allocate these costs to up to 8 different funding sources or users...." (Text)

Franklin P. Dixon, Director, Planetary Mission Studies, NASA OMSF, outlined manned Mars landing program for consideration as U.S. national space goal. Proposal called for seven-step program based principally on extension of mission-duration capability and including earth-orbital "Space Station/Mission Module," Mars and Venus flybys, and Mars landing. Discussing long-range NASA planning, he noted: "At the present time planning for the Saturn Apollo Applications Program includes subsystem developments for long duration manned flights.... This provides a logical growth pattern toward manned planetary travel with the easier flyby missions within technological reach by the 1975 to 1980 time period for actual operational flights." (Text)

November 17: NASA announced schedule changes in manned Apollo flights because of launch vehicle and spacecraft development problems: second manned flight of an Apollo spacecraft (AS-205), scheduled for spring 1967, was postponed and no new date set; AS-205 prime crew Walter M. Schirra, Donn F. Eisele, and Walter Cunningham became backup crew for AS-204 mission—first manned Apollo flight—scheduled for first quarter of 1967. (NASA Release 66-295)

- Some 1,000 meteors a minute were recorded at height of Leonid shower by Northwestern Univ.'s Observatory near Las Cruces, N. Mex., Dr. J. Allen Hynek reported. (Sullivan, *NYT*, 11/18/66, 39M)
- Referring to successful completion of Gemini series of space flights as a "historic triumph for mankind," editorial in *Washington Post* suggested that "as the conquest of space proceeds it ... gives the world a nobler and higher destiny and purpose than fratricidal strife for nationalistic objectives." (*Wash. Post*, 11/17/66, A22)

November 18: Maj. William J. Knight (USAF) flew X-15 No. 2 to 4,250 mph (mach 6.33) and 98,900-ft. altitude during 9-min. flight, successfully using full external propellant tanks for first time and setting new speed record for X-15 aircraft. Previous 4,104-mph (mach 6.059) record had been set by the late Joseph A. Walker (NASA), June 27, 1962. (X-15 Proj. Off.; UPI, *NYT*, 11/19/66, 24)

- NASA Nike-Apache launched from WSMR to 100-mi. (161-km.) altitude carried NASA-Dudley Observatory (Albany, N.Y.) experiment to collect micrometeorite particles at time of 33-yr. maximum of Leonid meteor shower and GSFC experiment to determine vehicle slant range. Rocket and instrumentation performed satisfactorily, but payload was not recovered and no data were obtained. (NASA Rpt. SRL)
- U.S. had notified Government of Nigeria that, with conclusion of Gemini program, NASA would close tracking station at Kano. (NASA Release 66-296)

November 18: George Cooper, ARC's Chief Test Pilot and Chief of Flight Operations, received Admiral Luis de Florez Flight Safety Award at International Air Safety Seminar in Madrid. He was cited for "his contribution to the safety of air transportation through improved understanding of the relation between pilot and aircraft." (ARC Release 66-17)

November 19: U.S.S.R. launched COSMOS CXXXII for continued space research into earth orbit with 280-km. (174-mi.) apogee; 207-km. (129-mi.) perigee; 89.3-min. period; and 65° inclination. Equipment was functioning normally. (Tass, 11/19/66)

November 20: U.S.S.R. began new carrier rocket tests in the Pacific that would last until Dec. 30. Tass said area with 56-km. (35-mi.) radius had coordinates of 0°5′ south latitude and 163°45′ west longitude. Ships and aircraft were requested not to enter this area daily between noon and midnight local time. (Tass, 11/19/66)

- ESRO successfully launched France's Centaure rocket from range at Paksuniemi, Sweden, to study the aurora borealis. (Reuters, *Wash. Post*, 11/21/66, A14)

- High military potential of space was discussed by Hanson W. Baldwin in *New York Times*: ". . . the advent of the military man in space is inevitable. For man can supply the judgment that no machine can offer; he can monitor and correct and interpret the instruments and aids essential to space flight; he will be able to repair satellites and assemble space platforms. He will pilot tomorrow's fighting machines in the cosmos. And in time he may actually be able, with a whole host of scientific aids, to control the weather with all the tremendous military and social consequences this implies." (Baldwin, *NYT*, 11/20/66, A7)

November 21: NASA successfully conducted second rocket-launched test [see Nov. 5] of a Mars entry parachute as part of an advanced technology effort to investigate possible parachute landing systems for Voyager program. Two-stage Honest John-Nike rocket launched from WSMR ejected parachute at 120,000-ft. altitude. Parachute descended to earth 40 min. later carrying 200-lb. payload with instrumentation for measuring shock of parachute opening and its oscillation characteristics. Experiment series was managed by LaRC. (NASA Release 66-298)

- MSFC awarded North American Aviation, Inc., a $141-million, cost-plus-incentive-fee contract to provide 30 F-1 rocket engines, beginning in November 1967, and varied support services. (NASA Release 66-297)

- NASA Aerobee 150 launched from WSMR to 102-mi. (164-km.) altitude carried GSFC experiment to obtain ultraviolet spectral scans of bright stars from 1110 A to 4000 A. Rocket and instrumentation performed satisfactorily. (NASA Rpt. SRL)

- Commenting on success of Project Gemini in developing techniques and experience vital for manned lunar landing, William J. Coughlin said in *Technology Week*: "Gemini technology will . . . be applied to the manned scientific space stations still to come. This fanning out of technology from the direct-line effort to effect a manned lunar landing is a dramatic demonstration of the rapid maturing of the national space program." (Coughlin, *Tech. Wk.*, 11/21/66, 50)

- Final glide flight of M2-F2 lifting body vehicle before installation of XLR11 8,000-lb.-thrust rocket engine [see Dec. 30] was made at Edwards AFB with Capt. Jerauld R. Gentry as pilot. Purposes of flight, 14th in unpowered series, were testing of stability and control and determination of vehicle performance characteristics. (NASA Proj. Off.)

November 21: Reported that USAF Titan II launch vehicle, which launched 12 out of 12 missions in the Gemini series, was expected to win a USAF incentive fee of over $2 million for Martin Co., the prime contractor for two-stage booster. (*Av. Wk.*, 11/21/66, 28)

- U.S.S.R. had orbited an atomic clock of the ammonia type in COSMOS XCVII, launched Nov. 26, 1965, a brief article in *Pravda* disclosed. Soviet Academician N. G. Basov, explaining possible uses of an ammonia frequency standard in a satellite, said it "permits carrying on communications with space devices, control over them, and transmission of telemetric information for very great distances. In addition, there is in this case a considerable increase in the operating precision of program timer devices and systems for determining the trajectory of the satellite's movement." (*Av. Wk.*, 11/21/66, 36)
- Beryllium ball floating in electrostatic field and spinning at some 60,000 rpm was part of new electrostatic gyroscope (Esg) being flight-tested by USAF at Wright-Patterson AFB aboard C-124 Globemaster aircraft. Instability formerly found in the spinning gyroscope had been eliminated in Esg since there was virtually no friction between the sphere and its cavity. Sphere would spin for three years without power source. (AFSC Release 221.66)
- Pressure for early action to establish Canadian domestic comsat system was rising, both within the Canadian government and in telecommunications and broadcasting industries, *Aviation Week* reported. Government advisory team was planning to complete report by end of 1966 making recommendations to Canadian Cabinet on technical aspects of system and on controversial policy issues. Team was under direction of Dr. J. H. Chapman, deputy chief superintendent, Defence Research Telecommunications Establishment, Dept. of National Defence. (*Av. Wk.*, 11/21/66, 35)
- Guy Warner Vaughn, president and chairman of Curtiss-Wright Corp. and among first to apply mass-production techniques to aviation industry, died in New Rochelle, N.Y., at 82. (*NYT*, 11/22/66, 39M)

November 22: GSFC awarded Hughes Aircraft Co. a $4.275-million, cost-plus-incentive-fee contract for reconfiguration of three gravity-gradient Applications Technology Satellites scheduled for launch in 1967 and 1968. (NASA Release 66-299)

- NASA awarded Perkin-Elmer Corp. and Chrysler Corp. nine-month, $250,000 contract extensions to continue evaluating optical experiments for possible future extended Apollo space flights. (NASA Release 66-300)
- AFSC presented first USAF Craftsmanship Awards in Industrial Zero Defects Program to RCA's Astro-Electronics Div., Communications Systems Div., and Missile and Surface Radar Div.; GE's Flight Propulsion Div. and Evendale Facility; Lockheed Missiles and Space Co.; Aerojet-General Corp., Sacramento Plant; and Douglas Aircraft Co., Inc., Missile and Space Systems Div. (AFSC Release 245.66)

November 23: GEMINI XII pilot Edwin E. Aldrin, Jr., told MSC news conference that with proper equipment, like the waist straps and foot restraints he used during Eva, a man could work outside with no fear of fatigue—if he were well trained. Following press conference, Astronauts James A. Lovell, Jr., and Aldrin together with a group of MSC, KSC, NASA Hq., and industry Gemini program participants flew to LBJ Ranch for visit with President Johnson, who said: "The splendid performance of man and machine in Gemini has been a product of the American system at its best.

"The months ahead will not be easy as we reach toward the moon.... But with Gemini as the forerunner, I am confident that we will overcome the difficulties and achieve another success." Group returned to MSC for formal awards ceremony.

Receiving NASA Exceptional Service Medal were GEMINI XII crewmen James A. Lovell, Jr., and Edwin E. Aldrin, Jr.; Col. John G. Albert (USAF), ETR; Osro H. Covington, GSFC; and John D. Hodge, MSC.

NASA Distinguished Service Medal: Dr. George E. Mueller, OMSF; and Charles W. Mathews, MSC.

Outstanding Leadership Medal: Robert F. Thompson, MSC; John J. Williams, KSC; and M/G Vincent G. Houston (USAF), ETR.

NASA Exceptional Scientific Achievement Award: James A. Chamberlin, MSC.

NASA Public Service Award: John F. Yardley, McDonnell Aircraft Corp.; Bastian Hello, Martin Co.; Bernhard A. Hohmann, Aerospace Corp.; Walter D. Smith, Martin Co.; Walter F. Burke, McDonnell; Louis D. Wilson, Aerojet-General; Lawrence A. Smith, Lockheed; William B. Bergen, Martin Co.; George M. Bunker, Martin Marietta; B/G Paul T. Cooper, AFSC Space Systems Div.; Daniel J. Haughton, Lockheed; Roger Lewis, General Dynamics; James S. McDonnell, McDonnell Aircraft; R. I. McKenzie, Aerojet-General; L. Eugene Root, Lockheed; David S. Lewis, McDonnell; and Jack A. Bowers, Martin Marietta.

NASA Superior Achievement Award: Arthur W. Vogeley, Richard J. Allen, LeRoy E. Day, John A. Edwards, Eldon W. Hall, Vearl N. Huff, and William A. Summerfelt, NASA Hq.; and Anthony L. Liccardi.

NASA Group Achievement Award: Gemini Astronaut Team, Manned Spaceflight Network Team, Gemini Spacecraft Launch Team, Gemini Program Office at MSC, Gemini Program Office at NASA Hq., and Gemini Support Team at MSC. (MSC *Roundup*, 11/25/66, 1, 8; Sheehan, *NYT*, 11/24/66, 36; *Pres. Doc.*, 11/28/66, 1718-9)

November 23: XB-70 No. 1 supersonic aircraft, piloted by North American Aviation, Inc., test pilot Van H. Shepard and L/Col. Joseph Cotton (USAF), reached mach 2.51 and 60,000-ft. altitude in flight from Edwards AFB for the national sonic boom program. (XB-70 Flight Log)

November 24: European conference on satellite communication, meeting at The Hague, Netherlands, requested ESRO to study feasibility and desirability of European comsat. (*NYT*, 11/25/66, 4)

- Airport built on artificial island on Goodwin's Sandbank—12 mi. off coast of England in English Channel—was London consulting engineer G. E. Bratchell's answer to noise problem anticipated from projected Anglo-French Concorde Sst and other planned supersonic jet aircraft, Reuters reported. Takeoffs and approaches would be made over sea, avoiding danger of crashes in populated areas. By time aircraft was over land, it would be at height where noise was no longer a problem. A natural gas fog-dispersal system would make the airport accessible every day of the year. (Reuters, *Wash. Post*, 11/24/66, K10)

November 26: President Johnson established White House committee headed by Dr. Donald F. Hornig to study ways to help West European nations improve their technology. Serving with Dr. Hornig would be representatives of Depts. of State, Defense, and Commerce; NASA; AEC; and

Council of Economic Advisers. (AP, *NYT*, 11/27/66, 41; *Wash. Post*, 11/27/66, A6)

November 26: U.S.S.R.'s LUNA XII, launched into lunar orbit Oct. 22, had completed 220 orbits, taken "unique photographs," and was continuing its mission, Tass reported. Satellite had transmitted data on lunar x-ray and gamma radiation and micrometeoroids. (Tass, 11/26/66)

- U.S. and U.S.S.R. were reported to have reached basic agreement on text of treaty for free and peaceful exploration of moon and outer space. Text would be presented within next two or three weeks for approval of U.N. General Assembly. U.S. was said to have agreed that all countries —not just U.N. members—should be allowed to sign treaty. Communist China, East Germany, North Vietnam, and North Korea would then be free to adhere to it if they chose. (AP, *NYT*, 11/27/66, 78)
- High-quality photographs of earth taken during Nov. 11-15 GEMINI XII mission had revealed substantial inaccuracies and omissions in geologic maps, GSFC scientist Paul D. Lowman, Jr., said in press interview. He added that space photography might someday help solve such geologic problems as continental drift, the relationship of volcanic activity to mountain building, and the formation of regional sand dune fields. (AP, *NYT*, 11/26/66, 26)
- U.S. was studying feasibility of defense against nuclear missiles for all or part of Western Europe, *New York Times* reported. Administration officials said that although there would be political, financial, and technical problems, missile shield could be developed at estimated cost of $3-$12 billion. (*NYT*, 11/27/66, 9)

November 27: Two propulsion developments emerging from engineering research might open Jupiter, the regions outside the solar system, and even the sun to closeup inspection during 1970s, said Dr. Homer Joe Stewart, manager of JPL's Advanced Studies Office. Developments were use of planetary gravitational drag and solar-electric propulsion. When a spacecraft neared a planet, it would gain speed from gravitational pull. While traveling under planet's gravitation attraction, it would pick up planet's own sun-orbit speed. This, added to spacecraft's momentum, would be enough to propel spacecraft past planet's gravitational field and toward another planet. Travel between gravitational fields would be powered by spacecraft's solar-electric system. Solar-electric propulsion would double useful payload for a Mars-orbiter mission, Dr. Stewart said, and it would offer the advantage of high specific impulse for a Jupiter flyby. (JPL Release; *A&A*, 12/66, 26-31)

- Gen. Bernard A. Schriever (USAF, Ret.) said on ABC-TV's "Issues and Answers" that the U.S. should concentrate on developing better missiles rather than antimissile defense system. He said there was "a grave question with respect to the effectiveness" of an antimissile system aimed at stopping the sophisticated weapons the U.S.S.R. probably had. On the other hand, the U.S. could build a defense against the type of missile the Communist Chinese were presently capable of launching. Schriever said he believed U.S. missiles would get through the antimissile system reportedly being deployed in U.S.S.R. (AP, *NYT*, 11/28/66, 53)
- Analysis of NASA LUNAR ORBITER I's orbit led to assumption that moon is homogeneous, not layered, Dr. C. L. Goudas of Boeing Scientific Research Laboratories said in an interview with Walter Sullivan of the *New York Times*. He said its shape—flattened at the poles with several

bulges at the equator and slightly pear-like—was not "dominated by a 'frozen tide.'" Gamma ray recordings by U.S.S.R.'s LUNA X spacecraft indicated similar conclusion; lunar surface lacked emissions typical of granites, which are associated with layering. (Sullivan, *NYT*, 11/27/66, E7)

November 28: NASA's MARINER IV had completed two years in space, flown 1,025,082,830 mi., and continued to operate properly, reporting its condition to earth three times a week. Launched Nov. 28, 1964, spacecraft had completed primary mission Aug. 2, 1965, after transmitting to earth 22 pictures of Martian surface. Photos were made when it flew within 6,118 mi. of planet July 14, 1965. (NASA Release 66-304; JPL Release)

- NASA awarded contracts totaling $825,000 for study of methods to explore the planets and to design advanced launch vehicles: Lockheed Missiles and Space Co., $200,000; Boeing Co., $150,000; Lockheed-California Co., $250,000; and North American Aviation, Inc., $225,000. Contract management would be performed by NASA Hq. Mission Analysis Div. (NASA Release 66-302)
- White House announced agreement for establishment of an ESRO station near Fairbanks, Alaska, to receive telemetry from and send commands to ESRO scientific satellites. This would be first foreign ground station on U.S. soil. (*Pres. Doc.*, 12/5/66, 1740)
- Partial nationalization of U.K.'s two large airframe manufacturers—British Aircraft Corp. and Hawker Siddeley Aviation Co.—was being planned on grounds no real competition existed between the two firms, *Aviation Week* reported. Also, U.K.'s Ministry of Aviation would merge with Ministry of Technology and be represented on Cabinet by Minister of Technology Anthony Wedgewood Benn. (*Av. Wk.*, 11/28/66, 23)
- U.S.S.R. launched COSMOS CXXXIII into earth orbit with 232-km. (144-mi.) apogee; 181-km. (112-mi.) perigee; 51.8° inclination; and 88.4-min. period. Equipment was functioning normally. (Tass, 11/28/66)
- U.S.S.R. was showing increasing signs of having conceded manned lunar landing race to U.S. as part of vastly revamped space program, Donald C. Winston postulated in *Aviation Week*. Significance of new program was that it reflected emphasis Premier Alexey Kosygin had fomented in Soviet planning. Brezhnev and Kosygin, who came to power while three-man VOSKHOD I was still in orbit in October 1964, ordered the spacecraft to earth prematurely as one of their first moves even before the overthrow of Nikita Khrushchev was made public. It was not long before Kosygin grounded the rest of the Voskhod series, which was to have continued with at least five launches through 1965. Only VOSKHOD II was permitted off the ground after Kosygin took power. The March 1965 mission was conducted because planning had reached an advanced stage by the time Khrushchev was ousted and because Cosmonaut A.A. Leonov's Eva—first in history—had high propaganda value. Winston theorized that the new space philosophy could produce a much less complex manned circumlunar mission without landing within the next year. (*Av. Wk.*, 11/28/66, 22)
- NASA selected Bendix Corp. for negotiation of a $6.9-million MSFC contract to develop and produce three pointing-control-system (Pcs) units for the Apollo Telescope Mount (Atm). (NASA Release 66-309)

Week of November 28: French space officials reported partial success for first firing of an experimental Coralie rocket from Hammaguir Range.

Rocket, second stage of ELDO's Europa booster, underwent nominal countdown, launch, and flight until T+62 sec., when circuit malfunction prevented monitoring to T+100 sec. as planned. (*Av. Wk.*, 12/5/66, 30)

November 29: President Johnson said at news conference at Texas White House that there would be $60-million program reduction in NASA FY 1967 budget requiring some contract cancellations. About $30 million in expenditures was involved. Asked if changes in the NASA budget would cause postponement of target dates for any of the space programs, Johnson said: ". . . $30-million out of some several billion dollars wouldn't cause a change in the target date of the moon schedule, if you are talking about that." (*Pres. Doc.*, 12/5/66, 1741; Transcript, *NYT*, 11/30/66, 18)

- X-15 No. 3 was flown by Maj. Michael J. Adams (USAF) to 3,204.5 mph (mach 4.65) and 92,100-ft. altitude in pilot familiarization flight. As secondary mission, pilot performed maneuvers to evaluate aircraft's controllability and natural stability. (X-15 Proj. Off.)
- Canada's ALOUETTE II had traveled about 135 million mi. in 4,037 orbits since launch Nov. 28, 1965, Defence Research Board announced. The 32-lb. spacecraft had obeyed 11,200 commands; all experiments had been carried out as planned. (AP, *NYT*, 11/30/66, 9)
- Eighth Uprated Saturn I booster was static-fired successfully at MSFC for 145 sec. First stage performed as expected, developing 1.6 million lbs. thrust. (MSFC Release 66-288)

November 29-December 2: AIAA's Third Annual Meeting and Technical Display was held in Boston. In keynote address, NASC Executive Secretary Dr. Edward C. Welsh described the manned spaceflight program as the most successful technological feat ever undertaken by the Nation. Panel discussion followed on future of U.S. space program.

Dr. Gordon J. F. MacDonald, Institute for Defense Analyses, predicted that man would be needed in space primarily as a "repairman" for scientific instruments. "The science . . . will be carried out on the ground where enormous facilities for computing, analysis, and discussion are available."

Gen. Bernard A. Schriever (USAF, Ret.) said emphasis on space activities should not be only on peaceful purposes but also on preservation of peace; this would inevitably require sending military men into space. With advances in technology, U.S. must convert from "throwaway" hardware concept to use of maneuverable reentry vehicles and recoverable boosters, he added.

This idea was echoed by NASA Associate Administrator for Manned Space Flight Dr. George E. Mueller, who said: ". . . all of the hardware developed for future projects would reflect our concern over . . . greater cost effectiveness, through the development of reusable launch vehicles and reusable spacecraft." Mueller said there would be manned space stations that could stay in orbit a year or longer but said there was no need for crews to stay aloft for such periods. A ferry vehicle would be used for resupply of both equipment and crew. He listed potential benefits of space technology in the next decade, describing air traffic control towers in space, direct-to-home international TV broadcasting systems, and possibilities for predicting and controlling the weather.

Rep. George P. Miller (D-Calif.) said one of the problems of the House Committee on Science and Astronautics—of which he was chairman—was educating the people of the Nation to the benefits of the space program. It was a difficult task, he said, to justify space expenditures against the costs of Vietnam, foreign aid, and Great Society projects. But, he insisted, "it is the translation of space technologies into useful benefits that is the overall value of the space program." (Texts; AIAA *Daily*, 11/30/66, 1, 3; Wash. *Eve. Star*, 11/30/66, A7)

Results from NASA's EXPLORER XXXIII had shown for first time that tail of earth's magnetosphere "extends more than 75,000 miles beyond the orbit of the moon," reported GSFC scientist Dr. Norman F. Ness. Findings were contrary to those of Soviet scientist Dolginov and colleagues who had reported they were unable to detect magnetospheric tail at lunar distances on basis of data obtained from LUNA X between April 3 and May 4, 1966. Ness suggested as reason for this that LUNA X data were probably obtained when spacecraft was near or within neutral sheet region of magnetosphere where very low magnetic fields would be expected. (GSFC Release 20-66)

In session on Future Large Subsonic Transports, John Borger of PAA said problems in air transport industry "are being created by the phenomenal growth rate of our business." Since the growth rate showed no signs of slackening in the coming years, continued development and extensive use of large subsonic jet aircraft were not only desirable but essential. Borger pointed out that the engine for the Boeing 747—which would carry 2½ times more passengers than present intercontinental transports—was being developed with only commercial airlines as potential customers. This was the first time in air history that a new aircraft powerplant had been undertaken on this basis. Borger noted that a major requirement for the "jumbo jets" was that they be available at least 5 to 6 yrs. ahead of the proposed Sst so that initial development and operation costs could be recovered.

H. F. Klump, general manager of cargo traffic and sales for Deutsche Lufthansa Airlines, sounded a warning about the subsonic jet aircraft's cargo capacity: unless airports expanded their freight handling facilities enormously, they would not be able to handle volume of material "dumped on them."

W. D. Perreault, program director for Lockheed 500—a giant commercial cargo aircraft for the 1970s—discussed evolution of the 500 from C-5A transport under development for USAF. After reviewing three proposed commercial versions of the aircraft, he said one of them "will provide for the first time a cargo plane with direct operating costs of less than 2¢ per ton statute mile." (AIAA *Daily*, 11/30/66, 1, 31; Hines, Wash. *Eve. Star*, 11/29/66, A10)

Commenting on the report of the AIAA Launch Vehicle and Missile Technical Committee, "The Next Generation of Launch Vehicles—Evolution or Bold Step?" N. E. Golovin of the Office of Science and Technology said: ". . . if a family of reusable boosters is to be developed for reasons of economy then development cost per booster must be so low that economical amortization must be possible over a relatively small number of flights for each type; or, if only a single reusable booster is to be developed its use must indeed be so cheap that it would pay to employ it for payloads which are but a small fraction of its orbiting capacity. This reasoning . . . casts considerable doubt on the Committee's suggestion that a 'bold step' leading to the development of reusable boosters is justified at this time." (Text)

NASA Deputy Associate Administrator for Space Science and Applications Dr. Edgar M. Cortright, participating in a panel discussion on "Do We Need New Propulsion Systems for Lunar and Planetary Flight?", concluded that unmanned missions during the next ten years would probably "not require either new propulsion systems or new launch vehicle stages. They probably will require adding a modified Centaur as an upper stage on the Saturn IB or the Titan IIIC. They may require and certainly could use a 260 inch solid first stage under the S-IV-B Centaur. . . ." Cortright recommended that current launch capability be put "to effective use in exploring space and in developing practical applications of space flight." (Text)

During panel on Space and Public Policy, "poor communications" between scientists and the general public was suggested reason for death of Project Mohole—killed by Congress because it would have cost at least three times the original estimate. NAS president Frederick Seitz said such escalations in cost were "not unusual" for such scientific ventures and revealed the project would soon be renewed under a different name such as "deep ocean drilling."

"The goals of space exploration are primarily those of science, not technology," asserted panelist Hunter Dupree, science historian and history professor at Univ. of California, Berkeley. Chairman of the President's Science Advisory Committee's Panel on Space Science Louis Branscomb replied that no one wanted to see the space program purely as a "spectator sport," and that space programs should be conceived to "maximize their scientific returns." Commenting on the same subject, NASA Deputy Administrator Robert C. Seamans, Jr., said: "We must grapple with the problem of what we will do with the moon. This is not an esoteric problem." Concerned about the lack of AIAA advice to Congress on questions of scientific policy, Rep. Emilio Q. Daddario (D-Conn.), member of House Committee on Science and Astronautics, said the loudest critics of the space program were scientists in other disciplines eying the space money. "I call on AIAA to come to Congress with a sharpened viewpoint. You can speak with the consensus of 35,000 members."

Avco Corp. vice president Arthur Kantrowitz feared that "AIAA is not equipped" to do the suggested job because unanimity within the society could not be achieved on policy questions. And unless it could, the assignment "would induce strains within AIAA and change its complexion." (AIAA *Daily*, 12/2/66, 1, 29)

Four major AIAA awards were made at Honors Convocation: Dr. W. Randolph Lovelace II (posthumously), the Louis W. Hill Award "to honor his many outstanding contributions to the biomedical sciences in their applications to space and in particular his pioneering development of protective devices enabling the human body to function and operate safely in the aerospace environment"; Joe H. Engle of NASA FRC, the Lawrence Sperry Award for "outstanding contribution to aeronautics as the youngest pilot to fly the X-15 in various space-probing research programs since June 1963"; Warren J. North of MSC, the De Florez Training Award for "contribution to astronaut training and operational crew procedures development in support of the manned space flight program"; and Dr. A. K. Oppenheim, the G. Edward Pendray Award for "outstanding contributions to the field of aeronautical and astronautical literature in the area of gaseous detonation and gas wave dynamics." (AIAA *Daily*, 11/29/66, 30; *Av. Wk.*, 11/28/66, 19)

November 30: NASA launched two-stage, 30-ft.-long Nike-Tomahawk sounding rocket from Wallops Station in cosmic ray and ionospheric experiment for Univ. of New Hampshire. 120-lb. payload reached altitude of 223 mi. (359 km.) and impacted 142 mi. downrange in Atlantic Ocean. Measurements obtained were radioed back during flight and recorded at ground receiving stations. (Wallops Release 66-56; NASA Rpt. SRL)

- NASA Aerobee 150 launched from WSMR reached 136-mi. (219-km.) altitude in NASA-Univ. of Minnesota experiment to measure atmospheric temperature and composition from 62- to 124-mi. (100- to 200-km.) altitude. Rocket and instrumentation performed satisfactorily. (NASA Rpt. SRL)

- Charles W. Mathews, MSC Gemini Program Manager, was named Director of Saturn-Apollo Applications in NASA Hq. Office of Manned Space Flight. He would replace M/G David M. Jones (USAF), Deputy Associate Administrator for Manned Space Flight Programs, who had been Acting Director. (NASA Release 66-310)

- First AEC-NASA Tech Brief was published, launching joint program for announcing to U.S. industry useful technical innovations resulting from their research programs. Tech Briefs would be issued in five categories: electrical and electronic; energy sources; materials and chemistry; life sciences; and mechanical. (NASA Release 66-307)

- National Science Foundation and National Foundation for the Advancement of Arts and Humanities urged in comments filed jointly with FCC that satellite system be devised that would "permit the distribution of high-quality programs in the arts, humanities, and sciences to remote areas in the Nation where educational and cultural opportunities are often minimal." Comments were filed in response to FCC request for suggestions on establishment of comsat facilities by organizations other than Government and communications companies. (AP, *Wash. Post*, 12/1/66, A18)

During November: Two tests important to launch of NASA's first manned Apollo mission had been successfully completed at MSC. First was six-day manned Apollo systems test of spacecraft 008 which was conducted in Space Environmental Simulation Laboratory chamber. Second was entry-egress training in Gulf of Mexico for AS-204 prime and backup crews. (MSC *Roundup*, 11/11/66, 8; NAA S&ID *Skywriter*, 11/11/66, 1, 2)

- ComSatCorp conducted test transmissions via INTELSAT II-A ("LANI BIRD") comsat, launched by NASA Oct. 27, which included: first live color programs between U.S. mainland and Hawaii, featuring Notre Dame-Michigan State, Army-Navy, and Green Bay Packers-Chicago Bears football games; first live news and public affairs programs between U.S. and Japan; and first live telecast between Great Britain and Australia. Comsat had failed to achieve desired synchronous orbit Oct. 30, but its communications system was functioning well and could be used for short periods of time when in line-of-sight of ground stations. Orbit would be adjusted in early December to maximize use of comsat for transmissions between ComSatCorp earth stations. (ComSatCorp Release; Reuters, *Wash. Post*, 11/26/66, A4; *Wash. Eve. Star*, 11/25/66, A8; AP, *Wash. Post*, 11/28/66, A1)

- Management organization to direct Apollo Telescope Mount (ATM) project at MSFC was announced by MSFC Director Dr. Wernher von Braun:

Leland Belew, Program Manager; Reine Ise, Project Manager; William Keathley, Project Experiments Manager; E. B. Craig, Contracting Officer; and Dr. Ernst Stuhlinger, Project Scientist. MSFC's Astrionics Lab., headed by Dr. Walter Haeussermann, Director, and William Horton, Assistant Director, would be responsible for overall technical aspects of Atm development. (MSFC Release 66-270)

During November: Two experiments developed at ARC and flown on Nov. 11-15 GEMINI XII mission were highly successful. Five runs with hand-held sextant proved it could deliver required accuracy of 10 sec. of arc (1/360th of one degree of a circle) in actual space flight. In experiment to study effects of weightlessness on living cells using newly fertilized frog eggs, astronauts arrested cell growth at programed intervals during mission. Data were being analyzed. (ARC Releases 66-15, 66-16; ARC *Astrogram*, 11/23/66, 1, 4)

- NASA selected Westinghouse Electric Corp. for an $8-million contract to install semiautomatic systems in tracking stations to speed compilation of satellite data. The data acquisition system would be used in NASA's Applications Technology Satellite (Ats) program; it would make possible reduction of data within 90 days of satellite launch. Systems would be installed at Rosman, N.C., Goldstone facility, and Toowoomba, Australia, tracking stations. (*WSJ*, 12/1/66, 8)

- NASA Hq. OSSA issued report for NRC Space Science Board describing past achievements and outlining plans for future missions. (Text)

- Boeing Co. awarded subcontracts totaling more than $25 million to Lockheed Aircraft Corp., General Precision Equipment Corp., Litton Industries, Inc., North American Aviation, Inc., General Telephone & Electronics Corp., and Universal Match Corp. for help in developing a short-range attack missile (Sram). Boeing had received $142.3-million USAF contract Oct. 31. (*WSJ*, 11/8/66, 13)

- Human voice was successfully relayed via DOD's Initial Defense Communications Satellite Project (Idcsp) from terminal in Nhatrang, South Vietnam, to Camp Roberts, Calif. U.S. Ambassador to South Vietnam would probably use military system to talk to President Johnson in times of crisis. (Wilson, *Wash. Post*, 11/10/66, M7)

- French Minister of Research Alain Peyrefitte, commenting on the 1967 space budget, disclosed that France planned to launch two D-1-type satellites from Hammaguir Range in February 1967 and one D-2-type satellite from Guiana Range in 1969, *Technology Week* reported. He said a Super Diamant booster was under development to launch French-built satellites from Guiana, but stressed that main French effort was centered on continuing development of ELDO booster. (*Tech. Wk.*, 11/14/66, 3)

- Joint NAS-NAE Committee had been established to provide continuing advisory services in science and engineering to ESSA. Verner Suomi, professor of meteorology at Univ. of Wisconsin, had been named chairman. (NAS-NRC-NAE *News Report*, 11/66, 3)

- Biologist Dr. Colin S. Pittendrigh, Dean of Princeton Univ. Graduate School, said in *Astronautics & Aeronautics* that biological exploration of the planets would be "a milestone in modern science," but such an undertaking, he cautioned, would first require scientific readiness and organization. Without adequate preparation, outright commitment now would risk the political consequences that would attend failure of a major, costly project. "Could commitment now to the opportunities in mid-1970s fail to freeze an inadequately formulated strategy? Could it avoid outright assignment of responsibility and the authority for decision on design and tactical detail to one of several *single* laboratories, no one of which can marshal the necessary scientific talent to ensure a competent project? . . ." The task immediately confronting those of us anxious to see Voyager go in the 1970s is thus more organizational than scientific. . . ." (*A&A*, 11/66, 76-89)

During November: Department of Transportation would be the means and choice to begin a new era in transportation planning, Alan S. Boyd, Under Secretary of Commerce for Transportation, wrote in *Astronautics & Aeronautics.* ". . . transportation policy-making today necessitates a resolving of three different elements: The methodology for planning and programming; the organization to parallel such methods; and the leadership to carry them out. These three ingredients have never been combined in an effective manner. . . . If the nation takes this opportunity, benefits to the public and to the users of transportation will be tremendous. . . ." (*A&A*, 11/66, 112-116)

- DOD was successfully using SYNCOM II comsat 12 hrs. daily and SYNCOM III comsat 22 hrs. daily for military communications across the Pacific, *Technology Week* reported. DOD had assumed control of both satellites —launched by NASA July 26, 1963, and Aug. 19, 1964, respectively— July 8, 1965. (*Tech. Wk.*, 11/21/66, 9)

- "The outstanding achievements of air transportation and associated social and economic gains have been paced in the main by the continued development of the gas-turbine propulsion system," LRC Director Dr. Abe Silverstein wrote in *Astronautics & Aeronautics.* "Despite the outstanding developments of the past 20 years, however, there still remain large gains in performance and operational capability to be achieved." Potential for major advances, Silverstein continued, existed in several areas: noise reduction, engine cooling, engine-inlet matching, development of improved materials, determination of engine temperature values, and use of liquid-methane fuel. (*A&A*, 11/66, 96-104)

- Trendex poll indicated favorable attitude toward NASA's Project Apollo had increased 13% over past three years; negative attitude had decreased 40%. Sentiment to speed up program was greater than to slow down, but majority of respondents were content with present progress. While there was evidence of some increase in desire to reduce expenditures on space program, this attitude was in line with public's increasing concern with governmental expenditures of all kinds because of the defense buildup. Public appeared receptive to news and information of further space programs and plans. (Text)

- "In the last analysis, the only reason for having a space program is to satisfy the needs of society," editorialized AIAA President Dr. Raymond L. Bisplinghoff in *Astronautics & Aeronautics.* "Unless the space program serves people in a useful way, it will be transitory and fade into history as a curiosity. It will not even be enough for the program to serve people by measuring interesting scientific data. It must eventually serve the masses of people of the Earth, if it is to command substantial resources." Communications, geodetic, navigation, and meteorological satellites, and other satellites designed to study and manage the world's natural resources could, in Bisplinghoff's estimation, contribute to meeting man's needs, but international cooperation would be essential. (*A&A*, 11/66, 54-55)

During November: Viewing space program in "historical perspective," Solomon W. Golomb wrote in *Astronautics & Aeronautics:* "The Manhattan Project established that we could run a program for three years and two billion dollars. This was the size limit for major technological projects until Apollo escalated the level to ten years and thirty billion dollars. The next major goal will be the establishment of permanent colonies on other planets. This may mean a commitment of 50 to 90 years and a price tag measured in Terabucks." (*A&A*, 11/66, 14)

December 1966

December 1: Deadline for outlining major national space objectives as requested by House Science and Astronautics Committee's NASA Oversight Subcommittee on Aug. 9. NASA told subcommittee chairman Rep. Olin E. Teague (D-Tex.) that "any detailed report on future goals must be made in light of our fiscal 1967 operating and the President's fiscal 1968 budget request. Since these budgetary considerations are not resolved now . . . any elaboration on future goals cannot be made. . . ." Representative Teague agreed NASA was powerless: "Until the Bureau of the Budget and the President make a decision, there isn't much NASA can do. . . ." (NASA Hq. PAO; *SBD*, 12/2/66, 159; *Av. Wk.*, 12/5/66, 22)

- Successful 384-sec. captive firing of first flight model of Saturn V 2nd stage (S-II) was conducted at MTF by North American Aviation, Inc., prime contractor. State was powered by five Rocketdyne J-2 liquid-hydrogen-fueled engines capable of developing a total of 1 million lbs. thrust. (MSFC Release 66-290; UPI, *Wash. Post*, 12/1/66, A4; AP, *Wash. Eve. Star*, 12/1/66)

- Problems of construction industry were discussed by William Rieke, NASA Assistant Administrator for Industry Affairs, at Construction Industry Joint Conference in Washington, D.C. Rieke praised accomplishments such as KSC's Vertical Assembly Building but urged management to increase its effectiveness, labor to expand apprentice programs and control rate escalation, and Government to examine labor policies and maintain knowledgeable labor relations staffs. (Text)

- McDonnell Aircraft Corp. study prepared for FAA concluded that new family of 100-passenger V/Stol aircraft operating from small airports close to downtown city areas could offer both greater convenience and reduced ground travel time between city and airport than the present air travel system. (FAA Release 66-102)

December 1-6: LUNAR ORBITER II continued transmitting photos of the lunar surface [see Nov. 10-30]. On Dec. 6—one day early—spacecraft's photography readout terminated when high-power transmitter failed. Breakdown resulted in loss of three medium-resolution and two high-resolution photos of primary site I, but full coverage of site had been provided by medium-resolution photography readout. Environmental, engineering, and selenodesy tracking data continued to be received through low-power transmission. (NASA Proj. Off.)

December 2: NASA Aerobee 150 sounding rocket was launched from WSMR to 142-mi. (229-km.) altitude to measure atmospheric composition and temperature at 62- to 124-mi. (100- to 200-km.) altitude range. (NASA Rpt. SRL)

- NASA announced sponsorship during 1967 of seven specialized summer programs of research and study for about 245 college and university faculty members in engineering and science. Developed by NASA and

American Society for Engineering Education, programs would combine space research at NASA centers and parallel seminar-type studies at adjacent universities. Fellowships would be available for about 225 faculty members. Participating were: (1) Univ. of Alabama, Auburn Univ., and MSFC; (2) Case Institute of Technology and LRC; (3) Univ. of Houston, Texas A&M Univ., and MSC; (4) Univ. of Maryland, Catholic Univ., and GSFC; (5) Virginia Associated Research Center (Univ. of Virginia, Virginia Polytechnic Institute, and College of William and Mary) and LaRC. Also, NASA would sponsor five-week summer faculty program in space physics at Yeshiva Univ. in cooperation with Goddard Institute of Space Studies. (NASA Release 66-305)

December 2: The *Washington Post* commented on criticism of NASA spending: "What an uproar there was in 1957 when Sputnik showed us we were second in space! The Eisenhower Administration was roundly condemned. Congressmen and Senators clamored for a stepped-up space program and promised all the money needed. President Kennedy, in his 1960 campaign . . . said repeatedly: 'I am not satisfied to be second in space.'

"Well, we are no longer second in space. And the critics of the Johnson Administration are now not satisfied to be first in space— unless it can be done cheap. As a result of the space program we are not only going to be first on the moon and first in space. We are going to be first in a thousand fields of science, education, and industry.

"This country does not have to choose between a space program that will keep it in first place and a poverty program that will keep it first . . . in national exertion for the disadvantaged. It can be and ought to be first in both." (*Wash. Post*, 12/2/66, A16)

- First program to provide pilots with current satellite photos adapted for their use in flight planning began at Kennedy International Airport under ESSA direction. Photos transmitted by ESSA II and III and NIMBUS II were composited by computer into a single picture showing cloud patterns over eastern North America and the North Atlantic and analyzed by meteorologists. (ESSA Release)

- USAF awarded contracts totaling over $6 million to Lockheed-Georgia Co. and North American Aviation, Inc., to advance Vtol aircraft development. NAA received $5,671,000 to develop and demonstrate a Vtol integrated flight system; Lockheed received $975,000 to modify XV-4A "Hummingbird" to new system with direct lift and diverted thrust engines. (AFSC Release 288.66)

- AEC Chairman Dr. Glenn T. Seaborg believed man could modify weather in this century through a global center operating a vast satellite system, AP reported. With international cooperation, he said, a worldwide forecasting system could become operational in 20 yrs.; later it might be possible to actually control the weather. (AP, *NYT*, 12/2/66)

December 3: ComSatCorp's INTELSAT II-A ("LANI BIRD") comsat, launched by NASA from ETR Oct. 27, began commercial service under temporary FCC authorization which would expire Feb. 2, 1967. Satellite, which had failed to achieve planned synchronous equatorial orbit Oct. 30, had been moved into modified 12-hr. orbit Dec. 1, to provide maximum visibility: eight to nine hours daily between U.S. mainland and Hawaii; over seven hours between U.S. and Japan; and about four hours between U.S. and Carnarvon, Australia. Authorized communications carriers could use satellite for television and telephone communications about eight hours daily. (ComSatCorp Release; *Wash. Post*, 12/2/66, F1; AP, *WSJ*, 12/5/66, 15; Wash. *Eve. Star*, 12/3/66, A3)

December 3: U.S.S.R. launched COSMOS CXXXIV into earth orbit with 319-km. (197-mi.) apogee; 214-km. (133-mi.) perigee; 89.6-min. period; and 65° inclination. Equipment "for continuation of the exploration of outer space" was functioning normally. (Tass, 12/5/66)

• AEC conducted 350-ton nuclear detonation for DOD near Hattiesburg, Miss., in underground cavity formed by previous nuclear explosion. Purpose of test was to determine extent of decoupling—a reduction of ground shock and other seismic signals. (AEC Release J-273)

December 4: Secretary of the Air Force Dr. Harold Brown had indicaten that future U.S. military exploration of space would concentrate od existing and proved technologies, Dr. I. M. Levitt, director of Fels Planetarium, reported in the *Philadelphia Inquirer*. Dr. Brown had isolated three areas of vital interest to the military: (1) electronic systems which would ensure long life and reliability in orbit; (2) reusable materials for spacecraft, launch vehicles, and propulsion systems; and (3) space electrical power systems which would provide better power-to-weight ratios at reduced cost per kw for long duration missions. (Levitt, *Phil. Inq.*, 12/4/66, 5)

December 4-6: At 43rd annual convention, National League of Cities meeting in Las Vegas adopted resolution challenging priority given to U.S. space program and the Vietnam war and urged President Johnson to establish "a national rebirth of the American city." Detroit Mayor Jerome P. Cavanagh argued that if funds for low-income housing, poverty, and education were restored by cutting NASA's appropriation, space programs would be delayed only a few weeks. Budget requests for housing and urban development, he charged, had been cut 17 times as much as the NASA budget. (Janson, *NYT*, 12/5/66, 1; *Wash. Post*, 12/6/66, A8)

NASA Administrator James E. Webb responded to accusations that he was more concerned with life on Mars than with life in Chicago or Los Angeles: "Neither I nor any other leader in the space program has ever suggested that the space program should have any priority over the needs of the American City...." He agreed that cities needed more Federal funds, but suggested that they first offer greater assurance that their programs would work well. Benefits in medical, industrial, and other fields added dimensions to space program far beyond the value of the lunar-landing program alone, he said. (Transcript; Janson, *NYT*, 12/6/66, C34)

December 5: Dr. George E. Mueller, NASA Associate Administrator for Manned Space Flight, commented on national impact of U.S. space program in a speech before ARCS (Achievement Rewards for College Scientists) Foundation in Los Angeles: "Space exploration in its broadest meaning and in all of its ramifications has become a powerful force ... socially, economically, politically, and even morally.... It is the creator of new technologies, new techniques, and new methods of management. It has great significance for our national security. It is a stimulus for our economic and national growth. It is a catalyst to the achievement of the goals of our society. And it provides us with the dimensions of a great challenge—to explore space for the benefit of all mankind." (Text)

• Project Makai—program to construct manned laboratories at various underwater depths near Honolulu to increase man's ability to work productively under the sea—was described in *New York Times* interview by Wayne Collins of Oceanic Institute, which was cosponsoring project with several industries. First phase was $3-million, 3-yr. effort to establish two test ranges at 70 and 200 ft. below sea level. Each range would consist of multilevel habitat and laboratory linked to shore by an "umbilical cord" and capable of housing four to six men for at least one month. Work environments under consideration for aquanauts included a simulated oil field, an experimental mining area, an underwater "weather" station, and instrument and material testing sites. (Brody, *NYT*, 12/5/66, 56)

December 5: AFSC would offer research opportunities in 17 scientific disciplines through National Research Council postdoctoral resident research associateships during 1967 and 1968. Associateships would be supported by ten separate AFSC laboratories and research centers located in Ohio, New Mexico, Texas, New York, and Alaska. (AFSC Release 249.66)

- USAF launched unidentified satellite with Atlas-Agena D booster from WTR. (*U.S. Aeron. & Space Act.*, 1966, 159)

December 6: NASA released summary of report on observatory-class earth satellites. Recommendations included: (1) redesign of outmoded electronic equipment that had performed poorly in space; (2) rigorous preflight testing with prototype models under simulated space conditions; (3) stronger management of observatory program within NASA; and (4) increased surveillance of performance at contractors' plants. Report was compiled by five-man board appointed April 21 by NASA Associate Administrator for Space Science and Applications Dr. Homer E. Newell and chaired by his Deputy for Engineering Robert F. Garbarini. It studied OAO I mission, which failed two days after April 8 launch because battery overheated, and reviewed Orbiting Solar Observatory (Oso), Orbiting Geophysical Observatory (Ogo), and Nimbus meteorological satellite project practices. All programs were managed by GSFC. (NASA Release 66-313; UPI, *NYT*, 12/8/66, 53)

- U.S. must continue to press for advanced technology to ensure "the fundamental strength of our society for all purposes," Karl Harr, president of Aerospace Industries Assn., told Town Hall in Los Angeles. "Because technological advance is largely indivisible, a nation either opts for technology or it doesn't. As some sister nations have learned, you either move with the tide of technological advance or you fall behind on all fronts. . . .

  "Today the question becomes the same for us. It is not should we as a nation have space or other things, but rather can we have the 'other things' if we reject the technological reach required by space. . . ." (Text)

- USAF had successfully penetrated ion-sheath blackout phase of an Athena missile's reentry in test at WSMR, AP reported. (AP, *Wash. Eve. Star*, 12/7/66)

- All F-104G Starfighter aircraft in West Germany's Air Force were grounded indefinitely as result of Nov. 28 air crash which killed pilot. Since 1962, 65 of Luftwaffe's 700 Starfighters had crashed, killing 37 pilots. Aircraft was manufactured by Lockheed Aircraft Corp. (Shabecoff, *NYT*, 12/7/66, 15)

December 6-7: NASA's ATS I (Ats-B), first satellite in Applications Technology Satellite (Ats) program, was successfully launched from ETR by Atlas-

Agena D booster into elliptical transfer orbit with 22,831-mi. (36,758-km.) apogee; 114-mi. (183-km.) perigee; 650-min. period; and 31° inclination. At 16:30 GET, apogee motor was fired to place satellite into near-synchronous 22,300-mi.-altitude orbit. It would be allowed to drift 12 days to stationary position over the Pacific at 151° west longitude.

One of the most versatile satellites ever developed, 775-lb. ATS I carried 15 communications, technology, and scientific experiments, including camera systems for cloud-cover photos; two-way VHF transponder for voice communications between ground stations and aircraft in flight; electronically despun antenna to continuously direct a radio beam toward earth; and low-thrust resistojet which produced thrust of 450 millionths of a pound. Capable of multiple-access communications, satellite would be able to transmit TV and voice communications among stations in North America, Asia, and Australia. It would also be used to send weather data from ESSA facility at Suitland, Md., to Automatic Picture Transmission (Apt) stations in U.S., Japan, and Australia. Mission objectives were to inject spacecraft into synchronous orbit and place it on station; operate spacecraft for at least 30 days; and conduct experiments. Ats program—a five-satellite project managed by GSFC under OSSA direction—sought to investigate technology common to various spacecraft applications through flight experiments carried on spin-stabilized and gravity-gradient-stabilized spacecraft. (NASA Proj. Off.; NASA Release 66-308; *Marshall Star*, 12/7/66, 1, 4; *Wash. Post*, 12/8/66, A2; UPI, *NYT*, 12/9/66, 57; UPI, *Wash. Post*, 12/21/66, A15)

December 7: U.S. confusion and concern over U.S.S.R.'s space objectives had increased because of the "long gap" in Soviet manned space flights and three recent Soviet launches "surrounded by unusual secrecy," Evert Clark reported in the *New York Times*. Two of the spacecraft, secretly launched Sept. 17 and Nov. 2, had exploded in orbit; the third, Nov. 28, had been announced as COSMOS CXXXIII but orbital inclination usually given in Cosmos launch reports was not disclosed. Clark warned that U.S. experts might be "underestimating the Soviet Union's commitment to space exploration" and suggested U.S.S.R. was either: (1) preparing to secretly launch manned spacecraft with three- to six-man crew in early 1967; (2) conducting tests to precede new missions to Mars and Venus; (3) changing its emphasis from manned lunar exploration to manned earth-orbiting stations; or (4) trying to provide maximum coverage of U.S. by launching additional reconnaissance satellites. (Clark, *NYT*, 12/8/66, 1)

- FCC ruled that ComSatCorp must share ownership of its ground stations with U.S.-owned international common carriers. Effective until late 1969, decision—which gave ComSatCorp controlling interest in existing and future ground stations—modified FCC's May 12, 1965, interim ruling awarding ComSatCorp "sole responsibility" for stations at least during "critical early years." (UPI, *NYT*, 12/8/66, 52; *WSJ*, 12/8/66, 3)

- AFSC awarded $2.3-million, fixed-price-incentive-fee contract to Douglas Aircraft Co., Inc., for design, development, and fabrication of Titan III launch vehicle components. (DOD Release 1029-66)

December 8: President Johnson announced U.S. and U.S.S.R. had reached agreement on space law treaty under consideration by U.N. Committee on the Peaceful Uses of Outer Space since May 7. Calling agreement "the most important arms control development since the Limited Test Ban Treaty of 1963," the President said he would seek fast Senate approval so U.S. would be first nation to ratify it. Primary provisions: (1) moon and other celestial bodies should be free for exploration, use, and scientific research to all countries and international organizations exclusively for peaceful purposes; (2) no country should be permitted a claim of sovereignty or appropriation; (3) no country should be permitted to place weapons of mass destruction, establish military bases, or conduct military activities on moon and celestial bodies; (4) all outer space activities would be subject to international law and should be reported to U.N. Secretary General; (5) cooperative efforts should be made to avoid harmful contamination, conduct scientific research, and assist all astronauts in need of aid; and (6) all stations, installations, and space vehicles on moon and other celestial bodies should be open to other countries on reciprocity basis. Use of ICBM's, reconnaissance satellites, and USAF's Manned Orbiting Laboratory (Mol) were not affected.

Treaty would be considered by U.N. General Assembly Dec. 19. (Frankel, *NYT*, 12/9/66, 1, 18C; Gwertzman, Wash. *Eve. Star*, 12/9/66, A1; *Pres. Doc.*, 12/12/66, 1781-2; Roberts, *Wash. Post*, 12/9/66, A1, A8, A10; *NYT*, 12/20/66, 6)

December 8: LUNAR ORBITER II executed plane change—possibly the first ever accomplished by satellite in lunar orbit—to obtain information about moon's gravity field and gain experience in flight operations at high inclinations. (NASA Proj. Off.; NASA Release 66-316)

December 9: NASA's ATS I satellite took first U.S. high-quality photos of the earth from synchronous orbit altitudes. Photos showed changing cloud-cover pattern over 40 per cent of the earth's surface. (NASA Proj. Off.)

- NASA's second Project Scanner instrument package was successfully launched by three-stage solid propellant Trailblazer booster from NASA Wallops Station on suborbital trajectory to gather information on earth's horizon. Two-channel radiometers measured infrared energy emitted from earth's horizon by carbon dioxide and water vapor; star-mapper telescope provided data on attitude of spin-stabilized spacecraft. Project Scanner was a phase of horizon definition research conducted by LaRC. (Wallops Release 66-58)

- Complexities of modern society required a new concept of management, NASA Administrator James E. Webb told graduation class of Harvard Univ.'s Graduate School of Business Administration. "The kind of challenges that we in management are facing today ... call for new and experimental approaches to organization. One ... is the question of the chief executive function. In traditional thinking, the structure of an organization peaked in the chief executive. ... However, as organizations have become more complex and their challenges more interdisciplinary ... there has been an increasing tendency to experiment with the idea of the multiple executive. ... We saw this kind of need at the very beginning of NASA's history. We evolved, therefore, a partnership arrangement which included Dr. Hugh Dryden, Dr. Robert Seamans, and myself. We all had many common ideas, and yet each brought to our work on the critical decisions affecting the nation's space effort certain specialized experience. To do it any other way would have deprived us of the kind of mutual support and broadly-based leadership that I think we achieved." (Text)

December 9: French President Charles de Gaulle and Soviet Premier Aleksey Kosygin signed extension of scientific, technical, commercial, and cultural cooperation between U.S.S.R. and France provided for by June 30 agreement. Declaration followed eight-day visit to France by Kosygin who had toured the Center for Nuclear Studies and Air Liquide laboratory and inspected prototype of the Concorde supersonic aircraft. (Reuters, *NYT*, 12/10/66, 16; Nossiter, *Wash. Post*, 12/8/66, A32; Mooney, *NYT*, 12/6/66, C10)

- Report on national document-handling systems in science and technology prepared by Federal Council for Science and Technology's Committee on Scientific and Technical Information (COSATI) was summarized in *Science* by Launor F. Carter, vice president of System Development Corp.—a participant in the study. Report had recommended that Office of Science and Technology: (1) accelerate its efforts to organize an integrated national network of information and document-handling systems in science and technology; (2) collaborate with other Federal agencies and private organizations to develop comprehensive, coordinated program and sponsor effective legislation; (3) encourage private sector to formulate plans for its consideration; and (4) encourage Federal support of experiments in information technology, including prototype information systems. (Carter, *Science*, 12/9/66, 1299-1304)

December 10: Editorial comment on U.S.-U.S.S.R. agreement on space law treaty announced Dec. 8.

New York Times: "No one has yet demonstrated that there would be any military advantage to placing bombs either in orbit or on celestial bodies. But there is a clear need both for space law and for reducing the wasteful competition in space exploration. The ability of Moscow and Washington to reach agreement . . . offers hope of further East-West cooperation on space problems." (*NYT*, 12/10/66, 36C)

Washington Daily News: ". . . reason and peacefulness have prevailed. And we will all benefit. There will be no military bases or forts, no testing of weapons and tactics—nor their threat or expense. And no claims to territory in space, and thus no arguments and battles over conflicting claims." (*Wash. Daily News*, 12/10/66)

Washington Post: ". . . the treaty can be applauded for the good that it has achieved as of today; and it can be praised for the larger good that it will do tomorrow when the practical aspects of space relations have become infinitely more pressing. The world will surely have occasion to look back to this day as one that set the nations on the right path and on which the great powers made a wise decision to shun the military exploitation of celestial bodies and of outer space for narrow, nationalistic purposes." (*Wash. Post*, 12/10/66)

- NASA successfully completed third rocket-launched test [see Nov. 5 and 21] of LaRC-managed parachute landing system for Voyager spacecraft. Honest John-Nike rocket launched from WSMR deployed 30-ft.-dia. disc gap band parachute at approximately 130,000-ft. altitude. Voyager program was managed by JPL. (NASA Proj. Off.)

December 11: USAF successfully launched OV1-IX and OV1-X research satellites pickaback from Vandenberg AFB with Atlas D booster. Each 220-lb. spacecraft carried 12 radiation-measuring experiments. (AP, *Wash. Eve. Star*, 12/12/66, A1; *Tech. Wk.*, 12/19/66, 13)

- Ford Foundation President McGeorge Bundy announced $10-million grant to show potential of educational television through a series of national programs. Series would be carried by Nation's 125 noncommercial educational TV stations joined by coaxial cable and microwave relay for the demonstration. In a brief to be filed with FCC, Bundy defended Foundation's Aug. 1 request for formation of nonprofit nationwide satellite system that would carry extensive schedule of educational programs financed by transmission of commercial programs. (*NYT*, 12/12/66, 1, 95; *Wash. Post*, 12/12/66, A1, A4; AP, *Wash. Eve. Star*, 12/12/66, A3)

December 11: Photograph of earth from ATS I, in 22,300-mi. synchronous orbit. This is one of seven pictures which together show changing cloud pattern over the earth for an entire day. ATS I began its photography Dec. 9.

December 11: Aerospace rather than shipbuilding companies were competing for $1-billion Fast Deployment Logistic Fleet (Fdl) program because of USN's new contract policy, Robert Wright reported in the *New York Times*. Under new policy, firms were required to bid for total package—concept, design, and construction—instead of construction only. Vernon A. Johnson, a Lockheed Aircraft Corp. vice president, believed Lockheed, General Dynamics Corp., and Litton Industries, Inc., were competing exclusively for contract because shipbuilding industry lacked resources and experience for total-package systems design. Goal of Fdl program was fleet of 30-40 ships capable of remaining at sea for indefinite periods and making fast delivery of heavy supplies to fight small wars. (Wright, *NYT*, 12/11/66, 12)

December 12: XB-70 No. 1 experimental research aircraft, piloted by NASA test pilot Fitzhugh Fulton and North American Aviation, Inc., pilot Van Shepard reached mach 2.52 and 60,000-ft. altitude during national sonic boom program flight. (NASA Proj. Off.)
- Howard H. Haglund, formerly Deputy Project Manager and Spacecraft System Manager for Surveyor at JPL, had been named Manager of JPL's Surveyor Project Office. He succeeded Robert J. Parks, who returned to his former position of JPL Assistant Director for Lunar and Planetary Projects. (JPL Release 426)
- Efficient management in both industry and Government was the key to a continuing successful space effort, Dr. Edward C. Welsh, NASC Executive Secretary, told the Management Club at Patrick AFB. "The space challenge is unique. To face up to its imperatives calls for new ways of doing things. The pacing item is as much managerial competence as it is technology or money. . . .

  "Under the stress and pressure of getting the job done, there is imposed a harsh necessity which breeds innovation. So long as purposeful men accept such challenges, this nation will flourish and its economic and political systems will continue to lead the way. On the other hand, should we slip into the path of the familiar and decline to accept the challenge of the new, this nation will suffer a grievous loss and so will the peoples throughout the world. . . ." (Text)
- MSFC Director Dr. Wernher von Braun discussed U.S. space objectives and accomplishments in *U.S. News & World Report* interview. Immediate objective, he explained, was "to land an American on the moon in this decade and bring him back alive"; far-reaching goal was "to develop a broad, national spaceflying capability." He described Gemini program as "a smashing 100 per cent success. . . . The purpose of Gemini was to learn the kind of things you can't learn on the ground . . . and we did make some surprise discoveries, for which we are more than grateful. . . ." Dr. von Braun said there was a "distinct possibility" U.S. might land a man on the moon in 1968. (*U.S. News*, 12/12/66, 62-7)
- U.S.S.R. launched COSMOS CXXXV into earth orbit with 662-km. (411-mi.) apogee, 259-km. (161-mi.) perigee, and 48.5° inclination. (*U.N. Public Registry*, 1/9/67)
- New Soviet launch site near Archangel was discovered by British schoolboys after they had tracked several Cosmos satellites and fed information, including data on orbital intersections, into a computer. U.S. later confirmed U.S.S.R. had been launching satellites from large military base south of Archangel. Only two Soviet launch sites—Baikonur and Kapustin Yar—had previously been publicized. Schoolboys' information, which appeared in *Flight International*, represented important break in secrecy surrounding Soviet reconnaissance satellites. (*Wash. Post*, 12/13/66; 12/19/66, A1, A8; *NYT*, 12/21/66, 25; *Time*, 12/30/66)

December 13: Versatility of ATS I satellite, launched by NASA from ETR Dec. 6, was demonstrated in program at GSFC. In first test, satellite relayed color TV show from Goldstone, Calif., to GSFC via Rosman, N.C., ground station. In second test, voice signals transmitted from GSFC to Goldstone and beamed by dish antenna to satellite were successfully relayed to three aircraft in flight, demonstrating publicly for the first time that 50,000-mi. earth-to-space circuit could be used with existing airline radio equipment in VHF range. Concurrent with these tests, satellite transmitted high-quality photos of weather over eastern Asia, Pacific Ocean, and western U.S. It had begun relaying photos Dec. 9. NASA Deputy Administrator Dr. Robert C. Seamans, Jr., told the press ATS I was "a major step forward" toward practical satellites and would have "an impact on everyone in the world." (Clark, *NYT*, 12/14/66, 38; Hines, Wash. *Eve. Star*, 12/14/66, A4; Reistrup, *Wash. Post*, 12/14/66, A12)

December 13: NASA had eliminated final three Surveyors because of good results from Ranger series, SURVEYORS I and II, and LUNAR ORBITERS I and II. It had been determined that five remaining Surveyor spacecraft would provide necessary additional information about the moon to support Apollo manned lunar landing mission. (NASA Release 66-318)

- Aerobee 150 sounding rocket launched by NASA and Brazilian Space Commission (CNAE) from mobile facility at Natal, Brazil, carried 225-lb. payload of four x-ray detectors on ballistic trajectory to 120-mi. (193-km.) altitude. Payload, which impacted 66 mi. downrange in the Atlantic, telemetered scientific data to ground stations. First NASA x-ray astronomy experiment on a sounding rocket in the Southern Hemisphere, flight searched for specific galactic x-ray sources, observed Magellanic clouds for x-ray emissions, and obtained spectral information on known x-ray sources. (NASA Release 66-311)

- Sen. Clinton P. Anderson (D-N.Mex.), Chairman of the Senate Aeronautical and Space Sciences Committee, told AP he saw "no urgency" in landing on the moon first. Discounting the theory that a manned lunar landing would have military significance, Senator Anderson said "had the United States and the Soviet Union cooperated from the start in space exploration, more knowledge at less expense might have resulted." (AP, *NYT*, 12/15/66, 34)

- Informed sources said U.S.S.R. had cut its nonmilitary space budget by one third and increased general military spending, according to AP. Report could not be confirmed. (AP, *NYT*, 12/14/66, 3; AP, Wash. *Eve. Star*, 12/14/66, A7)

- There might be a connection between period of inactivity in Soviet manned space flight and Soviet efforts to create an effective antimissile system, speculated Crosby S. Noyes in the Washington *Evening Star*. ". . . The Russians, in assessing the present state of the world, [might] have radically readjusted their priorities in favor of earthly defenses over celestial exploration. . . ." Noyes suggested U.S. might have to choose soon "between the astronomical costs of continuing exploration of space, and those of the best possible antimissile defense system. If the Russians have, in fact, already made their choice, our own dilemma should be easier to resolve. But in any event, the new Russian defense system raises military and political problems which cannot be simply ignored." (Noyes, Wash. *Eve. Star*, 12/13/66, A11)

December 14: U.S.S.R. had successfully completed series of carrier rocket tests in the Pacific 16 days early, Tass reported. Nov. 20 Tass statement had said that tests would continue until Dec. 30. (AP, Wash. *Eve. Star*, 12/14/66, D11)

- USAF launched unidentified satellite with Titan III-B-Agena booster from WTR. (*U.S. Aeron. & Space Act., 1966*, 159)

- NASC Executive Secretary Dr. Edward C. Welsh spoke at the dedication of Boeing Co.'s Industrial Research Center in Huntsville, Ala. "I believe that the next decade will find us marrying more closely the major features of aeronautics and astronautics so that manned spacecraft will take off and return to spaceports with reusable propulsion systems and reusable spacecraft. The near future will find us engaged in the exploration of the lunar surface, the development of manned earth orbiting permanent space stations, unmanned and then possibly manned travel to the planets, as well as in the progressive utilization of space technology in such fields as communications, navigation, meteorology, and observation.

"In addition, I believe the next decade holds great promise for the utilization of space systems technology in the solution of many of the economic and social problems of this complex civilization. . . ." (Text)

December 14: NASA Hq. released LUNAR ORBITER II photo of Ocean of Storms which confirmed moon's history of volcanic activity. Taken Nov. 25 from an oblique angle at 28-mi. altitude, photo showed an array of lunar dunes 2-10 mi. in diameter and about 1,000-1,500 ft. high which resembled volcanic domes of northern California and Oregon. Scientists believed they had been caused by upwelling of molten rock from moon's interior "which warped the overlying rock and in some cases spilled out on the surface as lava." Dr. Lawrence Rowan, U.S. Geological Survey, commented: "It is the most spectacular evidence of volcanism that we've had yet." (O'Toole, *Wash. Post*, 12/15/66, E2; UPI, *NYT*, 12/15/66, 34; *Tech. Wk.*, 12/9/66, 19)

- Skin from three Soviet researchers had been orbited in special sterilized bottles onboard LAMBDA satellite Aug. 19, 1960, and successfully grafted back three days later to test effects of space travel, *Trud* reported. Experiment had convinced Soviet scientists that human organisms would not be dangerously affected by cosmic rays. (Reuters, *Wash. Post*, 12/15/66, E1)

December 14-17: NASA's BIOSATELLITE I (Biosatellite A) carrying more than 10-million tiny living organisms was successfully launched from ETR by two-stage Thrust-Augmented Improved Thor-Delta booster on mission to study effects of weightlessness and space radiation on growth of plants and animals. Orbital parameters: 197-mi. (317-km.) apogee; 191-mi. (308-km.) perigee; 91 min. 16 sec. period; and 33.5° inclination. After 47 orbits, satellite had been scheduled to reenter earth's atmosphere over the Pacific, deploy a parachute, and be recovered aerially for return to NASA laboratories. Retrorocket failed to fire, however, and satellite remained in orbit.

First of three spacecraft in NASA's Biosatellite program, 936.5-lb. BIOSATELLITE I carried 13 biological experiments to study physiological effects of weightlessness at three different levels: (1) growth and form of entire plants and animals; (2) structure and growth of cells and tissues; and (3) basic biochemistry of the cell. Mission also studied effects of radiation on organisms in weightless environment. Biosatellite program was managed by ARC under OSSA direction. (NASA Proj. Off.; NASA Release 66-312; AP, Wash. *Eve. Star*, 12/15/66, A3; AP, *NYT*, 12/18/66, 86; UPI, *NYT*, 12/19/66, 66)

December 15: Vice President Hubert H. Humphrey presented 1966 Harmon International Aviator's Trophy in Washington, D.C., to GEMINI VI Astronauts Walter M. Schirra, Jr., and Thomas P. Stafford and GEMINI VII Astronauts James A. Lovell, Jr., and Frank Borman for achieving first rendezvous in space by two separately launched maneuverable spacecraft Dec. 15, 1965. After the ceremony, the Vice President told newsmen he was one of those "manning the ramparts" against those who failed to see any direct public benefit from space exploration. He thought the space program had resulted in a superior educational system, increased industrial competence, new products, "and if it hasn't done anything else it has taught us to manage" vast undertakings. (Clark, *NYT*, 12/16/66, 21)

December 15: President Johnson's waning interest in space was postponing NASA's decision on a post-Apollo program, William Hines suggested in the Washington *Evening Star*. "NASA . . . is more to be pitied than censured for this silence. The agency has been led down the primrose path by . . . [President Johnson] who suddenly lost interest in the now shopworn ingenue.

"NASA has no plans because . . . [he] has permitted it to make none; aspirations yes, but plans no. . . . " (Hines, Wash. *Eve. Star*, 12/15/66, A23)

- NASA would negotiate $15-million fixed-price incentive contract with Douglas Aircraft Co. for 14 improved 2nd stages for Delta launch vehicle. GSFC would manage contract. (NASA Release 66-321)
- AFSC announced awards to four aerospace companies: Lockheed Missiles and Space Co., $2-million initial increment to a $14-million fixed-price contract for production of Agena launch vehicles; Douglas Aircraft Co., $2.5-million initial increment to a $17.5-million fixed-price letter contract for production of Thor launch vehicles; and Aerojet-General Corp., $2.9-million initial increment to a $54.5-million contract, and TRW Inc., $2.9-million initial increment to a $50-million contract, both for R&D in an unmanned space technology program. (DOD Release 1048-66; *WSJ*, 12/16/66, 4)

December 15-16: ComSatCorp and AT&T filed separate plans with FCC to establish domestic comsat system. AT&T would orbit three synchronous satellites starting in 1969, eventually replacing them with four satellites of advanced design by 1976; under AT&T plan, ComSatCorp would own the satellites, and common carriers, the earth stations. ComSatCorp, in its filing, proposed launching four-satellite system beginning in 1969; more powerful comsats would be launched into synchronous orbit over equator and spaced over Nation's time zones in 1973 and 1978. Both AT&T and ComSatCorp objected to Ford Foundation's Aug. 1 proposal for nonprofit comsat system whose income would be used to support educational TV. (ComSatCorp Release; *NYT*, 12/16/66, 85M; 12/17/66, 59; *Wash. Post*, 12/17/66, C16; Wash. *Eve. Star*, 12/16/66, A24)

- News briefing on Project Apollo was held at MSC. Dr. Joseph Shea, manager of NASA Apollo Spacecraft Office at MSC, said NASA hoped to launch three manned Apollo missions in 1967 involving "essentially every test that needs to get done in order to insure that the lunar operation is proper." He outlined the flights: (1) AS-204—three-man, earth-orbital mission of up to 14 days to check out crew, spacecraft, and ground equipment; (2) AS-205/208—10- to 12-day flight in which Uprated Saturn I boosters would launch an Apollo spacecraft with three-man crew and an unmanned Lunar Module (LM) on successive days for rendezvous mission; and (3) AS-503—full-duration lunar mission rehearsal in earth orbit in which single Saturn V would launch manned Apollo spacecraft and LM. Shea warned against undue optimism about a manned lunar landing in 1968. "It is literally true that there is no planned date at the present time for the lunar operation . . . and I don't see how there can be until we have had hardware in orbit to understand what vestigial problems still remain."

Donald Slayton, MSC Director of Flight Crew Operations, said that third, nonflying crew had been added to each Apollo mission for additional ground support. He also revealed that training for astronauts selected for Apollo lunar missions would be 40 weeks—14 weeks longer than Gemini training course.

Comparing Mercury and Apollo training programs, AS-204 Command Pilot Virgil I. Grissom said: "When we started training for Mercury, we didn't know what to train for. So we did everything. . . . Training for Apollo has been pretty straight forward. We know what we need to know and we know that we have to learn about the spacecraft, how to operate it and get our flight plan and our mission rules squared away. Back early in Mercury . . . we were worried about whether a man could live and survive . . . so we did a lot of things like spinning around in 3 axis machines that we don't feel you really need anymore. . . ." Grissom confirmed that AS-204 mission would carry camera for live television coverage. (Text; Hines, Wash. *Eve. Star*, 12/7/66, A3)

December 16: XB-70 No. 1 experimental research aircraft, piloted by North American Aviation, Inc., test pilot Van Shepard and NASA test pilot Fitzhugh Fulton, reached mach 2.55 and 60,300-ft. altitude in national sonic boom program flight. (NASA Proj. Off.)

- Almost all orbiting West Ford dipoles (launched by USAF May 10, 1963) had reentered the atmosphere during period of several months centering on predicted Jan. 1 reentry date, Irving I. Shapiro wrote in *Science*. Some dipole clusters had remained in orbit, but these, with few exceptions, were expected to return to earth within two years. Experiments performed while dipoles were in orbit had verified calculations that predicted interference with optical and radio astronomical observations would be negligible. (Shapiro, *Science*, 12/16/66)

- *New York Times* correspondent Evert Clark was selected for AAAS-Westinghouse science writing award for reporters on newspapers with 100,000 or more circulation. Clark was cited for nine articles about NASA's SURVEYOR I published between May 31 and July 14. (*Science*, 12/16/66, 1431)

- U.S.S.R. would cooperate with "any country in developing space communications," Deputy Communications Minister Nikolay Talyzin told Tass in a Moscow interview. He claimed that four Soviet Molniya I comsats could ensure multi-channel telephone, telegraph, and phototelegraph links between countries in the Northern Hemisphere such as U.S., Japan, Cuba, and Southeast Asia. (*Wash. Post*, 12/17/66, C16)

- DOD's Advanced Research Projects Agency (ARPA) announced establishment of an Office of Advanced Engineering under Dr. Chiao Jen Wang, former Booz-Allen Applied Research, Inc., executive. (DOD Release 1053.66)

- AEC had selected Weston, Ill., site near Chicago for proposed 200-billion-electron-volt (bev) proton accelerator. Six sites recommended by NAS had been under study since March. (AEC Release J-282)

December 17: Dr. J. Allen Hynek, head of Northwestern Univ.'s Dept. of Astronomy and USAF consultant on Ufo's, writing in *Saturday Evening Post*, offered four possible explanations for the Ufo phenomena: (1)

hoaxes or hallucinations; (2) military weapons being tested in secret; (3) something from outer space; (4) unknown natural phenomena. Of the 15,000 cases that had come to his attention, Hynek considered several hundred "puzzling" and perhaps one in 25 of puzzling incidents "bewildering." Because of these arresting cases, he urged serious study of Ufo's and suggested, as two initial steps to help solve the problem, computerization of data and improved photography of Ufo's. (Hynek, *Sat. Eve. Post*, 12/17/66)

December 17: 63rd anniversary of first powered flights by Orville and Wilbur Wright from Kitty Hawk, N.C.

December 18: USAF awarded Boeing Co. $6-million, 2½-yr. contract to develop automatic flight control system that could double useful lifetime of large, flexible aircraft such as B-52, XB-70, and C-5A by dampening structural oscillations and reducing stresses from wind gusts and maneuvering loads, all of which caused metal fatigue. (AFSC Release 234.66)

- Dr. Jan H. Oort, recipient of Columbia Univ.'s 1966 Vetlesen Award for his work in earth sciences, told *Washington Post* that Univ. of Leiden Observatory, which he directed, was building world's largest radio-telescope. A one-mile-long row of twelve 83-ft.-dia. parabolic disks in a remote area of the Netherlands, device would enable Oort and his colleagues to extend their explorations beyond earth's galaxy. 1968 was scheduled completion date. (*Wash. Post*, 12/18/66, B19)

- U.S. and Soviet scientists had advanced two related theories to explain radio emissions that made Venus seem lethally hot, Walter Sullivan reported in the *New York Times*. Soviet scientists V. M. Vakhnin and A. I. Lebedinsky, writing in *Zemlya i Vselennaya*, proposed that the two halves of Venus' atmosphere, the sunlit and the dark, were of opposite electric charges, and, where the two halves met, lightning flashes could account for emissions. U.S. scientists Dr. Paul Harteck, Dr. Robert R. Reeves, Jr., Dr. Barbara A. Thompson, and D. C. Appleton of Rensselaer Polytechnic Institute, in *Journal of Geophysical Research*, reported on effects of electric discharges, such as lightning, in a simulated Venus atmosphere of carbon dioxide and sulphur dioxide. Investigation revealed that such discharges produced intense radio emissions in same region of microwave spectrum as observed emissions from Venus. (Sullivan, *NYT*, 12/18/66, 35)

- FAA announced allocations of $72.5 million in Federal matching funds to help local communities construct and improve 341 civil airports under Federal-Aid Airport Program for FY 1967. (FAA Release 66-107)

- 25th anniversary of Reaction Motors, Inc. Formed to develop James H. Wyld's liquid-fueled rocket, corporation became in 1958 Reaction Motors Div. of Thiokol Chemical Corp. (Thiokol, *Aerospace Facts*, 11-12/66, 2)

December 19: U.N. General Assembly unanimously approved space law treaty agreed upon by U.S. and U.S.S.R. Dec. 8. Treaty would become effective when ratified by five nations which must include U.S., U.S.S.R., and U.K. (*NYT*, 12/20/66, 6)

- NASA personnel changes: (1) Samuel H. Hubbard, Director, Gemini Flight Operations, had been redesignated Special Assistant to the Director, Gemini Program Office; and (2) General Counsel Walter D. Sohier had resigned effective Dec. 31, to enter private practice. Deputy

General Counsel Paul G. Dembling would succeed him. (NASA Ann.; NASA Release 66-322)

December 19: U.S.S.R. launched COSMOS CXXXVI for space research into earth orbit with 305-km. (190-mi.) apogee, 198-km. (123-mi.) perigee, and 64.6° inclination. (*U.N. Public Registry*, 1/9/67)

December 20: XB-70 research aircraft, flown by Col. Joseph Cotton (USAF) and NAA pilot Van H. Shepard, reached mach 2.53 and 60,800-ft. altitude in flight to generate sonic booms over instrumented sites and test structures at Edwards AFB. (NASA Proj. Off.; UPI, *NYT*, 12/22/66, 20)

- NASA announced sponsorship of four institutes in the summer of 1967 to acquaint about 160 nationally selected undergraduates with problems of space science and engineering. The institutes will be carried out by the following universities and NASA field centers: Columbia Univ. and Goddard Institute for Space Studies (space physics), USC and JPL (spacecraft technology), Univ. of Miami and KSC (environmental and planetary sciences), and UCLA and ARC (space physiology). (NASA Release 66-324)
- MSFC awarded $7.2-million contract modification to Chrysler Corp. to begin procurement of long-lead-time items for 12 additional Uprated Saturn I 1st stages (S-IB). (MSFC Release 66-295)
- Japan's second attempt to orbit a satellite failed when 4th stage of Lambda 4S-2 booster failed to ignite. First failure was Sept. 26. Both launches were conducted from Uchinoura Range, Kyushu Island. (*Wash. Post*, 12/20/66, A13; UPI, *Wash. Post*, 12/21/66, A2, A24)
- Karl Harr, president of Aerospace Industries Assn., spoke before the Aviation/Space Writers' Assn.: "As we stand at the end of one record year of achievements and at the beginning of a year that we confidently predict will establish new records on all fronts, will historians view this point of time as having any particular significance from an aerospace point of view? After all we're not going to the moon next year and we didn't inject man into space for the first time last year....

    "I believe historians may come to regard about this point of time as being the point when the aerospace business completed its transition from a matter of only peripheral concern to the total public interest, into the mainstream of American life...." (Text)
- West German Defense Ministry ordered limited resumption of flights by F-104G Starfighters—Lockheed-manufactured aircraft which had been grounded since Dec. 6 after 37 pilots had been killed in series of 65 crashes. (AP, *NYT*, 12/21/66)

December 21: USAF SV-5D lifting body was successfully launched from WTR by Atlas booster on reentry mission, but "due to an undetermined malfunction in the final phases of descent" was not recovered. Released at desired altitude, the 900-lb. vehicle had deployed parachute to slow its descent rate and reentered planned recovery area using lifting body principle. Goals of mission—first of four in USAF's Precision Recovery Including Maneuvering Entry (Prime) program—were to study aerodynamic characteristics of SV-5D shape, heat shield performance, guidance and flight control systems, instrumentation, and recovery system. (UPI, *Wash. Post*, 12/22/66, A4)

- U.S.S.R. launched two unmanned spacecraft. LUNA XIII, on mission to study the "moon and its close environments," was following trajectory close to calculated one. COSMOS CXXXVII, carrying instruments for further investigation of outer space, was in orbit with 1,672-km. (1,032-mi.) apogee; 225-km. (140-mi.) perigee; 104-min. period; and 49° inclination. (UPI, *Wash. Post*, 12/22/66, A3; Anderson, *NYT*, 12/22/66, 20)

December 22: NASA named crews for second and third manned Apollo missions, scheduled for 1967. Prime crewmen for AS-205/208 were James A. McDivitt, David R. Scott, and Russell Schweickart; backup crew was Thomas P. Stafford, John W. Young, and Eugene A. Cernan. Prime crewmen for AS-503, first manned flight using Saturn V launch vehicle, were Frank Borman, Michael Collins, and William A. Anders; backup crew was Charles Conrad, Jr., Richard F. Gordon, Jr., and C. C. Williams, Jr. (NASA Release 66-326)

- First free flight of NASA's HL-10 lifting body vehicle was successfully conducted at Edwards AFB. Air-launched from B-52 bomber flying at 45,000 ft. and 450 mph, vehicle was maneuvered by NASA test pilot Bruce A. Peterson to 200-mph landing three minutes later. Purpose of flight was complete design and systems checkout and evaluation of basic stability and control. NASA and AFFTC were studying lifting body concept inflight to help establish the technological base for design of future manned spacecraft capable of maneuvering inflight to pilot-controlled ground landing. (FRC Release 29-66)
- NASA announced sponsorship in 1967 of three specialized summer programs in systems engineering design for engineering faculty members. Undertaken as cooperative efforts between NASA research centers and adjacent universities, programs would enable faculty members from various engineering disciplines to work together as a team to design a complex space system. Fellowships would be available for about 70 faculty members. Participating universities and cooperating NASA centers: Stanford Univ. and ARC; Auburn Univ., Univ. of Alabama, and MSFC; and Univ. of Houston, Rice Univ., and MSC. (NASA Release 66-325)
- U.S. R&D funding in 1967 would total about $23.8 billion—$500 million more than in 1966—Battelle Memorial Institute said in its annual forecast. For the first time in a decade, the Government would not be the dominant factor in the growth; 80 per cent of predicted increase would come from industry, universities, and nonprofit institutions. (BMI Release 124-66)
- Static test of Saturn V 2nd stage (S-II) at MTF was postponed 2 min. 31 sec. before ignition when electrical cable in one of its two engines short-circuited. (UPI, *NYT*, 12/23/66, 12; 12/25/66, 32; AP, *Wash. Post*, 12/23/66, A2)
- Tests showed that as little as one or more inches of continuous snow cover could be reliably identified from satellite photos, ESSA's Office of Hydrology reported. Associate Director of the Weather Bureau William E. Hiatt commented: "An increased capability of assessing snow accumulations, especially in remote sections where observation stations are limited, could result in faster appraisal of river conditions and contribute to more timely release of river forecasts and flooding warnings." (ESSA Release 66-71)
- Inactivity in Soviet manned space program could be explained by "strict economy" and preparations for ambitious new manned missions, Radio Prague's Moscow correspondent reported. The new missions—"flights of much higher standards"—would include "a flight of man to the

moon and the creation of a large orbiting laboratory." Correspondent also predicted that the U.S.S.R. would launch a "group of rockets"—presumably unmanned ones—toward Mars about Jan. 4. (*NYT*, 12/23/66, 12)

New York Times later editorialized: "... The possibility is raised that the Soviet strategy for a manned lunar landing may be based on a direct earth-moon trip. In that case there would be no need for the rendezvous and docking maneuvers.... Enormously powerful rockets would be required ... but if Moscow has them they might make possible the spectacular achievement of a goal the Kremlin must want very much: a manned lunar landing on the moon by the fiftieth anniversary of the Bolshevik Revolution next Nov. 7." (*NYT*, 12/26/66, 20)

December 23: NASA announced Atlas-Centaur booster would replace Atlas-Agena for launching Orbiting Astronomical Observatory (Oao) satellites and Applications Technology Satellites (Ats) D and E. Launches were scheduled to begin in 1968. Centaur stage—first to use liquid hydrogen as a fuel—was capable of boosting about 40 per cent greater payload than the Agena into near-earth orbit and about three times the payload on lunar trajectory. (NASA Release 66-328)

- NASA Administrator James E. Webb appointed B. L. Dorman, vice president of Aerojet-General Corp., Assistant Administrator for Industry Affairs, effective Jan. 16, 1967, and Gen. Jacob E. Smart (USAF, Ret.) Acting Administrator for Administration, effective Jan. 1, 1967. Both positions were formerly held by William B. Rieke, who resigned to return to Lockheed Missiles and Space Co. (NASA Release 66-327)

- NASA awarded General Electric Co. $22.3 million to continue in its fourth year as general support contractor for NASA's Mississippi Test Facility. Award was for period November 1966 to July 1967. (MSFC Release 66-298)

- Bell Telephone Laboratories had developed experimental method of impressing over 57,000 two-way telephone conversations on single laser beam—a process essential for the commercial use of the laser in communications, *Wall Street Journal* reported. Researchers hoped to eventually replace coaxial cables, which could carry only about 32,000 telephone conversations, with cables of laser beams. (*WSJ*, 12/23/66, 12)

December 24: Tass reported that detailed chart of earth's magnetic field which would enable cosmonauts to determine radiation conditions in earth's environs had been compiled by scientists using data from Cosmos satellites. (Reuters, *Wash. Post*, 12/25/66, A3)

- Subrahmanyan Chandrasekhar, Univ. of Chicago astrophysicist, was one of 11 scientists named by President Johnson to receive 1966 National Medal of Science. (*Pres. Doc.*, 1/2/67, 1832-3)

December 24-31: U.S.S.R. achieved her second soft-landing on the moon when LUNA XIII spacecraft landed an instrument package near Ocean of Storms to photograph and test the lunar surface.

Landing sequence began when spacecraft was oriented vertically to the moon. At 43-mi. (69-km.) altitude, radar system aimed at lunar surface and turned on retrorocket, and shock-absorbing system was prepared. Less than one second before impact, after spacecraft had been slowed to 18 mph, instrument capsule was ejected and softlanded. Four minutes later, its petal-like covering opened, its antenna opened, and radio transmissions began.

18 hrs. after touchdown, LUNA XIII began photographing a panorama of the moon's surface. Pictures showed barren, heavily pitted landscape littered with stones ranging upward in size from one inch, which Soviet scientists concluded had resulted from either volcanic eruptions or impacting meteoroids. There was no evidence of a thick layer of dust.

In addition to photographic equipment, spacecraft also carried two mechanical manipulators to test firmness and density of lunar soil: (1) a rod which was driven into surface with specific force; and (2) a meter which measured soil's resistance to radiation. Analysis of data indicated that lunar surface could bear manned spacecraft and that soil from 8- to 12-in. depth was similar to medium-density terrestrial soil. Also onboard were a dynograph which tested firmness of lunar surface by recording duration and force of impact impulse, and a radiation meter which revealed that moon's surface reflected about 25 per cent of electrically-charged particles which struck it.

Tass commented on LUNA XIII's successful mission: "There is no doubt now that it is these cosmic stations that can supply the information permitting a new approach to solve such important problems of science as the origin of the solar system, the origin and development of life on other planets, and the internal structure of celestial bodies." (*NYT*, 12/25/66, 1, 40; 12/26/66, 19; 12/31/66, 6; AP, Wash. *Eve. Star*, 12/26/66, A3; 12/27/66, A4; UPI, *Wash. Post*, 12/27/66, A1; 12/31/66, 1, 7)

December 25: Nationalist China would build $10-million ground communications station by early 1969 to participate in Intelsat's planned worldwide satellite communications system, Reuters reported. (Reuters, *NYT*, 12/25/66, 40)

December 26: NASA, Japanese Science and Technology Agency, and Japanese Meteorological Agency agreed to launch jointly 10 Japanese MT-135 sounding rockets beginning in March 1967 to study rocket and payload performance and gain information on diurnal cycles of wind and temperature in the stratosphere. MT-135 could boost 6.6-lb. payload to 200,000-ft. (322,000-km.) altitude. (NASA Release 66-331)

December 27: Clear air turbulence (CAT)—unexplained disturbances encountered by pilots flying at high altitude in seemingly calm weather—should be studied by NASA, DOD, FAA, and Dept. of Commerce, National Committee for Clear Air Turbulence recommended. One committee official estimated that CAT had been "highly suspect" in six recent air crashes. (*Balt. Sun*, 12/28/66)

- NASA Administrator James E. Webb, speaking at Univ. of New Mexico, urged establishment of a seminar where experts could exchange scientific and industrial knowledge to bridge "the tremendous gap between physical development and social inventions." Mr. Webb also recommended a national rocket foundation to provide facilities for amateur rocket enthusiasts. (AP, *NYT*, 12/28/66, 4)

- "The American Space Outlook" for 1967 was summarized by Albert Sehlstedt, Jr., in the Baltimore *Sun*. ". . . 1967 should be a year of genuine accomplishments in space for men and machines.

". . . astronauts will begin to fly around the earth in the kind of spacecraft that will carry them to the moon in 1968 or 1969. . . .

"New unmanned spacecraft will climb into orbit for additional views of the earth's weather. Other craft will fly away to study the science of the sun; make further on-the-spot examinations of the moon's surface, probe the nature of the stars and take another close look at Venus.

"... the Defense Department will lay the groundwork for the 1969 launching of a manned orbiting laboratory to study possible military roles of men in space ... the people who make airplanes will be planning the big jumbo jets and, in Europe, fashioning airliners that will go faster than the speed of sound. ..." (Sehlstedt, Balt. *Sun*, 12/28/66)

December 27: Communist China exploded its fifth nuclear device at Lop Nor test site in Sinkiang Province. (Roberts, *Wash. Post*, 11/29/66, A17)

- Astronaut Bruce McCandless II, discussing role of manned space flight in science in address before Washington, D.C., Junior Academy of Sciences' annual meeting, concluded: "Man is flexible; he is capable of responding to unforeseen situations intelligently and effectively. He has a limited and on-the-spot redesign capability built in and contributes to mission success through improved reliability. He is capable of evaluating and screening information and making decisions accordingly and independently. As long as these elements are required, man's participation will be beneficial." (Text)

December 29: Aerospace industry believed 4,000-mph hypersonic transport (Hst) could be ready for commercial service in 1990's—but only if Government would support its development, Richard Cooke reported in the *Wall Street Journal*. "Aerospace ... officials agree that the rate at which new developments in air transportation come along will hinge in large measure on Government policy. They note that the military has sponsored the development of every large American jet engine. ... A military version of the craft will almost certainly have to precede the commercial one—but right now there's no driving military sponsor at high level behind these advanced materials and engines." Development of Hst would require new heat-resistant materials, a more efficient fuel, and higher-capacity engine. (Cooke, *WSJ*, 12/29/66)

- American Astronautical Society (AAS) held symposium on the moon in Washington, D.C., as part of American Association for the Advancement of Science's (AAAS) annual meeting.

  Dr. Eugene Shoemaker, U.S. Geological Survey, said NASA's LUNAR ORBITER I photos had revealed "mass wasting" on the moon—a process similar to "downhill creep" which produced slopes of fragmented dirt and rock debris on earth. Accelerated by small local earthquakes caused by impacting meteorites, wasting process was rapidly covering all of moon's small craters, Shoemaker said. He believed majority of lunar craters were caused by high-velocity explosive impacts of meteorites.

  Dr. Donald H. Menzel, Harvard Univ., told the press that man would make profound changes on the moon in 15-20 yrs.—particularly if lunar rock were found to contain water of crystallization. "There is a real possibility for man ... to establish self-support expeditions on the moon. Using solar or atomic energy for melting rocks, man could get water and then use the debris for building blocks. Lunar gardens could be established and animals could be raised there." (Hines, Wash. *Eve. Star*, 12/29/66, A1, A6; Clark, *NYT*, 12/30/66, 8)

- USAF launched unidentified satellite with Thor-Agena D booster from WTR. (*U.S. Aeron. & Space Act., 1966*, 160)

December 29: National Aviation Hall of Fame announced six additions: Glenn L. Martin and William F. Boeing, both aircraft designers; Adm. John H. Towers, commander of first attempt at trans-Atlantic flight; B/G William Mitchell, pioneer advocate of aerial bombing; Robert Hutchings Goddard, U.S. rocket pioneer; and Lincoln Beachey, early designer of dirigibles. All were deceased. (AP, *NYT*, 12/29/66, 14)

- Robert L. Hallett, Head of Communications and Control Section at NASA Wallops Station, retired after 20 yrs. with NACA and NASA. (Wallops Release 67-1)

December 30: NASA and ESRO signed first agreement under which foreign country or organization would purchase satellite launch services from U.S. on reimbursable basis. ESRO would furnish flight-ready scientific spacecraft and NASA would provide launch vehicle, range and launching facilities, and other support including initial tracking and telemetry reception. ESRO would reimburse NASA for launch vehicles and all identifiable direct costs of equipment and services furnished by or through NASA. First launch—ESRO's Heos-A interplanetary physics satellite—was scheduled for late 1968. (NASA Release 66-332)

- Successful full-duration static firing (364 sec.) of the S-II-1, first flight model of Saturn V's 2nd stage, was conducted at MTF. This was second and last full-duration acceptance firing for the stage, which would now be shipped to KSC. (MSF Historian)
- NASA announced plans to begin powered flights of M-2 and HL-10 manned lifting bodies at Edwards AFB in early 1967. Equipped with 8,000-lb.-thrust rocket engine, vehicles would be flown at altitudes up to 80,000 ft. and speeds of 1,000 mph in series of 8-min. test flights designed to simulate spacecraft reentry conditions. (NASA Release 66-329)
- John A. Edwards, NASA Deputy Director, Gemini Program, had been designated Director of Operations for Saturn/Apollo Applications in OMSF. (NASA Ann., 12/30/66)
- MSFC awarded Douglas Aircraft Co. a $300,000 9-month contract to analyze needs of an orbital support facility for future astronomical research from space. (MSFC Release 66-302)
- NASA awarded one-year, $360,094, cost-plus-fixed-fee contract to Philco Corp. for engineering support services at Wallops Station. (Wallops Release 66-60)
- Dr. Robert M. Page, retired as chief scientist and director of research at Naval Research Laboratory after nearly 40 yrs. service. One of foremost U.S. experts on radar and a pioneer in space exploration and rocket astronomy, Dr. Page warned that U.S. research was being threatened by "dangerous trends and controls. . . . A complete lack of confidence is being shown in the integrity and judgement of responsible operating officials which will certainly lead to bad management of Federal programs." (Naval Research Lab.; Kluttz, *Wash. Post*, 1/8/67, A2)

December 31: Boeing Co. and General Electric Co. had been selected to continue development and refinement of designs for first U.S. supersonic transport (Sst), FAA announced. Decision climaxed 30 mos. of intense competition between Boeing Co. and Lockheed Aircraft Corp. for airframe design; and GE and United Aircraft Corp.'s Pratt & Whitney Div. for engine design. Boeing's 350-passenger, 1,800-mph, variable-sweep-wing model would be powered by four 4.5-ton GE engines capable of generating 600,000 lbs. thrust. Approval for prototype construction was expected in early 1967. (FAA Release; AP, *Wash. Post*, 1/1/67, A1; Wash. *Sun. Star*, 1/1/67, A10)

December 31: New York Times characterized 1966: "In all future histories of lunar exploration the year 1966 will hold an honored and important place. In effect, it was the year of the moon, the year in which a dazzling array of Soviet and American instrumented capsules photographed, measured and probed the moon to an extent never before known. As a result of the successful flights of the Soviet Luna vehicles and of the American Surveyor and Lunar Orbiters, it is probable that more detailed information about earth's natural satellite was gained in 1966 than in all the rest of human history. . . ." (*NYT*, 12/31/66, 18C)

- Fourth earth-orbiting object [see Sept. 30] not identified "with any launching or country of origin" was reported by GSFC in *Satellite Situation Report*. Object apparently decayed Dec. 30. (GSFC *SSR*, 12/31/66)

During December: NRC's Space Science Board was studying possible harmful effects to astronomy and other sciences of proposed NASA project to orbit large reflecting mirror that could illuminate land masses at night, NAS announced. NASA's support for the study was expressed by Associate Administrator for Space Science and Applications Dr. Homer E. Newell in a letter to Dr. Harry H. Hess, Chairman of NAS Space Science Board: "I am gratified that the committee has promptly undertaken work on this task and am looking forward to their review for guidance with regard to the possible adverse effects of such a project." (NAS-NRC-NAE *News Report*, 12/66, 7)

- Proposal for fleet of 25 stand-by spacecraft which could be launched within three hours of a distress signal to rescue astronauts stranded in orbit was submitted to NASA by MIT aeronautics students. Each 13,000-lb., four-man Nero (Near Earth Rescue Operation) spacecraft, designed with wing-like lift similar to M2-F2 lifting-body vehicle, would be launched with Titan III-C booster to rendezvous with target; it would glide to a landing at a jet airport. Vehicles could also repair crippled satellites, resupply manned orbiting laboratories, inspect unidentified satellites, and remove orbiting debris. System could be implemented with current technology and be in operation by early 1970's at total cost of $2 billion. (Wilford, *NYT*, 12/6/66, 60C)

- NAS published *History of the Proceedings of the National Academy of Sciences*; 40-page study was prepared by Edwin B. Wilson, managing editor of *Proceedings* from 1915 until his death in 1964. (NAS-NRC-NAE *News Report*, 1/67, 8)

- "Love as an emotion will play a very important role in the future in the populating of outer space," Soviet scientist Igor Zabelin wrote in *Moskva*. "In the future when mankind will fulfill its mission of populating outer space, men and women will be sent in spacecraft somewhere for many years." Noting that "when a couple lives together for five to seven years, they may get sick of each other and very often there is a divorce," Zabelin warned that such coolness could cause grave problems in outer space. (UPI, *Wash. Post*, 12/4/66, M7)

During 1966: In 1966, NASA attempted 36 major launches with 34 successes. Of 29 NASA missions, 22 were successes. DOD orbited 70 payloads in 43 launches. U.S.S.R. launched 43 payloads and France, one.

Highlighting NASA space achievements were closeup photographs of the moon provided by SURVEYOR I and LUNAR ORBITERS I and II, the soft-landing on the moon by SURVEYOR I, and the five two-man Gemini space flights. SURVEYOR I transmitted more than 10,000 photographs of the moon; the two LUNAR ORBITERS returned a total of 635 lunar photographs. Engineering data from these spacecraft indicated Apollo Lunar Module (Lm) could land and operate safely on the lunar surface. In Project Gemini, NASA conducted five two-man space flights, successfully concluding the project [see Nov. 15] and bringing U.S. manhours in space to 1,993 hrs. 34 min. (compared to 507 hrs. 16 min. for U.S.S.R.).

Major ground and flight tests in Project Apollo accelerated preparatory to 1st three-man Apollo/Saturn spaceflight, planned for early 1967. Three Uprated Saturn I's were launched; Saturn V stages for the 1st flight model passed their final ground test firings.

Atlas-Centaur liquid-hydrogen-fueled booster completed its development testing, demonstrating restart capability in space.

Applications satellites launched were NIMBUS II meteorological satellite, three ESSA satellites inaugurating ESSA's operational meteorological satellite system, and INTELSAT II-A communications satellite for ComSatCorp. The first Applications Technology Satellite, ATS I, was launched into synchronous orbit. Scientific achievements included orbiting of OGO III, PAGEOS I, EXPLORERS XXXII and XXXIII satellites, and PIONEER VII interplanetary spacecraft.

More than 300 meteorological sounding rockets and 100 scientific sounding rockets were launched. A total of 20 flights of the X-15 research aircraft were conducted; X-15 No. 2 piloted by Maj. William J. Knight set a new record speed of 4,233 mph (mach 6.33). NASA-USAF flight research investigating sonic booms got underway in six flights with XB-70 aircraft; research for the U.S. supersonic transport program continued. M2-F2 lifting-body vehicle made 14 unpowered flights, and HL-10 lifting body made its first flight.

Among the highlights in propulsion technology was static test firing of the 260-in.-dia. solid-propellant rocket motor, producing 3.5 million lbs. of thrust. Two test series of Nerva nuclear reactors were conducted. One, in 110 min. of power operation, demonstrated stability and controllability of the hot-bleed cycle nuclear rocket engine. The other involved first successful full-power operation and a restart operation, ending the program with an accumulation of more than 105 min. of reactor operation at full or near-full power.

USAF flight tested its powerful Titan III booster and used it to orbit the first seven Idcsp satellites of an eventual worldwide military comsat system.

Of the 43 Soviet payloads, 34 were Cosmos satellites and five, Luna spacecraft. Notable among these were COSMOS CX, orbiting two dogs for prolonged period; COSMOS CXXII, apparently first Soviet weather satellite; LUNA IX, achieving first soft-landing on the moon; and LUNA X, achieving first orbit of the moon. Also orbited were two MOLNIYA I comsats and the 12-ton PROTON III unmanned spacecraft used for cosmic-ray research; Yantar I ion-engine payload was flown in suborbital spaceflight test.

France orbited her first instrumented scientific satellite, DIAPASON I, the second satellite built and launched by France. (NASA Release 66-319; *Major NASA Launches;* NASC; Tass; *U.S. Aeron. & Space Act., 1966, passim; A&A 1966, passim*)

During 1966: International space events were highlighted by the U.N.'s agreement on a space treaty to govern "activities of states in the exploration and use of outer space, including the moon and other celestial bodies." The draft treaty—suggested by President Johnson May 7—was negotiated by the Legal Subcommittee of the U.N. Committee on the Peaceful Uses of Outer Space. On December 19 the U.N. General Assembly adopted a resolution endorsing the treaty, subject to ratification by five nations including U.S., U.S.S.R., and U.K.

By the end of 1966, 72 countries had cooperated with the U.S. in space research activities.

Six countries joined the International Telecommunications Satellite Consortium (Intelsat), bringing the total membership to 54.

France and U.S.S.R. entered into agreement on cooperative space research, including experiments in astronomy and communications. (*U.S. Aeron. & Space Act., 1966, passim; A&A 1966, passim*)

Appendix A

SATELLITES, SPACE PROBES, AND MANNED SPACE FLIGHTS

A CHRONICLE FOR 1966

The following tabulation was compiled from open sources by Dr. Frank W. Anderson, Jr., Deputy NASA Historian. Sources included the United Nations Public Registry; the *Satellite Situation Report* issued by the Space Operations Control Center at Goddard Space Flight Center; public information releases of the Department of Defense, NASA, and other agencies; and the *Report to the Congress from the President of the United States: United States Aeronautics and Space Activities, 1966*. Russian data are from the U.N. Public Registry, the *Satellite Situation Report*, translations from Tass News Agency, statements in the Soviet press, and international news services' reports.

It might be well to call attention to the terms of reference stated or implied in the title of this tabulation. This is a listing of payloads that have (a) orbited, (b) as probes, ascended to at least the 4,000-mile altitude that traditionally has distinguished probes from sounding rockets, etc., or (c) conveyed one or more human beings in space. Furthermore, only flights that succeeded —or at least are not known to have failed—in doing one of the above are listed. Date of launch is referenced to local time at the launch site. An asterisk by the date marks those dates that are one day earlier in this tabulation than in listings which reference to Greenwich time.

This was the year in which world space activity leveled out. After a decade of virtual doubling of activity each year, 1966 showed only a slight rise in total launches—116 against 102 in 1965—and a decline in the total number of payloads orbited—143 against the record 160 in 1965. At first glance there seems to be an inconsistency in a rise in number of launches and a drop in number of payloads; the answer is found in the abrupt switch away from multiple payloads that was one of the notable changes of direction in 1966 (DOD still flew 16 multipayload flights totaling 42 individual payloads, including the 8-satellite Titan IIIC, but the Soviets launched none, compared with 7 flights for a total of 23 payloads in 1965, and NASA none, compared with 3 flights for a total of 6 payloads in 1965). Of the 1966 total, the U.S. launched 92 boosters carrying 100 payloads (compared with 61/94 in 1965), the U.S.S.R. launched 43 (compared with 50 launches for a total of 66 payloads in 1965), and France joined the space club with one satellite in 1966. Of the U.S. total, DOD accounted for 70 payloads and 42 launches (compared with 67/39 in 1965), NASA 30 (compared with 27 payloads in 22 launches in 1965; one of the 30 was a non-NASA mission, INTELSAT II-A, launched by NASA for ComSatCorp). Two areas of concentration were again significant: the U.S. continued to dominate manned space flight, with 5 Gemini manned flights during the year, while the Soviets had no manned flights (compared with 5/1 in 1965); the Soviets continued their numerical edge in lunar exploration with 5 lunar flights to 4 for the U.S. (compared with 5/2 in 1965).

As we have cautioned in previous years, the "Remarks" column of these appendixes is never complete because of the inescapable lag behind each flight of the analysis and interpretation of scientific results.

Launch Date	Name	International Designation	Vehicle	Payload Data	Apogee (st. mi.)	Perigee (st. mi.)	Period (minutes)	Inclination	Remarks
Jan. 7	COSMOS CIV (U.S.S.R.)	1966-1A	Not available	Total weight: Not available. Objective: Continuation of Cosmos scientific satellite series. Payload: Not available.	248	126	90.2	65°	Reentered 1/15/66.
Jan. 19	DOD Spacecraft (United States) and DOD Spacecraft	1966-2A 1966-2B	Atlas-Agena D	Total weight: Not available. Objective: Develop spaceflight techniques and technology. Payload: Not available. Total weight: Not available. Objective: Develop spaceflight techniques and technology. Payload: Not available.	161 122	86 92	88.4 87.6	98.8° 98.8°	Reentered 1/25/66. One of 2 spacecraft launched with single vehicle. Reentered 1/28/66.
Jan. 22	COSMOS CV (U.S.S.R.)	1966-3A	Not available	Total weight: Not available. Objective: Continuation of Cosmos scientific satellite series. Payload: Not available.	201	127	89.7	65°	Reentered 1/30/66.
Jan. 25	COSMOS CVI (U.S.S.R.)	1966-4A	Not available	Total weight: Not available. Objective: Continuation of Cosmos scientific satellite series. Payload: Not available.	350	180	92.8	48.4°	Reentered 11/14/66.
Jan. 28	DOD Spacecraft (United States)	1966-5A	Scout	Total weight: 144 lbs. Objective: Develop spaceflight techniques and technology. Payload: Not available.	753	585	105.9	89.7°	Still in orbit.
Jan. 31	LUNA IX (U.S.S.R.)	1966-6A	Not available	Total weight: 220 lbs. for instrument capsule; 3,483 lbs. for total lunar descent spacecraft. Objective: Make softlanding on moon; take TV photos of lunar surface and measurements of cosmic radiation and transmit these data to earth. Payload: Lunar descent spacecraft, including retrorocket, landing radar, communications equipment; also carried ejectable airtight instrument capsule containing 3.3-lb. camera, cosmic radiation measuring equipment, telemetry transmitter, programmer, thermal control system, "energy supply sources," antennas, shock-absorbing system,	Soft-landed on moon				First softlanding on moon, made 2/3. LUNA IX began landing sequence 8,300 km. (6,165 mi.) above the moon; retrorocket was turned on at 47 mi., 48 sec. before impact. Less than 1 sec. before impact instrument capsule was ejected; it landed safely near Ocean of Storms west of Reiner and Marius craters. During planned 48-hr. operating period, LUNA IX took 10 panoramic photos of lunar surface and transmitted them to earth. Other scientific measurement was of cosmic radiation.

Date	Name	Designation	Launch Vehicle	Payload/Objective	Weight (lbs)	Period (min)	Inclination	Remarks
Feb. 2	DOD Spacecraft (United States)	1966-7A	Thor-Agena D	petal-shaped metal screens as protective covering during landing. Total weight: Not available. Objective: Develop spaceflight techniques and technology. Payload: Not available.	252	112	90.4	75.03°	Reentered 2/27/66.
Feb. 3	ESSA I (United States)	1966-8A	Thor-Delta	Total weight: 305 lbs. Objective: Begin Tiros Operational Satellite (Tos) system; photograph complete world weather system daily. Payload: 22" x 42" 18-sided hatbox-shaped polygon, with 18" receiving antenna and 4 22" transmitting whip antennas; contains 2 wide-angle vidicon cameras; 2 tape recorders; 2 spin-control systems (magnetic coil; small solid-propellant rockets); 2 infrared horizon sensors; transmitters; 63 nickel-cadmium batteries; 9,100 n-on-p solar cells.	523	433	100	97.9°	First satellite in operational weather satellite system. ESSA I began taking photos on global basis on 2/5/66; each day 18 orbits of TV-photo data were read out and data used for operational and research purposes. Quality of TV data was excellent. Still in orbit, still transmitting.
Feb. 9	DOD Spacecraft (United States)	1966-9A	Thor-Agena D	Total weight: Not available. Objective: Develop spaceflight techniques and technology. Payload: Not available.	316	314	94.8	82.07°	Still in orbit.
Feb. 10	COSMOS CVII (U.S.S.R.)	1966-10A	Not available	Total weight: Not available. Objective: Continuation of Cosmos scientific satellite series. Payload: Not available.	200	127	89.7	65°	Reentered 2/18/66.
Feb. 11	COSMOS CVIII (U.S.S.R.)	1966-11A	Not available	Total weight: Not available. Objective: Continuation of Cosmos scientific satellite series. Payload: Not available.	587	141	95.3	48.9°	Reentered 11/21/66.
Feb. 15	DOD Spacecraft (United States)	1966-12A	Atlas-Agena D	Total weight: Not available. Objective: Develop spaceflight techniques and technology. Payload: Not available.	154	89	96.5	88.8°	Reentered 2/22/66. Tracking report showed some 39 other fragments associated with this satellite.
	and DOD Spacecraft	1966-12B		Total weight: Not available. Objective: Develop spaceflight techniques and technology. Payload: Not available.	159	73	96.48	88.1°	Reentered 2/16/66.
	and DOD Spacecraft	1966-12C		Total weight: Not available. Objective: Develop spaceflight techniques and technology. Payload: Not available.	183	92	96.5	88.6°	Reentered 2/22/66.

Launch Date	Name	International Designation	Vehicle	Payload Data	Apogee (st. mi.)	Perigee (st. mi.)	Period (minutes)	Inclination	Remarks
Feb. 17	DIAPASON I (D-1A) (France)	1966-13A	Diamant	Total weight: 49 lbs. Objective: Orbit instrumented satellite; make measurements of earth's magnetic field; test French tracking stations. Payload: 19.7" (dia.) x 7.8" cylinder, from which extend 4 29.5" whip antennas on top and 4 16.5" X 17.2" solar paddles folded downward at 45° angle; 2,304 n-on-p solar cells; cylinder contains transmitter and receiver, very stable oscillator, 8-element nickel-cadmium battery, yo-yo despin device; on outside of cylinder are attached 3 sets of solar cells for radiation-effect experiment.	1,710	312	118.6	34.04°	First instrumented French satellite; orbital parameters very near the intended ones indicated accuracy of guidance system. Satellite performed well, as did tracking stations. Two kinds of geodetic experiments were attempted, one using Doppler effect, the other attempting to photograph the satellite against star background. Still in orbit, still transmitting.
Feb. 19	COSMOS CIX (U.S.S.R.)	1966-14A	Not available	Total weight: Not available. Objective: Continuation of Cosmos scientific satellite series. Payload: Not available.	192	180	89.5	65°	Reentered 2/27/66.
Feb. 22	COSMOS CX (U.S.S.R.)	1966-15A	Not available	Total weight: Not available. Objective: Orbit animal payload for prolonged period; conduct biomedical and radiation experiments; recover animal payload. Payload: Spacecraft with 2 dogs in it; life-support system; radiation sensors; camera; telemetry; retromotor and recovery system.	562	116	96.3	51.64°	Dogs Veterok and Ugolyek were orbited for 22 days in orbit that took them through lower Van Allen belt on each revolution. Dogs were heavily instrumented with biosensors and were photographed in orbit. Reentered safely on 3/16; were said to have experienced heartbeat irregularity that became more pronounced toward end of flight.
Feb. 28	ESSA II (United States)	1966-16A	Thor-Delta	Total weight: 290 lbs. Objective: Add Automatic Picture Transmission (Apt) capability to Tiros Operational Satellite system; provide daily coverage of local weather systems for weather stations around the world. Payload: 22" x 42" 18-sided hatbox-shaped polygon, with	885	843	118.6	101°	First operational weather satellite to offer cloud photos to local Apt stations; as it passes overhead, ground station can command readout of current local weather photo. Second half of 2-satellite Tos system which provided pictures of earth's entire 200-million-sq.-mi. area daily. ESSA II's

ASTRONAUTICS AND AERONAUTICS, 1966

Date	Name	ID	Launch vehicle	Description					Remarks
				18' receiving antenna and 4 22' transmitting whip antennas; containing 2 wide-angle Apt vidicon camera systems, FM transmitters, 2 spin-control systems (magnetic coil; small solid-propellant rockets), 2 infrared horizon sensors; 63 nickel-cadmium batteries; 9,100 n-on-p solar cells.					photos were "of superior quality." Still in orbit, still transmitting.
Mar. 1	COSMOS CXI (U.S.S.R.)	1966–17A	Not available	Total weight: Not available. Objective: Continuation of Cosmos scientific satellite series. Payload: Not available.	140	119	88.6	51°	Reentered 3/8/66.
Mar. 9	DOD Spacecraft (United States)	1966–18A	Thor-Agena D	Total weight: Not available. Objective: Develop spaceflight techniques and technology. Payload: Not available.	260	112	90.5	75.02°	Reentered 3/29/66.
Mar. 16	GATV VIII (United States)	1966–19A	Atlas-Gatv	Total weight: 7,000 lbs. Objective: Act as target vehicle for Gemini VIII rendezvous and docking exercise. Payload: 26' x 5' cylinder, containing adapter system radar transponder, command control system, main engine, 2 secondary engines, attitude control system.	186	184	90.4	28.88°	Still in orbit.
Mar. 16	GEMINI VIII (United States)	1966–20A	Titan II	Total weight: 8,351 lbs. (for Gemini, including reentry and adapter modules). Objective: Rendezvous and dock with target vehicle; perform extravehicular activity for 1 orbit. Payload: 18'5" x 10' 2-module bell-shaped spacecraft, containing 2 astronauts, guidance and control equipment, rendezvous radar, cameras, 1 HF and 1 UHF transceiver, high and low frequency telemetry transmitters, tracking and recovery communications, fuel cell, environmental control system, reentry and recovery systems.	167	99	88.88	29.07°	GEMINI VIII made world's first docking in space, at 6:33 hrs. into flight. At 7 hrs. unexpected yaw and roll motion caused Astronauts Neil A. Armstrong and David R. Scott to undock, use their reentry control system to stabilize their spacecraft, and reenter on revolution 7 of planned 44 (10 hrs. 42 min. of planned 72 hrs. 50 min.); landed in stipulated emergency area in western Pacific, were hoisted aboard U.S.S. *Leonard F. Mason* in their spacecraft. Flight's 2nd primary objective—Eva for 1 orbit — was not achieved. Motion problem had been caused by 1 yaw thruster which fired constantly because of a short-circuit.

Launch Date	Name	International Designation	Vehicle	Payload Data	Apogee (st. mi.)	Perigee (st. mi.)	Period (minutes)	Inclination	Remarks
Mar. 17	COSMOS CXII (U.S.S.R.)	1966-21A	Not available	Total weight: Not available. Objective: Continuation of Cosmos scientific satellite series. Payload: Not available.	351	188	92.1	72°	Reentered 3/25/66.
Mar. 18	DOD Spacecraft (United States)	1966-22A	Atlas-Agena D	Total weight: Not available. Objective: Develop spaceflight techniques and technology. Payload: Not available.	187	91	89	101°	Reentered 3/24/66. One of two satellites launched with single vehicle.
	and DOD Spacecraft	1966-22B		Total weight: Not available. Objective: Develop spaceflight techniques and technology. Payload: Not available.	146	85	88.2	100.9°	Reentered 3/23/66.
Mar. 21	COSMOS CXIII (U.S.S.R.)	1966-23A	Not available	Total weight: Not available. Objective: Continuation of Cosmos scientific satellite series. Payload: Not available.	203	130	89.6	65°	Reentered 3/29/66.
Mar. 25	DOD Spacecraft (United States)	1966-24A	Scout	Total weight: 144 lbs. Objective: Develop spaceflight techniques and technology. Payload: Not available.	699	553	105.3	89.72°	Still in orbit, still transmitting.
Mar. 30	OV1-IV (United States)	1966-25A	Atlas D	Total weight: 193 lbs. Objective: Report effects of weightlessness on photosynthetic organisms and small vascular plants; report environmental effects on thermal-control coatings. Payload: 27" x 55" cylinder with hemispheric ends, containing chlorella algae and multicell duckweed, with photocells to measure cell division; wafers and coatings for thermal equipment; telemetry transmitter; 5,000 solar cells.	627	550	104.1	144.5°	OV1-IV completed its biological and thermal missions. Still in orbit. One of 2 satellites orbited with single vehicle.
	OV1-V and	1966-25B		Total weight: 252 lbs. Objective: Measure optical radiation of earth, of its background, and of space as basis for surveillance techniques. Payload: 27" x 55" cylinder with hemispheric ends, contain-	662	607	105.6	144.6°	OV1-V stabilized successfully, returned optical radiation data. Still in orbit.

Date	Name	Designation	Launch vehicle	Payload/Objective	Weight (lbs)	Perigee/Apogee	Period (min)	Inclination	Remarks
Mar. 30	DOD Spacecraft (United States)	1966-26A	Thor-Altair	ing 5 optical sensors (3 pointing at earth, 2 at horizon); stabilizer; telemetry; 5,000 solar cells. Total weight: Not available. Objective: Develop spaceflight techniques and technology. Payload: Not available.	580	392	100.5	98.61°	Still in orbit.
Mar. 31	LUNA X (U.S.S.R.)	1966-27A	Not available	Total weight: 540 lbs. for lunar satellite; 3,530 lbs. for total lunar spacecraft. Objective: Maneuver spacecraft into vicinity of the moon; test systems for putting satellite into orbit around the moon. Payload: Spacecraft including retromotor, maneuvering equipment, and containing ejectable lunar satellite—an airtight package containing radio equipment, telemetry system, instruments for studying the moon and near-lunar space, thermal and power systems, small jet engines for stabilizing the satellite.	(Lunar orbital data:) 621	217	180		First artificial satellite in lunar orbit; on 4/4 the main spacecraft fired retrorocket 4,969 mi. above the moon, then lunar satellite was separated into lunar orbit. Scientific observations by LUNA X of near-lunar space noted larger number of meteoroids than in interplanetary space, rise in electron flux (possibly caused by moon's passing through earth's magnetic tail), lack of radiation differentiation in lunar crust equivalent to earth's granite, indicating earth and moon were of separate origins. Still in lunar orbit.
Apr. 6	COSMOS CXIV (U.S.S.R.)	1966-28A	Not available	Total weight: Not available. Objective: Continuation of Cosmos scientific satellite series. Payload: Not available.	232	130	90.1	73°	Reentered 4/14/66.
Apr. 7	DOD Spacecraft (United States)	1966-29A	Thor-Agena D	Total weight: Not available. Objective: Develop spaceflight techniques and technology. Payload: Not available.	187	118	89.5	75.07°	Reentered 4/26/66.
Apr. 7	Centaur-Surveyor (United States)	1966-30A	Atlas-Centaur	Total weight: 1,780 lbs. Objective: Place dummy Surveyor spacecraft on simulated lunar transfer trajectory using 1-burn, indirect-ascent mode. Payload: Surveyor mass model, ballasted to simulate Surveyor's retrorocket and solar panels and antennas; S-band transponder; operational-type separation system.	208	113	90	31°	Atlas-Centaur went into 100-mi. parking orbit but 2nd burn by Centaur, intended to put dummy Surveyor into lunar transfer trajectory, failed after 8 sec. Improved RL-10 engines performed well in this 1st flight, as did modified guidance and vent systems. Reentered 5/26/66.

Launch Date	Name	International Designation	Vehicle	Payload Data	Apogee (st. mi.)	Perigee (st. mi.)	Period (minutes)	Inclination	Remarks
Apr. 8	OAO I (United States)	1966-31A	Atlas-Agena D	Total weight: 3,900 lbs. (including 1,000 lbs. of scientific instrumentation). Objective: Prove operational status of Oao system; stabilize and control spacecraft so scientific data can be obtained. Payload: 10' x 7' octagonal cylinder, from which extend 6 solar panels for total width of 21'. Radio command system includes 2 pairs of command receivers; 2 transmitters for wideband telemetry, 2 for narrowband; 2 radio tracking beacon transmitters; attitude control system; 4 experiment packages; 3 nickel-cadmium batteries; 74,618 n-on-p solar cells.	505	493	100.8	35°	OAO I orbited, functioned normally for 36 hrs.; then on 4/10 battery failure ended communication. No experiments had a chance to return data. Spacecraft's complex stabilization and control system did operate with intended accuracy. Still in orbit.
Apr. 19	DOD Spacecraft (United States)	1966-32A	Atlas-Agena D	Total weight: Not available. Objective: Develop spaceflight techniques and technology. Payload: Not available.	233	86	89.6	116.9°	Reentered 4/26/66.
Apr. 20	COSMOS CXV (U.S.S.R.)	1966-33A	Not available	Total weight: Not available. Objective: Continuation of Cosmos scientific satellite series. Payload: Not available.	183	118	89.3	65°	Reentered 4/28/66.
Apr. 22	OV3-I (United States)	1966-34A	Scout FW-4S	Total weight: 152 lbs. Objective: Measure angular distribution and energies of charged particles in magnetosphere and upper ionosphere, the pitch angle of particles being of primary interest. Payload: 29" x 29" octagonal cylinder, from which extend 2 54" plasma probe booms; carries proton and electron spectrometers, electrostatic analyzer, Geiger counter; transmitter; 2,560 n-on-p solar cells; nickel-cadmium batteries.	3,557	219	151.7	82.46°	Still in orbit. Performed well.
Apr. 25	MOLNIYA I-3 (U.S.S.R.)	1966-35A	Not available	Total weight: Not available. Objective: Develop and further improve a satellite radio and	24,545	310	710	64.5°	MOLNIYA I-3 on 5/18 took TV photos of earth, transmitted them successfully to earth.

ASTRONAUTICS AND AERONAUTICS, 1966

Date	Name	Designation	Launch Vehicle	Description				Remarks
Apr. 26	COSMOS CXVI (U.S.S.R.)	1966–86A	Not available	TV communication system (Sov-ComSat). Payload: Satellite with transmitter, command system, orientation system, orbit correction devices, power supply.	297	183	92.1	48.4°	Made a number of earth photos with varying focal lengths, lenses, and filters to observe earth's surface under differing illumination. First photo of earth from synchronous orbit distance. Still in orbit.
May 6	COSMOS CXVII (U.S.S.R.)	1966–87A	Not available	Total weight: Not available. Objective: Continuation of Cosmos scientific satellite series. Payload: Not available.	191	129	89.5	65°	Reentered 12/3/66.
May 11	COSMOS CXVIII (U.S.S.R.)	1966–88A	Not available	Total weight: Not available. Objective: Continuation of Cosmos scientific satellite series. Payload: Not available.	398	398	97.1	65°	Reentered 5/14/66.
May 14	DOD Spacecraft (United States) and DOD Spacecraft	1966–89A	Atlas-Agena D	Total weight: Not available. Objective: Develop spaceflight techniques and technology. Payload: Not available.	222	83	89.4	110.5°	Still in orbit.
		1966–89B		Total weight: Not available. Objective: Develop spaceflight techniques and technology. Payload: Not available.	344	322	95.4	109.94°	Reentered 5/21/66. One of 2 satellites orbited with single vehicle.
May 15	NIMBUS II (United States)	1966–40A	Thrust-Augmented Thor-Agena B	Total weight: 912 lbs. Objective: Test instruments for studying structure of the atmosphere; observe weather data in untried portions of electromagnetic spectrum while simultaneously observing cloud cover; test basic technology for weather satellites. Payload: 10′-tall structure consisting of 56″-dia. sensory ring forming the base, connected by truss structure to smaller hexagonal-shaped package, and flanked by 2.8′ x 3′ solar paddles covered with 10,500 n-on-p cells; containing 3 TV cameras; Apt vidicon camera system; Hrir instrumentation; Mrir instrumentation; active 3-axis control system; 8 nickel-cadmium batteries; 3 transmitters; 2 receivers; tape recorders; temperature control system.	730	678	108	80°	Still in orbit. NIMBUS II, in good near-polar orbit, provided excellent data from all experiments, including photos of weather globally. Achieved final test objective of 800-orbit continuous operation on 7/15/66. TV tape recorder froze 9/2/66, making picture storage impossible. Still in orbit, still transmitting.

Launch Date	Name	International Designation	Vehicle	Payload Data	Apogee (st. mi.)	Perigee (st. mi.)	Period (minutes)	Inclination	Remarks
May 19	DOD Spacecraft (United States)	1966-41A	Scout	Total weight: 144 lbs. Objective: Develop spaceflight techniques and technology. Payload: Not available.	552	534	103.4	90°	Still in orbit.
May 23	DOD Spacecraft (United States)	1966-42A	Thor-Agena D	Total weight: Not available. Objective: Develop spaceflight techniques and technology. Payload: Not available.	151	104	88.6	66.02°	Reentered 6/9/66.
May 24	COSMOS CXIX (U.S.S.R.)	1966-43A	Not available	Total weight: Not available. Objective: Continuation of Cosmos scientific satellite series. Payload: Not available.	811	136	99.8	48.5°	Reentered 11/30/66.
May 25	EXPLORER XXXII (United States)	1966-44A	Thor-Delta FW-4S	Total weight: 495 lbs. Objective: Measure temperatures, densities, and pressures in the upper atmosphere on a global basis. Payload: 35"-dia. hermetically sealed sphere fitted with 2,064 solar cells, with a canted turnstile antenna projecting from the bottom and 2 18" electrostatic probes from the sides; sphere contains ion mass spectrometer, 2 neutral particle mass spectrometers, 3 magnetron density gages, optical aspect sensors and switch detectors, 2 Pcm telemetry systems, command receiver, tracking transmitter, tape recorder and timer clock, magnetic spin-axis orientation dipole, silver-zinc batteries.	1,685	179	116	64.6°	Apogee was higher than planned (750 mi.) because of excessive burn by 2nd stage engine. Programming adjustment of onboard sensors compensated for high orbit and the aeronomy satellite returned good data. Still in orbit, still transmitting.
May 30	SURVEYOR I (United States)	1966-45A	Atlas-Centaur	Total weight: 2,194 lbs. (weight at launch, including 1,377-lb. retromotor, propellants, etc.; weight of Surveyor lander on the moon, 596 lbs.). Objective: Demonstrate capability of vehicle, spacecraft, and ground equipment to fly a lunar-intercept trajectory, maneuver and communicate, and softland Surveyor on the moon. Payload: 10' high × 14'	Soft-landed on moon				SURVEYOR I made 1st U.S. soft-landing on moon (it was also 1st U.S. attempt) on 6/2/66; landing sequence began 2,000 mi. above lunar surface, when spacecraft positioned itself with retromotor facing downward; retromotor slowed spacecraft from 6,000 mph to 400 mph, then vernier engines continued slowing spacecraft down until it reached 3½

Date	Name	Designation	Launch Vehicle	Payload/Objective	Weight (lbs)			Remarks
						99	90	30°	Reentered 6/11/66.

mph at 14 ft.; from there it free-fell to lunar surface, landing at 8 mph. Scanning the horizon from its location in the Ocean of Storms, the spacecraft took and transmitted to earth some 10,888 photos as well as engineering data indicating a lunar surface bearing strength of 5 psi, good enough for a manned landing. SURVEYOR I proved its ruggedness by surviving the cold of a 2-week lunar night; transmitted last photos 7/14/66.

(around 3 extended landing gear) spacecraft, consisting of triangular aluminum frame to which are attached: a mast supporting rotatable planar array antenna and solar panel (with 3,960 solar cells); 2 folding booms deploying conical omnidirectional antennas; survey TV camera; retromotor with propellants and equipment; 3 vernier motors.

Jun. 1 — ATDA (United States) — 1966-46A — Atlas D

Total weight: 1,748 lbs.
Objective: Serve as target for Gemini rendezvous and docking operation.
Payload: Spacecraft containing ascent shroud, target docking adapter, equipment section, reaction control section, battery.

174 | 99 | 90 | 30° | Reentered 6/11/66.

Jun. 3 — GEMINI IX-A (United States) — 1966-47A — Titan II

Total weight: 8,268 lbs. (including reentry and adapter modules).
Objective: Conduct rendezvous and docking with Augmented Target Docking Adapter; conduct extravehicular activity.
Payload: 18'5" x 10' (dia. at base) 2-module spacecraft, containing 2 astronauts; guidance and control equipment, cameras, 1 HF and 1 UHF transceiver; rendezvous system computer; high and low frequency telemetry transmitters, tracking and recovery communications; 2 fuel cells; environmental control system; reentry and recovery systems.

166 | 99 | 88.79 | 28.91°

GEMINI IX-A mission began on 6/1 with the launch of ATDA, a substitute for the usual Agena target vehicle; problems in guidance system delayed launch of Gemini 9A until 6/3; on 3rd revolution rendezvous was achieved as planned but ATDA was in "angry alligator" configuration because its shroud had not separated; docking was canceled, but 2 more rendezvous were made, one in the downward direction. Eva took place on 3rd day, with Astronaut Thomas Stafford piloting and Eugene Cernan doing Eva; after 72 min., as Cernan attempted to don and operate Astronaut Maneuvering Unit, his faceplate fogged and Eva was called off after 2 hrs. Reentry came on 6/6 after 72 hrs. 21 min. in flight, landing 2 mi. from target in view of live TV carried by EARLY BIRD I satellite. Spacecraft and astronauts were hoisted aboard U.S.S. Wasp.

Launch Date	Name	International Designation	Vehicle	Payload Data	Apogee (st. mi.)	Perigee (st. mi.)	Period (minutes)	Inclination	Remarks
Jun. 8	DOD Spacecraft (United States) and DOD Spacecraft	1966-48A 1966-48B	Atlas-Agena D	Total weight: Not available. Objective: Develop spaceflight techniques and technology. Payload: Not available. Total weight: Not available. Objective: Develop spaceflight techniques and technology. Payload: Not available.	127 75	89 75	88.4 88.3	86.90° 87°	Reentered 6/9/66. One of 2 spacecraft launched with single vehicle. Reentered 6/9/66.
Jun. 6*	OGO III (United States)	1966-49A	Atlas-Agena B	Total weight: 1,135 lbs. Objective: Operate 3-axis-stabilized spacecraft for 1 mo.; make correlated geophysical measurements within magnetosphere and interplanetary space. Payload: 67" x 32" x 31" box-shaped satellite; from the sides extend 2 rotatable solar panels (covered with 83,000 solar cells) and 2 sun-oriented experiments, one of them with a 30' antenna; from the ends of the box extend 2 22' booms and 4 shorter booms, all of them supporting experiments; total of 21 experiments; attitude control system; 3 tracking beacon transmitters; 2 telemetry transmitters; 2 tape recorders; 2 nickel-cadmium batteries.	75,874	188	2,916	31°	OGO III achieved planned highly elliptical orbit; all 21 experiments returned good data; attitude control system exceeded primary objective of 1-mo. operation, functioning more than 6 weeks, then going into spin-stabilized mode. Still in orbit, still transmitting.
Jun. 8	COSMOS CXX (U.S.S.R.)	1966-50A	Not available	Total weight: Not available. Objective: Continuation of Cosmos scientific satellite series. Payload: Not available.	186	124	89.4	51.8°	Reentered 6/16/66.
Jun. 9	DOD Spacecraft (United States) and SECOR VI	1966-51A 1966-51B	Atlas-Agena D	Total weight: Not available. Objective: Develop spaceflight techniques and technology. Payload: Not available. Total weight: 46 lbs. Objective: Continuation of geodetic measurements intended to tie the Pacific Island chain to North America and to provide an equatorial net. Payload: 9" x 11" x 13" rectangle covered with solar cells	2,206 2,248	115 106	124.3 124.9	90.04° 90.05°	Reentered 12/8/66. One of 3 spacecraft launched with single vehicle. Still in orbit.

Date	Name	Designation	Vehicle	Description	Orbital data	Perigee/Apogee	Period	Inclination	Status
	and ERS XVI	1966-51C		and containing solid-state transponder; transmitter. Total weight: 11 lbs. Objective: Test cold-welding experiments for metal-to-metal adhesion of space propulsion materials. Payload: 9" octahedron, containing 5 experiment actuators, including 4 with seat-and-poppet valves and 1 cyclical metal-to-metal contactor with 8 metal combinations; telemetry transmitter; solar cells.	2,238	110	124.7	90.02°	Still in orbit.
Jun. 10	OV8-IV (United States)	1966-52A	Scout	Total weight: 173 lbs. Objective: Measure spectral and depth dose of the inner Van Allen belt. Payload: 29" x 29" octagonal cylinder, with 5 18" booms and containing tissue-equivalent ionization chambers, linear-energy-transfer spectrometer, electron and proton spectrometers, solid-state charged particle spectrometer and triaxial magnetometer, transmitter, 2,560 n-on-p solar cells, nickel-cadmium batteries.	2,252	804	144	40.79°	Still in orbit.
Jun. 16	GGTS I (United States)	1966-53A	Titan IIIC	Total weight: 104 lbs. Objective: Test gravity-gradient stabilization at high orbital altitude; achieve steady-state pointing accuracy of ±8° in pitch and roll within 60 days. Payload: 32" x 36" symmetrical polyhedron with 24 faces; 2 52' booms for gravity-gradient test; 2 earth-albedo sensors; 5 sun sensors; telemetry transmitter; 1,824 n-on-p solar cells.	21,004	20,857	1,888.8	0.17°	Still in orbit. One of 8 spacecraft launched with single vehicle.
	and IDCSP I, II, III, IV, V, VI, VII	1966-53B, C, D, E, F, G, H		Total weight: 700 lbs. (100 lbs. each). Objective: Place in synchronous orbit the 1st 7 of 22 satellites in an interim defense communications satellite system. Payload: 32" x 36" symmetrical polyhedrons with 24 faces; communications relay equipment; 8,000 n-on-p solar cells producing 40 watts of primary power.	(Orbital range) 21,004–21,295	20,857–20,905	1,884.4–1,847.4	0.04°–0.40°	The 7 communications satellites were successfully distributed in their planned orbits and activated for communications operations. Still in orbit.

Launch Date	Name	International Designation	Vehicle	Payload Data	Apogee (st. mi.)	Perigee (st. mi.)	Period (minutes)	Inclination	Remarks
Jun. 17	COSMOS CXXI (U.S.S.R.)	1966-54A	Not available	Total weight: Not available. Objective: Continuation of Cosmos scientific satellite series. Payload: Not available.	220	130	98.9	72.9°	Reentered 6/25/66.
Jun. 21	DOD Spacecraft (United States)	1966-55A	Thor-Agena D	Total weight: Not available. Objective: Develop spaceflight techniques and technology. Payload: Not available.	221	121	90	80.09°	Reentered 7/4/66.
Jun. 23	PAGEOS I (United States)	1966-56A	Thrust-Augmented Thor-Agena D	Total weight: 247 lbs. (including Pageos, inflation powders, canister, and adapter; Pageos itself, 125 lbs.). Objective: Place 100' sphere into circular orbit to act as passive portion of worldwide geodesy system. Payload: 100' (when inflated) Mylar sphere; inflation powders; canister; adapter.	2,645	2,603	182	87°	PAGEOS I went into orbit remarkably close to planned one (2,646/2,636 mi.). Still in orbit.
Jun. 25	COSMOS CXXII (U.S.S.R.)	1966-57A	Not available	Total weight: Not available. Objective: Continuation of Cosmos scientific satellite series. Payload: Not available.	388	388	97.1	65°	COSMOS CXXII was launched from Baikonur cosmodrome, Kazakhstan, and launching was witnessed by President Charles de Gaulle of France, first Westerner allowed to watch a Soviet space launching. Still in orbit.
Jul. 1	EXPLORER XXXIII (United States)	1966-58A	Thrust-Augmented Thor-Delta	Total weight: 206 lbs. (including 81-lb. 4th-stage motor). Objective: Using lunar-anchored spacecraft, study magnetic tail and magnetohydrodynamic wake of earth once a month; measure interplanetary magnetic fields, solar plasma, and energetic particles. Payload: 28" (dia.) x 8" octagonal spacecraft, with 2 7' booms deploying flux-gate magnetometers; 3 radiation monitors; solar-wind analyzers; 4 solar paddles mounting 6,144	278,990	18,975	13.5 days	28.99°	EXPLORER XXXIII's intended lunar orbit was changed to an alternative highly elliptical earth orbit when launch vehicle developed excess speed. All experiments returned good data. Early results indicated tail of earth's magnetosphere extended more than 75,000 mi. beyond moon's orbit. Still in orbit, still transmitting.

Date	Name	Designation	Launch vehicle	Payload/Objective					Remarks
Jul. 5	Saturn (United States)	1966-59A	Uprated Saturn I	Total weight: 58,500 lbs. (includes S-IVB stage, instrument unit, and nose cone and 19,000 lbs. of hydrogen) Objective: Evaluate S-IVB/IU in orbit, including behavior of liquid hydrogen in orbit. Payload: 92'-long S-IVB/IU/nose cone combination; liquid-hydrogen and liquid-oxygen fuel; 2 cameras; telemetry transmitter.	117	117	88.21	31.94°	Heaviest weight orbited by U.S. to date. S-IVB stage went into good orbit; fuel-positioning system worked and engine made simulated restart in orbit. Intended 3-day flight ended during 4th orbit, when deliberately induced internal overpressures ruptured bulkhead and fragmented the payload. Fragments began reentering 7/5/66.
Jul. 6	PROTON III (U.S.S.R.)	1966-60A	Not available	Total weight: 24,000 lbs. Objective: Orbit unmanned space station to study cosmic radiation. Payload: Cylindrical satellite, containing scientific instrumentation "more sensitive than the old equipment"; 4 solar paddles; batteries; telemetry.	391	118	92.5	63.5°	PROTON III was said to be instrumented to attempt to experimentally detect "quarks," particles thought to appear during collisions of cosmic rays with atoms in space. Reentered 9/16/66.
Jul. 8	COSMOS CXXIII (U.S.S.R.)	1966-61A	Not available	Total weight: Not available. Objective: Continuation of Cosmos scientific satellite series. Payload: Not available.	329	163	92.2	48.8°	Reentered 12/10/66.
Jul. 12	DOD Spacecraft (United States)	1966-62A	Atlas-Agena D	Total weight: Not available. Objective: Develop spaceflight techniques and technology. Payload: Not available.	162	93	88.7	95.53°	Reentered 7/20/66.
Jul. 14	OV1-VIII (United States)	1966-63A	Atlas D	Total weight: 74 lbs. Objective: Test new design of passive communications satellite intended to be more efficient in performance and less subject to solar wind. Payload: 30' (dia.) (when inflated) balloon, structured with fine wire grid.	627	619	105.3	144.2°	Experiment was successful. Still in orbit.
Jul. 14	COSMOS CXXIV (U.S.S.R.)	1966-64A	Not available	Total weight: Not available. Objective: Continuation of Cosmos scientific satellite series. Payload: Not available.	188	129	89.4	51.8°	Reentered 7/22/66.

Launch Date	Name	International Designation	Vehicle	Payload Data	Apogee (st. mi.)	Perigee (st. mi.)	Period (minutes)	Inclination	Remarks
Jul. 18	GATV X (United States)	1966-65A	Atlas-Gatv	Total weight: 7,184 lbs. Objective: Serve as target for Gemini rendezvous and docking operation. Payload: 26' x 5' cylinder, containing adapter system, radar transponder, command control system, main engine, 2 secondary engines, attitude control system.	187	180	90.5	28.87°	Reentered 12/9/66.
Jul. 18	GEMINI X (United States)	1966-66A	Titan II	Total weight: 8,248 lbs. Objective: Conduct rendezvous and docking maneuvers with GATV X. Payload: 18'5" x 10' (dia. at base) 2-module bell-shaped spacecraft, containing 2 astronauts; guidance and control equipment; cameras; 1 HF and 1 UHF transceiver; rendezvous system; computer; high and low frequency telemetry transmitters, tracking and recovery communications; 2 fuel cells; environmental control system; reentry and recovery systems.	165	99	88.79	28.80°	GEMINI X achieved rendezvous and docking with GATV X; while docked, Astronauts John W. Young and Michael Collins used the Gatv engine to maneuver the combination into proper orbit for rendezvous with GATV VIII, undocked from GATV X and rendezvoused with passive GATV VIII. Collins performed 2 periods of Eva: standup Eva was terminated after 49 min. because of eye irritation and umbilical Eva after 28 min. because of spacecraft fuel shortage. Reentry occurred on schedule (7/21) in 44th revolution after 70 hrs. 47 min. of flight. Astronauts were picked up by helicopter, flown to U.S.S. *Guadalcanal*.
Jul. 20	COSMOS CXXV (U.S.S.R.)	1966-67A	Not available	Total weight: Not available. Objective: Continuation of Cosmos scientific satellite series. Payload: Not available.	155	155	89.5	65°	Reentered 8/2/66.
Jul. 28	COSMOS CXXVI (U.S.S.R.)	1966-68A	Not available	Total weight: Not available. Objective: Continuation of Cosmos scientific satellite series. Payload: Not available.	223	182	90	51.8°	Reentered 8/6/66.
Jul. 28	DOD Spacecraft (United States)	1966-69A	Titan IIIB-Agena D	Total weight: Not available. Objective: Develop spaceflight techniques and technology. Payload: Not available.	146	94	88.3	94.14°	Reentered 8/6/66.

Aug. 4	OV3-III (United States)	1966-70A	Scout	Total weight: 165 lbs. Objective: Measure hazards from charged particles for space payloads. Payload: Hexagonal cylinder, containing triaxial flux-gate magnetometer; VLF experiment for power density of plasma waves; magnetic fluctuation experiment; omnidirectional proton and electron spectrometers; high and low energy hydrogen-helium nuclei telescopes; Faraday cup and electron and proton spectrometer; induction coil to measure magnetic field; electron field antenna; solar cells; telemetry system.	2,781	223	136.9	81.47°	Returned intended data. Still in orbit.
Aug. 8	COSMOS CXXVII (U.S.S.R.)	1966-71A	Not available	Total weight: Not available. Objective: Continuation of Cosmos scientific satellite series. Payload: Not available.	173	127	89.2	51.9°	Reentered 8/16/66.
Aug. 9	DOD Spacecraft (United States)	1966-72A	Thor-Agena D	Total weight: Not available. Objective: Develop spaceflight techniques and technology. Payload: Not available.	176	118	89.2	100.1°	Reentered 9/11/66.
Aug. 10	LUNAR ORBITER I (United States)	1966-73A	Atlas-Agena D	Total weight: 850 lbs. Objective: Place 3-axis-stabilized Lunar Orbiter in lunar orbit; obtain high-resolution photos of various types of lunar terrain related to Projects Apollo and Surveyor. Payload: 5'6" x 5' (dia.) (when deployed, 18'6" across the solar panels) conical spacecraft, with body containing attitude control system, retromotor, S-band transmitter, duel-lens (24" and 3" focal length) camera system, 2 radiation dosimeters; high and low gain antennas.	(Lunar orbital data:) 1,152	119	218	12.2°	LUNAR ORBITER I on 8/14 became 1st U.S. probe to achieve lunar orbit; on 8/21 perilune was reduced to 36 mi.; on 8/25 to 30 mi.; photographed all 9 primary Apollo landing sites, including the one containing SURVEYOR I, and 7 other Apollo sites and 11 areas on back side of moon. Total of 207 frames (sets) of photos were taken and relayed back to earth; a malfunction caused image smear in high-resolution camera. On 10/29 LUNAR ORBITER I was crashed into the moon's far side to avoid complications in radio control of next Lunar Orbiter.

Launch Date	Name	International Designation	Vehicle	Payload Data	Apogee (st. mi.)	Perigee (st. mi.)	Period (minutes)	Inclination	Remarks
Aug. 16	DOD Spacecraft (United States) and DOD Spacecraft	1966-74A	Atlas-Agena D	Total weight: Not available. Objective: Develop spaceflight techniques and technology. Payload: Not available.	205	89	89.2	98.2°	Reentered 8/24/66. One of 2 spacecraft launched with single vehicle.
		1966-74B		Total weight: Not available. Objective: Develop spaceflight techniques and technology. Payload: Not available.	325	316	94.9	98.16°	Still in orbit.
Aug. 17	PIONEER VII (United States)	1966-75A	Thrust-Augmented Delta	Total weight: 140 lbs. Objective: Obtain scientific data on magnetic field, solar plasma, and cosmic ray density at points farther from the sun than earth is and for a period of at least 3 mos. Payload: 37" (dia.) x 35" cylindrical spacecraft, with a 4'4" boom protruding from the top containing communications antennas and another boom-antennas extending from the bottom to receive radio propagation experiments; 3 5'4" booms protrude from the sides, 2 for control jets, 1 for magnetometer; the sides covered with 10,368 solar cells except for a narrow band in which are located the experiments and 4 orientation sun sensors; data storage; transmitter; batteries; total of 6 experiments: for magnetic field study, single-axis magnetometer; for solar-wind study, plasma cup detector, quadrispherical plasma analyzer; and radio propagation detector; for cosmic ray study, cosmic ray anisotropy detector, cosmic ray telescope.	(Heliocentric orbital data:) 1.125 au	1.01 au	402.95 days	.0946°	PIONEER VII went into planned orbit outside that of earth and away from the sun. All experiments returned data, exceeded 3-mo. operating requirement. On 9/25 PIONEER VII appeared to have detected the tail of the earth's magnetosphere at 8.25 million mi. from earth as earth passed between the sun and the probe. Still in orbit, still transmitting.

ASTRONAUTICS AND AERONAUTICS, 1966

Date	Name	Launch vehicle	Designation	Description	(col 6)	(col 7)	(col 8)	Inclination	Status
Aug. 18	DOD Spacecraft (United States)	Scout	1966-76A	Total weight: 144 lbs. Objective: Develop spaceflight techniques and technology. Payload: Not available.	689	649	106.8	88.85°	Still in orbit.
Aug. 19	DOD Spacecraft (United States)	Atlas–Agena D	1966-77A	Total weight: Not available. Objective: Develop spaceflight techniques and technology. Payload: Not available.	2,294	2,283	167.6	90.18°	Still in orbit. One of 3 spacecraft launched by single vehicle.
	and SECOR VII		1966-77B	Total weight: 46 lbs. Objective: Continue geodetic measurement program. Payload: 9" x 11" x 18" rectangle covered with solar cells and extending 1 antenna from top, 8 from sides; containing solid-state transponder; telemetry system.	2,295	2,282	167.6	90.02°	Still in orbit.
	and ERS-XV		1966-77C	Total weight: 11 lbs. Objective: Develop spaceflight techniques and technology. Payload: 9" octahedron faced with solar cells; 1 dipole antenna.	2,295	2,275	167.4	90.11°	Still in orbit.
Aug. 24	LUNA XI (U.S.S.R.)	Not available	1966-78A	Total weight: 3,608 lbs. Objective: Orbit the moon; measure lunar radiation. Payload: Spacecraft with main power system; braking engine; 4 vernier engines; radiation detection instruments.	(Lunar orbital data;) 745	99	178	270°	LUNA XI entered lunar orbit on 8/27; made observations and had transmitted 137 times when batteries gave out on 10/4; observed gamma rays and x-rays emitted from lunar surface "to determine its chemical composition more exactly." Still in lunar orbit.
Aug. 27	COSMOS CXXVIII (U.S.S.R.)	Not available	1966-79A	Total weight: Not available. Objective: Continuation of Cosmos scientific satellite series. Payload: Not available.	226	182	90	65°	Reentered 9/4/66.
Sep. 12	GATV XI (United States)	Atlas–Gatv	1966-80A	Total weight: 7,199 lbs. Objective: Serve as target for Gemini rendezvous and docking operation. Payload: 26' x 5' cylinder, containing adapter system, radar transponder, command control system, main engine, 2 secondary engines, attitude control system.	185	174	90.5	28.88°	Reentered 12/30/66.

Launch Date	Name	International Designation	Vehicle	Payload Data	Apogee (st. mi.)	Perigee (st. mi.)	Period (minutes)	Inclination	Remarks
Sep. 12	GEMINI XI (United States)	1966-81A	Titan II	Total weight: 8,509 lbs. Objective: Conduct rendezvous and docking maneuver with GATV XI during 1st revolution. Payload: 18'5" x 10' (dia. at base) 2-module bell-shaped spacecraft, containing 2 astronauts; guidance and control equipment; cameras; 1 HF and 1 UHF transceiver; rendezvous system; computer; high and low frequency telemetry transmitters, tracking and recovery communications; 2 fuel cells; environmental control system; reentry and recovery systems.	178	100	88.99	28.85°	GEMINI XI spacecraft rendezvoused successfully with its GATV, docked with it 1 hr. 34 min. into flight. Astronauts Charles Conrad and Richard F. Gordon each docked twice. While still docked, they used the GATV propulsion to maneuver to a record 851/180-mi. orbit, then to lower orbit again. After undocking they conducted a tether exercise, followed by another rendezvous. Eva was performed twice: umbilical Eva, scheduled for 1 hr. 45 min., was ended after 44 min. because of Eva astronaut fatigue; standup Eva lasted 2 hrs. 11 min. Automatic reentry in 46th revolution, occurring in normal; splashdown was within 2 mi. of predicted point; flight had lasted 71 hrs. 17 min. Astronauts were picked up by helicopter and flown to U.S.S. *Guam*.
Sep. 15	DOD Spacecraft (United States)	1966-82A	Thor-Burner II	Total weight: Not available. Objective: Develop spaceflight techniques and technology. Payload: Not available.	559	432	100.8	98.5°	Still in orbit.
Sep. 16	DOD Spacecraft (United States)	1966-83A	Atlas-Agena D	Total weight: Not available. Objective: Develop spaceflight techniques and technology. Payload: Not available.	208	91	89.2	98.9°	Reentered 9/28/66. One of 2 spacecraft launched with single vehicle.
	and DOD Spacecraft (United States)	1966-83B		Total weight: Not available. Objective: Develop spaceflight techniques and technology. Payload: Not available.	306	290	94.2	92.02°	Still in orbit.
Sep. 17	Soviet Spacecraft (U.S.S.R.)	1966-88A	Not available	Total weight: Not available. Objective: Not available.	448	85	93.6	49.59°	First unannounced Soviet satellite since 1/4/63. Perhaps the

Date	Name	Designation	Launch Vehicle	Payload / Remarks		Apogee	Perigee	Period (min)	Inclination	Remarks
Sep. 20	SURVEYOR II (United States)	1966-84A	Atlas-Centaur	Total weight: 2,194 lbs. (weight at launch, including 1,377-lb. retromotor, propellants, etc.; weight of Surveyor lander on the moon, 596 lbs.) Objective: Demonstrate capability of launch vehicle, spacecraft, and ground equipment to fly a lunar-intercept trajectory, maneuver and communicate effectively, and softland the Surveyor spacecraft on the moon. Payload: 10'-high x 14' (around 3 extended landing gear) spacecraft, consisting of triangular aluminum frame to which are attached: a mast supporting rotatable planar array antenna and solar panel (with 3,960 solar cells); 2 folding booms deploying conical omnidirectional antennas; thermal compartment housing 2 transmitters, 2 receivers, signal processing decoder and silver-zinc battery; thermal compartment housing decoder and altitude-radar antennas; survey TV camera; retromotor with propellants and equipment; 3 vernier motors.	Impacted on the moon.					larger of the 2 configurations in the Cosmos series, the spacecraft disintegrated into 51 pieces identifiable by radar; these began reentering 9/20; 88A reentered 11/11/66. SURVEYOR II was launched into near-perfect lunar-transfer trajectory by Atlas-Centaur; at midcourse maneuver, 1 of 3 vernier engines failed to ignite, causing spacecraft to tumble; 39 other attempts were made to start balky engine, without success. Since softlanding was impossible in tumbling condition, ground control made many engineering checks, then on 9/22 fired the retromotor. All communication was lost 30 sec. later. SURVEYOR II crashed on moon SE of crater Copernicus.
Sep. 20	DOD Spacecraft (United States)	1966-85A	Thor-Agena D	Total weight: Not available. Objective: Develop spaceflight techniques and technology. Payload: Not available.		254	111	90.4	85.05°	Reentered 10/12/66.
Sep. 28	DOD Spacecraft (United States)	1966-86A	Titan IIIB-Agena D	Total weight: Not available. Objective: Develop spaceflight techniques and technology. Payload: Not available.		178	91	88.8	98.97°	Reentered 10/7/66.

Launch Date	Name	International Designation	Vehicle	Payload Data	Apogee (st. mi.)	Perigee (st. mi.)	Period (minutes)	Inclination	Remarks
Oct. 2	ESSA III (United States)	1966-87A	Thor-Delta	Total weight: 825 lbs. Objective: Orbit Tos satellite carrying 2 Advanced Vidicon Camera Systems (Avcs) to maintain satellite weather system capability of providing worldwide cloud-cover photos once a day for operational use by ESSA. Payload: 22" x 42" 18-sided hatbox-shaped polygon, with 18" receiving antenna and 4 22" transmitting whip antennas; containing 2 Avcs, FM transmitters, 2 spin-control systems (magnetic coil; small solid-propellant rockets), 2 infrared horizon sensors; 8 solar and terrestrial radiation sensors; 63 nickel-cadmium batteries; 9,100 n-on-p solar cells.	922	858	114.5	101°	ESSA III was put into near-polar orbit by 1st Delta vehicle launched from WTR. All systems functioned normally and photos were of excellent quality. Still in orbit, still transmitting.
Oct. 5	DOD Spacecraft (United States) and SECOR VIII	1966-88A see Sep. 17 1966-89A	Not available	Total weight: Not available. Objective: Develop spaceflight techniques and technology. Payload: Not available.	2,298	2,280	167.6	90.24°	Still in orbit. One of 2 satellites launched with single vehicle.
		1966-89B		Total weight: 46 lbs. Objective: Continue geodetic measurement program. Payload: 12" x 14" x 10" rectangle covered with solar cells; containing high-altitude transponder, batteries, telemetry system; power regulator.	2,299	2,282	167.6	90.2°	Still in orbit.
Oct. 12	DOD Spacecraft (United States) and DOD Spacecraft	1966-90A	Atlas-Agena D	Total weight: Not available. Objective: Develop spaceflight techniques and technology. Payload: Not available.	131	95	88.6	90.95°	Reentered 10/20/66. One of 2 spacecraft launched with single vehicle.
		1966-90B		Total weight: Not available. Objective: Develop spaceflight	103	108	88.9	90.88°	Reentered 10/21/66.

Date	Name	Designation	Launch Vehicle	Payload/Objective	Apogee	Perigee	Period	Inclination	Remarks
Oct. 14	COSMOS CXXIX (U.S.S.R.)	1966-91A	Not available	techniques and technology. Payload: Not available. Total weight: Not available. Objective: Continuation of Cosmos scientific satellite series. Payload: Not available.	191	126	89.4	65°	Reentered 10/21/66.
Oct. 20	MOLNIYA I-4 (U.S.S.R.)	1966-92A	Not available	Total weight: Not available. Objective: Develop and further improve a satellite radio and TV communication system (SovComSat). Payload: Satellite with transmitter, command system, orientation system, orbit correction devices, power supply.	24,658	301	718	64.9°	Still in orbit. On 12/16 U.S.S.R. offered other nations communications facilities via the 4-satellite Molniya I network for communications between the northern countries and Southeast Asia. Launching of MOLNIYA I-4 and COSMOS CXXX was witnessed by leaders of 8 Communist nations.
Oct. 20	COSMOS CXXX (U.S.S.R.)	1966-93A	Not available	Total weight: Not available. Objective: Continuation of Cosmos scientific satellite series. Payload: Not available.	211	131	89.9	65°	Reentered 10/28/66.
Oct. 22	LUNA XII (U.S.S.R.)	1966-94A	Not available	Total weight: Not available. Objective: Orbit the moon; take TV photos of the moon, scientific measurements of lunar radiation, meteoroids. Payload: Spacecraft with main power system; braking engine; 4 vernier engines; TV camera; radiation detection instruments; meteoroid detection system; telemetry; batteries.	(Lunar orbital data:) 1,081	62	205		LUNA XII entered lunar orbit 10/25, took TV photos of lunar surface and transmitted them to earth, where they were shown on Moscow TV; also sent data on x-ray and gamma radiation and on micrometeoroids. Still in orbit.
Oct. 25	Centaur (United States)	1966-95A	Atlas-Centaur	Total weight: 1,627 lbs. Objective: Demonstrate capability of Centaur vehicle to perform 2-burn, indirect lunar ascent mission. Payload: Surveyor mass model, ballasted to simulate Surveyor's retrorocket, solar panels, and antennas; S-band transponder assembly; operational-type separation system.	Completed part of 1 geocentric orbit				Atlas-Centaur attained 100-mi. parking orbit; Centaur's liquid-hydrogen engines reignited, put Surveyor mass model in simulated lunar transfer trajectory. Surveyor reentered 11/6/66. Eighth and last Centaur development flight.

Launch Date	Name	International Designation	Vehicle	Payload Data	Apogee (st. mi.)	Perigee (st. mi.)	Period (minutes)	Inclination	Remarks
Oct. 27	INTELSAT II-A (United States)	1966-96A	Thor-Delta	Total weight: 357 lbs. (with apogee motor; 192 lbs. in synchronous orbit). Objective: Launch satellite and apogee motor into proper transfer orbit, provide tracking and telemetry and backup calculations through the transfer orbit so the satellite can be injected into synchronous orbit. Payload: 56" (dia.) x 26½" circular spacecraft, with apogee motor and whip antennas beneath, sleeve rising above and covering the antennas; containing 2 frequency translation mode repeaters; 4 traveling wave tube amplifiers; control system; 2 batteries; 12,756 n-on-p solar cells.	23,892 After apogee motor firing: 23,330	186 2,072	669.6 728	26.4° 17.6°	NASA launched INTELSAT II-A for ComSatCorp, the Delta launch vehicle putting the spacecraft into an even better transfer orbit than hoped for; test signals were excellent; on 10/30 ComSatCorp fired the apogee motor intending to put the spacecraft into synchronous orbit but the engine failed after 5 sec. Spacecraft remained in elliptical orbit, transmitted occasional commercial messages. Still in orbit, still transmitting.
Oct. 28	OV3-II (United States)	1966-97A	Scout	Total weight: 201 lbs. Objective: Measure charged particle environment near earth. Payload: Spacecraft containing electrostatic analyzer; ion mass spectrometer; plasma probe; retarding potential analyzer; standing wave impedance probe; 2 aspect sensors (magnetometer and solar sensor); transmitter; power supply.	991	198	104.2	81.99°	Still in orbit. Returned desired data.
Nov. 2	DOD Spacecraft (United States)	1966-98A	Atlas-Agena D	Total weight: Not available. Objective: Develop spaceflight techniques and technology. Payload: Not available.	181	100	89	90.95°	Reentered 11/15/66. One of 2 spacecraft launched with single vehicle.
	and DOD Spacecraft	1966-98B		Total weight: Not available. Objective: Develop spaceflight techniques and technology. Payload: Not available.	161	114	88.8	90.6°	Reentered 11/16/66.

ASTRONAUTICS AND AERONAUTICS, 1966

Date	Name	Launch vehicle	Description					Remarks
Nov. 2	Soviet Spacecraft (U.S.S.R.)	1966–101A	Not available	Total weight: Not available. Objective: Not available. Payload: Not available.	470	87	94.6	49.58°	Second Soviet unannounced satellite of the year; like the 1st one (Sep. 17), the satellite disintegrated in orbit, this one into 34 pieces identifiable by radar; these began re-entering 11/6; 101A reentered 11/17/66.
Nov. 3	OV4-III (United States)	1966–99A	Titan IIIC	Total weight: 21,300 lbs. Objective: Perform experiments, study aerodynamics of its long shape during launch and exit from the atmosphere. Payload: 34' x 10' modified Titan II oxidizer tank, including a fuel cell, biocell experiment testing cell growth in weightlessness; 18 corner reflectors for laser experiments, 2 micrometeoroid detectors, heat-transfer experiment; 35' ionospheric sounding antenna.	184	183	90.4	32.85°	Still in orbit. One of 4 spacecraft launched with single vehicle.
and OV4-IR		1966–99B		Total weight: 300 lbs. Objective: Test transmission of intersatellite messages at low power through ionosphere ducts of F layer; measure cosmic noise, electron and air density. Payload: 55" x 17" cylinder with 1 end domed, containing receiver, telemetry system; silver oxide-zinc batteries.	182	180	90.3	32.88°	Still in orbit.
and OV1-VI		1966–99C		Total weight: 445 lbs. Objective: Develop spaceflight techniques and technology. Payload: 68" x 27" cylinder with hemispheres at each end covered with solar cells.	180	179	90.3	32.88°	Reentered 12/31/66.
and OV4-IT		1966–99D		Total weight: 240 lbs. Objective: Develop spaceflight techniques and technology. Payload: 55" x 27" cylinder with 1 end domed; transmitter; silver oxide-zinc batteries.	197	179	90.6	32.84°	Still in orbit.

412 ASTRONAUTICS AND AERONAUTICS, 1966

Launch Date	Name	International Designation	Vehicle	Payload Data	Apogee (st. mi.)	Perigee (st. mi.)	Period (minutes)	Inclination	Remarks
Nov. 6	LUNAR ORBITER II (United States)	1966-100A	Atlas-Agena D	Total weight: 850 lbs. Objective: Place 3-axis-stabilized Lunar Orbiter in lunar orbit; obtain high-resolution photos of various types of lunar terrain related to Projects Apollo and Surveyor. Payload: 5'6" x 5' (dia.) (when deployed, 18'6" along the antenna booms and 12' across the solar panels) conical spacecraft, with body containing attitude control system, retromotor, S-band transmitter, dual-lens (24" and 3" focal length) camera system, 2 radiation dosimeters; high and low gain antennas.	(Lunar orbital data:) 1,163	122	218	12.0°	LUNAR ORBITER II on 11/10 deboosted, went into lunar orbit; on 11/15 another firing of the retromotor dropped the perilune to 31.4 mi., then to 26. Some 211 photos were taken, including 13 primary Apollo sites and 17 secondary ones. Readout of photos stopped 12/6 with failure of high-power transmitter; several photos of 1st Apollo site photographed were lost. Engineering and selenodesy data continued to be transmitted. On 12/8 the spacecraft was commanded to fire its engine for the 4th time and made the 1st known plane change in lunar orbit. Still in orbit, still transmitting.
Nov. 8	DOD Spacecraft (United States)	1966-101A See Nov. 2 1966-102A	Thor-Agena D	Total weight: Not available. Objective: Develop spaceflight techniques and technology. Payload: Not available.	193	107	89.3	100°	Reentered 11/29/66.
Nov. 11	GATV XII (United States)	1966-103A	Atlas-Gatv	Total weight: 7,000 lbs. Objective: Serve as target for Gemini rendezvous and docking operation. Payload: 26' x 5' cylinder, containing adapter system, radar transponder, command control system, main engine, 2 secondary engines, attitude control system.	177	156	90	28.85°	Reentered 12/23/66.

Date	Name	Designation	Launch Vehicle	Remarks				
Nov. 11	GEMINI XII (United States)	1966-104A	Titan II	Total weight: 8,297 lbs. (including reentry and adapter modules). Objective: Conduct rendezvous and docking maneuver with GATV XII in 3rd orbit. Payload: 18'5" x 10' (dia. at base) 2-module bell-shaped spacecraft, containing 2 astronauts; guidance and control equipment; cameras; 1 HF and 1 UHF transceiver; rendezvous system; computer; high and low frequency telemetry transmitter; tracking and recovery communications; 2 fuel cells; environmental control system; reentry and recovery system.	167	100	88.87	28.87°	GEMINI XII rendezvoused and docked with GATV XII in 3rd orbit as planned, but trouble with GATV prevented Astronauts James A. Lovell, Jr., and Edwin E. Aldrin, Jr., from using GATV propulsion for further maneuvers. Aldrin had 3 Eva periods, 2 standup and 1 umbilical, for total of 5 hrs. 58 min. On 11/12 GEMINI XII rendezvoused with total solar eclipse and photographed it. Reentry occurred in 59th revolution on 11/15; splashdown was in the Atlantic, 4 mi. from aiming point, after 94 hrs. 31 min. 31 sec. of flight. Astronauts were picked up by helicopter and landed on U.S.S. Wasp. Last of Gemini flights.
Nov. 12	COSMOS CXXXI (U.S.S.R.)	1966-105A	Not available	Total weight: Not available. Objective: Continuation of Cosmos scientific satellite series. Payload: Not available.	351	133	89.8	72°	Reentered 11/20/66.
Nov. 19	COSMOS CXXXII (U.S.S.R.)	1966-106A	Not available	Total weight: Not available. Objective: Continuation of Cosmos scientific satellite series. Payload: Not available.	174	129	89.3	65°	Reentered 11/30/66.
Nov. 28	COSMOS CXXXIII (U.S.S.R.)	1966-107A	Not available	Total weight: Not available. Objective: Continuation of Cosmos scientific satellite series. Payload: Not available.	144	112	88.4	51.8°	Reentered 11/30/66.
Dec. 3	COSMOS CXXXIV (U.S.S.R.)	1966-108A	Not available	Total weight: Not available. Objective: Continuation of Cosmos scientific satellite series. Payload: Not available.	197	133	89.6	65°	Reentered 12/11/66.
Dec. 5	DOD Spacecraft (United States)	1966-109A	Atlas-Agena D	Total weight: Not available. Objective: Develop spaceflight techniques and technology. Payload: Not available.	226	86	89.7	104.63°	Reentered 12/14/66.

Launch Date	Name	International Designation	Vehicle	Payload Data	Apogee (st. mi.)	Perigee (st. mi.)	Period (minutes)	Inclination	Remarks
Dec. 6	ATS I (United States)	1966-110A	Atlas-Agena D	Total weight: 1,550 lbs. (Including 76-lb. apogee motor system and 768 lbs. of apogee-motor propellant; 775 lbs. after apogee-motor firing). Objective: Place into synchronous orbit and on station the spin-stabilized Applications Technology Satellite; operate spacecraft for minimum of 30 days; obtain useful data from the various applications, technology, and scientific experiments. Payload: 58" (dia.) x 53" cylinder, surfaced with 22,000 n-on-p solar cells; from the bottom end protrude 8 VHF antennas and apogee motor; from the top end protrude 8 whip antennas for telemetry and the microwave antenna; within the spacecraft are the guidance and control system; 2 VLF traveling-wave-tube transponders; 4 telemetry transmitters; communications and weather experiments; technology experiments; and 7 scientific experiments for measuring phenomena in the spacecraft's environment; 6 nickel-cadmium batteries.	22,881 after apogee-motor firing: 22,905	114 22,258	650.4 1,465.9	81.29° 0.28°	ATS I was launched into synchronous transfer orbit, 1st performed by an Atlas-Agena; on 12/7 the apogee motor was fired, circularizing orbit at synchronous altitude; by 12/16 ATS I was on station; all experiments performed well; by 12/18 it had photographed a 1-day weather pattern over 1 spot on earth, relayed color TV from 1 U.S. coast to the other; relayed earth-to-space-to-airliner conversation using existing VHF equipment on airliner; relayed 1st multiple-access satellite communications, handling more than 1 ground station at a time. By 1/19/67 all primary objectives had been met. Still in orbit, still transmitting.
Dec. 11	OV1-IX (United States)	1966-111A	Atlas D	Total weight: 300 lbs. (approx.) Objective: Study radiation hazards in space. Payload: 55" x 27" cylinder with hemisphere at each end covered with solar cells and containing experiment on altitude determination and spin rate; tissue equivalents for use in 3 experiments on cosmic radiation and biohazards.	2,290	295	142.2	99.14°	Still in orbit. Experiments performed as expected. One of 2 spacecraft launched with single vehicle.
	OV1-X and	1966-111B		Total weight: 300 lbs. (approx.) Objective: Study space environment around the earth.	477	397	98.8	98.79°	Still in orbit. Experiments performed as expected.

Date	Name	Designation	Launch Vehicle	Description	Col6	Col7	Col8	Col9	Remarks
Dec. 12	COSMOS CXXXV (U.S.S.R.)	1966-112A	Not available	Payload: 55″ x 27″ cylinder with hemisphere at each end covered with solar cells and containing 8 experiments on Lyman-alpha, dayglow, nightglow, solar x-ray, cosmic radiation, magnetic fields, and thermal coatings; vertistat system of 6 controllable stabilizing booms. Total weight: Not available. Objective: Continuation of Cosmos scientific satellite series. Payload: Not available.	397	158	93.6	48.41°	Still in orbit.
Dec. 14	DOD Spacecraft (United States)	1966-113A	Titan IIIB-Agena	Total weight: Not available. Objective: Develop spaceflight techniques and technology. Payload: Not available.	228	86	89.5	109.5°	Reentered 12/24/66.
Dec. 14	BIOSATELLITE I (United States)	1966-114A	Thrust-Augmented Delta	Total weight: 986.5 lbs. (includes 276-lb. experiment capsule). Objective: Investigate effect of controlled gamma radiation plus weightlessness and of weightlessness alone on living organisms aboard an attitude-controlled spacecraft with less than 10⁻⁴g effect for 3 days; maintain on-board environmental control to keep the organisms alive; expose radiation sources for duration of flight; recover the reentry experiment capsule. Payload: 72″ x 56″ (dia. at base) cylinder-cone adapter section, covered with thermal insulating material, and containing attitude control system (nitrogen jets, IR sensors, magnetometer), telemetry (transmitter, 2 receivers, decoders, tape recorder, tracking beacon), silver-zinc batteries and power controller; antenna; 40″ (dia. at base) reentry vehicle, containing experiment capsule with 13 biological experiments, strontium 85 radiation source and dosimeter; life support system; separation and entry systems; de-orbit telemetry transmitter, programmer, tracking beacon, tape recorder; recovery system.	197	191	91	33.5°	BIOSATELLITE I functioned well for its stipulated 3 days in orbit; on 12/17 the spacecraft was positioned for retrofire and the experiment capsule was separated from the spacecraft, but retrofire apparently did not take place. Capsule reentered 2/15/67 and was thought to have landed in the vicinity of Australia. Search was abandoned 2/22.

Launch Date	Name	International Designation	Vehicle	Payload Data	Apogee (st. mi.)	Perigee (st. mi.)	Period (minutes)	Inclination	Remarks
Dec. 19	COSMOS CXXXVI (U.S.S.R.)	1966-115A	Not available	Total weight: Not available. Objective: Continuation of Cosmos scientific satellite series. Payload: Not available.	190	123		64.6°	Reentered 12/27/66.
Dec. 21	LUNA XIII (U.S.S.R.)	1966-116A	Not available	Total weight: Not available. Objective: Softland on the moon; photograph lunar surface; study lunar surface characteristics; measure lunar radiation. Payload: Lunar descent spacecraft, including retrorocket, landing, radar, communications equipment; also carried ejectable airtight instrument capsule containing camera, cosmic radiation measuring equipment, 2 mechanical manipulators for testing firmness and density of lunar soil, telemetry transmitter, programmer, thermal control system, "energy supply sources," antennas, shock-absorbing system, petal-shaped metal screens as protective covering during landing.	Soft-landed on the moon				LUNA XIII made 2nd U.S.S.R. softlanding on the moon on 12/24 near Ocean of Storms; photographed lunar surface in panorama, showing barren, pitted landscape with stones down to 1 in. in size and no evidence of thick lunar dust; tests by soil density manipulators indicated soil to 1-ft. depth was similar to medium-density soil on earth. Radiation measurements indicated moon's surface reflected about 25% of electrically charged particles which struck it.
Dec. 21	COSMOS CXXXVII (U.S.S.R.)	1966-117A	Not available	Total weight: Not available. Objective: Continuation of Cosmos scientific satellite series. Payload: Not available.	1,032	140	104	49°	Still in orbit.
Dec. 29	DOD Spacecraft (United States)	1966-118A	Thor-Agena D	Total weight: Not available. Objective: Develop spaceflight techniques and technology. Payload: Not available.	329	300	94.4	75.02°	Still in orbit.

Appendix B

CHRONOLOGY OF MAJOR NASA LAUNCHINGS
JANUARY 1, 1966, THROUGH DECEMBER 31, 1966

This chronology of major NASA launchings in 1966 is intended to provide an accurate and ready historical reference, one compiling and verifying information previously scattered over several sources. It includes launchings of all rocket vehicles larger than sounding rockets launched either by NASA or under "NASA direction" (e.g., NASA provided vehicle, launch facilities, and performed the launch for the ComSatCorp INTELSAT II-A).

An attempt has been made to classify the performance of both the launch vehicle and the payload and to summarize total results in terms of primary mission. Three categories have been used for vehicle performance and mission results—successful (S), partially successful (P), and unsuccessful (U). A fourth category, unknown (Unk), has been provided for payloads where vehicle malfunctions did not give the payload a chance to exercise its main experiments. These divisions are necessarily arbitrary, since many of the results cannot be neatly categorized. Also they ignore the fact that a great deal was learned from shots that may have been classified as unsuccessful.

Date of launch is referenced to local time at the launch site. Sources used were all open ones, verified where in doubt from the project offices in NASA Hq. and from the NASA Centers. For further information on each item, see Appendix A of this volume and the entries in the main chronology as referenced in the index. Prepared January 1967 by Dr. Frank W. Anderson, Jr., Deputy NASA Historian (EH).

Date	Name (NASA Code)	General Mission	Launch Vehicle (site)	Performance Vehicle	Performance Payload	Performance Mission	Remarks
Jan 20	Apollo (A-004)	Suborbital Apollo capsule test.	Little Joe II (WSMR)	S	S	S	All priority mission objectives accomplished; Launch Escape Vehicle demonstrated satisfactory performance for an abort in power-on, tumbling flight.
Feb 3	ESSA I (OT-3)	Operational weather satellite.	Thor-Delta (ETR)	S	S	S	First of the Tiros Operational Satellite (Tos) series, 11th straight success of Tiros. Funded by Environmental Science Services Administration (ESSA); built, launched, and initially operated for them by NASA.
Feb 9	Reentry E	27,000 fps reentry test.	Scout (WS)	S	S	S	Vehicle performance was good, flight trajectory nominal. Reentry speed was near the planned 26,950 fps. Failure of part of tracking equipment did not prevent receipt of real-time telemetry in Bermuda.
Feb 26	Saturn IB (AS-201)	Launch vehicle development test.	Saturn IB (ETR)	S	S	S	First launch of Saturn IB; problem with low nitrogen pressure in S-IB stage caused automatic engine cutoff at launch, but count was recycled and launch and suborbital flight were successful. Apollo spacecraft (AFRM 009) was recovered from Atlantic 48 n.m. from planned impact point.
Feb 28	ESSA II (Tos OT-2)	Operational weather satellite.	Thor-Delta (ETR)	S	S	S	First operational weather satellite to offer cloud photos to local Automatic Picture Transmission (Apt) stations. Second of 2 experimental operational satellites funded by ESSA; built, launched, and initially operated by NASA. In near-polar orbit.
Mar 16	GEMINI VIII (Gemini VIII)	Orbital manned space flight.	Atlas-Gatv Titan II (ETR)	S	P	P	First docking in space, made on time 6 hrs. 38 min. into flight; at 7 hrs. unexpected yaw and roll motion caused Astronauts Neil A. Armstrong and David R. Scott to undock, use their reentry control system to stabilize the Gemini spacecraft, and reenter on revolution 7 of planned 44 (10 hrs. 42 min. of planned 72 hrs. 50 min.); landed in stipulated emergency area in Western Pacific, 3 mi. from intended impact point, were picked up by U.S.S. Mason. Other primary objective—Eva for one orbit—was not achieved. Flight trouble was electrical short in a yaw thruster.
Apr 7	Atlas-Centaur (AC-8)	Launch vehicle development test.	Atlas-Centaur (ETR)	P	Unk	U	2nd burn of Centaur engine was not sufficiently prolonged—because of propellant starvation—to inject the 1,780-lb. mass model into a lunar intercept trajectory.
Apr 8	OAO I (Oao-A)	Scientific satellite, astronomy.	Atlas-Agena D (ETR)	S	U	U	Shortly after orbiting, OAO I began having battery heating problems and other electrical malfunctions; on 4/10, after 1½ days in orbit, the battery failed; OAO I ceased transmitting without having returned any scientific data.

Date	Name	Purpose	Launch vehicle				Remarks
May 15	NIMBUS II (Nimbus C/Second Spacecraft)	Meteorological earth satellite.	Thrust-Augmented Thor-Agena B (WTR)	S	S	S	All sensors returned good data for both R&D and operational purposes. Stipulated 800-orbit continuous operation was reached 7/15/66.
May 17	GEMINI IX (Gemini IX)	Orbital manned space flight.	Atlas-Gatv (ETR)	U	Unk	U	Some 121 sec. after launch of the Atlas-Gatv, the No. 2 engine on the Atlas went hard over; vehicle was unable to correct; Gatv stage failed to orbit. Launch of Titan II with Gemini spacecraft was not attempted.
May 25	EXPLORER XXXII (Atmosphere Explorer-B)	Scientific satellite, aeronomy.	Thor-Delta (ETR)	S	S	S	2nd stage burned too long, giving higher apogee than planned. Experiments performed well; higher orbit would be used for studies of density drag. Completed 90-day-minimum operating life 8/25/66.
May 30	SURVEYOR I (Surveyor A)	Lunar probe, soft-landing.	Atlas-Centaur (ETR)	S	S	S	First U.S. attempt at soft landing on the moon was successful; landed on Ocean of Storms. Bearing strength of surface about 5 psi; TV photos excellent; no loose dust. Survived 2-wk. lunar night, continued to operate.
Jun 1	GEMINI IX-A (Gemini IX-A)	Orbital manned space flight.	Atlas-Atda Titan II (ETR)	S U	P U	U	Atlas and Titan launches were successful, 6/1 and 6/3; Astronauts Thomas Stafford and Eugene Cernan rendezvoused with ATDA on 3rd orbit, but docking was not possible because ATDA's shroud had not separated; Cernan performed 2-hr. Eva but fogged visor terminated attempt to exercise the Astronaut Maneuvering Unit; reentered and landed after 72 hrs. 21 min., were hoisted aboard U.S.S. Wasp.
Jun 6*	OGO III (Ogo-B)	Scientific satellite, geophysical.	Atlas-Agena B (ETR)	S	S	S	First highly elliptical orbit satellite to be 3-axis stabilized. All 21 experiments operated, returned good data.
Jun 23	PAGEOS I (Pageos A)	Scientific satellite, geodetic.	Thrust-Augmented Thor-Agena D (WTR)	S	S	S	Sphericity excellent; being tracked by radar and observed optically.
Jul 1	EXPLORER XXXIII (Imp-D)	Scientific probe, radiation.	Thor-Delta (ETR)	P	S	S	Anchored Interplanetary Monitoring Platform (Aimp) did not achieve lunar-anchored orbit because 2nd stage overspeed; went into alternative elliptical earth orbit; all experiments operated, returned desired scientific data.
Jul 5	Apollo-Saturn (AS-203)	Launch vehicle development test.	Uprated Saturn I (ETR)	S	S	S	Orbited an s-IVB stage for a U.S. weight record (58,500 lbs.); its capability for orbital restart checked out successfully. In final test of s-IVB structure (in 4th orbit) the stage fragmented after well exceeding design values.
Jul 18	GEMINI X (Gemini X)	Orbital manned space flight.	Atlas-Gatv Titan II (ETR)	S	S	S	First manned flight to rendezvous with 2 spacecraft; longest docking (38 min. 42 sec.); 1st use of another spacecraft to provide primary and secondary power for docked manned Spacecraft; 2 Eva's. Astronauts John W. Young and Michael Collins landed on 7/21, 4 mi. from U.S.S. Guadalcanal, after 70 hrs. 47 min. (43 revolutions).

Date	Name (NASA Code)	General Mission	Launch Vehicle (site)	Performance			Remarks
				Vehicle	Payload	Mission	
Aug 10	LUNAR ORBITER I (Lunar Orbiter A)	Lunar probe, orbital.	Atlas-Agena D (ETR)	S	P	S	First of 5 lunar orbiter missions was successful; spacecraft went into lunar orbit on 8/14; high-resolution camera returned smeared images after only one good frame, but medium-resolution camera obtained good photos of Apollo landing sites. Photos included back side of the moon and first view of earth from the moon. On 10/29, was crashed into moon's far side.
Aug 17	PIONEER VII (Pioneer B)	Scientific probe, sun-orbiting.	Thrust-Augmented Delta (ETR)	S	S	S	PIONEER VII went into orbit around the sun, successfully locked on earth for communication and stabilization; would provide radiation and magnetic field measurements from widely dispersed points in its solar orbit and for an extended period.
Aug 25	Apollo-Saturn (AS-202)	Launch vehicle development test.	Uprated Saturn I (ETR)	S	S	S	3rd flight test of Uprated Saturn I (Saturn IB) and 2nd flight test of Apollo heat shield. In this suborbital flight, service module motor fired 4 times, sent command module into reentry at 19,900 mph; heat shield withstood high heat load of reentry, and command module was recovered in good condition 500 mi. SE of Wake Island by U.S.S. *Hornet*.
Sep 12	GEMINI XI (Gemini XI)	Orbital manned space flight.	Atlas-Gatv Titan II	S	S	S	Achieved 1st-orbit rendezvous (94 min. GET) and docking, new manned spaceflight altitude record (851 mi.). Astronauts Charles Conrad and Richard Gordon landed on 9/15 after 71 hrs. 17 min. of flight (44 revolutions), were picked up in S. Atlantic by helicopters from U.S.S. *Guam*.
Sep 20	SURVEYOR II (Surveyor B)	Lunar probe, soft-landing.	Atlas-Centaur (ETR)	S	U	U	SURVEYOR II was launched on good trajectory for a lunar landing, but failure of 1 of 3 vernier engines to fire during midcourse maneuver set spacecraft spinning and made soft landing impossible; many engineering experiments were then made, including retrofire. Communications were lost 30 sec. after retrofire, probably from antenna failure. On 9/22, hit the moon SE of the crater Copernicus.
Oct 2	ESSA III (Tos-A)	Operational weather satellite.	Thor-Delta (WTR)	S	S	S	3rd Tiros Operational Satellite and 1st to include Avcs; achieved nearly polar, sun-synchronous orbit to provide daily cloud-cover photos. Launch was 1st from WTR for a Thor-Delta booster. Launched by NASA for ESSA.
Oct 26	Atlas-Centaur (AC-9)	Launch vehicle development test.	Atlas-Centaur (ETR)	S	S	S	Carried out 1st full-thrust restart in space of liquid-hydrogen engine, injecting mass model of Surveyor spacecraft on simulated lunar transfer trajectory. Last of 8 Centaur development test flights.

					S	P	P**	
Oct 27	INTELSAT II-A	Commercial communications satellite.	Thrust-Augmented Delta (ETR)		S			Orbit elliptical instead of geostationary over the Pacific because of apogee motor malfunction. Satellite, nicknamed "LANI BIRD," was functioning and usable for transmissions when in line of sight of ground stations. Transmitted 1st live color program between U.S. mainland and Hawaii. Launched by NASA for ComSatCorp.
Nov 6	LUNAR ORBITER II (Lunar Orbiter B)	Lunar probe, orbital.	Atlas-Agena D (ETR)		S	S	S	2d of 5 Lunar Orbiter missions; by 11/25 had taken all planned 211 medium and high resolution photographs of 13 potential Apollo landing sites. Readout of photos was interrupted near its end on 12/6 by failure of high-powered transmitter. Continued transmitting engineering data.
Nov 11	GEMINI XII (Gemini XII)	Orbital manned space flight.	Atlas-Gatv (ETR) Titan II (ETR)		S	S	S	Astronauts James A. Lovell, Jr., and Edwin E. Aldrin, Jr., rendezvoused and docked with GATV; Aldrin made 2 standup EVA's and one tether EVA, performing work tasks in space. After flight of 94 hrs. 34 min. 31 sec., GEMINI XII landed in Atlantic about 3 mi. from recovery ship U.S.S. *Wasp*.
Dec 6	ATS I (Ats-B)	Applications technology satellite.	Atlas-Agena D (ETR)		S	S	S	Launched into elliptical transfer orbit; on 12/7 onboard apogee-kick motor was fired to place ATS 1 in near-synchronous equatorial orbit, drifting westward at 7° per day toward preselected station of 151° west longitude. First satellite to photograph weather at same spot on earth continuously; 1st to relay messages directly to inflight aircraft using existing VHF equipment. Included spacecraft technology and science experiments.
Dec 14	BIOSATELLITE I (Biosatellite A)	Scientific satellite, biological.	Thrust-Augmented Delta (ETR)		S	U	U	Spacecraft systems and biological experiments performed normally, but spacecraft was not recovered. Command to initiate retrofire and reentry of experiment capsule was sent on 12/17 but retrofire did not occur. Spacecraft reentered 2/15/67 but was not found.

**Non-NASA mission.

Appendix C

CHRONOLOGY OF MANNED SPACE FLIGHT, 1966

This chronology contains basic information on all manned space flights in 1966. The information was compiled by William D. Putnam, Assistant NASA Historian for Manned Space Flight.

The year 1966 was marked by the last five missions of the United States' Gemini program and continuation of the hiatus in Soviet manned space flights (since VOSKHOD II in March 1965). Major historic milestones in manned space flight were rendezvous and docking with another space vehicle on GEMINI VIII, the use of a propulsion unit placed separately into orbit to propel the assembled combined spacecraft to new altitude and speed records on GEMINI X and XI, and extensive astronaut extravehicular activity on GEMINI X, XI, XII. Numerous technological and scientific experiments conducted in Gemini made significant contributions to man's knowledge of the space environment, the earth as observed from the vantage point of space, and astronomical phenomena.

By the end of 1966, the United States had conducted a total of 16 manned space flights, 14 of these orbital, with a total of 19 different crewmen. Seven of the 19 American astronauts had participated in two flights. The Soviet Union had conducted a total of eight manned flights, all orbital, with 11 different crewmen. No Soviet cosmonaut had yet experienced two space flights. Cumulative totals for manned spacecraft hours in flight had reached 1,023 hours 45 minutes for the United States and 432 hours 33 minutes for the Soviet Union. Cumulative total manhours in space were 1,993 hours 36 minutes, and 507 hours 9 minutes, respectively.

Data on United States flights are the latest available to date within NASA. Although details are subject to modification as information is refined, the major aspects of United States manned flights have been subject to direct observation by the interested citizens of the world.

Weight given is the weight of the total spacecraft as placed in orbit. Maximum altitude is chosen from many possible performance measurements because it represents a world-record category as recognized by the Fédération Aéronautique Internationale.

Date Launched	Date Recovered	Designation	Crew	Weight (lbs.)	Revolutions	Max. Apogee (st. mi.)	Duration	Remarks
Mar. 16	Mar. 16	GEMINI VIII	Neil A. Armstrong David R. Scott	8,351	6½	186	10 hrs. 41 min.	Gemini spacecraft launched into orbit by modified Titan II booster, 1 hr. 41 min. after Gemini Agena Target Vehicle (Gatv)–Atlas booster combination. Rendezvous with the GATV was accomplished in fourth orbit, 6 hours after launch, and first docking of two vehicles in space was achieved 33 min. later. After 28 min. of flight in the docked configuration a thruster in the orbital attitude maneuvering system stuck open and spacecraft began to spin. Gemini spacecraft was undocked from GATV and, after 25 min., was stabilized by use of reentry control system. Mission terminated early with landing in secondary recovery area in Western Pacific. Scheduled extravehicular activity (Eva) was not accomplished. GATV was placed in a "storage" orbit for later use as a rendezvous target.
Jun. 3	Jun. 6	GEMINI IXA	Thomas P. Stafford Eugene A. Cernan	8,268	45	194	72 hrs. 21 min.	First mission attempt made on 5/17 when an Atlas booster engine failed at 121 sec. after liftoff, and the Gatv failed to reach orbit. A backup mission utilizing Augmented Target Docking Adapter (Atda) as the target was scheduled for 6/1. The Atda was placed into orbit on 6/1 with an Atlas booster, but inability of Gemini spacecraft to receive final guidance information caused postponement of Gemini launch to 6/3; then spacecraft was launched into orbit by its modified Titan II booster. Rendezvous with ATDA achieved on third orbit. Crew confirmed that shroud covering the docking collar had failed to separate and docking was canceled. Two additional rendezvous were performed as planned, one using visual techniques, the last from above the ATDA. Eva was delayed from 6/4 to 6/5 due to crew fatigue. After 1 hour of Eva, Cernan's visor accumulated fog and communications between Stafford and Cernan were poor as Cernan checked out the Astronaut Maneuvering Unit (Amu) in the adapter section of the spacecraft. Planned use of the self-contained life support and propulsion systems to maneuver with the Amu was canceled. Total 2 hr. 7 min. Eva. Radio polarization and photographic experiments were conducted in addition to micrometeoroid collection. Controlled reentry to .42 mi. of target.

July 18	GEMINI X	John W. Young Michael Collins	8,248	43	476	70 hrs. 47 min.	GATV-Atlas booster combination launched 100 min. before Gemini spacecraft launched into orbit by modified Titan II booster. Rendezvous and docking with GATV X accomplished in fourth orbit. GATV engine used to propel the docked combination to record altitude, then proper orbit to rendezvous with GATV VIII. After separation from GATV X, GEMINI X effected rendezvous with the passive target. First period of standup EVA was terminated at 50 min. when both crewmen suffered eye irritation. Second EVA period, after rendezvous with passive GATV, consisted of Collins on umbilical moving to GATV and recovering micrometeorite experiment package. Umbilical EVA terminated after 39 min. to conserve maneuvering fuel. Of 14 scheduled experiments, data were obtained on 12. First rendezvous with two different spacecraft, first extensive test of docked spacecraft, first post-docking maneuvers using propulsion unit and fuel of target vehicle, first crewman to touch another spacecraft. New altitude and speed (476 mi. and 17,700 st. mph) records for manned space flight.
Sep. 12	GEMINI XI	Charles Conrad, Jr. Richard F. Gordon, Jr.	8,509	44	853	71 hrs. 17 min.	GATV-Atlas booster combination launched 1 hr. 37 min. before Gemini spacecraft was launched into orbit by modified Titan II booster within 2-sec. launch window. Rendezvous accomplished on first orbit using onboard information exclusively. Docking at 1 hr. 34 min. after launch. Docking accomplished twice by each crewman. During umbilical EVA, Gordon removed the nuclear emulsion experiment package and attached tether between Gemini spacecraft and GATV. Excessive fatigue from these activities caused early termination of EVA at 33 min. After returning to, and repressurizing spacecraft, the hatch was opened for 3 min. to jettison umbilical EVA equipment. Using GATV primary propulsion system, docked configuration was propelled to new altitude and speed (853 mi. and 17,943 st. mph) records. Hatch opened third time for 2 hr. 10min.standup EVA during which scheduled photography was accomplished. The two spacecraft undocked and station keeping by use of 100-ft. nylon tether was exercised with a slow spin rate imparted to the tethered combination. After separation of the tether a second rendezvous with GATV was conducted from 25-mi. separation. Reentry sequence was fully automatic with impact 2.9 mi. from the aiming point. Ten of 11 scheduled experiments were conducted as planned, power tool evaluation canceled due to shortened umbilical EVA.

Date		Designation	Crew	Weight (lbs.)	Revolutions	Max. Apogee (st. mi.)	Duration	Remarks
Launched	Recovered							
Nov. 11	Nov. 15	GEMINI XII	James A. Lovell, Jr. Edwin E. Aldrin, Jr.	8,297	59	187	94 hrs. 35 min.	GATV-Atlas booster combination launched 1 hr. 38 min. before Gemini spacecraft was launched into orbit by modified Titan II booster. Rendezvous was accomplished in third orbit with docking 4 hrs. 14 min. after Gemini launch. Scheduled boost of docked GEMINI-GATV combination into higher orbit was canceled due to pressure fluctuations in GATV primary propulsion system. The docked combination was maneuvered to obtain photographs of solar eclipse during 10th revolution. Two standup EVA periods, on first and third days, totaled 3 hrs. 34 min. On second day, Aldrin conducted 2 hr. 6 min. umbilical EVA, testing restraint devices to overcome body positioning problems experienced on prior flights. Resting frequently, Aldrin used portable hand rails, foot restraints, and various tethers. He completed 19 tasks, demonstrating that useful EVA work was feasible with proper planning and restraint devices. GEMINI and GATV spacecraft undocked, remaining joined by 100-ft. tether for station keeping and gravity-gradient stabilization experiment. Fourteen experiments were conducted. Automatic reentry sequence was used for second time.

Appendix D

ABBREVIATIONS OF REFERENCES

Listed here are abbreviations for sources cited in the text. This list does not include all sources provided in the chronology, for some of the references cited are not abbreviated. Only those references which appear in abbreviated form are listed below. Abbreviations used in the chronology entries themselves are cross-referenced in the Index.

A&A	AIAA's magazine, *Astronautics & Aeronautics*
A&A 66	NASA's *Astronautics and Aeronautics 1966* [this publication]
ABC	American Broadcasting Company
AEC Release	Atomic Energy Commission News Release
AF Info. Pol. Ltr.	Air Force Information Policy Letter for Commanders
AF/SD	*Air Force and Space Digest* magazine
AFOSR Release	Air Force Office of Scientific Research News Release
AFSC Release	Air Force Systems Command News Release
AGU *Transactions*	American Geophysical Union *Transactions*
AIAA Release	American Institute of Aeronautics and Astronautics News Release
AIP News	*American Institute of Physics News*
AMR	*Applied Mechanics Reviews*
AP	Associated Press
ARC Release	NASA Ames Research Center News Release
Atlanta J/Const.	*Atlanta Journal and Constitution* newspaper
Atomic Energy Programs, 1966	AEC, *Major Activities in the Atomic Energy Programs, 1966*
Av. Daily	*Aviation Daily* newsletter
Av. Wk.	*Aviation Week and Space Technology* magazine
Balt. Sun	Baltimore *Sun* newspaper
CBS	Columbia Broadcasting System
Chic. Trib.	*Chicago Tribune* newspaper
ComSatCorp Release	Communications Satellite Corporation News Release
CR	*Congressional Record*
CSM	*Christian Science Monitor* newspaper
CTNS	Chicago Tribune News Service
DAC Release	Douglas Aircraft Co. News Release
DJNS	Dow Jones News Service
DMSSD *Apogee*	Douglas Missile and Space Systems Div. *Apogee*
DOD Release	Dept. of Defense News Release
EH	NASA Historical Staff (formerly EPH)
EPH	NASA Historical Staff (now EH)
ESSA Release	Environmental Science Services Administration News Release
FAA Release	Federal Aviation Agency News Release
FonF	*Facts on File*
FRC Release	NASA Flight Research Center News Release

FRC X-Press	NASA Flight Research Center's *FRC X-Press*
GE Forum	*General Electric Forum* magazine
Goddard News	NASA Goddard Space Flight Center's *Goddard News*
GSFC Release	NASA Goddard Space Flight Center News Release
GSFC SSR	NASA Goddard Space Flight Center's *Satellite Situation Report*
Houston Chron.	*Houston Chronicle* newspaper
HTNS	New York Herald Tribune News Service
Int. Science and Tech.	*International Science and Technology* magazine
J/Armed Forces	*Journal of the Armed Forces*
JPL Release	Jet Propulsion Laboratory News Release
KB	NASA Inventions and Contributions Board
KSC Release	John F. Kennedy Space Center, NASA, News Release
Langley Researcher	NASA Langley Research Center's *Langley Researcher*
LaRC Release	NASA Langley Research Center News Release
L.A. Times	*Los Angeles Times* newspaper
Lewis News	NASA Lewis Research Center's *Lewis News*
LRC Release	NASA Lewis Research Center News Release
M&R	*Missiles and Rockets* magazine
Marshall Star	NASA George C. Marshall Space Flight Center's *Marshall Star*
Minn. Trib.	*Minneapolis Tribune* newspaper
MSC Release	NASA Manned Spacecraft Center News Release
MSC Roundup	NASA Manned Spacecraft Center's *Space News Roundup*
M/S Daily	*Missile Space Daily* newsletter
MSFC Release	NASA George C. Marshall Space Flight Center Release
NAA Skywriter	North American Aviation, Inc., *Los Angeles Skywriter*
NAA Release	North American Aviation, Inc., News Release
Nat'l. Aeron. Assn. Release	National Aeronautical Association News Release
NANA	North American Newspaper Alliance
NASA Ann.	NASA Announcement
NASA Auth. Hearings	*NASA Authorization* [FY 1967] *Hearings*
NASA Hq. Bull.	NASA Headquarters' *Weekly Bulletin*
NASA Hq. PB	NASA Headquarters Personnel Bulletin
NASA LAR V/50	NASA Legislative Activities Report, Vol. V, No. 50
NASA Off. Int. Aff.	NASA Office of International Affairs
NASA Proj. Off.	NASA Project Office
NASA Release	NASA (Hq.) News Release
NASA Rpt. SRL	NASA Report of Sounding Rocket Launching
NASA SP-4006	NASA Special Publication #4006
NASA X-15 Proj. Off.	NASA (Hq.) X-15 Project Office
NASC Release	National Aeronautics and Space Council News Release
NAS-NRC Release	National Academy of Sciences-National Research Council News Release
NAS-NRC-NAE *News Report*	National Academy of Sciences-National Research Council-National Academy of Engineering *News Report*
NBC	National Broadcasting Company
NMI-	NASA Management Instruction-
NN	NASA Notice
NSC Release	National Space Club News Release
N.Y. Her. Trib.	*New York Herald Tribune* newspaper
N.Y. J/Amer.	*New York Journal-American* newspaper
NYT	*New York Times* newspaper
NYTNS	New York Times News Service
OAR Release	Office of Aerospace Research (USAF) News Release
Orl. Sent.	*Orlando Sentinel* newspaper
P&W Release	Pratt & Whitney Div. News Release
Phil. Eve. Bull.	Philadelphia *Evening Bulletin* newspaper
Phil. Inq.	*Philadelphia Inquirer* newspaper
Pres. Doc.	National Archives and Records Service's *Weekly Compilation of Presidential Documents*
Sat. Eve. Post	*Saturday Evening Post* magazine
SBD	*Space Business Daily* newsletter
Sci. Amer.	*Scientific American* magazine
Sci. Serv.	Science Service

S.F. Chron.	*San Francisco Chronicle* newspaper
SR	*Saturday Review* magazine
Tech. Wk.	*Technology Week* magazine (formerly *Missiles and Rockets*)
Testimony	Congressional testimony, prepared statements
Text	Prepared report or speech text
Transcript	Official transcript of news conference or congressional hearing
UPI	United Press International
U.S. Aeron. & Space Act., 1966	President's Report to Congress, *United States Aeronautics and Space Activities, 1966*
U.S. Naval Inst. Proc.	*U.S. Naval Institute Proceedings* magazine
U.S. News	*U.S. News and World Report* magazine
USS–T	Translation by NASA Scientific and Technical Information Div., Translators
Wallops Release	NASA Wallops Station News Release
Wash. Daily News	*Washington Daily News* newspaper
Wash. *Eve. Star*/Wash. *Sun. Star*	Washington *Evening*/*Sunday Star* newspaper
Wash. Post	*Washington Post* newspaper
WBE Sci. Serv.	World Book Encyclopedia Science Service
WSJ	*Wall Street Journal* newspaper

INDEX

A-1 (French satellite), 56, 94
Aa. See Apollo Applications program.
AAAS. See American Association for the Advancement of Science.
AAS. See American Astronautical Society.
ABC. See American Broadcasting Co.
Abelson, Dr. Philip H., 1
Abhau, R/Adm. Conrad (USN), 332
Able, Charles R., 167
Abraham, Karl, 18
Abres. See Advanced Ballistic Reentry Systems.
Academia Sinica, 339
Accelerator, 376
Accident (see also Aircraft, accident), 25, 55, 129, 183, 281
Acv. See Air-cushion vehicle.
Adam (Air Deflection and Modulation) II (V/Stol aircraft), 309
Adams, Rep. Brock, 90
Adams, Dr. Mac C., 1, 62, 92, 111
Adams, Maj. Michael J. (USAF), 313, 357
Adams, Dr. Thomas W., 55
Adcsp. See Advanced Defense Communications Satellite Project.
Adelaide, Australia, 162
Adler, Fred, 250
Administrative History of NASA, 1958-63, 231
Advanced Ballistic Reentry Systems (Abres), 54
Advanced Defense Communications Satellite Project (Adcsp), 200
Advanced Manned Strategic Aircraft (Amsa), 72, 341
Advanced Research Projects Agency (ARPA), 15, 322, 376
Advanced Vidicon Camera System (Avcs), 178, 283, 310
AEC. See Atomic Energy Commission.
AEDC. See Arnold Engineering Development Center.
Aegean Sea, 87
Aerial Coast Patrol (U.S. Coast Guard), 278
Aerobee (sounding rocket)
150
    booster test, 297
    micrometeoroid sampling, 326, 328, 350
    radar test, 171
    solar astronomy, 156, 187, 337, 347
    ultraviolet astronomy, 145, 190, 241, 242, 295, 352
    upper atmosphere data, 60, 138, 277, 360
  x-ray astronomy, 89, 219, 319
150-A
    ultraviolet astronomy, 18, 182
    upper atmosphere data, 79, 84, 276
350—269
Aeroflot, 50, 339
Aerojet-General Corp., 62, 103, 110, 235, 259, 283, 295, 353, 375
Aeronautical and Space Research Institute (IIAE), 342
Aeronautical Research Associates of Princeton, Inc., 151
Aeronautics, 66, 208, 223, 280, 323
Aeronomy satellite (see also EXPLORER XXXII), 192, 289
Aerospace, 200
Aerospace Corp., 354
Aerospace Education Foundation, 36
Aerospace Industries Assn., 97, 122, 160, 230, 295, 367, 378
Aerospace industry, 97, 230, 295, 350, 372, 378, 382
Aerospace Medical Research Laboratories, 203
Aerospace medicine. See Space biology.
Aerospace Sciences Meeting, Third, 26
Aerospace Systems Conference, 238
AFA. See Air Force Assn.
AFCRL. See Air Force Cambridge Research Laboratories.
AFFTC. See Air Force Flight Test Center, Edwards AFB, Calif.
AFL-CIO. See American Federation of Labor–Congress of Industrial Organizations.
AFSC. See Air Force Systems Command.
Agaltsov, Air Marshal V. A., 270
Agena (booster) (see also Atlas-Agena, Thor-Agena), 54, 67, 92
  High-reliability, 213
Agena Target Vehicle. See Gemini Agena Target Vehicle.
Agency for International Development (AID), 34
AGM-69A (Short Range Attack Missile), 334
Agreement, 179, 208
  FAA-AID, 34
  international, 125, 155, 198, 220, 236, 285, 330
  NAS-Czechoslovak Academy of Sciences, 231
  NASA
    -Argentina, 253
    -Brazil, 14, 161, 310, 320
    -ComSatCorp, 235
    -Department of Commerce, 103
    -DOD, 7, 29, 43, 110, 150, 198, 305, 334

-ESRO, 383
-France, 195
-Naval Oceanographic Office, 19
-Spain, 15, 137–138
-USAF, 128–129
-USN, 8
-U.S.S.R., 90, 156
-Vocational Rehabilitation Administration, 222, 328
-West Germany, 289, 303, 304
U.S.–U.S.S.R., 221, 247, 250, 293, 298, 339, 355, 368, 370
U.S.S.R.–France, 227, 261, 281, 334, 370, 386
Agriculture, Dept. of, 3, 117
AGU. See American Geophysical Union.
Ahmedabad, India, 235
AIAA. See American Institute of Aeronautics and Astronautics.
AIAA-USN Marine Systems Conference, 271
AID. See Agency for International Development.
Aimp. See Anchored Interplanetary Monitoring Platform.
Air cargo, 197
Air-cushion vehicle (Acv), 168
Air et Cosmos, 22, 82
Air Force Academy, 139
Air Force Assn. (AFA), 36, 58, 60, 113, 294
Air Force Avionics Laboratory, 330, 342
Air Force Cambridge Research Laboratories (AFCRL), 79, 84, 124, 146, 161, 337, 350
Air Force Flight Test Center (AFFTC), Edwards AFB, Calif., 3, 180, 264
Air Force Man of the Year, 60
Air Force Systems Command (AFSC), 6, 57, 66, 77, 87, 111, 129, 172, 188, 285, 295, 316
   Aeronautical Systems Div., 114
   arresting gear, aircraft, 180
   associateships, 367
   award, 168, 268, 294, 315, 353
   contract, 72, 88, 159, 177, 220, 225, 284, 335, 348, 368, 375
   Electronic Systems Div., 283
   experiment, 342
   Flight Dynamics Laboratory, 87
   personnel, 146, 148
   Space Systems Div., 255, 268
   Systems Engineering Group, 284
Air-Launched Air-Recoverable Rocket (Alarr), 118–119
Air Line Pilots Assn., 64
L'Air Liquide, 370
Air Museum Act, 49
Air pollution, 142, 194
Air Products and Chemicals, Inc., 114, 257
Air Proving Ground Center (Eglin AFB, Fla.), 131
Air Research and Development Command, 316
Air traffic control, 18
Air Transport Association of America (ATA), 165, 197, 250, 265
Air University Review, 253

Aircraft (see also individual aircraft, such as B-52, X-15, etc.), 139–150, 267, 282, 329
   accident, 3, 28, 58, 64, 76, 159, 211, 214, 250, 263, 367, 378
   liability, 43, 165
   prevention, 159, 262, 265, 284
   anniversary, 259, 377
   bomber, 61, 72, 153, 270, 341
   cargo, 164, 333, 358
   carrier, 228
   fighter, 94, 111, 116, 119, 168, 195, 256, 262, 288, 323, 367, 378
   foreign, 43, 50, 58, 118, 196, 226, 235, 270, 333
   general-aviation, 250, 295
   helicopter, 60, 132, 187, 194, 290, 291, 326, 340, 344
   hypersonic, 66, 382
   interceptor, 27
   personal, 37, 134
   reconnaissance, 27, 28, 251, 253
   record, 62, 132, 134, 187, 250, 251, 253, 291, 302, 333, 351, 385
   research, 5, 16, 47, 62, 114, 129, 158, 211, 223, 225, 338, 378, 385
   statistics, 172, 210
   tracking, 128
   training, 76, 85
   transport (see also Aircraft, V/Stol, Vtol; Supersonic transport)
      air-bus, 323, 330
      jet, 58, 137, 143, 145, 149, 153, 235, 250, 285, 330, 334, 358
      military, 57, 281
   variable-sweep-wing, 61, 116, 168, 215, 383
   V/Stol, 72, 106, 131, 185, 194, 253, 309, 330, 364
   Vtol, 94, 284, 340, 365
Airlessness, 61
Airlines, 180, 223, 272, 334, 339
   anniversary, 159
   forecast, 11, 172, 197
   statistics, 165, 209, 250
   strike, 236, 254, 257, 272
Airlock, 272
Airports, 91, 155, 217, 243, 330, 354, 364, 377
AiResearch Manufacturing Div., 156
Alabama, 341
Alabama, Univ. of, 365, 379
Alarr. See Air-Launched Air-Recoverable Rocket.
Alaska, 161, 216, 221, 226, 260, 265, 310, 356, 367
Albert, Col. John G. (USAF), 354
Albuquerque, N. Mex., 221
Aldrin, L/C Edwin E., Jr. (USAF), 110, 219, 327, 340, 344, 349, 353
Alexander, George, 261
Alfven Propulsion Engine (Ape), 152
All-weather landing system (Awls), 66
Allen, Alice S., 154
Allen, H. Julian, 145
Allen, Richard J., 354

Allentown, Pa., 325
"Alleviation of Jet Aircraft Noise Near Airports" (report), 107
Allied Research Associates, Inc., 214, 275
Allmanna Svenska Elektriska, A.B., 239
ALOUETTE II (Canadian satellite), 63, 357
Alpha Lyrae (star), 241
Alpha Virginis (star), 40
Alsep. See Apollo Lunar Surface Experiment Package.
AMA. See American Medical Association.
America Illustrated, 304
American Academy of Political and Social Science, 141
American Association for the Advancement of Science (AAAS), 22, 382
American Association of School Administrators, 52
American Astronautical Society (AAS), 61, 121, 134, 160, 189, 221, 225, 314, 382
American Broadcasting Co. (ABC), 46, 256, 287, 355
American Chemical Society, 322
American Federation of Government Employees (AFGE), 33, 258
American Federation of Labor-Congress of Industrial Organizations, 271
American Geophysical Union (AGU), 38, 79, 144, 215
American Institute of Aeronautics and Astronautics (AIAA), 74, 159, 307
award, 323
fellows, 235, 289
meeting, 26-27, 89-90, 119-120, 134, 162, 235, 271, 314, 338, 357-359
American Institute of Physics, 46-47
Center for History and Philosophy of Physics, 229
American Legion, 137
American Medical Assn. (AMA), 8
American Meteorological Society, 325
American Optical Co., 209
American Ordnance Assn., 168, 341
American Petroleum Institute, 350
American Philosophical Society, 147
American Physical Society, 31-32, 150-152
American Science and Engineering, Inc., 89, 219, 229, 319
American Society for Engineering Education, 365
American Society for Microbiology, 166
American Society for Public Administration (ASPA), 21-22, 140
American Society of Civil Engineers, 315
American Society of Mechanical Engineers (ASME), 166
American Society of Tool & Manufacturing Engineers, 108
American Telephone & Telegraph Co. (AT&T), 114, 126, 131, 152, 256, 278, 375
Ames, Milton B., Jr., 99
Ames Research Center (ARC), 258, 378
award, 20, 145, 352
Hypervelocity Free Flight Facility, 246
probe, 79, 269, 301
Project Luster, 326
research, 74, 152-153, 271, 347, 361, 374

Amsa. See Advanced Manned Strategic Aircraft.
AMTRAN. See Automatic Mechanical Translator.
Amu. See Astronaut Maneuvering Unit.
An-2 (U.S.S.R. airliner), 256
An-10 (U.S.S.R. airliner), 256
An-22 (U.S.S.R. airliner), 50
An-154 (U.S.S.R. double-decked airliner), 50
Anaheim, Calif., 189
Anchored Interplanetary Monitoring Platform (Aimp), 231
Anders, Edward, 144
Anders, William A., 110, 379
Anderson, Sen. Clinton P., 157, 208, 373
Anderson, Roy E., 321
Anderson, Wendell, 82
Andoeya, Norway, 111, 124, 322
Andover, Me., 122, 278, 314, 331
Andrews AFB, Md., 131
Andrews, Rep. Mark, 104
ANL. See Argonne National Laboratory.
Ann Arbor, Mich., 110, 113, 116
Anna (geodetic satellite), 162
Anniversary, 15, 193, 278, 304
flight
  manned, 107, 135, 259
  unmanned, 103, 240
Government, 123, 217, 224, 302, 310
industry, 159, 241, 347
institution, 217, 329
LRC, 23, 312
satellite, 28, 35, 106, 227, 312, 322
Anosov, Aleksey, 291
Antaeus (U.S.S.R. cargo aircraft), 333
Antarctica, 157, 170, 348
Antenna, 88, 157, 197, 372
Antimissile missile, 72, 170, 343, 355
Antioch, Calif., 237
Antiparticles, 225
Antonov, Oleg K., 50
AP. See Associated Press.
Ape. See Alfven Propulsion Engine.
Apollo (program), 238, 246
astronaut, 57, 110, 144, 241, 243, 269, 306, 375, 379
experiment, 15, 29, 255, 322
  hydrogen tank conversion, 167, 272
facilities, 122, 176, 311, 331
funds for, 24, 288, 301
launch (see also Apollo spacecraft)
  AS-201, 67
  AS-202, 274
  AS-203, 233
lunar exploration, 5, 11, 25, 314
  landing site, 96, 296, 313
management, 268, 289, 291
plans for, 5, 64, 72, 75, 93, 127, 167, 229, 332, 350, 375
progress, 95, 167, 179, 193, 362
test, 127, 167, 375, 384-385
tracking, 30, 81, 128
training, 40, 144, 360, 376
Apollo (spacecraft), 25, 193, 332, 351
Command Module (Cm), 21, 25, 40, 67, 93, 156, 272, 274, 298, 313

heat shield, 67, 112, 274
launch
  AS-201, 51, 67, 93
  AS-202, 274
launch vehicle. See Saturn.
Lunar Module (Lm), 25, 40, 96, 272, 375, 385
Service Module (Sm), 25, 132, 272, 274, 328
test, 328, 360, 385
  Launch Escape System (Les), 19, 21, 42
recovery, 35
Apollo Applications (Aa) program
  contract, 195, 315
  funds for, 23, 138, 149, 171, 278
  management, 235, 239
  plans for, 13, 52, 59, 261
  proposed merger, 111, 137, 307
Apollo Lunar Surface Experiment Package (Alsep), 6, 29, 83, 104, 221
Apollo/Range Instrumentation Aircraft (A/Ria), 7
Apollo Telescope Mount (Atm), 239, 289, 303, 325, 356, 360
Appleton, D. C., 377
Applications Technology Satellite (Ats) (see also ATS I), 77, 84, 295, 353, 361, 372, 380, 385
  launch, 367
Applied Physics Laboratory (APL) (Johns Hopkins University), 162
Apt. See Automatic Picture Transmission.
Aquanaut, 316, 367
ARC. See Ames Research Center.
Arcas (sounding rocket) (see also Boosted-Arcas), 15, 65, 87, 117, 173, 342
Archangel, U.S.S.R., 227, 372
Arches of Science Award, 312
Arco, Ida., 276
ARCS (Achievement Rewards for College Scientists) Foundation, 366
Arctic Circle, 299
Arencillo, Spain, 322, 325
Arendt, Dr. Paul R., 254
Arenstorf, Richard F., 315
Argentina, 14, 60, 154, 253, 297, 342, 346
Argentine Space Commission (CNIE), 154, 253, 342
Argo D-4 (sounding rocket). See Javelin.
Argonne National Laboratory (ANL), 142, 240
A/Ria. See Apollo/Range Instrumentation Aircraft.
ARINC Research Corp., 165
Arizona, 12, 339
Arizona, Univ. of, 44, 96, 116, 187, 224, 325
Arlington National Cemetery, 85
Armed Forces Communications and Electronics Assn., 212
Armed Forces Management, 335
Armed Forces Management Assn., 172
Armstrong, Neil A., 110, 137, 208, 214, 297, 316
  award, 117
  GEMINI VIII (flight), 99–101, 104, 105, 106, 107, 112, 117, 131
  training, 90, 131
Army Aviation Materiel Laboratories (AVLABS), 156
Army Corps of Engineers, 11, 114, 154, 332
Army Electronic Command Labs., 254
Arnold Air Society, 129
Arnold Engineering Development Center (AEDC), 8, 53, 54, 67, 86, 121, 126, 132, 189, 221
  Rocket Propulsion Research Laboratory, 111
Arnold, Gen. Henry H. ("Hap"), 168
Arnold, Henry H., Gold Medal, 168
Arons, Daniel M., 282
Arp, Dr. Halton C., 93
ARPA. See Advanced Research Projects Agency.
Asahi Shimbun, 109
Ascension Island, 69
Asia, 54, 61, 95, 152
Asia-Pacific Seminar, 25
Asia, Southeast, 376
ASME. See American Society of Mechanical Engineers.
Aso, Japan, 299
Asp (booster), 236
ASPA. See American Society for Public Administration.
Assembly of the Western European Union, 210
Associated Electrical Industries, Ltd., 136
Associated Press (AP), 55, 149, 161, 264, 322, 327, 332, 365, 367, 373
Associated Universities, 257
Asteroid, 69, 317
Astronaut (see also Extravehicular activity), 25, 98, 109, 189, 203, 213, 217, 246, 248, 313, 323, 381, 382
  Apollo mission, 106, 110, 351, 360, 376, 379
  contract, life story, 161
  death, 72, 76, 85
  flights
    GEMINI IV, 9, 105, 208, 232
    GEMINI V, 9, 105, 304
    GEMINI VI-A (Gemini VI), 8, 30, 54, 86, 226, 259, 374
    GEMINI VII, 12, 54, 86, 87, 170, 203, 208, 226, 259, 374
    GEMINI VIII, 17, 53, 86, 96, 99, 100, 104, 105, 145, 162, 226
    GEMINI IX-A (Gemini IX), 85, 105, 110, 192, 178, 180, 206–207, 208, 209, 213, 218, 236
    GEMINI X, 27, 226, 240, 243–244, 247, 248, 249, 255
    GEMINI XI, 110, 290, 293, 294, 297, 301, 304
    GEMINI XII, 218, 327, 340, 344, 347
  former, 54, 60, 105, 169, 191, 242, 314, 328
  good-will tour, 54, 60, 69, 72, 169, 297
  honors, 9, 12, 16, 22, 30, 69, 87, 191, 208, 209, 225, 232, 236, 255, 259, 304, 374

performance, 7, 69, 243
physiology, 166, 243, 273, 285, 308
press conference, 92, 218, 257, 301, 353, 375
record, 251
rescue, 78, 111, 145, 164, 192, 250, 269, 297, 336, 384
scientist-astronaut, 302
selection, 7, 127, 134
training, 14, 76, 122, 130, 198, 240, 375–376
vision, 7, 57, 78, 236, 294, 342
women as, 7, 107, 324
Astronaut Maneuvering Unit (Amu), 206, 304
Astronautics & Aeronautics, 37, 73, 159, 200, 229, 253, 281, 308, 361, 362, 363
Astronautics Engineer Award, 103
Astronomical clock, 169
Astronomisches Rechen Institut, (Heidelberg, W. Germany), 220
Astronomy (see also Radioastronomy; Star), 39, 40, 63, 65, 95, 121, 125, 339, 384
 gamma-ray, 151, 263
 optical, 95, 219
 solar, 11, 79, 87, 137, 156, 174, 187, 274, 278, 337, 341, 346, 347
 star catalog, 107
 ultraviolet, 18, 79, 138, 145, 182, 190, 241, 242, 295, 352
 x-ray, 79, 89, 96, 219, 229, 319, 328, 347, 373
Astrophysical Journal, 141, 229
ATA. See Air Transport Association of America.
AT&T. See American Telephone & Telegraph Co.
Atacama Desert, Chile, 202
Atda. See Augmented Target Docking Adapter.
Athena (missile), 342, 367
Athens, Greece, 206
Atlantic City, N. J., 52, 136, 233
Atlantic Ocean, 235, 247, 269, 290, 308
Atlantic Research Corp., 117
Atlas (booster), 92, 96, 121, 180, 182, 193, 202, 213, 243, 290, 378
 D, 320
Atlas (missile), 54
 D, 108
 F, 261
Atlas-Agena (booster), 92, 380
 Gatv 9, 180–181
 launch, 180
  Agena B (ogo III), 209
  Agena D
   ATS I, 367
   Ers satellites, 212, 272
   Lunar Orbiters, 257, 263, 340
   OV1-IV, V, 121
   Secor satellites, 212, 272, 312
   unidentified, 20, 54, 106, 121, 132, 143, 178, 208, 238, 268, 293, 319, 337, 367
  high-reliability Agena, 213

Atlas-Centaur (booster) (see also Centaur), 58, 238, 380, 385
 AC-7, 296
 AC-8, 120, 131, 164
 AC-9, 329
 AC-10, 164, 196
Atm. See Apollo Telescope Mount.
Atmosphere (see also Ionosphere)
 artificial, 12, 13, 285, 335
 meteorological experiments, 3, 221, 260, 265, 276, 342, 360, 364
 grenade, 23, 39, 51, 161, 164, 216, 226, 260, 307, 310
 upper, study of, 59, 79, 237, 267, 276, 277
Atomic bomb, 232, 246
Atomic clock, 353
Atomic Energy Commission (AEC) (see also NASA-AEC Space Nuclear Propulsion Office, Snap, Nerva, Rover, and Vela programs), 19, 41, 128, 147, 165
 accelerator, 10, 376
 annual report, 35
 Argonne National Laboratory, 142, 240
 award, 259
 Brookhaven National Laboratory, 10, 225, 257
 budget, 24, 171
 contract, 112
 cooperation, 42, 82, 240, 336, 360
 National Reactor Testing Station, 13, 276
 nuclear reactor, 13, 56–57
 nuclear rocket engine, 42, 52, 82, 103, 115, 145, 211, 336, 385
 nuclear test, 49, 342, 366
 Oak Ridge National Institute, 243
 radioisotope generator, 83, 112, 227
 space nuclear power, 27, 56, 273
Atomic Industrial Forum, 336
Ats. See Applications Technology Satellite.
ATS I (Ats-B) (Applications Technology Satellite), 367, 369, 372, 385
Auburn, Mass., 103, 263, 350
Auburn Univ., 365, 379
Augmented Target Docking Adapter (Atda), 8, 181, 202, 206
AURORA 7 (spacecraft), 107
Aurora borealis, 95, 137, 299, 352
Austin, Tex., 287
Australia, 288, 315
 good-will tour, 54, 95
 INTELSAT II-A transmissions, 360, 365
 international cooperation, 123, 191, 267, 329
 tracking stations, 64, 162, 234
Automatic Mechanical Translator (AMTRAN), 81
Automatic Picture Transmission (Apt), 70, 178, 284, 368
Avco Corp., 230, 264, 327, 359
Avcs. See Advanced Vidicon Camera System.
Aven, Alex P., 173
Aviation/Space Writers' Assn., 166, 191–192, 223, 275, 378

Aviation Week and Space Technology
  aviation, 143, 197, 308
  defense, 49, 72, 213, 243
  Government, 19, 128, 317
  international, 220, 285, 334, 356
  space, 2, 11, 96, 149, 150, 171, 232–233, 250, 261, 273, 295
Aviatsiya i Kosmonavtika, 55
AVLABS. See Army Aviation Materiel Laboratories.
Awards
  civic, 331
  Government, 43, 50, 57, 77, 117, 218, 223, 259, 304, 314, 380
  institutions, 22, 140, 324, 377
  society
    aeronautics, 20, 166, 188, 208, 235, 298, 301, 308, 323, 374
    astronautics, 103, 225, 235, 321
    engineering, 145, 213, 230, 296
    medical, 215
    military, 168, 294
    science, 376, 380
Awls. See All-weather landing system.
Ax-4 (balloon), 319
Aykovlev, Col. Gen. Alexander (U.S.S.R.), 118

B

B-52 (Stratofortress), 168, 377
  M2-F2 flights, 114, 237, 244, 264, 274, 284, 293, 297, 298, 306, 312, 319, 330, 348
Bacchus, Utah, 226
Backus, George E., 52
Bahrein Island, 219
Baikonur, U.S.S.R., 121, 223, 324, 372
Bailey, F. John, 149
Baja California, 297
Bak-11 (barrier arresting component device), 180
Baldwin, Robert H. B., 13
Ball, Robert S., Memorial Award, 166
Balloon, 185, 251
  accident, 161
  award, 225
  high-altitude, 152, 153, 242, 263, 275
  record, 40, 237, 319
  weather, 3, 89, 141, 168, 195, 287, 333
Baltimore, Md., 134, 167, 325
Baltimore *Sun*, 5, 210, 381
Bangkok, Thailand, 72
Barabashov, Nikolay, 47
Barber, Richard J., 165
Barbon, Dr. Roberto, 287
Barium, 299, 301
Barnes, Tracy, 289
Barron, Paul A., 139, 396
Barstow, Calif., 211, 263
Bartholemew, C. S., 134
Bascom, Willard N., 154
Bassett, Capt. Charles A., II (USAF), 14, 72, 76, 85, 110
Bastin, J. A., 316
Battelle Memorial Institute, 379

Baum, Dr. Werner A., 293
Bazykin, Prof. Victor, 12
BBC. See British Broadcasting Corp.
Beachey, Lincoln, 383
Beale AFB, Calif., 9
Beaudry, Col. Emil G. (USAF), 78
Becker, H. S., 284
Becklin, Eric E., 277
Beecher, William, 200
Belai, U. E., 318
Belew, Leland F., 235, 315
Belgium, 125, 191, 210, 220
Belka (dog), 154
Bell Aerospace Corp., 78, 106, 225
Bell Aerosystems Co., 168
Bell, Rep. Alphonzo, 128
Bell Telephone Laboratories, 31, 51, 329, 380
Beller, William S., 232
Belossov, Dr. Vladimir V., 251
Belton, M. J. S., 9
Belyayev, Col. Pavel (U.S.S.R.), 92, 107
Bendix Corp., 104, 210, 244, 261, 288
Bendix Field Engineering Co., 315
Bennett, Dr. Ivan L., Jr., 111, 328
Bennett, Robert A., 251
Benton, Dr. George S., 111
Bergeron, Prof. Tor, 210
Berkeley, Calif., 221, 283
Berkner, Dr. Lloyd V., 146, 251, 315
Berlin, Germany, 324
Berman, Col. Robert A. (USAF) 294,
Blue Book, Project, 326
Berry, Dr. Charles A., 8, 134, 166, 208, 225, 294, 318, 323
Berry, William R., 237
Beverly Hills, Calif., 299
Bhabha, Dr. Homi J., 46, 85
Bhagavantam, Dr. S., 106
Binder, Alan B., 187
Bioastronautics. See Space biology.
Biology and the Exploration of Mars (report), 199
Biosatellite (program), 84, 187
BIOSATELLITE I (Biosatellite A), 374
Bioscience. See Space biology.
Birmingham, Ala., 312
BIS. See British Interplanetary Society.
Bisplinghoff, Dr. Raymond L., 26, 74, 128, 159, 223, 280, 307, 362
Bjurstedt, H. B., 318
Black and Decker Tool Co., 78
Blackman, A. W., 27
Blagonravov, Dr. Anatoli A., 28, 46, 173, 175, 287
Blanchard, Gen. William H. (USAF), 199
Blessley, L/C R. C. W. (USAF), 116
Bleymaier, B/G Joseph S. (USAF), 113
Bloom, Harold L., 192
Blossom Point, Md., 307
Blue Book, Project, 112, 116, 130, 325, 326
Blue Streak (British rocket), 190
BMWF. See German Ministry of Scientific Research.
BNS. See Broadcasters Nonprofit Satellite Service.

BOAC. See British Overseas Airways Corp.
BOB. See Budget, Bureau of.
Bochum Observatory (W. Germany), 45, 104, 109, 126
Boeing Co., 134
 anniversary, 241
 booster, 223
 contract, 3, 85, 115, 194, 244, 249, 332
 jet engine inlet, 199
 Lunar Orbiter, 49
 supersonic transport, 85, 215, 225, 264, 286, 296, 299, 307, 323, 334, 341, 383
 Vertol Div., 17
Boeing 707 (jet aircraft), 143, 145
Boeing 720 (jet aircraft), 143
Boeing 727 (jet aircraft), 64
Boeing 747 (jet aircraft), 137, 143, 250
Boeing, William F., 383
Bogard, Dr. Howard M., 88
Boggs, Rep. Hale, 224
Bogotá, Colombia, 316
Bolivia, 297
Bollerud, Col. Jack (USAF), 54
Bolling AFB (D. C.), 123
Bomarc (missile), 275
Bomber aircraft, 61, 72, 153, 168
Bonn, W. Germany, 303
Bonney, Walter T., 36
Bono, Phil, 318
Boone, Adm. William F., 1, 198
Boosted Arcas (sounding rocket), 15, 87, 178, 185, 187, 249, 253, 277,
Boosted Dart (sounding rocket), 14, 15, 253
Booz-Allen Applied Research, Inc., 376
Borman, L/C Frank (USAF), 87, 144, 203, 306, 379
 award, 208, 226, 259, 374
 good-will tour, 54, 61, 66, 69, 72, 95
 interview, 12, 16, 170
Boron fiber, 77
Boston Globe, 21
Boston, Mass., 235, 330, 332
Boulder, Colo., 133
Boushey, Homer A., 259
Bow, Rep. Frank T., 84
Bowers, Jack A., 354
Boy Scouts of America, 169
Boyd, Alan S., 362
Boynton, Melbourne W., Award, 225
Bozajian, John, 250
Bradner, Hugh, 52
Bradt, Hale, 229
Branch, M/G Irving H. (USAF), 3
Brand, Vance D., 127
Bratchell, G. E., 354
Brazil, 14, 22, 154, 161, 252, 260, 265, 297, 310, 320, 347, 373
Brazilian Space Commission (CNAE), 14, 161, 252, 310, 320, 347, 373
Brenner, Edward J., 319
Bretigny Tracking Center (France), 56
Brevard (County) Building and Construction Trades Council, 240
Brewer, Frank G., Trophy, 210
Brewer, Jack D., 330

Brewer, L/C Martin H. (USAF), 294
Brewster, Sen. Daniel B., 20
Brewster Flat, Wash., 179
Brezhnev, Leonid I., 94, 356
Brighton, U. K., 189
Bristol Siddeley Engines, Ltd., 235
British Aircraft Corp., 356
British Assn. for the Advancement of Science, 285
British Broadcasting Corp. (BBC), 287
British Interplanetary Society (BIS), 45
British Meteorological Office, 199
British National Space Research Committee, 114, 118
British Overseas Airways Corp. (BOAC), 143, 145
British Research Council, 329
Brizendine, John C., 330
Broadcasters Nonprofit Satellite Service (BNS), 256
Bromberg, Jack L., 317
Brookhaven National Laboratory, 10, 116, 225, 257
Brooks AFB, Tex., 203
Brooks, Col. Allison (USAF), 164
Brooks, Melvin, 115
Brown, Ed, 111, 183
Brown, Gov. Edmund, 275
Brown Engineering Co., 130
Brown, Rep. George E., 246
Brown, Dr. Harold, 113, 122, 130, 138, 150, 199, 219, 232, 264, 335, 366
Brownell, Charles, 78
Brush Instruments Co., 216
Buchwald, Dr. Julius, 88
Buckley, Edmond C., 1, 83, 157
Budget, Bureau of (BOB), 112, 128, 258, 278, 364
Buenos Aires, Argentina, 60
Buffalo, N. Y., 330
Bulgaria, 214
Bull, Lt. John S. (USN), 127
Bundy, McGeorge, 270, 370
Bunker, George M., 354
Burcham, Dr. Donald P., 198
Bureau of Ships (USN), 13
Burke, Walter F., 354
Burma, 60
Burroughs, Richard Hansford, Test Pilot Award, 20
Bush, Edgar G., 208
Bush, Dr. Vannevar, 155
Butler, Herbert I., 43
Butler, Stuart Thomas, 251
Bykovsky, Valery F., 55
Byram, E. T., 80
Byrd, Sen. Harry F., Jr., 12

C

C-5A (cargo transport), 35, 58, 377
C-124 (cargo aircraft), 7
C-133 (Cargomaster), 333
C-141 Starlifter (cargo jet), 66
CAAGS. See Civil Aviation Assistance Groups.

CAB. See Civil Aeronautics Board.
Cabell, Rep. Earle, 90
Cable & Wireless, Ltd., 219
Caen, France, 348
Cahn, Robert, 28
Califano, Joseph A., Jr., 258
California, 215, 275
California Institute of Technology (Cal Tech), 2, 29, 90, 96, 124, 128, 152, 230, 236, 253, 260, 277, 329, 348
California, Univ. of, 10, 163, 329
  Berkeley, 85, 90, 221, 283, 286, 359
  La Jolla, 159
  Los Angeles (UCLA), 32, 128, 142, 320, 837
  San Diego, 7, 27
Callaghan, Richard L., 2
Cal. Tech. See California Institute of Technology.
Calvin, Melvin, 90
Cambridge Astrophysical Observatory. See Smithsonian Cambridge Astrophysical Observatory.
Cambridge Catalog of Radio Objects, Third, 253
Cambridge, Mass., 37, 342
Cambridge Research Laboratories. See Air Force Cambridge Research Laboratories.
Cambridge Univ., 57
Camera (see also Advanced Vidicon Camera System; Automatic Picture Transmission System; SURVEYOR I; LUNAR ORBITERS I and II)
  COSMOS CXXII, 308
  GEMINI VIII, 107
  K-24 ballistic, 114, 118
  reconnaissance, 289
  lunar, 11, 12, 252
  miniature, 234
Cameron, Roy E., 202
Camm, Sir Sydney, 94
Camp Roberts, Calif., 361
Canada, 43, 63, 94, 252, 353, 357
Canadian Aeronautics and Space Institute, 94
Canadian Defence Research Board Telecommunications Establishment, 313
Canadian National Research Council, 252, 348
Canadian Overseas Telecommunication Corp., 126
Canberra, Australia, 64
Cannon, Dr. Robert H., Jr., 298, 362
Canopus (star), 203, 262, 340
Cape Canaveral National Seashore, 15
Cape Kennedy, Fla. (see also Eastern Test Range and Kennedy Space Center), 244, 303
Carabela 4 (Spanish sounding rocket), 322
Carbon dioxide, 142, 273
Carlson, Loren D., 90
Carnarvon, Australia, 365
Carnegie Commission on Educational Television, 256, 278
Carpenter, Leslie, 301
Carpenter, Cdr. M. Scott (USN), 25, 89, 189, 316
Carr, Maj. Gerald P. (USMC), 127
Carrobio, Renzo di, 41
Carter, Launor F., 370
Case Institute of Technology, 365
Casey, Rep. Robert, 90
Cassidy, Dr. William A., 94
Cassino, Brazil, 347
Castiella, Fernando, 137
Castro, Fidel, 193
Castruccio, Dr. Peter A., 61
Cat. See Clear Air Turbulence.
Cate, Col. Albert M. (USAF), 264
Catholic Univ., 365
Causse, Jean-Pierre, 188
Cavanagh, Mayor Jerome P., 366
CBS. See Columbia Broadcasting System.
Cedar Crest College, 325
Celestial mechanics, 83
Centaur (booster) (see also Atlas-Centaur), 298
Centaure (French sounding rocket), 87, 124, 253, 299, 327, 352
Center for Nuclear Studies, 370
Centre National d'Etudes Spatiales (CNES), 188, 195, 323
Centre National d'Etudes des Télécommunications, 20
Centre National de la Recherche Scientifique, 264
CERN. See European Center for Nuclear Research.
Cernan, LCdr. Eugene A. (USN), 379
  award, 251
  GEMINI IX-A (Gemini IX), 110, 180, 206, 209, 213, 219, 236,
  training, 14, 178, 198
Cesium engine, 336
Cessna 206 (light aircraft), 134
Cetus (constellation), 287
Chaffee, Dr. R. R., 136
Chaffee, LCdr. Roger B. (USN), 110, 144, 332
Chamberlin, James A., 354
Chamical Rocket Range, Argentina, 154, 253
Champine, Robert, 253
Chandrasekhar, Subrahmanyan, 380
Chaparral (missile), 267
Chapman, Dr. J. H., 353
Charyk, Dr. Joseph V., 131, 172, 221
Chavanac, Mayor Albert, 191
Chemistry, 1
Cheng, Dr. Chu-yuan, 30
Cheng, Richard, 250
Chevalier, Roger, 188
Chew, Peter, 13
Cheyenne Mountain, Colo., 32
Chicago American, 294
Chicago Assn. of Commerce and Industry, 154
Chicago Daily News, 332
Chicago Ill., 330, 332, 366, 376
Chicago Sun Times, 261
Chicago Tribune, 16, 228
Chicago, Univ. of, 144, 151, 339, 380
Childs, J. Howard, 16

Chile, 94, 162, 202, 297
Chimpanzee, 336
China, 381
China, Communist, 325
  engineers, 30
  missile, 331, 355
  nuclear test, 172, 331, 382
  scientist exchange, 77
  scientists, 30
Christchurch, New Zealand, 89, 168
Christian Science Monitor, 28, 122
Chrysler Corp., 5, 22, 122, 178, 250, 255, 263, 267, 353, 378
Chubb, T. A., 80
Churchill Research Range, Canada, 252
  launch
    Aerobee, 60, 150
    Nike-Apache, 50, 55, 95, 237, 245, 260, 274, 284, 292
    Nike-Cajun, 23, 39, 51, 161, 164, 216, 221, 226, 260, 265
    Nike-Tomahawk, 137, 283, 293
Churchman's Club, Baltimore, 167
Cincinnati, Univ. of, 252
"City of San Francisco" (balloon), 237
Civil Aeronautics Board (CAB), 159, 180, 227
Civil Aviation Assistance Groups (CAAGS), 34
Claggett, C. Thomas, Jr., 314
Claremont, Calif., 232
Clark, Evert, 57, 95, 294, 333, 348, 368, 376
Clark, Dr. John F., 44, 166, 264, 315
Clark, Sen. Joseph, 190
Clark Univ., 208, 314
Clear air turbulence (Cat), 23, 381
Clear Lake (Tex.) Chamber of Commerce, 325
Clearinghouse for Federal Scientific and Technical Information, 235
Cleaver, Arthur V., 189
Clemence, Dr. Gerald M., 252
Clock, atomic, 353
Cloud, 267
  noctilucent, 252, 260, 266
  photographs, 7, 310, 365, 369
Cloud, James D., 250
Cm. See Command module.
CNAE (Comissao Nacional de Atividades Espaciais). See Brazilian Space Commission.
CNES. See Centre National d'Etudes Spatiales.
CNIE (Comisión Nacional de Investigaciónes Espaciales). See Argentine Space Commission.
Cobb, Dr. James C., 116
Cobb, Roland W., 151
Cocoa Beach, Fla., 89
COESA. See Committee on Extension to the Standard Atmosphere.
Coggan, B. F., 34
Colbert, Ralph, 250
Coleman Bros. Corp., 332
Coleman, Paul J., Jr., 151

Colgate Univ., 117
College of William and Mary, 365
College Park, Md., 273
Collier, Robert J., Trophy, 324
Collins, Maj. Michael (USAF), 27, 240, 246, 248, 249, 255, 306, 379
Collins, Wayne, 367
Collision, 263–264, 284, 317–318, 320
Colloquium of Law of Outer Space, 317
Colombia, 297, 316
Colorado, 341
Colorado Springs, Colo., 139, 215
Colorado, Univ. of, 315
Colorado Women's College, 107, 138
Columbia, Mo., 350
Columbia Broadcasting System (CBS), 256, 287, 327
Columbia Univ., 10, 57, 150, 163, 258, 279, 286, 302, 324, 343, 377, 378
  Nevis Laboratory, 225
Columbus, Christopher, 298
Columbus Day, 298
Columbus, Ohio, 134
Colwell, Arch T., Merit Award, 213
Comet, 235, 260, 277, 287
Comisión Nacional de Investigación del Espacio (CONIE) (Spain), 15
Comisión Nacional de Investigaciónes Espaciales (CNIE). See Argentine Space Commission.
Comissao Nacional de Atividades Espaciais (CNAE). See Brazilian Space Commission.
Command module (Cm), 21, 25, 40, 67, 93, 156, 272, 274, 313
Commerce, Dept. of, 80, 103, 111, 323, 354, 381
Committee on Extension to the Standard Atmosphere (COESA), 121
Committee on Space Research (COSPAR), 173
Communications.
  deep space, 3, 9, 84, 157, 263
  global, 67, 73, 83, 96, 172
  international, 118, 120, 139, 152, 162, 198, 210, 221, 235, 281, 386
  laser use in, 51, 126, 209
  military use, 35, 200, 220, 323, 362
Communications satellite (see also individual satellites: EARLY BIRD I, MOLNIYA I, INTELSAT II-A, etc.), 180, 223
  aeronautical, 91, 120
  agreement, 40, 65, 235
  contract, 20, 142
  cooperation, 65, 331
    international, 40, 155, 237, 354
  military, 40, 52, 200, 216, 220, 321, 323, 362, 376
  system, 20, 83
    domestic, special purpose, 81, 131, 146, 256, 270, 278, 288, 375
    foreign, 149, 271, 325, 353
    global, 67, 172, 287, 331, 376
  transmission via, 46, 152, 154, 162–163, 182, 287, 360
  launch
    INTELSAT II-A, 331

Initial Defense Communications Satellite Program, 216
MOLNIYA I (3), 149
MOLNIYA I (4), 325
use of, 98, 118, 162–163, 278, 338
Communications Satellite Corp. (ComSatCorp), 83, 118, 125, 165, 199
contract, 20, 142, 179, 232
EARLY BIRD I, 65, 67, 83, 122, 252, 331
revenue, 125, 252
transmission, 237, 278, 350
Gemini recovery, 86, 179, 207, 290, 332
ground station, 122, 131, 152, 199, 314, 331, 350, 368
INTELSAT II-A ("LANI BIRD"), 331, 341, 360, 365, 385
rates, 311, 327
Research and Development Center, 221, 337
satellite program, 46, 67, 73, 114, 136, 146, 197, 256, 278, 287, 288, 375
services, 237, 311, 365
test, 237, 360
Communications blackout, 88, 330, 367
Communist Party Congress, 125
Compania Telefónica Nacionale de Espana (CTNE), 97
Compagnie Francaise Thomson-Houston, 239
Comprehensive Designers, Inc., 288
Computer Sciences Corp., 169, 223
Computers, 348
foreign use, 372
NASA, 9, 81, 104, 165, 169, 223, 262, 265, 267, 273, 275, 291–292, 298
universities, 220
Concorde (U.K. - France) supersonic transport, 19, 21, 43, 123, 143, 189 194, 226, 233, 289, 354
Condon, Dr. Edward U., 315
Condor (missile), 227
Conference on Biomedical Knowledge, 328
Congress, 80, 83, 119, 131, 138
AEC report to, 35
budget request, President's 23–24
NASA budget, 23–24
State of Union message to, 14
Congress, House of Representatives, 12, 19, 49, 119, 171, 211, 215, 224, 227, 244, 274, 359
bills introduced, 84, 288
bills passed, 163, 171, 172, 228, 244, 246, 247, 279
Committee on Appropriations, 54–55, 166, 171, 182, 211, 213,
Committee on Armed Services, 87, 116, 130
Committee on Civil Service, Subcommittee on Manpower, 258–259
Committee on Government Operations, Subcommittee on Military Operations, 111, 323
Committee on Science and Astronautics, 10, 28–29, 59, 76, 90, 91, 92, 116, 127, 143, 232–233, 281, 291, 325, 357–358, 359
Subcommittee on Advanced Research and Technology, 62, 66, 81, 83–84 88–89
Subcommittee on Manned Space Flight, 59, 64–65, 273
Subcommittee on NASA Oversight, 180, 272, 280
Subcommittee on Space Science and Applications, 56, 62–64, 65–66, 71, 75–76, 113, 200, 244, 275, 308
Congress, Senate, 20, 32, 34, 89, 187, 212, 227, 247, 261, 266, 273
bills introduced, 21, 167
bills passed, 190, 227, 247, 263
Committee on Aeronautical and Space Sciences, 12, 70–71, 80–81, 91, 176, 208, 373
Committee on Appropriations, 122, 257, 263
Committee on Armed Services, 72, 122, 145
Commitee on Commerce, Subcommittee on Communications, 270
Committee on Foreign Relations, 271
Committee on Finance, 21
Committee on Government Operations, Subcommittee on Government Research, 179
Committee on Judiciary, 268
Congressional Record, 12, 119
CONIE. See Comisión Nacional de Investigación del Espacio.
Connes, Dr. Pierre, 264, 322
Connes, Dr. Janine, 264, 322
Connor, John T., 77, 111
Conrad, LCdr. Charles, Jr. (USN), 7, 54, 60, 110, 290, 293, 294, 304, 379
Construction Industry Joint Conference, 364
Contamination, 75
Contract (see also under agencies, such as NASA, USAF, etc.)
cost-plus-award-fee, 223, 235, 257, 315, 330
cost-plus-fixed-fee, 224, 257, 262, 383
cost-plus-incentive-fee, 130, 257, 165–166, 169, 224, 255, 352, 353
fixed-price, 85, 142, 145, 176, 195, 215, 220, 223, 298, 338, 375
fixed-price-incentive-fee, 334, 368, 375
fixed-price letter, 375
incentive–fee, 85, 90, 244, 353
sole source, 321
university, 247, 251, 315, 328
Control Data Corp., 157, 165, 215, 258
Convair 990 (jet aircraft), 153, 154
Cooke, Richard, 382
Cooper, George, 352
Cooper, L/C L. Gordon (USAF), 7, 54, 60, 105, 219
Cooper, B/G Paul T. (USAF), 268, 354
Copernicus (moon crater), 344
Coralie (rocket engine), 356–357
Cornell Univ., 10

Corona, 241, 272
Corps of Engineers. See Army Corps of Engineers.
Cortright, Edgar M., 56, 75–76, 120, 315, 325–326, 359
Cosmic dust, 9, 328
Cosmic rays, 85, 86, 133, 151, 264, 279, 336
"Cosmic rubble," 279
Cosmology, 47
Cosmonaut, 20, 55, 60–61, 70, 92, 169–170, 193, 306, 323
   anniversary, 135–136
   extravehicular activity, 20, 107, 318
   interview, 135–136, 214, 237, 330
   scientists as, 327
   space cooperation, 10, 109–110, 248–249
   space flight, effects of, 60, 285
   VOSKHOD II flight, 12–13, 107
Cosmonautics Day, 135–136
COSMOS LXXX (U.S.S.R. satellite), 2
COSMOS LXXXI, 2
COSMOS LXXXII, 2
COSMOS LXXXIII, 2
COSMOS LXXXIV, 2
COSMOS C, 2
COSMOS CIV, 8
COSMOS CV, 22
COSMOS CVI, 27
COSMOS CVII, 51
COSMOS CVIII, 51
COSMOS CIX, 59
COSMOS CX, 60–61, 69, 84, 91, 102, 138
COSMOS CXI, 75
COSMOS CXII, 104
COSMOS CXIII, 110, 121
COSMOS CXIV, 130
COSMOS CXV, 144
COSMOS CXVI, 152
COSMOS CXVII, 168
COSMOS CXVIII, 175
COSMOS CXIX, 190
COSMOS CXX, 211
COSMOS CXXI, 218
COSMOS CXXII, 223, 269, 271, 308
COSMOS CXXIII, 235
COSMOS CXXIV, 240
COSMOS CXXV, 246
COSMOS CXXVI, 252
COSMOS CXXVII, 261
COSMOS CXXVIII, 277
COSMOS CXXIX, 321
COSMOS CXXX, 325
COSMOS CXXXI, 347
COSMOS CXXXII, 352
COSMOS CXXXIII, 356
COSMOS CXXXIV, 366
COSMOS CXXXV, 372
COSMOS CXXXVI, 378
COSMOS CXXXVII, 378
COSPAR. See Committee on Space Research.
Cotton, Col. Joseph F. (USAF), 136, 158, 264, 338, 354, 378
Coughlin, William J., 25, 112, 162, 214, 272, 348, 352

Council of Economic Advisers, 354–355
Council on Higher Education (D. C.), 55
Courant, Dr. Richard, 52
Courter, Robert F., 78
Cousins, Frank, 51
Covington, Osro H., 354
Crab Nebula, 263
Craig, E. B., 361
Crater Lake, Ore., 299
Crawler-transporter. See Kennedy Space Center (NASA).
Cronkite, Walter, 17
Crowell, Gen. Benedict, Gold Medal, 341
Cryogenics, 55, 77, 114
Crystal fibers, 250
CTNE. See Compania Telefónica Nacional de Espana.
Cuba, 193, 334, 376
Cuddeback Dry Lake, Calif., 313
Culver City, Calif., 132
Cunningham, R. Walter, 144, 306, 351
Curtis, Sen. Carl T., 227–228
Curtiss-Wright Corp., 72, 353
Cygnus (constellation), radiation sources, 79–80, 151
Czechoslovak Academy of Sciences, 231
Czechoslovakia, 92

D

D-1 (French satellite), 361
D-2 (French satellite), 361
Daddario, Rep. Emilio Q., 64, 214, 359
Dade County, Fla., 62, 63, 110
Dallas, Tex., 14, 113, 297
Dana, William H., 47, 242, 258, 271, 292, 336
Daniel, L/C Walter F. (USAF), 251
Daniel Guggenheim School of Aeronautics, 319
Das. See Data automatic subsystem.
Data, 229, 281
Data automatic subsystem (Das), 337
Davies, M. Helen, 315
Davies, Merton E., 73
Davies, Richard W., 162, 202
Davis, Richard, 250
Day, LeRoy E., 232, 354
Day, Melvin S., 209, 243
Dayglow, 277
Dayton, Ohio, 383
Deep Space Facilities, Goldstone, Calif., 226
Debus, Dr. Kurt H., 198, 215, 283
Deep Space Network (DSN) (NASA), 3–4, 84, 157, 261, 263, 315
Defense Atomic Support Agency, 311
Defense Communications Agency (DCA), 222, 232
Defense, Dept. of (DOD), 22, 57, 90–91, 110, 134, 147–148, 178, 236, 251, 291–292, 332, 382
   Advanced Research Projects Agency, 322, 376
   aircraft, 131, 253, 288, 323
   budget, 23–24, 145, 246

communications satellite system, 40, 200, 306, 323, 361, 362
contract, 54
cooperation, 40, 295, 354
   NASA, 7, 29, 43, 110–111, 150, 198, 303, 334
   missile program, 170
   nuclear detection satellite, 322
   R&D, 154, 165, 220, 323
   space program, 153, 384
Defense Electronics, Inc., 128
DeFlorez Training Award, 359
DeGaulle, President Charles (France), 221, 223, 334, 370
Delamar Dry Lake, Nev., 168
Delorme, Jean, 307
Delta (booster) (see also Thor-Delta and Thrust-Augmented Improved Delta), 192, 374
Delta Scorpii (star), 241
Dembling, Paul G., 282, 378
Denmark, 125
Denver, Colo., 107, 177, 315
Denver Moonwatch, 284–285
Denver Post, 118
Detroit, Mich., 177, 296
Deutsche Lufthansa Airlines, 358
Devons, S., 10
Diamant (French booster), 52, 55–56, 261
DIAPASON I (D-IA) (French satellite), 55–56
Dick, John M., 78
Dineen, Col. Richard E. (USAF), 218
Disarmament, 84–85, 97, 302, 304
DISCOVERER XVII (satellite), 332
Distance Measuring Equipment-Instrument Landing System (Dme-Ils), 131
District of Columbia Teachers College, 225
Djerfisherite (mineral), 287
DMS, Inc., 172
Dmu. See Dual-purpose Maneuvering Unit.
Docking, 17, 99–102, 104, 105–106, 108, 202, 212, 243, 290, 291, 294, 314, 344
Documentation, Inc., 273–274
DOD. See Defense, Dept. of.
Dog experiment, 60–61, 84, 102, 179, 237, 318
Dolginov, A. I., 358
Doll, Ronald, 164
Doolittle, Gen. James H. (USAFR), 27
Doppler effect, 254
Dorman, B. L., 380
Doty, L. L., 285
Douglas Aircraft Co., 34, 384
   contract, 2, 5, 145, 167, 224, 263, 306, 316, 368, 375, 383
   Delta launch vehicle, 375
   Missile and Space Systems Div., 318, 353
   space equipment, 78, 383
   Sacramento Test Center, 107, 202
   Saturn V, 107, 202, 224, 252, 261, 306
Douglass, Howard W., 258
Dounreay, Scotland, 51
Dow Chemical Co., 262
Downey, Calif., 167, 215, 257, 328
Dragon (French sounding rocket), 322
Draper, Dr. Charles S., 26, 317
Drell, Dr. Sidney D., 111
Drinkwater, Fred J., III, 20
Drummond, Roscoe, 160
Dryden, Dr. Hugh, L., 20, 27, 50, 148, 193, 219, 225, 305, 315, 369
Dryden, Mrs. Hugh L., 305
Dryden, Dr. Hugh L., Memorial Fund, 305
DSN. See Deep Space Network.
Dual-purpose Maneuvering Unit (Dmu), 225
Duart, Calif., 221
DuBridge, Dr. Lee A., 29, 329
Dubrovin, Leonid I., 157
Dudley, Dr. Seibert, 7
Dudley Observatory (Albany, N. Y.), 267, 351
Duke, Capt. Charles M., Jr. (USAF), 127
Duke Univ., 25
Dulles International Airport, 217
Dunning, Dr. John R., 57
Dupree, Prof. A. Hunter, 359
Durant, Okla., 89
Duthie, J. G. M., 151
Dynatronics, Inc., 176

E

Eaker, L/G Ira C. (USAF, Ret.), 328
Early, Benjamin N., 103
EARLY BIRD I (INTELSAT I) (communications satellite), 65, 67, 83, 122, 252, 331
   revenue, 125–126, 252
   transmission, 237, 278, 350
   Gemini recovery, 86, 179, 207, 290, 332
Earth, 150, 302, 305
   infrared data, 176, 178, 251, 257, 283–284
   magnetic field, 86, 146
   magnetic tail, 86, 301, 358
   photographs of, 7, 265, 294, 297, 355, 369
East Central Florida Planning Council, 15
East Grand Forks, Minn., 228
Eastern Air Lines, 144, 194
Eastern Test Range (ETR) (see also Cape Kennedy and Kennedy Space Center), 334
   launch
      Apollo/Saturn
         AS-201, 67–69, 93
         AS-202, 274
         AS-203, 233
      Atlas-Centaur
         AC-7, 296
         AC-8, 131–132
         AC-9, 329
         AC-10, 196
      Augmented Target Docking Adapter (Atda), 202
      GEMINI VIII, 99–102
      GEMINI IX, 206
      GEMINI X, 243, 248

GEMINI XI, 248, 290
GEMINI XII, 340, 344
Minuteman, 255
PIONEER VI, 214, 301
satellite, 69–70, 132, 213, 249, 262, 265, 269, 296, 313, 331, 342, 344, 365, 367
Titan III-C, 216, 276
Eastman Kodak, 284
Easton, Pa., 140
Eckert, Dr. Wallace J., 150
Eclipse, solar, 87, 178, 187, 311, 320, 331–332, 344, 346
Ecole Normale Supérieure, 339
"Ecological Technology—Space, Earth, Sea" (symposium), 54
Economic Club of Detroit, 296
Ecs. See Environmental control system.
Ecuador, 297
Eddington Gold Medal, 82
Education, 28–29, 32, 35, 53, 238–239
Edwards AFB, Calif. (see also Air Force Flight Test Center)
flight
C-130B, 119
F-104, 111
F-111A, 116
M-2-F2 (lifting body vehicle), 114, 152–153, 237–238, 244, 264, 274, 284, 293, 298, 306, 312, 319, 330, 348, 352, 383
SR-71, 28
XB-70, 5, 16, 136, 158, 184, 211, 338, 342, 354, 372, 376, 378
strike, 35
Edwards, John A., 354, 383
Eglin AFB, Fla., 131
Egypt. See United Arab Republic.
Eisele, Donn F., 351
ELDO. See European Launcher Development Organization.
Electric propulsion. See Engine, electric.
Electro-Optical Systems, Inc., 62, 336
Electron, 237, 277, 287, 292, 293, 313, 328
auroral producing, 55, 95
measurement, 237, 277, 283, 287, 292, 293, 313, 328, 346–347
ELECTRON I (U.S.S.R. satellite), 2
ELECTRON III, 2
Electronics Research Center (ERC) (NASA), 37, 288, 332, 342
Electrostatic gyroscope (Esg), 353
Elliott-Automation, 239
Elms, James C., 288
Emerson Electric Co., 247
ENEA. See European Nuclear Energy Agency.
Engine (see also individual engines, such as F-1, M-1, etc.)
aircraft, 27, 77, 137, 143
gas turbine, 143, 326, 362
jet, 77, 137, 143, 199–200, 249, 273, 382
supersonic transport, 85–86, 109, 123, 199–200, 226, 250, 273, 286, 296, 383

cesium, 336
chemical, 258
electric, 27–28, 82, 119–120, 336, 343
ion, 89–90, 312, 339
solar-electric (see also solar cell), 89–90, 194, 355
hydromagnetic, 151–152
nuclear (see also Nerva), 62, 92, 145, 251, 336, 385
rocket, 27, 34, 39, 110, 114, 168, 241, 383
turbo-rocket, 285
vernier, 296
Engle, Capt. Joe H. (USAF), 47, 127, 246, 359
Environmental control system (Ecs), 332
Environmental Science Services Administration (ESSA), 43, 199, 240, 293, 361, 368
budget, 24
cooperation, 103, 127, 168, 267, 321
Institute for Telecommunications Sciences and Aeronomy, 328
Institutes for Environmental Research, 111
Natural Disaster Warning System, 77
Office of Hydrology, 379
Project Stormfury, 267
satellite, 41, 69–70, 103, 310, 321, 365, 385
solar eclipse expedition, 311
sounding rocket, 127, 328
Eole (meteorological forecasting project), 195
Epsilon Canis Majoris (star), 295
Epstein, L. M., 289
ERC. See Electronics Research Center (NASA).
Ercoupe (monoplane), 259
Eremea, George, 302
Erhard, Chancellor Ludwig (W. Ger.), 303, 311
Eros (Earth Resources Observation Satellite) project, 297
ERS-XV (research satellite), 272
ERS XVI, 212
Esg. See Electrostatic gyroscope.
Eshleman, Dr. von R., 199
ESRO. See European Space Research Organization.
ESSA. See Environmental Science Services Administration.
ESSA I (meteorological satellite), 41, 43, 82, 308
ESSA II, 70, 79, 103, 365
ESSA III, 310, 321, 365
ETR. See Eastern Test Range.
Euboea (Greek island), 87
Euratom. See European Atomic Energy Community.
Europa (booster), 348, 357
Europa I (ELDO booster), 190, 214, 236, 348, 357
Europe (see also European Launcher Development Organization; European

Space Research Organization; Eurospace), 15, 41, 239, 257, 275, 307
cooperation, 73
European Atomic Energy Community (Euratom), 133, 189, 285–286, 303, 307, 311
European Broadcasting Union, 287
European Center for Nuclear Research (CERN), 286
European Launcher Development Organization (ELDO), 41, 189, 285, 361
 launch, 190, 348, 356–357
 policy, 156, 211, 214, 220, 236
European Satellite Team, 239
European Space Research Organization (ESRO), 11, 87, 124, 133, 208, 299, 307, 352, 354, 356, 383
European Space Symposium, 189
Eurospace, 96, 307
Eva. See Extravehicular activity.
Evans, LCdr. Ronald E. (USN), 127
EXAMETNET. See Inter-American Experimental Meteorological Rocket Network.
Exceptional Service Medal (NASA), 218, 223, 255
Exhibit, 312
Exobiology, 198–199
Experimental Aircraft Assn., 308
EXPLORER XXVIII (Imp-C) (satellite), 150–151
EXPLORER XXX (IQSY Solar Explorer) (satellite), 174
EXPLORER XXXII (AE-B) (aeronomy satellite), 192, 276, 385
EXPLORER XXXIII (Imp-D) (satellite), 231, 358, 385
Explorer's Club, 333
Extraterrestrial life, 40, 74, 141, 171, 174, 189, 241, 264, 322, 366
Extra-terrestrial Research Agency, 154
Extravehicular activity (Eva), 104, 316, 324
 equipment, 12, 17, 78–79, 225, 251, 301, 304–305, 317, 353–354
 GEMINI VIII, 92, 100
 GEMINI IX-A, 202, 207, 218
 GEMINI X, 243–244, 248, 255
 GEMINI XI, 290, 293, 294, 301
 GEMINI XII, 344, 353–354
 hydrogen tank conversion experiment, 167, 272
 U.S.S.R., 20, 107, 318

F

F-1 (rocket engine), 115, 352
F-4B (Phantom II) (fighter aircraft), 211
F-4E, 323
F-4J, 195
F-104 (Starfighter) (aircraft), 111, 211, 263
F-104G, 183, 256, 367, 378
F-106 (Delta Dart), 198
F-111 (supersonic fighter), 61, 288
F-111A, 77, 116, 288
F-111B, 24
FAA. See Federal Aviation Agency.
Fabbrica Italiana Aparecchi Radio, 239
Faget, Dr. Maxime A., 166
Fairbanks, Alaska, 257, 356
Fairchild Hiller Corp., 77, 264
FAITH 7 (spacecraft), 105
Fairchild, Sherman, Technology Center, 322
Falcon (fan-jet aircraft), 302
Falmouth, Mass., 5
Farnsworth, Clyde H., 136
Fast Deployment Logistic Fleet (Fdl) program, 371
Faul, Dr. Henry, 139, 208
FB-111 (supersonic bomber), 72
FCC. See Federal Communications Commission.
Fdl. See Fast Development Logistic Fleet program.
Federal-Aid Airport Program, 377
Federal Aviation Agency (FAA), 22, 24, 80, 143, 227, 261, 340, 343
 air traffic control, 13, 18, 91, 233
 airports, 243, 364, 377
 annual report, 131
 contract, 233, 341
 cooperation, 284, 381
 criticism, 64
 landing system, 66
 noise, aircraft, 133
 statistics, 11, 91, 155, 209, 250
 transport, supersonic (see also Supersonic Transport), 24, 84, 112, 122–123, 128, 131, 143, 200–201, 296
 design and development, 85–86, 286, 299, 307, 323, 332, 334, 343, 383
Federal Communications Commission (FCC), 81, 179
 approvals, 126, 332, 365
 briefs filed with, 256, 288, 360, 371, 375
 ComSatCorp, 142, 146, 152, 176, 197, 199, 233, 235, 311, 314, 327, 331, 350, 365, 368, 375
 ground stations, 278, 314
 requests to, 46, 67, 73, 90, 122, 152, 176, 270, 321, 331, 341, 365
Federal Council for Science and Technology, 31, 34
Federal Electric Corp., 261
Federal Executive Board, 238
Federal Woman's Award, 77
Fédération Aéronautique Internationale (FAI), 187, 253
Fédorenko, Ambassador Nikolay T., 198
Fels Planetarium, 366
Feoktisov, Konstantin Petrovich, 136
Ferguson, Gen. James (USAF), 283
Fermi, Enrico, Award, 259
Fermi, Enrico, Institute, 151
Field Enterprises Educational Corp., 161
Finger, Harold B., 82, 83, 315
Fink, Daniel J., 111, 153, 220
Finland, 174
Fire
 aircraft, 159

electrical, 177
escape system, 250
Firebee (target missile), 267
"First Soft Step" (TV program), 198
Fisher, Adrian S., 85, 97
Fitch, John, 198
Flax, Alexander H., 225
Fleming, William A., 327
Flemming, Arthur S., Award, 57
Flight Dynamics Laboratory (AFSC), 87–88
Flight International, 372
Flight Research Center (FRC) (NASA) (see also Lifting body vehicle; X-15; XB-70) 14, 47, 200
Flight Safety Foundation, 20
Flight Safety Review Board, 180–181
Flight Test Center. See Air Force Flight Test Center.
Florida, 62, 63
Florida, Univ. of, 29–30
Flying saucers. See Unidentified flying objects (Ufo).
Forster, Wilhelm, Observatory, 121, 324
Fokker, 239
Ford Foundation, 229, 256, 270, 288, 327, 370
Ford, Rep. Gerald R., 116
Ford, Henry, II, 247
Foreign Office (U.K.), 211
Forman, Edward S., 124
Ft. Campbell, Ky., 131
Ft. Eustis, Va., 253
Ft. Greely, Alaska, 119
Ft. Monmouth, N. J., 254
Foster, Dr. John S., Jr., 29, 43, 90–91, 200
Foster, Leslie D., 288
Foster, Willis B., 61
Fourier spectroscopy, 264
France, 20, 43, 264
 atomic bomb, 232, 246, 312
 Concorde (France-U.K. supersonic transport) 43, 233
 cooperation, 63, 73, 125, 133, 190, 195, 210, 214, 220, 221, 239, 261, 299, 348, 370
  U.S.S.R., 227, 261, 281, 334, 348, 370, 385
 launch, 11, 52
  missile, 82, 201
  satellite, 52, 55–56, 94, 384, 385
  sounding rocket, 322, 327, 352
 satellite, 233
 space program, 334, 361
Franklin Institute, 82
Franzini, Dr. Juliet Lee-, 225, 286
Franzini, Dr. Paolo, 225, 286
Frascati, Italy, 37
FRC. See Flight Research Center (NASA).
Fredericka, Queen Mother (Greece), 317
Freeman, Orville, 117
Freeport, Tex., 196
Freitag, Capt. Robert F. (USN, Ret.), 143, 177
French Guiana, 82, 261
Frese, Col. Gregory C., Jr. (USAF), 294

Friedman, Dr. Herbert, 79–80, 109, 174
Friendly, Fred W., 270, 327
FRIENDSHIP 7 (spacecraft), 105
Frutkin, Arnold W., 2, 32, 74, 141
FS-3 (nuclear reactor power system), 27–28
Fuchs, Dr. Louis H., 287
Fuel, 55
 aircraft, liquid methane, 293, 362
 carbon dioxide-water, 120
 cesium, 336
 hydrogen-peroxide, 78
 liquid, 57–58
 liquid hydrogen, 55, 222, 385
 liquid hydrogen-liquid oxygen, 114, 120, 168, 252
 liquid oxygen-kerosene, 285
 plutonium, 83, 112
 slush hydrogen, 9–10
 solid, 42, 57–58, 62, 90, 92, 110, 176, 190, 219, 226, 241, 297, 385
 tank, 328
  liquid hydrogen, 167, 168, 195, 198, 210, 272, 283
  liquid oxygen, 272, 277
Fulbright, Sen. J. William, 32, 190
Fuller, Prof. Buckminster, 311
Fulton County (Georgia) Airport, 251
Fulton, Fitzhugh, 338, 372
Fulton, Rep. James G., 163, 214
Fulton Recovery System, 164
Funk, M/G Ben I. (USAF), 268

G

Gabriel, David S., 315
Gagarin, Col. Yuri A. (U.S.S.R.), 20, 60–61, 135, 193, 214
Galaxy, 79–80
Galileo (jet aircraft), 347
Galveston Island, 36
Galipeau, George, 308
Gamma ray, 136–137, 263
GAO. See General Accounting Office.
Garbarini, Robert F., 44, 149, 204, 250, 367
Gardner, Richard N., 163
Garmire, Gordon, 229
Garrard, Dr. W. C. J., 296, 314
Garrett Corp., 238
Gary, Ind., 16
Gas turbine. See Engine.
GATV. See Gemini Agena Target Vehicle.
Gaudry, Roger, 229
Gautschi, Theodore F., 250
Gay, Clarence C., Jr., 340
Gazenko, Dr. Oleg G., 321
GCA Corp., 19, 214, 239, 274, 292
GE. See General Electric Co.
GE 4 (turbojet engine), 273, 332, 383
Gell-Mann, Dr. Murray, 152, 329
Gemini (program), 16, 64, 218, 314, 318, 338, 352, 372
 award, 27, 218, 324
 completion, 344, 349, 384–85
 management, 43

Gemini (spacecraft), 7-8, 121, 164, 177, 338
  exhibit, 60, 312, 340
  GEMINI VIII, 7, 90, 94
GEMINI IV (flight), 105
GEMINI V (flight), 7, 105
GEMINI VI-A (flight), 8, 126
GEMINI VII (flight), 203
GEMINI VIII (flight), 8, 53, 86, 104, 118, 145, 162, 226
  launch, 99
  preparations, 7, 90, 94, 96
GEMINI IX-A (Gemini IX) (flight), 210, 215, 236
  crew, 72, 85, 110, 208, 219
  Gatv, 180
  launch, 206
  preparations, 105, 172, 178, 195
GEMINI X (flight), 27, 226, 240, 247, 248
  launch, 243
GEMINI XI (flight), 248, 294, 297
  crew, 110, 293, 301, 304
  launch, 290
GEMINI XII (flight), 327, 355, 361
  crew, 219, 340, 347, 349
  launch, 344
Gemini Agena Target Vehicle (Gatv), 8, 53, 86, 126
  GATV VIII, 99-100, 104, 115, 118, 243, 248, 255
  GATV 9, 180
  GATV X, 243-244, 248, 255
  GATV XI, 290, 297
  GATV XII, 344
Gemini Mid-Program Conference, 64
General Accounting Office (GAO), 202
General Aviation Jet Training Standards Board, 131
General Dynamics Corp., 11
  award, 354
  contract, 30, 268, 284, 315, 371
  Convair Div., 54
  F-111A, 116
  missile system, 119
  tracking communications vessel, 47, 214
General Electric Co. (GE), 111, 317, 321
  award, 103, 353
  contract, 72, 77, 90, 222, 343, 380
  Industrial Electronics Div., 157
  Missile and Space Div., 239
  Re-Entry Systems Div., 192
  research, 126, 174, 190
  supersonic transport engine, 86, 249, 273, 286, 296, 299, 332, 383
  XB-70 accident, 211
General Electric Co., Ltd., 136
General Electric Forum, 308
General Precision Equipment Corp., 337, 361
General Telephone & Electronics Corp., 361
Geneva, Switzerland, 135, 199, 221, 237, 248, 250, 258, 286, 287
Gentry, Capt. Jerauld R., 319, 348, 352
Geodetic satellite, 162, 212, 272, 312
Geological Society of America, 349

George Washington Univ., 287
Georgetown Univ., 18
Gerdes, Ron, 253
Germantown, Md., 322
Germany, East, 174, 355
Germany, West, 25, 210, 307
  aircraft, 53, 183, 256, 367, 378
  Bochum Observatory, 45, 104, 109, 126, 348
  cooperation, 73, 125, 190, 191, 220, 240, 299, 303, 304, 311
  Ministry of Scientific Research (BMWF), 299, 301
  satellite, 176, 342
  sounding rocket, 240, 299, 342
  Wilhelm Forster Observatory, 121, 324
Ghost. See Global Horizontal Sounding Technique.
Giacconi, Riccardo, 229
Giannini Corp., 221
Giberson, W. Eugene, 250
Gill, Dr. Jocelyn R., 77, 215
Gillespie, Rollin W., 26
Gilmore Creek, Alaska, 41, 310
Gilruth, Dr. Robert R., 93-94, 104, 117, 321
Ginter, Roll D., 315, 327
Giordmaine, Dr. Joseph A., 51
Givens, Maj. Edward G., Jr. (USAF), 127
GISS. See Goddard Institute for Space Studies.
Glenn, Col. John H., Jr. (USMC, Ret.), 54, 60, 105, 169, 191, 241, 314, 328
Glennan, Dr. T. Keith, 257, 315
Glider, 124
Global Horizontal Sounding Technique (Ghost), 2, 89, 168
Glod, G. D., 318
Goddard Institute for Space Studies (GISS), 159, 235, 365, 378
Goddard, Dr. Robert Hutchings, 103, 209, 263, 350, 383
Goddard, Robert H., Award, 27
Goddard, Robert H., Historical Essay Award, 103
Goddard, Dr. Robert H., Lecture, 97
Goddard, Robert Hutchings, Library, 314
Goddard, Robert H., Memorial Trophy Award, 102
Goddard Space Center (GSFC), 151, 187, 231, 260, 288, 355, 358, 375
  aeronomy satellite, 192
  award, 57, 315, 354
  contract, 216, 353
  Humphrey, Vice President Hubert H., visit, 79
  management, 264, 289
  personnel, 9, 86, 166, 208, 232
  satellite monitoring
    ATS I, 367, 372
    ESSAS I, II, III, 41, 69-70, 310
    NIMBUS II, 178, 185, 349
    Observatory class, 63-64, 181, 209, 218, 264, 367

Satellite Situation Report, 294, 307, 337, 384
  Tiros, 145
  sounding rocket experiments, 80–81
    astronomical, 18, 80–81, 185, 190, 295
    atmospheric data, 50, 55, 95, 111, 245, 260, 276, 277, 284, 313, 341
    meteorological, 19, 69, 161, 221, 241, 265, 276
      grenade, 23, 164, 216, 252
    micrometeorite collection, 328, 351
    radar, 171
Goethert, Dr. Bernhard H., 316
Gold, Dr. Thomas, 128
Goldberg, Arthur J., 237, 246, 258, 271, 291, 298
Goldstein, Dr. Richard, 175
Goldstone Tracking Station, 3, 7, 157, 194, 226, 372
Golomb, Solomon W., 363
Golovin, Dr. Nicholas E., 57, 358
Goodrich, B. F., Co., 87–88
Goodwin's Sandbank, U. K., 354
Goodyear Aerospace Corp., 78, 159
Gordon, LCdr. Richard F., Jr. (USN), 110, 297, 379
  award, 304
  GEMINI XI, 285, 290, 293
    Eva, 294, 301
Gordon, T. J., 317
Gorenstein, Paul, 229
Gorkin, Jess, 10, 248
Gorman, Harry H., 315
Goudas, Dr. C. L., 355
Gould, R. Gordon, 258
Government Printing Office (GPO), 107
Graduate Research Center of the Southwest, 127, 139
Grand Canary Island, 97
Grand Forks AFB, N. Dak., 165
Grants
  research, 28, 55
  facilities, 128, 147, 287
Great Silver Medal of Paris, 191
Greece, 87, 178, 187
Greek National Committee for Space Research, 87, 185
Green, Rep. William, 90
Green, William L., Jr., 202
Green River, Utah, 342
Greenstein, Dr. Jesse L., 329
Gregory, Prof. Richard L., 57
Grigorov, Prof. Naum, 140
Grissom, L/C Virgil I. (USAF), 105, 110, 144, 332
  press conference, 94, 257, 376
Gromyko, Andrey A., 227
Gross, Courtland S., 170
Gross, F. R., 159
Gross, Rep. H. R., 172
Grubbs, Haydon Y., Jr., 60, 91
Grumman Aircraft Engineering Corp., 145, 167
GSFC. See Goddard Space Flight Center.
Guggenheim Aeronautical Laboratory, 235
Guggenheim, Daniel, Medal, 94, 321
Guiana Range, Guiana, 361
Gulf of Mexico, 36, 287
Gunter, Richard, 250
Gurjian, A. A., 318
Gursky, Herbert, 229
Guthrie, Sir Giles, 143

H

H-1 (rocket engine), 130, 169, 227
Haas, Conrad, 231
Haeussermann, Dr. Walter, 361
Hagge, Dr. Donald E., 151
Haggerty, James J., Jr., 87, 134, 219
Haglund, Howard H., 372
The Hague, Netherlands, 354
Hahn, Dr. Otto, 259
Haise, Fred W., Jr., 127
Halaby, Najeeb E., 343
Haley, Andrew G., 281, 289
Hall, Eldon W., 354
Hall, Floyd D., 194
Hall of Aeronautical Propulsion, 236
Hallett, Robert L., 383
Halpern, Dr. Irwin P., 195
Hammaguir Range, Algeria, 52, 55, 94, 327, 344, 356, 361
Hammarskjold, Knut, 156
Hampton, Va., 121–122
Handler, Prof. Philip, 25
Haney, Paul P., 76
Hanscom Field, Mass., 294
Hansen, Grant L., 315
Harmon, Col. Clifford B., 259
Harmon International Aviator's Trophy, 259, 374
Harper, Charles W., 37, 66
Harr, Dr. Karl G., Jr., 97, 160, 367, 378
Harris, Sen. Fred R., 89
Harris poll, 249
Harris, Sydney, 242
Harteck, Dr. Paul, 377
Hartford Times, 248
Harvard Univ., 22, 131, 260, 382
Harvard Univ. Graduate School of Business Administration, 369
Hasp (meteorological rocket), 173
Haughton, Daniel J., 350, 354
Hausman, William, 322
Haut-Provence Observatory, 322
Hawaii, 83, 222, 311, 360
Hawaii Institute of Geophysics, 38
Hawaii, Univ. of, 39, 144
Hawker Siddeley Aviation Co., 356
Hawker Siddeley Dynamics, Ltd., 191
Haworth, Dr. Leland J., 34
Haworth, Michael, 33
Hayes International Corp., 130
HC-130H (cargo aircraft), 36, 164
Healey, Denis, 40, 61
Health, Education, and Welfare (HEW), Dept. of, 177, 284
Hechler, Rep. Ken, 214
Hedges, Fielding, 250
Heidelberg, W. Germany, 220

Helicopter, 132, 228, 290, 344
   commercial use, 340
   military use, 60, 187, 326
   record, 187, 291
Heliport, 155
Helium, 12, 237, 242, 245, 263
Hello, Bastian, 354
Henry, William, 118
Henty, Denham, 64, 123, 162
Heos (Highly Eccentric Orbiting Satellite), 208
Heos A (interplanetary physics satellite), 383
Hercules (constellation), 260
Hercules, Inc., 226
Herget, Dr. Paul, 252
Hess, Dr. Harry H., 280, 349, 384
Hess, Dr. Wilmot N., 57
HEW. See Health, Education, and Welfare, Dept. of.
Hewlett, William R., 111
Heydon, Rev. Francis J., 160, 341
Hiatt, William E., 379
Hibex (High-E Boost Experiment), 15
High Altitude Background and Signal to Noise (Hitab) program, 108, 115–116
High Resolution Infrared Radiometer (Hrir), 178, 284
Highly Eccentric Orbiting Satellite. See Heos.
Hilburn, Earl D., 1, 202, 270
Hill, Gladwin, 93
Hill, Louis W., Award, 359
Hill, William, 131
Hillsdale, Mich., 116
Himmel, Dr. Seymour C., 16
Hines, William, 7, 154–155, 166, 185
   boosters, 57–58, 241
   manned space flight, 145–146, 193, 295, 305
   NASA, 37, 123, 220, 375
   U.S.S.R., 106, 125, 138, 338–339
Hitab. See High Altitude Background and Signal to Noise program.
Hjornevik, Wesley L., 166
HL-10 (lifting-body vehicle), 19, 135, 379, 383, 385
Hodge, John D., 100, 115, 354
Hohmann, Bernhard A., 354
Holifield, Rep. Chet, 227
Holland, Sen. Spessard L., 190
Holloman AFB, N. Mex., 40, 175, 185, 283, 342, 336
Holmes, D. Brainerd, 53
Honest John-Nike (rocket), 275, 340, 352, 370
Honeywell, Inc., 60, 92, 264, 298
Hong Kong, 219
Honolulu, Hawaii, 77, 106, 134, 366
Hooker, Ray W., 315
Hootman, Dr. James A., 317
Hope, Bob, 129
Horn, Helmut J., 315
Horne, Walter B., 213
Horner, H. M., 308

Hornig, Dr. Donald F., 15, 28, 33–34, 57, 106–107, 308, 354
Hornick, Clay, 281
Horowitz, Dr. Norman H., 189
Horton, William, 361
Hospital Topics, 254
Hot Springs, Ark,. 330
Hotz, Robert, 19, 180, 220, 273, 317, 334
Hough, Richard, 270
House of Commons, U. K., 289
Housing and Urban Development, Dept. of, 80
Houston, Tex., 165, 213, 216, 332
Houston Post, 76, 220, 269
Houston, Univ. of, 365, 379
Houston, M/G Vincent G. (USAF), 354
Hovercraft, 168
Howick, George J., 12
Hrir. See High Resolution Infrared Radiometer.
HS-303A (communications satellite), 65
Hsinhua (Chinese press agency), 172
Hubbard, Samuel H., 377
Huff, Vearl N., 354
Hugh, W. F., 241
Hughes Aircraft Co., 39, 132, 176, 227, 260, 297, 320, 353
Humphrey, Vice President Hubert H., 28, 139
   Assistant for Aeronautics and Space Matters, 54
   award by, 215, 324, 374
   NASA visit, 43, 79, 125
   science, 239
   space program, 44, 103, 119, 125, 302, 307, 374
Humphrey, Mrs. Hubert H., 247
Hunn, Col. Spencer S. (USAF), 294
Hunter, Willson H., 315
Huntington Beach, Calif., 202
Huntsville, Ala., 373
Hurricane Beulah, 267
Hurricane control, 267
Hurricane Esther, 267
Hutchingson, Henry P., 254
Hydrogen (see also Fuel), 335
Hyman, William A., 237
Hynek, Dr. J. Allen, 112–113, 116, 122, 255, 326, 351, 376
Hypersonic flight, 66
Hypersonic transport (Hst), 382

I

IAF. See International Astronautical Federation.
IAM. See International Association of Machinists.
IATA. See International Air Transport Association.
IAU. See International Astronomical Union.
IBM. See International Business Machines Corp.
ICAO. See International Civil Aviation Organization.
Icarus (asteroid), 251

ICAS. See International Council of the Aeronautical Sciences.
Icm. See Improved capability missile.
ICSU. See International Council of Scientific Unions.
IDA. See Institute for Defense Analyses.
IDB. See Inter-American Development Bank
Idcsp. See Initial Defense Communications Satellite Project.
IEEE. See Institute of Electrical and Electronics Engineers.
IGY. See International Geophysical Year.
Ikeya-Seki (comet), 277
Il-18 (U.S.S.R. airliner), 256
Illinois, Univ. of, 111, 214
Improved capability missile (Icm.), 219
India, 106, 163
Indian Atomic Energy Commission, 46
Indian Institute of Technology, 253
Indian National Commission for Space Research (INCOSPAR), 234
Indian Radioastronomy Group, 46
Indiana, 16
Information technology, 31, 370
Infrared. See Earth, infrared data; Instrument, radiometer.
Initial Defense Communications Satellite Project (Idcsp), 40, 200, 216, 276, 361
Institute for Defense Analyses (IDA), 214, 357
Institute for the Study of Science in Human Affairs, 343
Institute of Electrical and Electronics Engineers (IEEE), 112, 238, 327
Institute of Geophysics and Planetary Physics (UCLA), 17
Institute of Navigation, 321
Instituto de Investigaciónes de Geológicas de Chile, 94
Instituto Nacional de Técnica Aeroespacial (INTA) (Spain), 15, 137
Instrument
  Langmuir probe, 234, 292
  magnetometer, 29, 56, 83, 129, 221, 234
  microscope, 260
  photometer, 83
  pitot-static probe, 260, 263, 276
  radiation detector, 83, 129
  radiometer, 2, 176, 178, 238, 257, 268, 369
  sensor, 21, 61, 175, 289
  spectrometer, 2, 18, 111, 129, 238, 274, 277, 347
INTA. See Instituto Nacional de Técnica Aeroespacial.
Intelsat. See International Telecommunications Satellite Consortium.
INTELSAT I. See EARLY BIRD I.
INTELSAT II-A ("LANI BIRD") (communications satellite), 341, 350, 360, 365, 385
  launch, 331
Inter-Agency Bird Hazard Committee, 284
Inter-American Experimental Meteorological Rocket Network (EXAMETNET), 154, 253
Interavia, 335

Interior, Dept. of, 263, 284, 295
International Academy of Astronautics (IAA), 317
International Air Safety Seminar, 352
International Air Transport Association (IATA), 156-157
International Airline Pilots Association, 165
International Assn. of Machinists (IAM), 144, 236, 242, 254, 257, 259, 272
International Astronautical Federation (IAF), 289
  Congress, 188, 317, 319, 320, 321
International Astronomical Union (IAU), 171, 324
International Business Machines Corp. (IBM), 61-62, 103, 165, 233, 257
  Federal Systems Div., 321, 348
International Civil Aviation Organization (ICAO), 43
International Conference of Social Work, 288
International Conference on Selenodesy, 196
International conference on outer space (proposed), 20, 28, 32, 287
International Congress on Air Technology, 330
International Congress on Oceanography, Second, 189
International Council of Scientific Unions (ICSU), 46
International cooperation, 84-85, 291
  aircraft, 188, 235
  astronomy, 40, 125, 171, 251, 329, 346
  communications, 40, 162-163, 376, 381
  meteorology, 8, 14, 15, 21, 89, 96, 133, 154, 185, 195, 210, 298, 308, 326, 346-347
  military, 40, 306
  oceanography, 240
  science and technology, 15, 49, 77, 133, 221, 231, 251, 303, 346-347
International cooperation, space (see also European Launcher Development Organization; European Space Research Organization; space law treaty), 10, 96, 111-112, 119
  conference (proposed), 20, 28, 32
  satellite, 11, 380
  Europe, 237, 239
  NASA-ESRO, 383
  -France, 195
  -West Germany, 176, 342
  U.S.-Canada, 63
  U.S.-U.K., 40
  U.S.-U.S.S.R., 308
  U.S.S.R.-France, 221, 226, 227, 261, 281, 334, 348, 370, 386
  sounding rocket
  ESRO, 124, 235, 299
  NASA-Argentina, 154, 342, 346-347
  -Brazil, 14, 161, 252, 310, 320, 347, 373
  -Canada, 313
  -France, 327

-Greece, 178, 185, 187
-India, 234
-Japan, 381
-Pakistan, 69, 114–115
-Spain, 15, 138
-West Germany, 191, 299
space research, 141, 155–156, 152
Europe, 37, 73, 189, 210, 214, 220, 285
NASA-Sweden, 51
-West Germany, 303
U.S.-Australia, 123
-New Zealand, 89
-U.K., 123
-U.S.S.R., 74, 90
tracking, 81
NASA-Spain, 137
U.S. -Australia, 64
-ESRO, 356
-U.S.S.R., 90, 293, 298, 299
International Council of Scientific Unions (ICSU) 46, 310, 316
International Council of the Aeronautical Sciences (ICAS), 291, 293
International Flat Earth Society, 282
International Galabert Astronautical Prize, 188
International Geophysical Year (IGY), 251, 260
International Institute of Space Law, 289, 317
International Scientific Radio Union, 199
International Telecommunications Satellite Consortium (Intelsat), 120, 126, 139, 159, 165, 197, 386
International Telecommunications Union (ITU), 180
International Telecommunications Union Conference, 135
International Telephone and Telegraph Corp. (ITT), 126, 145, 179, 210, 236
International Union of Electrical Workers (IUEW), 142, 144, 150
International Year of the Quiet Sun (IQSY), 64, 80, 251, 260
"Internationale" (U.S.S.R. anthem), 157, 171
Inventions, 16, 156, 199, 285, 288, 335
Ion propulsion. See Engine, electric, ion.
Ionosphere
composition, 111, 274, 277
properties of, 79, 127, 234, 301, 328, 341, 347
seasonal changes, 10
Iowa, 341
IQSY. See International Year of the Quiet Sun.
Iranian Poets Association, 49
Irwin, Maj. James B. (USAF), 127
Isaacs, John, 52
Ise, Reine, 361
Italy, 239
cooperation, 191, 210, 220
ground station, 306
satellite, 133, 174

ITT. See International Telephone and Telegraph Corp.
ITT World Communications, Inc., 86, 222, 332
ITU. See International Telecommunications Union.
IUEW. See International Union of Electrical Workers.
Iverson, Richard, 250
Izvestia, 44, 59, 82, 115, 157

J

J-2 (rocket engine), 2, 85, 168, 195, 233, 252, 257, 364
Jackass Flats, Nev., 52, 82, 104, 115, 211, 222
Jackson, Mich., 298
Jackson, Nelson P., Aerospace Award, 103
Jacksonville, Fla., 311
Jaffe, Leonard, 203, 228, 295
James, Jack N., 149
Japan, 45, 222, 224, 237, 311, 347, 360, 365
astronaut visit to, 54, 61, 66
booster, 333, 339
probe, 3
satellite, 3, 74, 123, 301, 333, 378
Satellite Navigation Research Office, 123
sounding rocket, 109, 149, 257
Japan-U.S.S.R. youth festival, 237
Jastrow, Dr. Robert, 235
Javelin (sounding rocket) (Argo D-4), 79, 161, 185, 237, 267, 299, 313
Jeffries, John, Award, 323
Jet-assisted takeoff (Jato), 259
Jet engine. See Engine, jet.
Jet Propulsion Laboratory (JPL) (Cal Tech), 2, 45, 177, 261, 378
anniversary, 124
award, 147, 166
contract, 227, 239, 261, 327, 337
Deep Space Network, 261, 263, 340
Goldstone Tracking Station, 3, 138, 157, 194, 203, 226, 372
Lunar Orbiter (see also LUNAR ORBITERS I, II), 265, 340
Mariner project (see also MARINER II, IV), 9, 175, 238
radar, 154, 171
research, 89, 162, 175, 189, 202, 253–254, 322, 348, 355
Surveyor project (see also SURVEYORS I, II), 39, 340, 372
Voyager project, 9, 198, 243, 279, 352, 370
"Jet shoes," 78
Jodrell Bank Experimental Station, 160, 209
Luna spacecraft, 41, 47, 115, 126, 278, 328, 330
-U.S.S.R. Venus experiment, 9, 47
VENUS III, 75
Johannesburg, S. Africa, 316, 325
Johns Hopkins Univ., 60, 111, 141, 145, 162

Johnson, Arthur W., 286
Johnson, Clarence L., 27, 135
Johnson, David S., 43
Johnson, Dr. Gerald W., 49
Johnson, President Lyndon B., 131, 159, 214, 298
  air pollution, 194
  appointment, 111, 293, 340
  astronaut, 54, 66, 117, 293, 345, 354
  awards by, 50, 251, 380
  budget, 23–24, 259, 286, 357, 364
  communications satellite system, 83, 361
  defense, 343
  disarmament, 97, 304
  GEMINI VIII, 101
  GEMINI XII, 344
  international cooperation, 10, 73, 258, 303, 311, 354
  LUNA IX, 42
  meteorology, 59, 113
  oceanography, 219, 240
  science and technology, 33, 303, 354
  space program, 35, 162, 169, 259, 297, 303, 375
  space treaty, 169–170, 276, 277, 368, 386
  State of the Union Message, 19
  supersonic transport, 112, 281, 334
  SURVEYOR I, 205, 212
  Transportation, Dept. of, 80, 321, 340
Johnson, Marshall, 250
Johnson, Vernon A., 371
Johnson, Dr. William G., 235, 315
Johnston, Richard, 226, 317
Joint Chiefs of Staff, 72
Jonash, Edmund R., 315
Jones, M/G David M. (USAF), 229, 360
Jones, Robert F., 199
Jones, William J., 309
Journal of Air Law and Commerce, 282
Journal of Geophysical Research, 377
Journal of Spacecraft and Rockets, 159
Journal of the Armed Forces, 87, 134, 219
Joyner, Upshur T., 213
JPL. See Jet Propulsion Laboratory.
JT9D (turbofan engine), 137, 143
JTF-17A (turbofan engine), 109
Junior Academy of Sciences, 382
Junkers Flugzeug-und Motorenwerke A. G., 208
Jupiter (planet), 29, 74, 150, 210, 281, 311, 355

K

Kai-shek, President Chiang, 72
Kakol, Stanley J., 106
Kaminski, Heinz, 45, 104
Kano, Nigeria, 351
Kansas City, Mo., 327
Kansas City Star, 303
Kansas, Univ. of, 200
Kantrowitz, Arthur, 359
Kapitsa, Dr. Pyotr L., 21, 30, 165
Kaplan, Dr. Lewis D., 264, 322
Kapustin Yar, U.S.S.R., 294, 372

Karo, V/Adm. H. Arnold (USN), 293
Karolinska Institute, 318
Karth, Rep. Joseph E., 19, 98, 113, 200, 214, 250, 275, 308
Kastler, Dr. Alfred, 339
Kazakhstan, U.S.S.R., 121, 223
KC-135 (Stratolifter), 106
Keathley, William, 361
Keldysh, Mstislav V., 20, 50, 82, 140, 224
Keller, K. T., 22
Kelley, Dr. Albert J., 149
Kemmerer, Dr. Walter, 64
Kennedy, Harvey M., Jr., 139
Kennedy International Airport, N. Y., 365
Kennedy, President John F., 193, 365
Kennedy, John F., Center for Special Warfare, 156
Kennedy, John F., Memorial Library, 314–315
Kennedy Space Center (KSC) (NASA), 125, 155, 172, 215, 258, 283
  Apollo/Saturn, 69, 134, 193, 383
  astronauts at, 112, 213, 247, 249, 293, 349
  award, 314–315
  contract, 223, 262, 321, 337
  crawler-transporter, 33, 86, 248
  facilities, 86, 220, 364
  personnel, 258
  spacecraft delivery, 49
  strike, 142, 144, 150, 242, 268, 271, 272
Kerguélen, 227
Kerr, Breene M., 2, 88, 195, 209
Kerster, George, 250
Keyhoe, Maj. Donald E. (USMC, Ret.), 119
Kilston (comet), 260
Kilston, Stephen, 260
Kincheloe, Iven C., Award, 301
King, Joseph C., 120
King, Ludlow, 249
King, Robert E., 6
King's Medal (U.K.), 2
Kings Point, N. Y., 321
Kirtland AFB, N. Mex., 119
Kitt Peak National Observatory, 9
Kitty Hawk, N. C., 377
Klass, Phillip J., 272
Klein, Milton, 336
Klein, Murray, 9
Kleinman, Dr. David A., 51
Kliore, Dr. Arvydas J., 175
Klump, H. F., 358
Knight, Maj. William J. (USAF), 237, 247, 257, 263, 278, 351, 385
Knowles, Gov. Warren P., 69
Kock, Dr. Winston E., 288
Koesterer, M. G., 174
Kok, Dr. Bessel, 174
Kolman, Prof. E., 201
Kolomiyets, Nikolay, 333
Komarov, L/C Vladimir (U.S.S.R.), 237
Kommunist, 40
Komsomolskaya Pravda, 21
Konecci, Dr. Eugene B., 54, 79, 221, 252
Konstantinov, Boris, 236
Korad Corp., 135

Korea, 54
Korea, North, 355
Korean War, 311
Korff, Dr. Serge A., 333
Kormanyos, Steven W., 142
Korolev, Sergey Pavlovich, 16
Koroni, Greece, 87, 178, 187
Kosygin, Premier Alexey N. (U.S.S.R.), 94, 356, 370
Kotelnikov, Vladimir, 9
Koval, Ivan, 46
Kraft, Christopher C., Jr., 166, 235
Krakatau, Indonesia, 299
Kratochvil, L/C C.H. (USAF), 336
Krasnaya Zvezda, 90, 118, 148, 333
Kreplin, Dr. Robert W., 174
Krieger, Robert L., 315
Krisch, Dr. Alan D., 142
KSC. See Kennedy Space Center.
Kuiper, Dr. Gerard P., 44, 96
Kuo, T. K., 152
Kuppermann, Dr. Aron, 128
Kuhrt, Wesley A., 98
Kurzweg, Dr. Hermann H., 81
Kyoto Univ., 45
Kyushu Island, Japan, 149, 301, 333, 378

L

Labor, Dept. of, 6
Labor relations (see also strike), 6, 259, 271, 364
Lafayette College, 140
Lagerfeld Base, Munich, Germany, 183
Lagos, Nigeria, 134
La Jolla, Calif., 159
Lakehurst, N. J., 134
Lam, S. H., 151
Lambda (Japanese booster), 74
　4S-1, 301
　4S-2, 378
Lambda Scorpii (star), 241
LAMBDA (U.S.S.R. satellite), 374
Lamont Geological Observatory, 94
Lancaster, Calif., 214
Lancaster, Dr. Otis E., 298
Les Landes (missile launching site), France, 82
Landing gear, aircraft, 158, 296, 314
"Landing on the Moon" (TV program), 198
Lange, Dr. Oswald H., 192
Langley Research Center (LaRC) (NASA), 309
　award, 213, 315
　computer, 165, 215
　contract, 19, 90, 92, 135, 177, 284, 316
　lifting body research (see also HL-10), 19, 135
　Lunar Orbiter program, 262, 265, 281, 313, 340, 344
　PAGEOS I, 222
　parawing, 156
　Planetary Reentry Parachute Program, 153, 242–243, 279, 284, 340, 352, 370
　contract, 275
　Project Scanner, 92, 268, 369
　Scout Reentry Heating Project, 49, 90
Langmuir probe, 234, 292
"LANI BIRD." See INTELSAT II-A.
Larraga, Gen. Jose Lacalle, 317
Las Cruces, N. Mex., 235, 351
Las Vegas, Nev., 145, 366
Laser, 135
　chemical, 221, 289
　patent, 126, 258
　use of, 51, 97, 126, 209, 310, 380
Lasilla, Chile, 125
Lasswell, Harold D., 190
Launch Complex 14, 243
Launch Complex 17A, 269
Launch Complex 37, 233
Launch Complex 39, 15, 268, 315
Launch Complex 41, 276
Launch Escape System (Les), 19, 21, 42
Launch Umbilical Tower, 86
Launch vehicle (see also individual launch vehicles such as Atlas-Centaur, Saturn, etc.)
　reusable, 99, 221, 358
Lausche, Sen. Frank J., 137
Lawrence Radiation Laboratories, 49
Learned, Dr. Edmund P., 131
Lebedinsky, Dr. Alexander I., 50, 174, 377
Lederberg, Joshua, 260
Lederman, Leon, 286
Lee, Capt. Chester M. (USN, Ret.), 267
Leiden, Univ. of, Observatory, 324, 377
Leighton, Dr. Robert B., 90, 236
Lem (Lunar excursion module). See Lunar module.
Lemke, George A., 268
Leningrad Institute of Cytology, 337
Leningrad Jet Study Group, 10
Leningradskaya Pravda, 10
Leonid meteor shower, 220, 267, 350, 351
Leonov, L/C Aleksey (U.S.S.R.), 92, 107, 330, 356
Le Poole, Rudolph S., 116
Les. See Launch Escape System.
Lesher, Dr. Richard L., 12
Levitt, Dr. I. M., 366
Levy, Lawrence, 275
Lewis, David S., 354
Lewis, Dr. George W., 23
Lewis Research Center (LRC) (NASA), 258, 281
　anniversary, 23, 312
　Atlas-Centaur, 197, 296, 329–330
　award, 315
　contract, 251, 283, 298
　research, 120, 194, 293
　rocket engine
　　hydrogen-fueled, 125
　　Sert, 312
　turbopump, 114
Lewis, Richard, 261
Lewis, Roger, 354
Li-2 (U.S.S.R. aircraft), 256
Libby, Dr. Willard F., 128
Library of Congress, 243

LIBERTY BELL 7 (spacecraft), 105
Liccardi, Anthony L., 354
Lick Observatory, 260
Life, artificial, 18–19
Life magazine, 125
Life science. See Space biology.
Life support system, 11, 94, 120, 273
Lifting body vehicle, 99, 383
   HL-10, 19, 135, 379
   M2-F2, 114, 237, 244, 264, 274, 284, 293, 297, 298, 306, 312, 319, 330, 352
   SV-5D, 378
   SV-5P, 152
Lin, Prof. Shao-Chi, 27
Lind, Dr. Don L., 127
Ling-Temco-Vought (LTV) Aerospace Corp., 194, 264, 267, 309
Lipski, Prof. Yuri, 305
Lisbon, Portugal, 302
Little Joe II (booster), 19, 21
Little Lake (NASA barge), 74
Litton Industries, Inc., 337, 361, 371
Livermore, Calif., 237
Lm. See Lunar Module.
Lockheed Aircraft Corp., 284
   award, 27, 135, 354
   C-5A (cargo transport), 35, 58, 377
   C-141 (cargo jet), 66
   contract, 5, 66, 77, 208, 361
   F-104G (Starfighter), 53, 256, 367
   500 (cargo aircraft), 358
   HC-130H, 164
   Launch Escape System rocket motor, 42
   supersonic transport, 85, 170, 224, 286, 296, 299, 323, 341, 343, 383
Lockheed-California Co., 111, 335, 356
Lockheed-Georgia Co., 58, 314, 365
Lockheed Missiles and Space Co., 179, 195, 213, 319, 353, 356, 375
Lockheed Propulsion Co., 16, 90
Loewy, Dr. Robert G., 298, 362
Loginov, Yevgeni F., 339
Logistics Management Symposium, First, 291
London Times, 251, 260
London, U.K., 41, 77, 170, 262, 293, 354
Long Tank Thor, 6
Look, 323
Loosbrock, John, 36
Lop Nor, Communist China, 382
LORHO (test facility), 221
Lorsch, H. G., 174
Los Angeles, Calif., 188, 221, 238, 330
   meeting in, 166, 241, 271, 273, 275, 314, 366
Los Angeles Times, 82, 142, 248
Louisiana, 341
Lousma, Capt. Jack R. (USMC), 127
Lovelace, Dr. W. Randolph, II, 36, 54, 359
Lovelace, Dr. W. Randolph, II, Award, 225
Lovell, Sir Bernard, 16, 44, 75, 115, 126, 160, 209, 330
Lovell, Capt. James A., Jr. (USN), 110
   award, 69, 226, 259, 354, 374
   GEMINI XII (flight), 219, 327, 340, 345, 349
Low, George M., 235
Lowenfeld, Andreas F., 43
Lowman, Paul D., Jr., 7, 355
LRC. See Lewis Research Center.
LTV. See Ling-Temco-Vought Aerospace Corp.
Luce, Adm. Sir David, 61
Luckey, Dr. T. D., 166
Ludlow King Associates, 249
Luedecke, M/G Alvin R. (USAF-Ret.), 9
Luftwaffe, 111
LUNA III (U.S.S.R. lunar probe), 333
LUNA IX, 41–42, 130, 203
   launch, 35
   photographs, 44, 47, 48, 49, 50, 59, 115, 116, 213, 228
   soft-landing, 42, 45, 50, 146, 385
LUNA X, 125, 146, 171, 279, 385
   launch, 122
   lunar orbit, 126–127
   results, 129, 133, 136, 159, 181, 356, 358
LUNA XI, 279, 330
   launch, 273
   lunar orbit, 278
   results, 292, 303, 312
LUNA XII
   launch, 326
   lunar orbit, 328, 330, 331
   results, 333, 355
LUNA XIII, 380
   launch, 378
Lunar Excursion Module (Lem). See Lunar Module.
Lunar Module (Lm), 6, 11, 96, 104, 167, 198, 203, 206, 221, 298, 375
   contract, 40
   test, 54, 189
Lunar Orbiter (program), 47, 49, 176, 246
Lunar Orbiter I (Lunar Orbiter A), 279, 282, 332, 384–385
   launch, 262
   lunar orbit, 265
   results, 275, 313, 355, 382
Lunar Orbiter II (Lunar Orbiter B), 333, 347, 373, 384
   launch, 340
   lunar orbit, 344, 364, 369
Lunar roving vehicle, 6
   simulated cabin, 60, 91, 138
Lunar Sample Receiving Laboratory (MSC), 163, 171, 172, 176, 255
Lund, Thomas, 250
Lurain. See Moon, photographs and surface.
Luster, Project, 326
Luxembourg, 210
Lyman-alpha radiation experiment, 267, 274, 347
Lyons, Dick, 289
Lytton Systems, 287
LyVere, Capt. Gerald T. (USAF), 164

M

M-1 (rocket engine), 114
M2-F2 (lifting-body vehicle), 384, 385
  flights, 114, 237, 244, 264, 274, 284, 293, 297, 298, 301, 306, 312, 319, 330, 352
M-87 (celestial body radiation source), 79
MA-25S (silicon ablator), 15
MacDonald, Dr. Gordon J. F., 32, 128, 142, 214, 357
McCall, G. J. H., 299
McCall, Dr. Jerry C., 1
McCandless, Lt. Bruce, II (USN), 127, 382
McCarthy, Sen. Eugene J., 288
McCarthy, Dr. John, 314
McCollom, John S., 264
McConnell, Gen. John P. (USAF), 72, 113, 282
McCormack, James, 146, 199, 270
McCormack, Rep. John W., 112
McCulloch, Rep. William M., 137
McDivitt, L/C James A. (USAF), 54, 93, 105, 110, 144, 216, 332, 379
McDivitt, Mrs. James A., 216
McDonald, Adm. David L. (USN), 251
McDonald, Dr. James C., 325
McDonnell Aircraft Corp., 7, 103, 145, 157, 167, 177, 272, 335, 354, 364
McDonnell, James S., 354
McGregor-Werner Co., 272
McInturff, Raymond M., 199
McKay, John B., 47, 168, 252, 263, 275, 288
McKee, Gen. William F. (USAF, Ret.), 128, 131, 296, 334
McKenzie, R. I., 354
McKnew, Dr. Thomas W., 232
McNamara, Secretary of Defense Robert S., 40, 43, 57, 72, 170, 264, 303, 341, 343
McWilliams, William J., 73
Madison, Wis. 69
Madras, India, 253
Madrid, Spain, 317, 319, 320, 321, 352
Magellanic Gold Medal, 147
Magnesium, 221
Magnetic field (see also Earth; Moon)
  earth, 86, 161, 320, 380
  interplanetary, 86
Magnetic hammer, 288
Magnetohydrodynamics, 222
Magnetometer, 29, 56, 83, 129, 221, 234
Magnuson, Sen. Warren G., 219, 286
Mahon, Rep. George H., 213
Maiman, Dr. Theodore, 135
Makai, Project, 366
Malaysia, 54, 165
Mali, 139, 174
Malina, Frank J., 124
Malinovsky, Rodion Y., 125
"Man and Materials" (symposium), 147
Management, 77, 172, 291, 294, 331, 350, 369, 372
Management Club, 372
Management Services, Inc., 130
Manchester, U. K., 196
Manhattan Project, 363
Manley, Oscar P., 229
Manned Orbital Operations Safety Equipment (Moose), 192
Manned Orbiting Laboratory (Mol), 87, 121, 150, 153, 167, 246, 323, 348, 369
  appropriations, 23–24, 87, 227
  contract, 157, 177
  cooperation, 111, 138, 307
  launch plans, 11, 243, 381
  test, 87, 111, 167, 338, 381
Manned space flight (see also Apollo program, Gemini program and flights; Astronaut; Cosmonaut; Manned Orbiting Laboratory; and Space Biology), 382
  achievements, 23, 99–100, 100–102, 105–106, 212, 248, 291, 324, 384
  award, 103, 259, 324
  cooperation, 10, 73, 111, 319–320
  criticism, 109
  Eva. See Extravehicular activity.
  hazards, 105, 107, 111–12, 120, 145, 182–183, 255
  long-duration, 120, 351
  lunar landing, manned. See Moon landing, manned.
  manpower required, 56, 149
  military potential, 352, 373, 381–382
  policy and plans for
    U.S., 49, 93, 229, 238, 262, 307–308, 323, 334, 338
    schedule, 93, 257, 269, 332, 351, 375
    U.S.S.R. 135–136, 303, 316, 319, 321, 338, 348, 356, 368, 373, 379–380
  space simulation testing, 130, 203, 273
    lunar roving vehicle cabin, 60, 91–92, 138
Manned Space Flight Policy Committee, 43
Manned Spacecraft Center (MSC) (NASA), 146, 293, 334, 349
  Apollo Spacecraft Office, 167, 375
  astronauts at, 22, 27, 104, 117, 218, 246, 249, 255
  award, 103, 255, 323, 354
  contract, 40, 104, 270
  Crew Systems Div., 317
  Gemini Mid-Program Conference, 64
  Lunar Sample Receiving Laboratory, 163, 171, 172, 176, 255
  management, 166, 235
  manned spacecraft missions. See Gemini flights.
  Mission Control Center, 244, 247, 327
  personnel, 64
  Philip, Prince, visit, 93
  Real Time Computer Complex, 258
  spacecraft test. See Apollo (spacecraft).
Marbarger, Dr. John P., 90
Marble, Prof. Frank E., 235
March Field, Calif., 259
Marcos, President Ferdinand E. (Philippines), 292, 311
Marine Resources and Engineering Development Act, 219
Marine Science, Engineering and Resources Commission, 219

Mariner (program), 32, 73, 281, 337
Mariner (spacecraft), 83, 238, 239, 264
MARINER II (Venus probe), 65, 75, 86, 151
MARINER IV (Mars probe), 33, 166
 photographs, 240, 241, 262, 299, 356
 results, 31–32, 86, 151, 175, 199, 236, 241
 signal, 3, 84, 194
Mariner E (Venus probe), 7
Marion Power Shovel Co., 33
Mark 12 (reentry vehicle), 255
Mark, Dr. Herman F., 52
Mars (planet) (see also MARINER IV; Voyager program), 86, 157, 173, 212, 239
 atmosphere, 80, 199, 236, 264, 314, 322
 craters, 187, 299
 exploration, 23, 73, 290, 334, 348, 368, 380
  manned, 69, 146–147, 229, 268, 327
  plans for, 26, 65, 281
  solar-electric propulsion, 89, 355
  spacecraft, 253–254, 268, 279
  unmanned, recommendation, 17, 198, 281
 life on, 53, 65, 189, 241, 366
 surface, 162, 175, 187, 202
 water on, 236
Marshall, Gen. George C., 22
Marshall Space Flight Center (MSFC) (NASA), 119, 168, 194, 241
 Apollo Telescope Mount, 239, 289, 325, 360
 Automatic Mathematical Translator, 81
 award, 315
 Computation Laboratory 169, 223
 contract, 5, 129, 165, 244, 247, 249, 268, 306, 328, 343, 383
 Saturn, 129, 165–166, 255, 378
 launch vehicle. See Saturn.
 management, 1, 125, 235, 360
 marine transportation, 74
 personnel, 1, 235, 284, 360
 Saturn Manufacturing Review meeting, 191
 space science exhibit, 312
 test (see also Saturn), 55, 60, 91, 138
Marshall, Mrs. Jane, 210
Martin Co., 14, 40, 78, 103, 112, 135, 152, 177, 195, 210, 264, 270, 275, 325, 354
Martin, Glenn L., 383
Martin, Dr. James S., Jr., 262
Martin, Joseph W., 214
Martin Marietta Corp., 5, 177, 220, 251, 354
Maryland, Univ. of, 241, 252, 293, 365
Massachusetts Institute of Technology (MIT), 26, 31, 47, 214, 223, 258, 280
 experiment, 89, 229
 research, 323, 384
Masursky, Dr. Harold, 313
Materials, aerospace, research, 77, 250, 382
Mathews, Charles W., 17, 64, 86, 103, 162, 342, 354, 360
Mattingly, Lt. Thomas K. (USN), 127
Maxwell, B/G Jewell C. (USAF, Ret.), 85, 122, 200–201
May, T. R., 58

May Day (U.S.S.R.), 154, 161
Medium Resolution Infrared Radiometer (Mrir), 178, 257
Mayhew, Christopher, 61
Mbelouson, Yuri, 46
Mdta. See Modulation, demodulation, terminal and associated equipment.
Meitner, Dr. Lise, 259
Memorandum of understanding, 15, 40, 43, 51, 138, 195
Menzel, Dr. Donald H., 382
Mercury (planet), 17
Mercury, Project, 7, 103, 321
Merz, Ernest J., 190
Metal fatigue, 377
"Metalaw," 289
Meteor, 149, 351
Meteorite
 crater, 94
 origin, 116, 144
 lunar impact, 299, 316, 382
Meteoroid detection satellite, 205, 247, 337
Meteorology, 3, 23
 award, 210
 balloon use in, 3, 89, 168, 287, 326
 cooperation, 8, 43, 89, 210
  ESSA-USN, 267
  NASA-Argentina, 154
  -Brazil, 154, 161
  -France, 195
  -Spain, 15
  U.S.-U.S.S.R., 96, 156, 269, 271, 308
 forecasting, 3, 79, 195
 National Natural Disaster Warning System, 77
 satellite. See Meteorological satellite.
 sounding rocket experiments (see also individual sounding rockets), 154, 173, 265, 360
  chaff, 14, 15
  grenade, 23, 39, 51, 114, 117, 152, 161, 164, 216, 226, 260, 307, 310
  Japan, 109, 149
  mylar sphere, 25, 42
  pitot-static probe, 69, 260, 276
  U.S.S.R., 174
  vapor cloud, 19, 69, 152, 239, 242, 292, 299, 301, 347
 Stormfury, Project, 267
 weather modification. See Weather modification.
 World Meteorological Day, 113
 World Weather Watch, 114, 133, 326
Meteorological satellite, 43, 70, 79, 173, 298, 385
 cooperation
  NASA-France, 195
  U.S.-U.S.S.R., 96, 156, 269, 271, 308
 Nimbus program, 367
  NIMBUS I, 3, 264
  NIMBUS II, 178, 185, 240, 257, 283, 347, 349, 365
 Tiros program, 24, 41, 69, 145, 220, 310, 321, 365
  contract, 176
  U.S.S.R., 269, 271, 385

Methane
  Martian atmosphere, 322
  liquid, 293, 362
"Metro Air Support '66" (air transport exercise), 340
Meudon Observatory, 323
MI-1 (U.S.S.R. helicopter), 291
Miami, Univ. of, 378
Michael, William, 313
Michigan, 110, 113, 116, 122
Michigan Assn. of the Professions, 177
Michigan Technological Univ., 22
Michigan, Univ. of, 25, 30, 42, 69, 142, 237, 251, 260, 276, 277
Michoud Assembly Facility (MSFC), 15, 22, 34, 191, 298
Microelectronics, 77, 177
"Microeye" (camera), 234
Microflora, 162
Micrometeorite, 86, 267, 328, 350, 351
Micrometeoroid, 288, 327, 336
  collection experiment, 252, 263, 275, 292
Micrometeoroid detection experiment
  GEMINI X, 243, 248
  LUNAR ORBITER I, 313
Microscope, 260
Midas (satellite), 172
Midway Island, 149, 233
Midwest Research Institute, 12, 200
Midwest Slavic Congress, 117
Midwest Space Exposition, 298
MiG-21 (U.S.S.R. fighter aircraft), 323
Mikhailov, Prof. Alexandr I., 140
Mikoyan, Anastas I., 94
Miles, Marvin, 82
Miles, Rufus E., Jr., 310
Milky Way (constellation), 292
Miller, Rep. George P., 64, 157, 171, 291, 297, 325, 357
Miller, James W., 13
Miller, Paul, Jr., 106
Miller, Rep. William E., 214
Millikan, Dr. Clark Blanchard, 2
Mills College, 332
Milwaukee, Wis., 69, 327, 337
Milwitzky, Benjamin, 250
Minderman, Peter A., 315
Minneapolis, Minn., 138, 287
Minneapolis Tribune, 34
Minnesota, Univ. of, 169, 360
Minuteman (ICBM), 22, 66
Minuteman II, 165, 255
Mirny Antarctic station, 157
Mirovaya Ekonomika i Mezhdunarodnyye Otnosheniye, 201
Missile, 97, 113, 153, 219
  air-to-surface, 334, 341, 361
  antiaircraft, 119, 267
  antimissile, 170, 355, 373
  ballistic, intercontinental (ICBM), 115, 165, 261, 369
  foreign
    Communist China, 331
    France, 82, 201
    India, 106
  U.K., 31
  U.S.S.R., 181, 261
  medium-range, 227
  nuclear, 355
  passenger, 34
  short range attack (Sram), 334, 341, 361
  launch, 22, 66, 255, 275, 342, 367
  submarine, missile carrying, 247, 302
Missiles and Rockets. See *Technology Week*.
Mississippi Test Facility (MTF) (MSFC), 114, 147, 181, 283, 364, 380
Missouri, Univ. of, 55, 136, 166, 350
MIT. See Massachusetts Institute of Technology.
Mitchell, LCdr. Edgar D. (USN), 127
Mitchell, Gen. William E., 383
Mitchell, Gen. William E., Memorial Award, 137
Miyamoto, Dr. Shotaro, 45
Mock, Mrs. Geraldine, 134
Mohole, Project, 150, 155, 182, 247, 359
Mol. See Manned Orbiting Laboratory.
Molniya I (U.S.S.R. communications satellite program), 376, 385
MOLNIYA I (3) (U.S.S.R. communications satellite), 149, 152, 154, 182, 224
MOLNIYA I (4), 325
Mombasa, E. Africa, 174
Monkey experiment, 187, 322
Montreal, Univ. of, 229
Moon
  atlas, 305
  crater, 187, 299, 305, 313, 333, 344, 382
  exploration of, 1, 29, 56–57, 95, 180, 198, 255, 307, 369
  equipment, 40, 244–245, 252
  gravitational field, 265
  landing
    manned
      U.S., 95, 96, 134, 219, 352
        plans for, 23, 229, 327, 334, 350, 375–376
      U.S.S.R., 50, 125, 157, 285, 319–320
    unmanned, soft
      equipment, 11, 39, 175, 203
      U.S., 203–205, 232, 249, 296, 384–385
      U.S.S.R., 41–42, 44–45, 50–51, 91, 380–381, 385
  magnetic field, 133, 173
  photographs
    LUNA IX, 44, 46, 47, 48, 59, 174, 203, 213, 228
    LUNA XI, 278, 279
    LUNA XII, 333
    LUNA XIII, 380–381
    LUNAR ORBITER I, 262–263, 265, 279, 281–282, 313, 382, 384
    LUNAR ORBITER II, 344, 364, 374
    SURVEYOR I, 203–204, 208, 226, 228, 234, 341
    ZOND III, 305
  probe. See individual probes: LUNAS III, IX, X, XI, XII, XIII, LUNAR ORBITERS I, II, RANGER IX, SURVEYORS I, II, ZOND III.

radiation, 51, 133, 173
shape, 275, 313, 355–356
surface
  composition, 136, 173, 228, 355
  texture, 49, 50, 96, 174, 203, 228
  topography, 47, 313, 316, 374
  temperature, 50, 203
  volcanism, 299, 313, 316, 374
Moorefield, W. Va., 122, 176, 314
Moorestown, N. J., 194
Moose. See Manned Orbital Operations Safety Equipment.
Moraes, Carlos de, 189
Morehead Planetarium (Univ. of North Carolina), 14
Morozov, Platon D. 218, 237, 248, 258, 291, 293
Morrison, Dr. Philip, 28, 47
Moscow (U.S.S.R.), 95, 161, 254, 379
  airline service, 339
  airport, 58
  comsat transmission to, 152, 154
  meeting in, 125, 189, 227
  press conference, 50, 82, 140
  television, 47, 48, 135, 147, 333
  weather data exchange, 156
Moscow State Univ., U.S.S.R., 221
Mosher, Rep. Charles A., 214
Moskva, 384
Motor, rocket (solid-propellant), 16, 62, 110, 177, 283, 297, 385
Motorola, Inc., 239
Mt. Palomar Observatory, 28, 93, 219 287
Mt. Wilson Observatory, 219, 287
Moyers, Bill, 212
Mrir. See Medium Resolution Infrared Radiometer.
MSC, See Manned Spacecraft Center.
MSFC, See Marshall Space Flight Center.
MT-135 (Japanese sounding rocket), 381
MTF. See Mississippi Test Facility.
Mu-1 (Japanese booster), 333, 339
Mueller, Dr. George E., 1, 311
  Apollo, 69, 275, 291
  Apollo Applications, 52
  award, 354
  Gemini, 108, 207, 318
  manned spaceflight, 30, 56, 64, 93, 110, 269
  space program, 56, 93, 162, 191, 196, 329, 337, 366
  technology utilization, 357, 366
Mulholland, Donald R., 188
Mullin, Jerome P., 339
Mullins, George C., 128
"Multiple Sclerosis Woman of the Year," 215
Mundt, Sen. Karl, 179
Murphy, Charles, 180
Murphy, Franklin D., 128
Murphy, Dr. Thomas P., 55
Murray, Dr. Bruce C., 73, 236
Mururoa Atoll, 232, 246, 312

N

N*3245 (nuclear particle), 142
NAC. See National Aviation Club.
NACA. See National Advisory Committee for Aeronautics.
Nadge. See NATO Air Defense Ground Environment Project.
NAE. See National Academy of Engineering.
Nagler, Kenneth, 7
Naples, Italy, 306
Naples, Univ. of, 319
Napolitano, Dr. Luigi, 319
Narodna Armiya, 214
NAS. See National Academy of Sciences.
NASA. See National Aeronautics and Space Administration.
NASA Headquarters
  Inventions and Contributions Board, 16
  Natural Resources Program Office, 33
  Office of Advanced Research and Technology (OART), 60, 66, 99, 232, 234, 238, 258, 327, 334
  Office of Manned Space Flight (OMSF), 18, 26, 53, 125, 143, 232, 233
  Office of Space Science and Applications (OSSA), 3, 18, 32–33, 43–44, 75, 113, 125, 178, 215, 222, 231, 261, 263, 269, 281, 296, 325, 340, 367, 374
  Office of Technology Utilization, 3
  Planetary Quarantine Committee, 308
  Scientific and Technical Information Facility, 273
NASA-AEC Space Nuclear Propulsion Office (SNPO), 42, 83, 336
NASA Contributions to Cardiovascular Monitoring, 309
NASA Pasadena Office (NAPO), 261
NASA Space Radiation Effects Laboratory, 19
NASC. See National Aeronautics and Space Council.
Nasher, Raymond D., 258
Natal, Brazil, 14, 161, 164, 252, 260, 265, 307, 310, 373
Nathan, Dr. Robert, 262
National Academy of Engineering (NAE), 145, 155
National Academy of Sciences (NAS), 39, 53, 85, 123, 133, 150, 173, 231, 302, 305, 310, 320, 361, 376, 384
  Physics Survey Committee, 95
  Space Science Board (see also National Research Council), 17, 53, 95, 198, 251, 280, 314, 349
National Advisory Committee for Aeronautics (NACA), 36, 270
National Aeronautics and Space Administration (NASA) (see also NASA centers, programs, satellites, and related headings, such as Ames Research Center, Apollo program, LUNAR ORBITER I, etc.)
  accomplishments, 23–24, 35, 47, 62–63, 65, 83–84, 88–89, 92, 99–100, 155–156, 203, 205, 207, 246, 248, 265, 294, 349, 384–385

agreement. See Agreement.
anniversary, 23, 35, 310
astronaut. See Astronaut.
awards and honors, 20, 22, 30, 43, 50, 57, 77, 103, 117, 140, 145, 188, 191, 208, 213, 215, 218, 223, 225, 232, 235, 236, 255, 259, 301, 304, 314–315, 321, 323, 324, 331, 352, 354, 359, 374
appropriations, 127, 143, 163, 166, 172, 176, 190, 214, 244, 247, 257, 259, 261, 263, 271, 273, 286, 328
budget, 23, 26–27, 28, 59, 64, 70–71, 75–76, 80, 91, 93, 113, 138, 143, 149, 171, 187, 262, 278, 317, 357, 364
conference, 5, 64, 191, 291
contract
  administration, 239
  engine, 2, 5, 16, 85, 90, 115, 145, 224, 238, 255, 257–258, 330, 332, 356, 375
  facilities, 255, 257, 258, 273
  space equipment, 54, 92, 104, 117, 175, 176, 177, 179, 194, 221, 234, 264, 271, 325, 337, 348, 353, 356
  spacecraft, 5, 77, 90, 176, 227, 275
  space medicine, 64
  study, 2, 5, 176, 179, 195, 356
  support services, 9, 165–166, 195, 223, 235, 244, 262, 298, 315, 321, 380, 383
  tracking, 261, 315, 361
cooperation, 284
  AEC, 42, 82–83, 211, 336, 360
  American Society for Engineering Education, 364–365
  ComSatCorp, 235
  DOD, 7, 29, 43, 110, 150, 198, 225, 304–305, 334
  ESSA, 43, 103
  FAA, 128
  NAS-NRC, 302
  Naval Oceanographic Office, 19
  Small Business Administration, 89
  USAF, 90–91, 111, 114, 128–129, 150, 203
  USDA, 3, 117
  USN, 8
cooperation, international. See International cooperation, space and Sounding rocket, international programs.
criticism of, 7, 37, 108, 109, 123, 165, 202, 220, 232–233, 258, 260, 264, 266, 273, 280–281, 319, 365, 366
exhibit, 60, 253, 312
facilities, 19, 23, 83, 220, 247, 257, 258, 273–274
  construction, 97, 255, 364
history, 231
Humphrey, Vice President Hubert H., visit, 43
information dissemination, 240
Inventions and Contributions Board, 317
labor relations, 6, 258, 271, 364
launch, 120, 131
  balloon, 89, 242
  booster, 131, 329
  failure, 162, 164, 180, 182, 193
  manned
    GEMINI VIII, 99
    GEMINI IX-A, 206
    GEMINI X, 243
    GEMINI XI, 290
    GEMINI XII, 344
  postponed, 61, 67, 96, 121, 180, 202
  probe
    PIONEER VII, 269
    SURVEYOR I, 196, 232
    SURVEYOR II, 296
  satellite
    ATS I, 367–8, 371
    BIOSATELLITE I, 374
    ESSAS I, II, and III, 41, 69, 310
    EXPLORERS XXXII and XXXIII, 192, 231
    INTELSAT II-A, 331
    LUNAR ORBITERS I and II, 262, 340
    NIMBUS II, 178
    OAO I, 132
    OGO III, 209
    PAGEOS I, 222
Scanner, Project, 268
test
  Apollo-Saturn, 67, 88, 233, 274
  Augmented Target Docking Adapter, 202
  Launch Escape System, 21
  Scout Reentry Heating Project, 50
sounding rocket
  Aerobee 150, 60, 89, 138, 156, 277, 364, 373
  Aerobee 150A, 18, 79, 84, 182, 276
  Aerobee 350, 269
  Arcas, 173
  Boosted Arcas, 185, 277
  Hasp, 173
  Javelin, 79, 161, 237, 267, 299
  Nike-Apache, 10, 19, 42, 55, 56, 95, 111, 117, 127, 132, 214, 234, 237, 239, 240, 242, 245, 276, 292, 307, 351
  Nike-Cajun, 39, 42, 51, 161, 307, 310
  Nike-Tomahawk, 276, 301, 360
  Orion II, 342
  Pacemaker, 333
  Wasp, 210
management, 9, 125, 258, 268, 289–290, 291, 331, 350, 360
manpower, 93
organization, 1, 16, 43–44, 139, 188, 235, 261, 327
patents, 16, 335
Observatory Class Spacecraft Review Board, 149
personnel, 9, 32, 34, 36, 105, 106, 127, 128, 132, 202, 203, 208, 209, 214, 232, 235, 249, 258, 264, 267, 268, 270, 283, 288, 289–290, 305, 316, 317, 327, 328, 336, 340, 360, 377, 383
  appointments, 6, 12, 54, 55, 73, 166, 173, 195, 243, 246, 249, 275, 279, 283, 284, 298, 310, 311, 315, 319, 380

programs (see also specific programs, such as Apollo, Gemini, etc. and Space program, national)
  aeronautics, 62, 66, 208
  astronomy, 39, 62–63, 65, 150–151, 278, 288
  international, 62–64, 80–81, 90, 133, 141, 155, 386
  manned space flight, 52, 64–65, 71, 93, 167, 238, 261, 269, 318, 324, 327, 332, 349, 351, 375–376
  meteorology, 3, 381, 384–385
  nuclear propulsion, 42, 62, 82–83
  sounding rocket, 65, 80–81, 310, 384–385
  space medicine, 187, 317–318, 361, 374
  space science, 59, 62–64, 71, 83, 133, 229, 238, 264, 359, 384–385
  technology utilization, 88–89, 281, 328, 334
  tracking and data aquisition, 81, 83
Saturn Manufacturing Review meeting, 191
Science Advisory Committee, 95
Space Science Steering Committee, 83
supersonic transport, 92, 128, 224
test
  Agena target vehicle, 67, 86
  booster, 14, 22, 64, 84, 107, 114, 122, 147, 181, 185, 195, 210, 227, 233, 252, 350, 357, 364, 382, 383
  motor, solid propellant, 62, 110
  nuclear, 42, 52, 82, 103–104, 115, 211, 222
  parachute, 279, 284, 340, 352, 370
  spacecraft, 19, 35–36, 39, 42, 175, 185, 328
universities, 53, 71, 238–239, 298, 364, 378, 379
  grants, 55, 97, 287
X-15. See X-15.
National Aeronautic and Space Engineering Meeting, 314
National Aeronautic Assn., 210, 237, 324
National Aeronautics and Space Act, 288, 310
National Aeronautics and Space Council (NASC), 35, 44, 46, 53, 79, 99, 102, 167, 180, 212, 252, 319, 324, 357, 372, 373
National Aerospace Education Council, 210
National Air and Space Museum, 49, 169, 227, 324
National Air Museum, 227
National Assn. of Broadcasters, 118
National Assn. of Educational Broadcasters, 327
National Assn. of Manufacturers, 256, 319
National Assn. of Science Writers, Inc., 47
National Aviation Club (NAC), 208, 268, 308
National Aviation Hall of Fame, 383
National Broadcasting Co. (NBC), 136, 256, 287
National Bureau of Standards, 9
National Center for Atmospheric Research and Information, 89, 133, 168, 263
National Committee for Clear Air Turbulence, 381
National Conference of Lieutenant Governors, 223–224
National Council of Marine Resources and Engineering Development, 219
National Cowboy Hall of Fame, 236
National Cybernetics Foundation, 54
National Dairy Council, 25
National Defense Transportation Assn. Award, 297
National Environmental Satellite Center, 96
National Foundation on the Arts and Humanities, 147, 360
National Geographic Society, 132, 155, 232
National Investigations Committee on Aerial Phenomena (NICAP), 119
National Labor Relations Board, 271
National League of Cities, 366
National Medal of Science, 50, 380
National Natural Disaster Warning System, 77
National Observer, 13
National Oceanography Assn. (NOA), 224
National Park Service, 15
National Pilots Assn. 308
National Radio Astronomy Observatory, 257
National Reactor Testing Station, Ida., 13, 276
National Research Council (NRC), 302, 336
  Space Science Board, 336, 384
National Referral Center, 243
National Satellite Center, 286
National Science Board, 10, 243
National Science Foundation (NSF), 10, 24, 30, 38, 54–55, 59, 96–97, 104, 223, 243, 247, 257, 360
  Special Commission on Weather Modification, 21
"The National Science Foundation - Its Present and Future" (report), 223
National security, 83, 130, 135, 196, 366
National sonic boom program (see also Supersonic transport; X-15), 338, 342, 354, 372, 376, 378
National Space Club (NSC), 19, 53, 102, 153, 221, 269, 297, 323
National Space Club Press Award, 103
National space program. See Space program, national.
National Survey of Historic Sites, 263
National Youth Science Camp program, 239
NATO. See North Atlantic Treaty Organization.
NATO. Air Defense Ground Environment (Nadge) Project, 227
NATO's 15 Nations, 199, 282
Nature, 289
Naugle, Dr. John E., 44, 326

Naval Electronics Laboratory, 13
Naval Laboratories, 49
Naval Missile Center, 327
Naval Research Laboratory, 79, 109, 156, 174, 242, 317
Naval Reserve Officers Assn., 316
La Nazione, 45
NBC. See National Broadcasting Co.
Neher, H. Victor, 152
Nelson, Clifford H., 313
Nero (Near Earth Rescue Operation) spacecraft, 384
Nerva. See Nuclear Engine for Rocket Vehicle Application.
Ness, Dr. Norman F., 86, 315, 358
Netherlands, 125, 190–191, 210, 220, 239, 306, 324, 347, 377
Nettles, J. Cary, 315
Nevada, 215, 339
New Delhi, India, 34
New Hampshire, Univ. of, 274, 360
New Mexico, 367
New Mexico, Univ. of, 89, 381
New Rochelle, N. Y., 353
New York Board of Trade, 331
New York, N. Y., 18, 97, 191, 246, 258, 316, 319, 330, 340, 367
New York Times, 57, 355, 376
  science and technology, 30, 260, 367, 377
  space law treaty, 250, 277, 299
  space program
    U.S., 44, 47, 52, 105, 182, 279, 347–348
    U.S.S.R., 20, 82, 95, 130, 157, 279, 294, 339
  space technology, 33, 69, 96, 109, 136, 339, 352, 384
  supersonic transport, 281, 289, 333
New York Univ., 52, 97, 293, 319, 333
New York, Univ. of (Stony Brook), 225
New York *World Telegram*, 110
New York Zoological Society, 96
New Zealand, 54, 89, 347
New Zealand Weather Service, 168
Newcastle-upon-Tyne, Univ. of, U. K., 150
Newell, Dr. Homer E., 1, 26, 44, 95, 110, 149, 235, 251, 325, 367
  space astronomy, 121, 384
  space programs, scientific, 59, 62, 65, 71, 121, 324
Newport News, Va., 87
News conference. See Press conference.
Ney, Edward P., 241
Nhatrang, South Vietnam, 361
NICAP. See National Investigations Committee on Aerial Phenomena.
Nigeria, 351
Nike-Apache (sounding rocket), 65, 342
  electron measurement, 50, 55, 95, 127, 214, 240, 292, 293, 347
  ionosphere experiments, 10, 56, 111, 127, 234, 274
  micrometeoroid sampling, 267, 351
  Polar cap absorption data, 245, 284
  upper atmosphere data, 19, 69, 117, 152, 237, 260, 276, 307
Nike-Cajun (sounding rocket), 65–66
  radar test, 25, 42
  upper atmosphere data, 23, 114, 328
  fall, 307
  summer, 161, 164, 216, 221, 226, 260, 265
  winter, 39, 51
Nike-Hydac (sounding rocket), 347
Nike-Javelin (sounding rocket), 108, 115
Nike-Tomahawk (sounding rocket), 137, 276, 277, 283, 287, 293, 341
Nike-X (antimissile missile system), 191, 343
Nikolayev, L/C Andrian G. (U.S.S.R.), 60
Nikolayeva-Tereshkova, Maj. Valentina (U.S.S.R.), 90
Nilsson, Carl, 288–289
NIMBUS I (meteorological satellite), 2, 264
NIMBUS II(Nimbus C) 2, 178, 185, 240, 257, 283, 347, 349, 365, 385,
  launch, 178
Nimbus B, 2, 322
Nininger, Dr. Harvey H., 232
Nippon Electric Co., 179
Nitrogen, 335
NOA. See National Oceanography Assn.
Nobel Prize, 339
Noervich, W. Germany, 256
Noise, aircraft, 164, 274, 279, 284, 288, 338
Norfolk, Va., 332
Normyle, William, 149
North American Air Defense Command (NORAD), 32, 43, 194
North American Aviation, Inc. 136, 158, 215, 268, 283–284, 302
  Apollo spacecraft, 93, 313–314, 328
  Atomics International Div., 320
  contract, 2, 5, 115, 130, 264, 268, 284, 306, 330, 335, 341, 356, 361
  labor, 35
  rocket engine
    F-1, 115, 352
    H-1, 130
    J-2, 2, 85, 168, 233, 252, 257
  Rocketdyne Div., 35, 111, 168, 233, 252
  Saturn, 5, 14, 306, 330, 364
North Atlantic Treaty Organization (NATO), 199, 227, 275, 306,
North Carolina, Univ. of, 14
North, Warren J., 359
Northrop Corp., 19, 130, 227, 264, 306
Northrop Space Laboratories, 270
Northwestern Univ., 255, 326, 351, 376
Norway, 227, 322
Norwegian Defense Research Establishment, 111
Noshiro, Japan, 339
Nottingham, U.K., 285
Novae, 339
Novosibirsk, U.S.S.R., 117
Novosti Press Agency (U.S.S.R.), 224
Noyes, Mrs. Blanche W., 22
Noyes, Crosby S., 373
NRC. See National Research Council.

NRDS. See Nuclear Rocket Development Station.
NRL. See Naval Research Laboratory.
NRX-A5 (nuclear reactor), 211, 222
NSC. See National Space Club.
NSF. See National Science Foundation.
Nuclear detection satellite (see also Vela), 24, 35, 322
Nuclear Engine for Rocket Vehicle Application (Nerva), 145, 385,
   test, 42, 52, 82–83, 103–104, 115, 145, 211, 222
Nuclear fission, 259
Nuclear power, 177
Nuclear propulsion, 83
Nuclear reactor, 35, 51, 57, 211, 222
Nuclear Rocket Development Station (NRDS), 42, 52, 82, 104, 115
Nuclear submarine, 87, 125
Nuclear test, 31, 35, 172, 312, 331, 342, 382
Nuclear test ban treaty, 58, 109, 259, 304, 342, 369

O

Oahu, Hawaii, 179
Oakland, Calif., 332
Oak Ridge National Institute, 243
Oao. See Orbiting Astronomical Observatory.
OAR. See Office of Aerospace Research.
OART. See NASA Headquarters Office of Advanced Research and Technology.
O'Brien, Edward W., 269
Observatoire de Meudon, 264
Observatory class satellite. See Orbiting Astronomical Observatory; Orbiting Geophysical Observatory; Orbiting Solar Observatory.
Ocean of Storms (Oceanus Procellarum) (moon), 48, 203, 313, 316, 325, 341, 380
Ocean Science and Engineering, Inc., 154
Oceanic Institute, 367
Oceanographer (oceanographic research ship), 240, 311
Oceanography, 13, 19, 154, 182, 219, 224, 240, 247, 271, 302, 366–367
Oceanus Procellarum (moon). See Ocean of Storms.
Ochao, Dr. Severo, 52
O'Connell, James D., 163
O'Connor, B/G Edmund F. (USAF), 315
Odishaw, Dr. Hugh, 337
OECD. See Organisation for Economic Cooperation and Development.
O'Fallon, John R., 142
Office of Aerospace Research (OAR), 326
Office National d'Etudes et de Recherches Aérospatiales (ONERA), 201
Office of Aircraft Noise Control and Abatement (proposed), 274, 279
Office of Emergency Planning, 340
Office of Naval Research (ONR), 13, 166
Office of Science and Technology (President's), 28, 33, 57, 106, 328, 358, 370

Offik, Wolfgang G., 250
O'Gallagher, J. J., 151
Oglesby, Capt. Charles, 194
Ogo. See Orbiting Geophysical Observatory.
OGO II (Ogo-C, Orbiting Geophysical Observatory), 289
OGO III (Ogo-B, Orbiting Geophysical Observatory), 209, 213, 218, 248
Ogonyok, 60
OH-6A (helicopter), 132, 187
Ohio, 142, 367
Okinawa, 106
Oklahoma City, Okla., 236, 328
Old Moore's Almanack, 281
Oliver, Dr. Bernard M., 112
OMSF. See NASA Headquarters. Office of Manned Space Flight.
"On Rockets" (manuscript), 321
ONERA. See Office National d'Etudes Recherches Aérospatiales.
ONR. See Office of Naval Research.
Oort, Dr. Jan H., 324, 377
Opik, Dr. Ernst J., 241
Oppenheim, Dr. A. K., 359
Optical Society of America, 97
Orbiting Astronomical Observatory (Oao) 5, 39, 47, 121, 132, 181, 264, 380
Orbiting Data Relay System (Ords), 179
Orbiting Geophysical Observatory (Ogo) (see also OGOs II, III), 5, 25, 63–64, 150–151, 248, 264
Orbiting Primate Spacecraft, 322
Orbiting Solar Observatory (Oso) (see also OSO II), 5, 39, 367
Ords. See Orbiting Data Relay System.
Oregon, 341
Orenburg region, Ural Mts., U.S.S.R., 291
Organisation for Economic Co-operation and Development (OECD), 15, 49
Orion (constellation), 11
Orion (NASA barge), 74
Orion II (sounding rocket), 342
Orionid meteor shower, 327
Orlando, Fla., 78
Orlov, Vladimir, 279
Ormand Beach, Fla., 132
Oso. See Orbiting Solar Observatory.
OSO II (Orbiting Solar Observatory), 212
OSSA. See NASA Headquarters. Office of Space Science and Applications.
Ousley, Gilbert W., 232
OV-1 Mohawk (surveillance aircraft), 253
OV1-IV (reseach satellite), 121
OV1-V, 121
OV1-VI, 338
OV1-VIIIT, 240
OV1-IX, 370
OV1-X, 370
OV3-I, 146
OV3-II, 331
OV3-III, 258
OV3-IV, 213
OV4-IR (communications satellite), 338
OV4-IT, 338
OV4-III, 338

Owens-Corning Fiberglas Corp., 249
Oxygen (see also Fuel), 11, 12 174, 285, 306, 335
Ozonide, 11

P

P-1127 (U.K. Vtol aircraft), 94, 131
Pacemaker (booster), 333
Pacific Ocean, 232, 247, 261, 311, 373
  communications satellite, 235, 287, 311, 321
  U.S.S.R. rocket test, 149, 193, 233, 275, 287, 302, 352, 373
Pacific Science Center, 312
Paffell, Col. Donald W. (USAF), 54
Pacific Area Travel Association Conference, 53
Pacific Missile Range (see also Western Test Range), 176, 227
Page Communications, Inc., 139
PAGEOS I (Pageos A Passive Geodetic Earth-Orbiting Satellite), 222, 385
Paige, Dr. Hilliard W., 317
Pakachoag Hill, Mass., 263, 350
Pakistan, 69, 114, 117–118, 174, 271
Pakistan Space and Upper Atmosphere Research Committee (SUPARCO), 69, 114, 152
Paksuniemi, Sweden, 352
Palaemon (NASA barge), 74
Palestine, Tex., 263
Pan American World Airways, Inc. (Pan Am), 137, 143, 297, 302, 339, 343
Parachute, 275, 278
  fatality, 278
  test, 40–41, 161
  Voyager, 153, 242–243, 340, 352, 370
Parade, 10, 109–110, 248–249
Paraguay, 297
Parawing, 156
Parin, Vassily V., 232, 318
Paris, France, 11, 15, 188, 220, 287, 339
Parks, Robert J., 197, 250, 372
Parry, Albert, 117
Parsons, John W., 124
Particles, charged, 225
Pastore, Sen. John O., 270
Patent Law Assn. of Chicago, 154
Patents, 126, 254, 268, 319, 335
Patrick AFB, Fla., 113, 372
Patuxent Naval Air Station, Md., 106, 131
Paul, Norman S., 213, 323
Pcs. See Pointing-control system.
Peach Tree Valley, Calif., 331
Pearl River (NASA barge), 74
Pearson, Drew, 314
Pecora, Dr. William T., 297
Peebles, P. J. E., 31
PEGASUS I (meteoroid detection satellite), 205–206, 337
PEGASUS II, 205–206, 337
PEGASUS III, 205–206, 247, 337
Pelly, Rep. Thomas M., 76
Pendray, G. Edward, Award, 359

Pennsylvania, 141
Pennsylvania State Univ., 298
Penrose Medal, 349
Pensacola, Fla., 234
Pensacola (Fla.) Junior College, 317
Pentagon, 251
Pépin, Dr. Eugene, 317
Perkin-Elmer Corp., 175, 353
Perm, U.S.S.R., 107
Peroxide, 11
Perreault, W. D., 358
Pert. See Program Evaluation Review Technique.
Peru, 297, 347
Peterson, Bruce A., 293, 298, 379
Peterson, L/C Harry W. (USAF, Ret.), 283
Petit Jean Air Service, 302
Petranek, Jan, 254
Petrocelli, Dr. A. W., 11
Petroleum Today, 281
Peyrefitte, Alain, 348, 361
Pezdirtz, George F., 315
Pfund, Edward, 250
Philadelphia, Pa., 82, 147
Philadelphia Evening Bulletin, 18, 282
Philadelphia Inquirer, 366
Philco Corp., 221, 233, 244, 247, 261 320, 382
Philip, Prince (U.K.), 93–94, 291
Philippines, 54, 222, 292–293, 311
Phillips, Col. Donald Boyer (USA, Ret.), 124
Phillips, M/G Samuel C. (USAF) 229, 350
Phipps, Clifford, 327
Phoenix, Ariz., 17, 242
Phoenix (missile), 176
Phoenix Gazette, 96
Photography. See Advanced Vidicon Camera System; ATS I; Automatic Picture Transmission System; Cloud, photographs; Earth, photographs; LUNAR ORBITERS I, II; MARINER IV; Moon, photographs; SURVEYOR I.
Photometer, 83
"Photostate I" (radiation measurement device), 337
Physical Research Laboratory (Ahmedabad, India), 234
Physical Review Letters, 143, 225
Physics, 85, 95–96, 229
Piantanida, Nick, 40–41, 161, 278
Piccard, Don, 237
Piccard, Dr. Jeanette Ridlon, 225
Pickering, Dr. William H., 147, 157, 166, 208–209, 317
Pierce, John R., 329
Piloted Low Speed Tests (Pilot) project, 153
Pioneer (program), 269
PIONEER VI (interplanetary probe), 32, 269
  results, 32, 79, 144, 183, 214–215, 301
PIONEER VII (Pioneer B), 269, 301, 385
Pitot-static probe, 260, 263
Pittendrigh, Colin S., 361–362
Pittsburgh, Pa., 336
Pittsburgh, Univ. of, 237
Planetarium, 12, 14

Planetary Reentry Parachute Progam, 153, 243, 340, 352, 370
Plasma probe, 83
Plessey Co., Ltd., 136
Plummer, William, 141
Pluto (planet), 17
PMR. See Pacific Missile Range.
Po, Shu-jen, 339–340
Pogo Stick, rocket propelled, 78–79
Pogue, Maj. William R. (USAF), 127
Point Barrow, Alaska, 38–40, 51, 161, 164, 216, 226, 252, 260, 265, 321
Point Mugu, Calif., 327
Pointing-control system (Pcs), 356
Poland, 231
Polar Cap Absorption, 245, 277, 284
Polaris (ICBM), 31, 247, 343
"Policy Planning for Aeronautical Research and Development" (report), 208
The Politics of Research, 165
Polytechnic Institute of Brooklyn, 52
Pomona College, 232
Pope, David P., 260
Popovich, Pavel R., 60
Porter, D. C., 134, 173
Porter, Dr. Richard W., 173
Poseidon (ICBM), 343
Poseidon (NASA barge), 15, 74, 298
Potrero, Calif., 16
Prather, Victor A., Award, 226
Pratt & Whitney Div. See United Aircraft Corp., Pratt & Whitney Div.
Pravda, 135–136, 279, 353
Presidential Medal of Merit, 2
President's Science Advisory Committee (PSAC), 25, 34, 111, 248
Press conference, 80, 106, 131, 194, 319
  Gemini flights, 16
    GEMINI VIII, 17, 92, 104, 117
    GEMINI IX-A, 207, 208, 218
    GEMINI X, 255
    GEMINI XI, 301
    GEMINI XII, 353–354
  LUNA IX, 50–51
  LUNA X, 126–127, 140–141
  LUNAR ORBITER I, 313
  NIMBUS II, 185
  U.S. space program, 14, 162, 196, 258, 334
  U.S.S.R. space activities, 14, 50–51, 82, 139, 140–141, 162, 232–233, 285, 316, 322
  SURVEYOR I, 203
Press, Harry, 2–3, 315, 349
Price, Don K., 22
Princeton Univ., 11, 40, 295, 349
Prishchep, A. G., 173–174
Probe (see also individual probes, such as MARINERS II and IV, PIONEERS VI and VII, VENUS II, III, etc.)
  contamination by, 75, 134
  interplanetary, 29, 50, 79, 144, 176, 183, 268, 269, 384–385
  lunar. See LUNAS IX, X, XI, XII, XIII; LUNAR ORBITERS I, II; SURVEYORS I, II.
  Mars, 173, 380

Venus, 7, 75, 86–87, 95
Program Evaluation and Review Technique (Pert), 262
Promise (NASA barge), 74
Propellant. See Fuel.
Propulsion (see also Engine; Motor), 73, 215, 258, 385
Propulsion Joint Specialist Conference, Second, 215
Propulsion Module. See Service Module.
Propulsive wing aircraft (V/Stol), 309
Proton, 287
PROTON III (U.S.S.R. space station), 233, 236, 285, 385
Protostar, 229
Proxmire, Sen. William F., 261
Pryor, Harold E., 139
PSAC. See President's Science Advisory Committee. 25, 34, 111,
Puerto Rico, 90, 152, 267, 350
Pulkovo Observatory (U.S.S.R.), 199, 224, 292
Punch, 32
Pushkov, N. V., 251

Q

Quark (matter theory), 236
Quasar (quasi-stellar object), 31, 92–93, 171, 230, 253

R

Radar, 18, 42–43, 51, 57, 95, 199, 203, 210, 337
  planetary sounding technique, 153, 171
Radar signature analysis (Rsa), 326
Radcliffe, Capt. Chester R., Jr. (USAF), 291
Radiation, 61
  cosmic, 85, 99, 133–134, 136–137, 151, 190, 295, 306, 374
  effects, 121, 243, 337, 374
  gamma, 129, 136–137, 151
  measurement, 51, 146, 199, 370
  resistant cell, 259–260
  shielding from, 306–307
  solar, 7, 79, 173, 199, 289, 337
  space, 12–13, 121, 213
  Van Allen belts, 102, 168, 213, 260, 332
Radio Corporation of America (RCA), 9, 126, 176, 177, 222, 255, 259, 338, 353
Radio Liberty, 226
Radio Moscow, 107
Radio Prague, 379
Radio signal, 9, 118
Radioactive fallout, 342
Radioactivity, 116–117, 263
Radioastronomy (see also Astronomy; Radio telescope), 46, 230
Radiometer, 2, 176, 178, 238, 257, 268, 284, 369
Radiotelescope, 112, 152, 292, 377
Radius, Dr. Walter A., 195
RAeS. See Royal Aeronautical Society. 15, 94

RAF. See Royal Air Force. 61
Raman, Dr. C. V., 253
Ramo, Dr. Simon, 329
Ramsey, Dr. Norman F., 132
Rand Corp., 73, 273
Randolph, Sen. Jennings, 227
Ranger (program), 262, 373
RANGER VII (lunar probe), 333
RANGER VIII, 333
RANGER IX, 333
Rangoon, Burma, 60
Ratner, Lazarus G., 142
Ray, Dr. Carleton, 96–97
Raytheon Corp., 53
Razumov, Vladimir Vasilyevich, 10
RCA. See Radio Corporation of America. 9, 126
RCA Service Co., 130
Rcs. See Reentry Control System. 100
Reaction Motors, Inc., 377
Reader's Digest, 37
Real Time Computer Complex (RTCC), 258
Reconnaissance satellite, 22, 61–62, 369, 372
Record
  aircraft, 134, 250, 253, 302, 351
  balloon, 319
  helicopter, 187, 291
  spacecraft, 207, 251, 294, 349
  speed, 201, 302, 351, 385
  women's, 134
Redeye (missile), 119
Reflector satellite, 249, 384
Redlands, Calif., 42
Redstone (rocket), 123, 267
Redstone Arsenal, Ala., 192
Reentry
  control, 105, 180, 207, 290–291
  Gemini, 99–100, 207, 244, 290–291, 344
  heating, 88, 258
  Scout Reentry Heating Project, 49–50
  test, 238, 367
  vehicle, 267, 271, 330, 378
Reentry Control System (Rcs), 100
Rees, Eberhard F. M., 315
Reeves, Dr. Robert R., Jr., 377
Reid, Rep. Ogden, 288
Reid, Sylvanus Albert, Award, 27
Reiff, Glenn, 32–33
Reinartz, Stanley R., 1
Reistrup, J. V., 294
Remote Maneuvering Unit (Rmu), 225
Rendezvous, 34, 212, 314
  GEMINI VI-A (Gemini VI), 226, 259, 374
  GEMINI VIII, 17, 99–102, 107–108
  GEMINI IX-A (Gemini IX), 202, 206–207
  GEMINI X, 243–244, 255
  GEMINI XI, 290, 291, 294
  GEMINI XII, 344
Rendezvous Docking Simulator, 130–131
Rensselaer Polytechnic Institute, 147, 377
Republic Supply Co., 73
Research, Inc., Barrier Arresting Component, 180
Research and development
  aeronautics, 291
  benefits, 97, 119, 154
  Federal support, 22, 30, 34, 55, 104, 165
  funds for, 33, 179, 257, 323
    geographic distribution, 104
  information, distribution of, 370
  management, 154
Research Institute for Advanced Studies, 174
Research Review, 229
Research satellite, DOD, 212, 213, 331–332, 370
Research Society of America, 123–124
Restrepo, President Carlos Lleras (Colombia), 316
Review of Popular Astronomy, 224
Reynolds, Dr. Orr E., 90
Rice Univ., 56, 139
Rickover, V/A Hyman G. (USN), 206
Rieke, William B., 188, 364, 380
Rigel (star), 18
Rio Grande, Brazil, 347
Rio Grande, N. J., 289
Risso, William P., 139
Ritchie, Donald W., 89
RL-10 (rocket engine), 125, 132, 169
Rmu. See Remote Maneuvering Unit.
Roberts, W. Cameron, Jr., 117
Roberts, Dr. Leonard, 258
Roberts, Walter Orr, 133
Robillard, Geoffrey, 9
Robledo de Chavela (Spain) Space Tracking Station, 265, 320
Rochester Conference, 286
Rochester, Univ. of, 151
Rock, Dr. Vincent P., 163
Rocketdyne Division. See North American Aviation, Inc.
Rocket engine. See Engine, chemical, electric, nuclear, rocket, etc., and individual rocket engines such as F-1, M-1, J-2, etc.
Rodana Research Corp., 40
Roderick, R. L., 250
Rogallo, Francis M., 156, 235
Rogallo, Gertrude S., 156
Rogers Dry Lake, Calif., 135
Rogers, Dr. Thomas F., 219
Rolls-Royce, Ltd. 189, 285
Roman, Dr. James A., 200
Roman, Dr. Nancy G., 107, 121
Romania, 165, 174, 231
Roney, Robert, 250
Roosa, Capt. Stuart A. (USAF), 127
Root, L. Eugene, 354
Rose, V/Adm. Rufus E. (USN, Ret.), 214
Rosenthal, Rep. Benjamin J., 274, 279
Rosholt, Dr. Robert L., 331
Rosman, N. C. 257, 361, 372
Ross, Stanley, 26
Rostow, Walt W., 258
Rota, Spain, 135
Rotary Club, 77
Roth, Herbert E., 285
Roth, J. Reece, 120
Rover (program), 24, 211
Row, Dr. Ronald, 88

Rowan, Dr. Lawrence, 313, 374
Royal Aeronautical Society (RAeS), 15, 94, 291
Royal Air Force (RAF), 61
Royal Aircraft Establishment (RAE), 262
Royal Astronomical Society, 82
Royal Crown Cola International, Ltd., 328
Royal National Foundation, 206
Royal Society of Great Britain, 77
Rsa. See Radar signature analysis.
RU Camelopardalis (star), 230
Ruby, Charles H., 165
Ruddick, Keith, 142
Ruderman, M. A., 152
Ruina, Dr. J. P., 214
Rumsfeld, Rep. Donald, 12
Runcorn, S. K., 150
Runyan, Thorne L., 78
Rushworth, L/Col. Robert A. (USAF), 47, 182, 223, 231, 315
Rusk, Secretary of State Dean, 137–138
Russell, Sen. Richard B., 145
Rust Engineering Co., 130

S

Sabha, Ebrahim, 49
Sabreliner (twin-jet), 302
Sacramento, Calif., 252
SADL. See Sterilization Assembly Development Laboratory.
SAE. See Society of Automotive Engineers.
Sagan, Dr. Carl, 98
Saigon, Vietnam, 302
St. Albans School, 313
St. Croix, V. I., 152
St. John's Newfoundland, 302
St. Louis, Mo., 95
St. Thomas, V. I., 152
Salinas, Dr. Luis Tapis, 317
Salto di Quirra Test Range, Sardinia, 235
Samos (satellite), 289
San Antonio Express, 328
San Diego, Calif., 7, 61, 160, 289
San Fernando State College, 302
San Francisco, Calif., 150, 322, 330, 349
San Francisco *Examiner and Chronicle*, 242
San Jose, Calif., 260
San Marco (Italian satellite), 174
Sanchez, Joaquin, 94
Sandri, G., 151
Santa Susana, Calif., 85
Santiago, Chile, 125, 334
Saros (French communications satellite), 233
Satellite. See individual satellites, satellite programs, and type of satellite, such as Aeronomy, Geodetic, Meteoroid detection, Meteorological, Navigation, Nuclear detection, and Reconnaissance satellite.
Satellite collision, 320–321
Satellite Educational and Informational Television (SEIT), 162–163
Satellite Situation Report, 337

Satellite, unidentified launch vehicle
  Atlas-Agena D, 20, 54, 92, 106, 143, 208, 212, 213, 238, 268, 272, 293, 312, 319, 337, 367
  Scout, 33, 116, 146
  Thor-Agena D, 40, 50, 91, 121, 132, 188, 221, 262, 297, 341, 382
  Thor-Altair, 9
  Thrust-Augmented Thor-Agena B, 178, 222
  Titan III-B-Agena D, 252, 305, 373
Satellite, weather. See Meteorological satellite.
Sato, Premier Eisaku, 66
Satterthwaite, Joseph C., 53
Saturday Evening Post, 376, 377
Saturday Review, 1
Saturn I, Uprated (Saturn IB) (booster), 144, 159, 177, 178, 194, 261, 328
  capability, 25, 58, 241
  contract, 2, 19, 255, 378
  engine
    H-1, 130, 169
    J-2, 2, 85, 168, 195, 233, 252, 257, 364
  launch, 51, 60–61, 67, 88, 385
    AS-201, 67
    AS-202, 274
    AS-203, 233
  program, 72, 92
  stage
    S-IB, 88, 255, 263, 378
      test, 227
    S-IVB, 88, 224, 263, 272
      test, 233
  test, 122, 350, 357
Saturn V (booster), 72, 86, 125, 193, 261, 272
  capability, 18, 25, 58, 119–120, 241, 317
  contract, 8, 9, 85, 115, 145, 224, 330, 332
  crawler-transporter, 33, 86, 248
  engine, 2
    F-1, 115, 210, 352
    J-2, 2, 85, 168, 192, 233, 252, 257, 364
  equipment, 9
  facilities, 272, 277
  stage
    1st (S-IC), 85, 115, 213, 292, 332
      test, 210
    2nd (S-II), 2, 213, 292, 306, 330
      explosion, 195, 198, 283
      test, 85, 147, 181, 185, 364, 383
    3rd (S-IVB), 2, 8, 145, 178, 202, 213, 224, 292, 306
      test, 107, 252
  transport, marine, 74
Saturn Manufacturing Review meeting, 191
Scanner, Project, 92, 268, 369
Schawlow, Dr. Arthur L., 97, 258
Scheer, Julian W., 1, 213
Scherer, Lee R., 176
Schirra, Capt. Walter M., Jr. (USN), 8, 95, 140

Apollo, 144, 306, 381
  award, 30, 225–226, 374
  good-will tour, 54, 61, 66, 69
Schlichter, Dr. Louis B., 128
Schmeck, Harold M., Jr., 33, 52
Schneider, William C., 195
Schneiderman, Dan, 9
Schreiber, Dr. Carl, 270
Schriever, Gen. Bernard A. (USAF, Ret.), 111, 148, 211, 215, 295, 355
  award, 129, 137, 168, 315
  military technology, 77, 191, 357
  retirement, 146, 153, 283, 285
Schurmeier, Harris M., 9
Schwartz, Murray L., 163
Schweickart, Russell L., 110, 144, 332, 379
Schwinghamer, Robert J., 288
Science, 98, 312, 343
  benefits, 167–168, 206, 308
  human needs, 74, 113, 147, 159, 223, 329
  policy, 287, 328
Science, 1, 33–34, 51, 85, 133, 370
  astronomy, 46, 92–93, 339–340
  geology, 150, 208
  interplanetary physics, 86, 133–134, 241, 288
  Mars, 187, 241–242, 264
  moon, 115, 187
  Ufo's, 255, 326
Science Advisory Committee (President's). See President's Science Advisory Committee.
"Science Frontiers" (seminar), 221
"Science Reporter" (TV series), 198
"Scientific and Engineering Manpower in Communist China, 1949–1963," 31–32
"Scientific Progress and Human Values" (Cal Tech convocation), 329
Scientist-astronaut, 302
Scientist-cosmonaut, 327
Scientists, 21, 34, 50–51, 327
Scorpius, 89, 188, 241
Scott, L/Col. David R. (USAF), 8, 144
  Apollo (mission), 110, 332, 379
  award, 117
  GEMINI VIII (flight), 99–102, 104–106, 107–108, 112, 117, 130–131
  plans for, 17, 78–79, 92, 94
  training, 90, 130–131
Scott, Col. James G. (USAF), 264
Sco X-1 (x-ray source), 219, 229
Scout (booster), 174, 176, 195, 223, 261
  launch
    Reentry E, 49–50
    USAF satellite, 183, 213, 258, 271, 331–332
  Reentry Heating Project, 50, 90
Scripps Institute of Oceanography, 52, 247
Scull, Wilfred E., 315
SDS Data Systems, 175
Seaborg, Dr. Glenn T., 18–19, 56–57, 147, 273, 365
Seal, 96–97
Seal Beach, Calif., 86
Sealab Project, 316

Sealab II, 13
Sealab III, 13
Seamans, Dr. Robert C., Jr., 29, 44–45, 157, 331, 359, 369
  Apollo Applications, 59, 171, 229, 278
  appointment, 1, 32, 43
  award given by, 117, 218, 255
  budget, 24, 26, 71
  space program, significance, 238–239, 241, 335
  U.S.S.R., 44–45
"Search" (deep-sea retrieval system), 154–155
"The Search for Extraterrestrial Life" (symposium), 189–190
Sears, R. E. 250
Seattle *Times*, 176
Seattle, Wash., 238
SECOR VI (geodetic satellite), 212
SECOR VII, 378
SECOR VII, 378
SECOR VIII, 312
Sedov, Prof. Leonid I., 319–320
See, Elliot M., Jr. (Cdr., USNR), 14, 72, 76 85, 110
Segrè, Emilio, 85
Sehlstedt, Albert, Jr.,
SEIT. See Satellite Educational and Informational Television.
Seitz, Dr. Frederick, 8, 85, 123–124, 305
Sekkujokki, Sweden, 299
Sell, Ted, 311
Sellman, Mrs. John, 57
Sells, Dr. S. B., 120
Senghor, President Leopold Sedar (Senegal), 131
Sensor, 3, 21, 61, 153, 175, 200
SEREB. See Société pour l'Etude et la Réalisation d'Engins Balistiques.
Service Module (Sm), 25, 272, 274, 328
SETP. See Society of Experimental Test Pilots.
Shanghai, China, 339
Shapiro, Prof. Ascher H., 235
Shapiro, Irving I., 376
Shapley, Willis H., 1
Sharp, Dr. Robert P., 329
Shastak, Dr. Arnold, 166
Shea, Dr. Joseph F., 17, 72, 93, 375
Sheely, Horace, 263
Sheldon, Dr. Charles S., II, 99
Shelton, Dr. Russel D., 120
Shen, C. S., 152
Shenton, Sam, 282
Shepanek, Capt. Raymond A. (USNR), 133
Shepard, Capt. Alan B., Jr. (USN), 76, 248–249
Shepard, Van H., 354, 376, 378
Shepherd, Dr. L. R., 45
Sherwin, Dr. Chalmers W., 154
Shoemaker, Dr. Eugene M., 228, 305, 382
Short Range Attack Missile (Sram), 334, 341
Shultze, Charles, 258
Sikorsky, Igor I., 60
Silverstein, Dr. Abe, 293–294, 362

Simmons, Henry, 74
Simon, Dr. Dorothy Martin, 230
Simons, Howard, 160
Simpson Desert, Australia, 348
Simpson, J. Wesley, 279
Simpson, Dr. John A., 151
Simpson, Wyatt C., 309
Sinkiang Province, Communist China, 382
Sioux Falls, S. Dak., 40
Sirius (star), 18, 203
Sisakian, Dr. Norair M., 94
Sky and Telescope, 230
"Skyhook," 52
Skylark (U.K. sounding rocket), 235, 299
SL-1 (solid rocket motor), 62
SL-2 (solid rocket motor), 62, 110
Sla. See Spacecraft Lem Adapter.
Slattery, Bart J., Jr., 322
Slayton, Maj. Donald K. (USAF, Ret.), 22, 69, 116, 376
Sloan, Alfred P., Foundation, 343
"Slush hydrogen," 9–10
Sm. See Service module. 25
Small Business Administration, 89
Smart, Gen. Jacob E. (USAF, Ret), 336, 380
Smirnov, Leonid, 20
Smith, Dr. Alexander G., 29–30
Smith, Francis B., 149
Smith, Lawrence A., 354
Smith, Sen. Margaret Chase, 187
Smith, Walter D., 354
Smith's Ranch, Nev., 288
Smithsonian Astrophysical Observatory (Cambridge, Mass.), 251, 260, 287
  catalog, 107
  research, 22, 144, 284–285
Smithsonian Institution, 49, 104, 169, 236, 253, 284, 324
Smull, Dr. Thomas L. K., 128, 264
Snap (Systems for Nuclear Auxiliary Power) program, 83, 273
Snap-8 (nuclear reactor), 24, 244
Snap-10A (nuclear reactor), 13, 28
Snap-27 (thermoelectric generator), 83
Snap-29 (radioisotope generator), 112, 325
SNAPTRAN-2 (AEC test), 13
Snow, Lord Charles (C. P.), 28–29
SNPO. See NASA-AEC Space Nuclear Propulsion Office.
Société pour l'Etude et la Réalisation d'Engins Balistiques (SEREB), 188
Society of American Scientists, 170
Society of Automotive Engineers (SAE), 213, 296, 314,
Society of Experimental Test Pilots (SETP), 299
Society of Women Engineers, 230
Sodium vapor experiment, 327
Sohier, Walter D., 377
Solandt, O. M., 229
Solar cell, 194, 259
Solar eclipse. See Eclipse, solar.
Solar plasma, 86
Solar radiation. See Radiation, solar.
Solar system, 9, 347, 381

Solar wind, 29, 32, 79, 144, 151, 183, 209, 221, 269, 301
Solid propellant. See Motor.
Sonic boom (see also Noise), 57, 332
  supersonic transport, 57, 203, 333, 341, 385
  XB-70, 338, 342, 354, 372, 376, 378
Sonmiani, Pakistan, 69, 114, 117, 152
Sonnichsen, Dede, 319
Sophia, Princess (Spain), 317
Sorlie, L/C Donald (USAF), 297, 306, 312
Sounding rocket (see also individual sounding rockets: Aerobee 150, Aerobee 150A, Aerobee 350, Arcas, Boosted Arcas, Hasp, Javelin, Nike-Apache, Nike-Cajun, Nike-Tomahawk, Nike-Hydac, Orion II, Wasp), 39–40, 65–66
  foreign
    France, 322, 327
    Japan, 109, 123, 149, 257
    Spain, 322, 325
  international programs, 155–156
    ESRO, 124, 235, 299
    NASA-Argentina, 154, 342, 347
    -Brazil, 14, 161, 252, 310, 320, 347, 373
    -France, 327
    -Greece, 178, 185, 187
    -India, 235
    -Japan, 381
    -Pakistan, 69, 114–115, 152
    -Spain, 15–16, 138
    -West Germany, 191, 299
South Africa, 7
South America, 7, 139, 152
Southeastern State College, Okla., 89
"Southern Cross" (USN program), 13
Southern Illinois Univ., 311
SOVCOMSAT, 149
Soviet Civil Aircraft Research Institute, 255
Soviet Academy of Sciences, 21, 50, 82, 140, 174, 201, 224, 236
  cooperation, 90, 171, 231
  new members, 52–53
Soviet Commission for Exploration and Use of Outer Space, 46
Space biology
  animal experiments, 232, 374
    dog, 60–61, 84, 102, 179, 237, 318
    mouse, 71
    primate, 187, 322, 336
  atmospheres, artificial, 11, 273, 285, 335
    helium in, 12, 13
  award, 226, 323
  biotelemetry, 91, 102, 254
  drugs, use of, 318
  environment, effects of, 60, 179, 187, 232, 236, 374
    Gemini program, 226, 294–295, 318, 346
    psychological, 60, 120
  fatigue, 206, 208, 317–318
  heart, 8, 102, 285, 294–295
  hibernation state, 96–97, 136

life support system, 11, 94, 120, 285
nutrition, 203, 335
physiological monitoring, 60, 254, 433–434
radiation, effects of, 12–13, 102, 213, 243, 337, 374
recycling, 54, 99
technology utilization, 200, 328–329, 349
U.S.-U.S.S.R. cooperation, 90
vision, 7, 57, 78, 236, 295
weightlessness
  effects of, 102, 179, 226, 294–295, 318, 322, 361, 374
  work during, 61, 170, 294–295, 317–318
Space Business Daily, 278
Space debris, 194, 225
Space Digest, 280
Space Electric Rocket Test (Sert), 312
Space General Corp., 177
Space law, 289, 317
Space Law and Government, 289
Space law treaty, 330
  U.N. consideration, 246, 248, 250, 276, 291, 368–369, 370, 377, 386
  U.S. draft proposal, 169, 171, 205, 216, 218
  U.S.S.R. draft, 312
Space medicine. See Space biology.
Space, military use of, 135–136, 147–148, 221, 248
  communications, 306
  manned space flight, 357, 381–382
  nuclear detection, 24, 35, 322
  potential uses, 200, 216, 220, 335, 370
  reconnaissance, 289
  space station, 368
Space Nuclear Propulsion Office. See NASA-AEC Space Propulsion Office.
Space, peaceful use of, 23, 60–61, 291–292, 357–358.
Space program, national (see also individual programs, such as Apollo program, etc.)
  accomplishments, 35, 53, 70, 125, 319, 324
    international, 141, 304
  management, 249
  manned space flight, 95, 246, 314, 344, 357, 372
  budget, 23–24, 26–27, 41, 70, 224
  cost of, 338–339, 362
  criticism, 7, 32, 109–110, 131, 210–211, 220, 232–234, 266–267, 366
  education, benefits to, 20, 35, 53, 54, 238–239
  employment, 13, 56, 93, 295
  international cooperation. See International cooperation.
  Johnson, President Lyndon B., 35, 162, 169, 258, 297, 303, 375
  lunar landing. See Moon, landing.
  manned space flight. See Manned space flight.
  military, 35, 111, 135, 153, 196, 200

  objectives, 26–27, 70–71, 167, 196, 372
    scientific and technological, 39, 53, 187–188, 359
    social, 12, 25, 160, 280, 362–363
  policy, 25, 95, 143, 179, 193, 199, 220
  post-Apollo, 54, 74, 248, 261, 269, 350
    budget decision, 70, 108, 210–211, 267, 272–273, 317, 364
    suggested programs, 17, 92, 99, 103, 281, 338
  significance, 32, 160, 167–168, 177–178, 187–188, 238–239, 297, 335, 343, 367
    international, 141, 148–149, 162, 241, 332
  U.S. vs. U.S.S.R. See Space race.
  Vietnam war, effect of, 24, 138, 275, 317, 323, 325
Space race, 109–110, 209, 212, 279, 286, 303, 373
  booster, 380
  criticism, 301, 328, 365
  manned space flight, 79, 316, 334, 356, 373
  moon, 44, 314
  post-Apollo, 70, 138, 162, 307
Space rescue, 262, 269
  need for, 111–112, 145–146, 273
  proposal, 78, 297
  test, 164–165, 192, 336
Space results (see also Earth; Moon; Mars; Venus; individual probes satellites, sounding rockets, and Apollo and Gemini flights)
  astronomy, 92, 151–152, 174, 188, 219, 229, 230, 348
  communications, 20, 88, 163, 372–373
  earth science, 144, 228, 236, 313
  medicine, 177, 309
  meteoroid detection, 205–206
  meteorology, 185, 257, 271
  oceanography, 33
  physics, 133–134, 140–141, 144,
  radiation, 51, 136–137, 173
  social science, 297, 329, 343, 366
  technology, 25, 88–89, 92, 209, 238–239, 329, 351–360
Space Research: Directions for the Future, 17, 39, 53
Space station (see also Manned Orbiting Laboratory), 180, 352, 357, 368, 373–374
Space suit, 203, 226, 301, 317–318, 336
Space tools, 17–18, 40, 78–79, 270
Space Tracking and Data Acquisition Network (STADAN), 70, 257, 307
Spacecraft (see also individual spacecraft, such as Apollo, Gemini, Lunar Orbiter, Luna, Mariner, Surveyor, etc.), 78, 170,
  braking, 279, 339
  design, 46, 127, 132, 167, 176, 192, 197, 204, 368
  development testing, 30, 39, 42, 132, 175, 185, 328
  electrical systems, 27, 82, 119–120, 134
  emergency equipment, 40, 78

environment, simulated lunar vehicle, 60, 91–92, 138
environmental control system, 96
equipment, 6, 11, 12, 57, 86–87, 129, 149, 178, 234,
extravehicular equipment, 29, 78, 246
heating, 112, 307
instrumentation, 117, 213, 236
landing system, 35–36, 39, 86, 88, 279, 284, 380
life support system, 11, 120
propulsion. See Engine; Motor, rocket; and individual launch vehicles, such as Saturn, etc.
reentry control system (see also reentry), 100, 105, 207, 290
reusable, 5, 313–314, 374
space tool, 17–18, 40, 78–79
window, 236
Spacecraft Lem Adapter (Sla), 177
Spacetrack, 225
Spaco, Inc., 41, 130
Spain, 15–16, 58–59, 137–138, 265, 322, 325
Spanish Astronautical Assn., 317
Spanish Space Commission. See Comisión Nacional de Investigación del Espacio (CONIE).
Sparkman, Sen. John J., 212
Sparta (reentry measurement program), 267
Spectrometer, 18, 111, 129, 274, 277, 347
Sperry, Lawrence, Award, 359
Sperry Rand Corp., 130
Univac Div., 157, 223
Spica (star). See Alpha Virginis.
Spilhaus, Dr. Athelstan F., 169
Spirit of St. Louis Medal, 166
Spohn, Dr. Clifford A., 286
SPUTNIK I (U.S.S.R. satellite), 17, 312
SPUTNIK V, 154
SR-71 (long-range reconnaissance aircraft), 129
SR-71B (training aircraft), 9
Sram. See Short Range Attack Missile.
Sreekantan, B. V., 229
SSRC. See Swedish Space Research Committee.
Sst. See Supersonic transport.
STADAN. See Space Tracking and Data Acquisition Network. 70
Stafford, Maj. Thomas P. (USAF), 72
Gemini IX-A, 206–207, 209, 213, 218, 236
preparation for, 178, 180–181, 198, 257
honors, 225–226, 236, 374–375
Stanford Univ., 97, 111, 199, 258, 298, 362
Stanton, Austin M., 61
Star, 11, 18–19, 40, 190, 203, 229, 230, 241, 295, 352
catalog, 107
Star Tracking Rocket Attitude Positioning (Strap) system, 241
Starr, Dr. Chauncey, 320
STATALTEX (French missile program), 201

State, Dept. of, 137, 195, 213, 341, 354
State Univ. of New York, 286
Stearns, John, 89
Stearns, John F., 243
Stepanzov, V. I., 318
Stephens, Col. Robert L. (USAF), 251
"Stepping Stones to Mars" meeting, 119–120
Sterilization Assembly Development Laboratory (SADL), 327
Sternberg, Sidney, 62
Stevenson, M/G John D. (USAF, Ret.) 311
Stewart, Dr. Homer Joe, 355
Stewart, Joseph, 151
Stoddard, Dr. David H., 254
Stoltenberg, Minister of Science (W. Germany), 303
Stoner, George H., 338
Stoolman, Leo, 250
Stormfury, Project, 267
Strang, W. J., 189
Strap. See Star Tracking Rocket Attitude Positioning.
Strassman, Dr. Fritz, 259
Strategic Air Command (SAC), 9, 165
Strelka (dog), 154
Strike 236, 257, 259
Edwards AFB, 35
Kennedy Space Center, 142, 144, 150, 239–240, 242, 268, 271
United Technology Center, 145, 150
Strom, Robert G., 116
Strong, Dr. John, 141
Strughold, Dr. Hubertus, 295
Stuhlinger, Dr. Ernst, 73, 120, 361
Submarine, missile-carrying, 247, 302
Sub-Aviation (France), 189
Sudden Ranch, Calif., 11
Suitland, Md., 268
Sullivan, Walter, 82, 109, 168, 355–356, 377
Sumi, Ukraine, U.S.S.R., 291
Summer Conference on Lunar Exploration and Science, 5–6
Summerfelt, William A., 354
Sun (see also Eclipse, solar; Solar cell; Solar wind), 236, 239, 287
IQSY, 251
radiation, 79, 173, 199, 289, 337
research, 109, 278, 336, 355
satellite data, 79, 269
Sun, K. H., 289
Sunspots, 129, 174, 199, 347
Suomi, Prof. Verner, 361
SUPARCO. See Pakistan Space and Upper Atmosphere Research Committee.
Super Diamant (French booster), 361
Super Guppy (cargo aircraft), 202
Supersonic combustion ramjet (Scramjet), 77, 335
Supersonic transport (Sst) (see also Concorde U.K. - France supersonic transport), 128, 131, 183
benefits, 170, 201, 296
contract, 249, 286, 296, 383
cost, 122–123, 289

criticism, 142, 185, 267, 281
design and development, 14, 16, 80, 85–86, 97, 122–123, 184, 215, 224–225, 226, 273, 297–298, 299, 307, 332, 334, 343, 350, 385
foreign, 19, 20–21, 43, 118, 143, 189, 194, 196, 226, 233, 289, 370
funds, for, 19, 24, 80, 84, 112, 167, 348
reservations, 118, 144, 272
sonic boom, 57, 203, 333, 341, 350.
Supersonic Transport Authority (proposed), 84, 167
"Sure-Fire," Project, 126
Surveyor (program), 5, 39, 75–76, 196–197, 202, 232–233, 340, 372, 373
Surveyor (spacecraft), 39, 44–45, 131–132, 175, 185, 330
SURVEYOR I (lunar probe), 209, 212, 215, 246, 262, 313, 333, 373
  instrumentation, 202, 296
  launch, 197–198
  lunar landing, 203–204, 384–385
  photograph enhancement, 263
  reactivation, 226, 316, 325, 341
  results, 228, 235, 249
"Surveyor I—A Preliminary Report," 235
SURVEYOR II (Surveyor B), 296
SV-5D (lifting-body vehicle), 378
SV-5P (manned lifting-body vehicle), 152–153
Sweden, 125, 174, 210, 239, 299, 318
Swedish Space Research Committee (SSRC), 51
Swetnick, Dr. Martin J., 344
Swigert, John L., Jr., 127
Sydney, Australia, 96
Sydney (Australia) Univ., 251
Sylvania Electronic Systems, 254
Symington, Sen. Stuart, 288
SYNCOM II (communications satellite), 133–134, 362
SYNCOM III, 362
Syracuse Univ., 237
System Development Corp., 370
Systems engineering, approach to social problems, 252
Syvertson, Clarence A., 258

T

T-38 (jet trainer), 211
Tabanera, Teófilo M., 253
Tad. See Thrust-Augmented Delta.
Taipei, 72
Taiwan, 54
Talyzin, Nikolay, 376
Tass
  cosmonaut, 307–308, 330
  Cosmos launchings, 60–61, 75, 110, 152, 223
  LUNAS IX, X, XI, XII, and XIII, 41–42, 47, 48–49, 122, 127, 129, 312, 326–327, 355, 381
  U.S. space program, 104
  U.S.S.R. space program, 16–17, 139, 233, 236, 287, 302, 306–307, 352, 373, 376
  U.S.S.R. weapons, 270
  VENUS II, 86–87
  ZOND III, 81
Tata Institute, Bombay, 46, 85
Taub, William P., 103
Taylor, Hal, 225
Taylor, Gen. Maxwell D. (USAF), 214
TD-1 (research satellite), 11, 239
TD-2 (research satellite), 11, 239
Teague, Rep. Olin E., 180, 211, 214, 273, 364
Technology, 28, 98, 187, 233, 329
  benefits, 66, 74, 113–114, 159, 191, 215, 253, 334
  gap, 15
  military, 77
  misuse of, 206
  sharing of, 355
Technology utilization, space, 61–62, 88–89, 167, 238–239, 281, 286, 287, 308, 328, 329
Technology Week (formerly *Missiles and Rockets*)
  Mol, 348
  Pert, 268
  space medicine, 335
    France, 361
    Japan, 3
    U.S., 11, 24, 72, 76, 112, 214, 272–273, 307–308, 334, 352
    U.S.S.R., 232, 285
  Webb, James E., 249
Tektite, 139, 208, 324
Teledyne Systems Co., 234
Telescope, 28, 39, 46, 47, 239, 241, 263, 268, 289–290, 303, 329
Television
  educational, 256, 270, 370–371
  laser beam use in, 310
  space probe, use of, 11, 59–60, 86–87
  via satellite, 12, 69, 86, 96, 162–163, 179, 287
Teller, Dr. Edward, 334
Tempel-Tuttle Comet, 220
Temperature, 61–62, 141
Il Tempo, 205
Tennessee, Univ. of, Space Institute, 316
Teplinskiy, M/G B. (U.S.S.R.), 147
Tepper, Dr. Morris, 43
Tereshkova, Valentina. See Nikolayeva-Tereshkova, Valentina.
Test ban treaty. See Nuclear test ban treaty.
Texas, 297
Texas A&M Univ., 365
Texas Atomic Energy Commission, 196
Texas Christian Univ., 120
Texas Instruments, Inc., 239.
Texas Technical College, 334
Texas, Univ. of, 56
Teweles, Sidney, 199
Thailand, 54, 72, 165, 222
Thiokol Chemical Corp., 112, 177, 265, 328
  Reaction Motors Div., 377

Thomas, Rep. Albert D., 54–55
Thomas, David F., 78
Thome, P. G., 190
Thompson, Dr. Barbara A., 377
Thompson, Col. John M. (USAF), 333
Thompson, Llewellyn E., 339
Thompson, Milton O., 47, 237–238, 244, 264, 274, 284, 301
Thompson, R. P., 253
Thompson, Robert F., 354
Thor (booster) (see also Long Tank Thor), 375
Thor-Agena (booster), 61
  B, 178
  D, 50, 91, 121, 132, 188, 221, 262, 297, 341, 382
Thor-Altair (booster), 9
Thor-Burner II (booster), 292
Thor-Delta (booster), 11, 41, 70, 310
Thrust-Augmented Improved Delta (Tad), 269, 331
Tiros (meteorological satellite), 24
TIROS VII (meteorological satellite), 220
Tiros Operational Satellite (Tos), 41, 70, 176, 310
Titan II (booster), 99, 164, 202, 206, 248, 318, 353
Titan III (booster), 24, 220, 305, 368
Titan III-B-Agena, 252, 373
Titan III-C (booster), 113, 216, 219, 276, 338, 359, 385
Titanium, 224, 308
Titov, L/C Gherman S. (U.S.S.R.),135
Tokyo, Japan, 179, 340
Tokyo Observatory, 219
Tokyo Univ., 109, 149, 257, 301, 339
Tonkin, Leo S., 55
Toowoomba, Australia, 361
Toronto, Canada, 194
Toronto, Univ. of, 229
Torrance, Calif., 156
Tos. See Tiros Operational Satellite.
Toth, Robert, 343
Towers, Adm. John H. (USN), 383
Townes, Dr. Charles H., 31, 97
Townes-Schawlow patent, 258
Tracy, Calif., 319
Traffic control, 247
Trailblazer (booster), 330, 369
Trans World Airlines (TWA), 159, 223, 236, 242
Thumba Equatorial Rocket Launching Station (India), 234
Thursday Island, 162
Tidbinbilla, Australia, 234
Time measurement, 166
Time, Inc., 161
Time (magazine), 326
Times of India, 45
Timiyasu, Dr. Kiyo, 126
Tipton, Stuart G., 197
Tracking, 216
  aircraft, 128
  deep space, 81, 83, 137–138, 157
  ship, 214
  Spacetrack, 225
  stations
    Australia, 162, 234, 361
    Cuba, 334
    Pakistan, 271
    Spain, 265
    U.S., 194, 299, 307, 361
    U.S.S.R., 299, 334
Transit (navigational satellite), 302
TRANSIT IV-A (navigation satellite), 227
Transportation, Dept. of, 80, 227, 274, 279, 321, 340, 348, 362
Trendex poll, 265, 362
Trimethyl-aluminum (TMA), 117, 152, 292, 347
Trippe, Juan T., 137, 297
Trud, 196, 285, 294, 374
Truszynski, Gerald M., 81
TRW, Inc., 235, 375
TRW Systems, Inc., 24-25, 142, 267, 306
Tsarapkin, Semyon K., 85, 97
Tu-104 (U.S.S.R. airliner), 256
Tu-114 (U.S.S.R. transport aircraft), 58
Tu-144 (U.S.S.R. supersonic transport), 196, 226
Tu-154 (U.S.S.R. airliner), 256
Tulsa, Okla., 177
Tupolev, Andrei N., 196
Turkey, 227
Twiss, Robert L., 176
Tydings, Sen. Joseph D., 12
Tyler, Gordon H., 283
Tyura Tam, U.S.S.R., 294

U

U-2 (reconnaissance aircraft), 27
Uchinoura Range, Kyushu, Japan, 109, 149, 301, 333, 378
UCLA. See California, Univ. of, at Los Angeles.
Udall, Secretary of the Interior Stewart L., 297, 350
Ufo. See Unidentified flying objects.
Ugolyek (dog), 60–61, 69, 84, 91, 102, 141, 147, 179, 318
U.K. See United Kingdom.
U.N. See United Nations.
Unidentified flying objects (Ufo), 160, 176–177, 299, 333, 341, 377
  Georgia, 251
  Michigan, 112–113, 116, 122
  Ohio, 142
  Pennsylvania, 200
  USAF, 130, 255, 272, 325, 326, contract, 172, 315
Union Carbide Corp., 135
United Air Lines, 159
United Aircraft Corp., 16, 27, 286, 296, 299, 308, 383
  Pratt & Whitney Div., 72, 235
  supersonic transport engine, 109, 123, 143, 226, 249, 273, 286, 296, 299, 332, 343, 383–384
United Aircraft Research Laboratories, 98
United Arab Republic (U.A.R.), 139
United Auto Workers (UAW), 35

United Kingdom (U.K.), 42, 43, 44, 46, 115, 136, 330, 347
  aircraft, 61, 356
    Concorde, 20–21, 43, 285, 289
  cooperation, defense, 40
  cooperation, space, 73, 133, 155–156, 239, 267, 287, 299, 322, 329, 377
  ELDO, 156, 189, 190, 211, 214, 220, 236, 285
  House of Commons, 31
  Ministry of Aviation, 356
  Ministry of Technology, 28, 356
  nuclear reactor, 51
  nuclear test, 31
  science and technology, 25–26, 28–29
United Nations (U.N.), 135, 169, 171, 216, 218, 221, 277, 299, 312
  Commission to Study the Organization of Peace, 181
  Committee on the Peaceful Uses of Outer Space, 20, 28, 32, 94, 258, 282, 287, 291, 386
    Legal Subcommittee, 237, 246, 248, 250
  General Assembly, 298, 355, 369, 377
  World Meteorological Organization, 89, 113-114
United Press International (UPI), 76, 134, 233
United States (see also appropriate Government agencies).
  Asian summit conference, 311
  budget, 19, 23–24, 33
  communications, 83
  contract policy, 128
  criticism, 85–86, 131, 288
  defense, 40, 170, 191–192, 341, 342, 343, 355
  disarmament, 304
  international cooperation, 15–16, 113–114, 133, 226, 231, 240, 303
  nuclear weapons, 58, 84–85, 342
  research and development, 15, 55, 119, 154, 179
  science and technology, 1, 13, 22, 28–29, 33–34, 70–71, 97–98, 99, 133, 150, 182, 219, 240, 348–349, 370
  space law treaty. See Space law treaty.
  space program. See Space program, national.
  transportation (see also Supersonic transport), 80, 340, 348
United Technology Center, 150
Univac Div. See Sperry Rand Corp., Univac Div., 223
Universal Match Corp., 361
Universe, origin, 31, 46–47
UPI. See United Press International.
Uprated Saturn I. See Saturn I, Uprated.
Ural Mountains, U.S.S.R., 107
Urban Renewal Administration, 37
Urey, Dr. Harold C., 159
Uruguay, 297
USA. See U.S. Army.
U.S. Air Force (USAF) (see also individual bases, centers, and commands, such as Air Force Systems Command, Arnold Engineering Development Center, Edwards AFB, etc.)
  aircraft (see also individual aircraft, such as C-5A, F-111A, X-15, XB-70, etc.), 9, 113, 164–169, 253, 282, 309, 323, 341, 365
    accident, 28, 211, 214
  anniversary, 302
  award, 113, 129, 137, 168, 232, 251
  booster, 6, 94, 216, 218, 385
  contract, 88, 122, 138–139, 195, 255, 268, 315, 328, 334, 335, 377
  V/Stol and Vtol aircraft, 194, 365
  cooperation, 128–129, 203, 281, 385
  Flight Safety Review Board, 180, 193
  launch
    booster, 213, 276, 329–330, 338
    Hitab test, 108, 115
    missile, 255, 261, 342–343, 367
  lifting-body vehicle, 152–153, 378
  missile program, 22, 108, 113, 165, 261, 275, 355
  Mol. See Manned Orbiting Laboratory. 177
  motor, solid propellant, 90, 177
  Office of Aerospace Research, 326
  personnel, 54, 57, 127, 146, 199, 259, 283, 298, 311, 336, 345
  research, 113, 209, 263
  satellite (see also Satellite, unidentified), 121, 146, 216, 331–332, 338, 370
  sounding rocket, 342
  Scientific Advisory Board, 172
  space program, 90–91, 283, 320–321, 335, 366
  spacecraft recovery, 100
  Ufo, 116, 119, 130, 142, 172, 255, 272, 315, 325, 376–377
  Vietnam war, 311
USAF. See U.S. Air Force.
USAF Aero Propulsion Laboratory, 78
USAF Aerospace Medical Div., 295
USAF Air Rescue Service, 78
USAF School of Aerospace Medicine, 12, 273, 335
U.S. Arms Control and Disarmament Agency, 339
U.S. Army (USA), 156, 284
  aircraft, 106, 131, 187, 253, 281, 309
  missile, 119, 267
  satellite, 212, 272, 312
  training, 150
U.S. Coast Guard, 114, 278
U.S. Comptroller General, 236
U.S. Congress. See Congress.
U.S. Court of Customs and Patent Appeals, 258
U.S. Geological Survey, 297, 313, 382
U.S. Information Agency (USIA), 304
U.S. Junior Chamber of Commerce, 9
U.S. Marine Corps (USMC), 150, 195
U.S. Merchant Marine Academy, 321
U.S. Military Academy, 203

U.S. Naval Aerospace Medical Institute, 234
U.S. Naval Oceanographic Office, 14
U.S. Navy (USN), 206, 207, 228
  agreement, 8, 179
  aircraft, 106, 131, 195
  contract, 30, 77, 338
  cooperation, 150, 267, 284
  missile, 176, 275
  satellite, 227, 302, 338
  spacecraft recovery, 207, 244, 251, 290–291, 332, 344
U.S. News & World Report, 372
USNS *Croatan*, 80–81
USNS *Kingsport*, 134–135
USNS *Point Barrow*, 74, 86
USNS *Range Recoverer*, 87, 178, 185, 187
U.S. Public Health Service, 64
U.S.S. *Benjamin Franklin*, 87
U.S.S. *Bennington*, 185
U.S.S. *Boxer*, 86
U.S.S. *Guadalcanal*, 244, 247
U.S.S. *Guam*, 290, 293
U.S.S. *La Salle*, 338
U.S.S. *Leonard F. Mason*, 100, 106
U.S.S.R. (Union of Soviet Socialist Republics) (see also Soviet Academy of Sciences, etc.) 219–220, 289, 294
  aircraft, 50, 58, 226, 251, 255–256, 291, 323, 333
  antiaircraft defense, 125, 161
  astronomy, 171, 174–175, 305, 377
  booster, 91, 138, 212, 290, 316
  communications satellite, 149, 152, 154, 324, 353, 376
  cooperation, 10, 251, 269, 330, 376
    France, 221, 226, 227, 261, 334, 348, 370, 386
    U.S., 90, 96, 109–110, 155–156, 271, 303, 308, 339
  cosmonaut. See Cosmonaut.
  Cosmonaut's Day, 135–136
  disarmament, 58–59, 84–85, 97
  launch
    aerostat, 336
    Luna spacecraft, 35, 122, 273, 326–327, 378–379
    satellite
      Cosmos, 8, 22, 27, 51, 59, 60–61, 75, 104, 110, 130, 144, 152, 175, 190, 218, 223, 235, 240, 246, 252, 261, 277, 294, 321, 325, 337, 347, 352, 356, 366, 372, 378–379
      Molniya, 149, 325
      Proton, 233
      Yantar I (ion propulsion system test), 339
  launch site, 372
  life science, 11, 12–13, 60–61, 69, 84, 90, 91, 94, 102, 136–137, 154, 179, 232, 285, 306–307, 318, 374
  lunar exploration (see also LUNA IX, etc.),
    manned, 55, 125, 130, 131, 162, 214, 285, 286, 319–320
    unmanned, 41–42, 44–45, 46, 47, 48, 157, 204–205, 380–381
  meteorology, 96, 174, 269, 271, 385
  missile and rocket program, 10, 181, 270, 302, 341, 355
  planetary exploration, 2, 40, 75, 81–82, 86–87, 126, 173, 236
  rocket test, 147–148, 193, 233, 275, 287, 302, 352, 373
  science, 21, 25–26, 117, 146, 337, 377
  space law treaty. See Space law treaty.
  space program. (see also individual probe and satellite programs, such as Cosmos, Luna, Molniya, Proton, Venus, Voskhod, etc.), 16–17, 20, 70, 130, 173, 201, 212, 237, 303, 312, 318, 320–321, 368, 373, 385
    manned space flight, 79, 92, 135–136, 138, 162, 303, 314, 338–339, 356, 379–380
  spacecraft, 17, 254, 285, 306, 321
  submarines, 125
  supersonic transport, 118, 196, 226
  tracking station, 9, 16, 48, 139, 334
  weapons, 84–85, 97, 161, 181, 191–192, 198, 270, 341, 355, 373
U.S.S. *Redstone*, 214
U.S.S. *Silas Bent*, 189
U.S.S. *Wasp*, 179, 207, 209, 332, 334
U.S. Weather Bureau, 5, 41
U Thant, U.N. Secretary Gen., 198

V

Vaccaro, Michael J., 60, 91
Vacuum Technology and Space Simulation, 334
Van Allen, Dr. James A., 134
Van Allen radiation belts, 102, 120, 213, 260
Van Dusen Aircraft Supplies, 308
Vandenberg AFB, Calif. (see also Western Test Range), 11, 195, 294
  launch, 108, 115
    missile, 275
    rocket, Arcas, 342
    satellite launch vehicle
      Atlas-Agena, 213
      Atlas-Agena D, 54, 92, 121, 272, 337
      Atlas D, 240, 370
      Scout, 146, 183
      Thor-Agena D, 40, 341
      Thor-Altair, 9
      Titan III-B-Agena D, 252, 306
  Missile, 22, 66, 115–116
VANGUARD I (satellite), 106
Varo, Inc., 61
Vashkov, V. I., 173
Vassilyev, P. V., 318
Vaughn, Guy Warner, 353
Vecchietti, George J., 139, 315
Veda a Technika Mladezi, 92
Veis, Dr. George, 22
Vela (nuclear detection satellite), 35, 322
Vema (research ship), 302
Venezuela, 297

Venus (planet), 142, 145, 157
  atmosphere, 251, 377
  exploration, 17, 26, 73, 75, 82, 86, 95, 147, 159, 268, 281, 351, 368
  life on, 26, 141
  rotation, 175
  surface, 16, 175
  temperature, 26, 159
VENUS II (U.S.S.R. interplanetary probe), 51, 75, 86–87, 95
VENUS III, 51, 75, 82, 86–87, 95,
Vershinin, Air Marshal Konstantin (U.S.S.R.), 136–137
Vertical and short takeoff and landing aircraft. See V/Stol and Vtol aircraft.
Veterok (dog), 60, 69, 84, 91, 102, 141, 147, 179, 318
Vetlesen Award, 324, 377
Vetlesen, G. Unger, Foundation, 324
Vienna, Austria, 171, 287
Vietnam, 216
Vietnam, North, 355
Vietnam, South, 361
Vietnam war, 228, 311, 325
  aircraft, 326
  effects on space budget, 24, 138, 275, 317, 323, 325, 328
  effects on space law treaty, 237, 277
  opposition to, 246, 366
Vine, Allyn C., 52
Vinogradov, Aleksandr, 50, 140
Virgin Islands, 152
Virginia Associated Research Center, 19, 365
Virginia Polytechnic Institute, 365
Virus control, 64
Vitro Corp. of America, 130
Vivian, Rep. Weston E., 110
Vladivostok, U.S.S.R., 154
Vocational Rehabilitation Administration, 222, 328
Vogel, Col. Lawrence W. (USA), 1
Vogeley, Arthur W., 354
Volcano, 299, 313, 316, 374
Volunteer Flight Officers Network, 284
Von Braun, Dr. Wernher, 1, 241, 298, 322, 372
Von Eckardt, Wolf, 311
Von Kármán, Dr. Theodore, 124, 259
Von Markatz, Dr. Hans-Joachim, 210
Von Ohain, Hans J. P., 27
VOSKHOD I (U.S.S.R. spacecraft), 285, 356
VOSKHOD II, 92, 107, 321, 339, 356
VOSTOK VI, 90
Voyager (program), 113, 171, 198, 244, 279, 281, 308, 352, 362
  Planetary Reentry Parachute Program, 153, 242–243, 279, 340, 352
  contract, 275
Voyager (spacecraft), 327
Vreeland, Thad, Jr., 260
V/Stol aircraft, 106, 131, 185, 194, 253, 364
Vtol aircraft, 94, 284, 309, 323, 365

W

WA-50 (light aircraft), 118
Walker AFB, N. Mex., 279, 284
Walker, Cliff W., 58
Walker, Dr. Eric A., 155
Walker, Joseph A., 47, 211, 214, 351
Wall Street Journal, 28, 219, 250, 350, 380
Wallops Station (NASA), 41, 312, 315, 383
  launch
    Pacemaker, 333
    satellite
      OV3-IV, 213
    Scanner, Project, 268, 369
    Scout (Reentry E), 49
    sounding rocket, 252
      Aerobee 150A, 79, 84, 182, 276
      Aerobee 350, 269
      Boosted-Dart, 14
      Javelin, 79, 185, 237, 267, 299, 313
      Nike-Apache, 10, 19, 56, 127, 132–133, 214, 239, 240, 274, 276, 293, 307
      Nike-Cajun, 25, 39–40, 42–43, 51, 161, 260, 265, 307, 310
      Nike-Tomahawk, 276, 277, 301, 341, 360
      Orion II, 342
      Wasp, 210
    Trailblazer, 330
  personnel, 382
Wang, Dr. Chiao Jen, 376
Wapakoneta, 137
Warrior Constructors, Inc., 255
Warsaw Convention, 43, 157
Washington Board of Trade, 319
Washington, D.C., 140, 172, 271, 289, 293, 311, 313, 330, 334
  awards presented at, 208, 215, 232, 294, 314–315
  meetings, 19, 56, 136, 140, 143, 144, 150, 162–163, 172, 199, 212, 214–215, 221, 269, 271, 288, 297, 338, 382
  museum, 49, 236
  news conference, 80, 119
Washington, D.C. Business Council, 56
Washington Daily News, 370
Washington *Evening Star* (see also Hines, William), 373
  editorial, 76, 106, 109, 170, 183, 210–211
Washington (D.C.) National Airport, 13, 149, 217
Washington Post, 82, 143, 225, 260–261, 289, 343
  editorial, 105, 183, 246, 320–321, 351, 365, 370
  interview, 294–295, 311, 341
Washington Society of Investment Analysts, 199
Washington *Sunday Star*, 267, 294, 301
Washington (state), 83
Washington, Univ. of, 287
Wasp. See Weightless Analysis Sounding Probe.
Wassmer Aviation, 118
Water, 105, 274

Waters, John R., 229
Watertown (N.Y.) Daily Times, 248
Watson, James Craig, Medal, 150
Watson Laboratory, 150
Weapon systems, 72, 125, 147–148, 161
Weather modification, 8, 59, 62, 99, 261, 326
Webb, Alvin B., 178
Webb, James E., 1, 60, 73, 79, 132, 140, 147, 381–382
   Apollo applications, 70, 138, 196, 210–211
   appointments by, 166–167, 173, 275, 319, 336
   award by, 304, 315
   award to, 314, 341
   labor agreement, 33
   manned space flight, 43
   nuclear propulsion, 145
   space cooperation, 133, 138, 289, 303, 304
   space program, 95, 108, 147, 297, 332, 366
   U.S.S.R., 70, 301, 316, 323
   tribute to, 102, 249
Weidner, Herman K., 315
Weightless Analysis Sounding Probe (Wasp), 210, 214
Weightlessness
   effects of, 102, 179, 210, 226, 295, 318, 322, 361, 374
   work during, 17, 61, 170, 294–295, 299–301, 317, 318
Weitz, L/Cdr. Paul J. (USN), 127
Welsh, Dr. Edward C., 46, 53, 79, 98, 119, 167–168, 180, 212, 319, 324, 357, 372, 373–374
West Ford, Project, 35, 376
West Palm Beach Facility, Fla., 123, 226
West Virginia, 341
Westbury, N.Y., 108
Western Air Lines, 159
Western Operations Office (WOO) (NASA), 215
Western Support Office (WSO) (NASA), 215
Western Test Range (WTR) (see also Vandenberg AFB, Calif.)
   launch
      lifting body, SV-5D, 378
      missile, 66
      satellite, 153, 178, 222, 284, 310, 321, 349
      launch vehicle
         Atlas-Agena D, 20, 54, 106, 143, 178, 208, 212, 238, 268, 293, 312, 319, 367
         Scout, 33, 116, 258, 271
         Thor-Agena D, 50, 91, 121, 132, 188, 208, 212, 262, 297, 382
         Thrust-Augmented Thor-Agena B, 178, 222
         Titan III-B-Agena D, 373
Western Union International, Inc., 176, 321
Western Union Telegraph Co., 341

Westinghouse Electric Corp., 103, 222, 249, 361, 376
Westinghouse Research Laboratories, 289, 309
Weston, Ill., 376
Westphal, James A., 277
Whitaker, Ewen A., 116
White, Alvin S., 36, 158, 211
White, L/C Edward H., II (USAF), 92, 144
   Apollo flight, 110, 332
   European visit, 54
   GEMINI IV flight, 17, 104, 216–217
   honors, 9, 208, 232
White, George C., Jr., 336
White House, 50, 77, 101–102, 146, 212, 293, 301, 303, 328, 354, 356
White House Committee on Meteorology, 3
White House Conference on International Cooperation, 258
White, John Michael, 128
White, L/C Robert M. (USAF), 47
White Sands Missile Range (WSMR), N. Mex.
   launch
      Aerobee 150
         atmospheric data, 360, 364
         booster test, 297
         dayglow measurement, 138
         micrometeoroid sampling, 326–327, 350
         radar test, 171
         solar astronomy, 187, 331
         ultraviolet astronomy, 40, 145, 190, 241, 242, 295, 352
         x-ray astronomy, 89, 188, 219, 319
      Air-Launched, Air-Recoverable Rocket (Alarr), 118–119
      Boosted Arcas, 249, 277
      Nike-Apache, 267, 351
   test
      Athena, 342–343, 367
      Chaparral, 267
      Voyager parachute, 279, 284, 340, 352, 370
White, Gen. Thomas D., Space Trophy, 232
Whitten, James R., 126
Wildt, Rupert, 82
Wilford, John Noble, 44–45
Will Rogers (missile-carrying submarine), 247
Williams, Clifton C., Jr., 379
Williams, G. P., 348
Williams, George K., 296, 314
Williams, John J., 354
Williams, B/G Robert R. (USA), 281
Williams, William W., 296, 314
Wilson, Andrew, 143
Wilson, Edwin B., 384
Wilson, George C., 128
Wilson, Prime Minister Harold, 31.
Wilson, Herbert A., 315
Wilson, Louis D., 354
Wilson, R. R., 10
Wilson, Robert W., 31
Wimer, Arthur G., Jr., 315–316
Wind tunnel, 2, 97, 246, 309

Wing, aircraft
  delta, 225
  flexible, 156
  variable-sweep, 116, 215, 307, 383
Wingert, Lowell T., 114
Wings Club, 137
Winston, Donald C., 356
Winter, Frank H., 103
Wisconsin, 69
Wisconsin State Chamber of Commerce, 337
Wisconsin, Univ. of, 132, 182, 328, 361
WMO. See World Meteorological Organization.
Wolff, Rep. Lester L., 165
Women as astronauts, 7, 107, 324–325
WOO. See NASA Western Operations Office.
Wood, Clotaire, 232, 322
Wood, David S., 260
Woodcock, Gordon R., 22
Woodland, Ga., 176
Woods Hole, Mass., 39, 53
Woods Hole Oceanographic Institute, 52
Woodward and Lothrop, Inc., 77
Wooldridge, Dean E., 235
Woollard, Dr. George P., 38, 39
Woomera Rocket Range, Australia, 123, 190, 348
Worden, Capt. Alfred M. (USAF), 127
World Affairs Council, 241
World Federation of Democratic Youth congress, 214
World Journal Tribune, 311
World Magnetic Survey, 64
World Meteorological Day, 89, 113, 133
World Meteorological Organization (WMO), 113–114, 210, 326
World War II, 311
World Weather Watch, 114, 133, 326
Worthington, Minn., 161
Wright Brothers Lecture, 26
Wright Brothers Medal, 296, 314
Wright, Orville, 377
Wright-Patterson AFB, Ohio, 27, 203, 284, 294, 309, 326, 353
Wright, Robert, 371
Wright, Wilbur, 377
WSMR. See White Sands Missile Range.
WSO. See NASA Western Support Office.
Wyatt, DeMarquis D., 1, 351
Wydler, Rep. John W., 108, 164, 279
Wyld, James H., 377
Wyndham, Dr. John D., 230, 253

X

XV-4A ("Hummingbird") (Vtol aircraft), 365
XV-5A (V/Stol aircraft), 253
XV-6A (V/Stol aircraft), 131
X-15 (rocket research aircraft), 15, 129
  flight, 47, 62, 282, 385
    No. 1, 168, 238, 252, 263, 275, 288, 313
    No. 2, 182, 231, 247, 257, 263, 278, 351
    No. 3, 242, 258, 271, 292, 351, 357
  pilots, 47, 211, 246, 359
  record, 62, 211, 351, 385
  test, 182
    accelerometer, 279, 336
    base drag, 247, 278–279
    electrical loads, 237, 275
    glass, nonglare, 237, 252
    heat, 278
    horizon scanner, 252, 263, 275
    micrometeoroid, 252, 263, 275, 292
    pilot checkout, 258, 313, 357
    radiometer, 292, 336
    stability, 182, 279, 357
    star tracker, 247, 263
X-22A (V/Stol aircraft), 106
XB-70 (supersonic aircraft), 114, 385
  accident, 211, 263–264
  flight, 5, 16, 136, 158–159, 183–184, 338, 342, 354, 372, 376, 378
XB-70A, 183
XC-142A (V/Stol aircraft), 182
XE (nuclear rocket engine), 336
Xi, Ze-zong, 339
X-ray (see also Astronomy, x-ray), 79, 174, 187, 219, 373, 385
  flux, 87
  sources, 89, 188, 229

Y

Yak-40 (U.S.S.R. airliner), 256
Yale Club, 314
Yale Univ., 190, 252
Yale Univ. Observatory, 220
Yangel, Mikhail K., 224
Yantar I (U.S.S.R. automatic ionospheric laboratory), 339, 385
Yardley, John F., 354
Yarymovich, Dr. Michael, 150
Yegorov, Dr. Boris B., 84, 147, 249, 285
Yegorov, Pyotr, 327
Yegorychev, Nikolay, 20
Yeremin, A. V., 318
Yeshiva Univ., 365
YF-12A (reconnaissance aircraft), 28, 251
Yomiuri Shimbun, 45
Young, Dr. C. Gilbert, 209
Young, Cdr. John W. (USN), 240, 246, 249, 255, 299, 379
Young, John D., 27, 188
Young Men's Business Club of Greater New Orleans, 247
Young, Sen. Milton R., 167
Young, Robert B., 235
Yttrium-aluminum-garnet crystal (Yag), 209
Yugoslavia, 165, 231
Yurovsky, Vladimir, 46

Z

Zabelin, Igor, 384
Zahner, Donald, 224

Zemlya i Vselennaya, 377
Zenith Radio Corp., 310
Zero Defects Achievement Award, 131
Zero Defects Program, 353
Zero (zero gravity) (see also Weightlessness), 121, 179, 210, 299
Zero Length Launch (Zell) platform, 111
Zeta Ophiuchi (star), 190, 241

Zia Corp., 235
Zimmerman, Fred L., 350
Zimmet, Donald, 250
Zond (U.S.S.R. spacecraft), 73
ZOND III (U.S.S.R. space probe), 81–82, 305, 333
Zuckert, Eugene M., Trophy, 129
Zwayer, James T., 28

NASA HISTORICAL PUBLICATIONS

HISTORIES:

Robert L. Rosholt, *An Administrative History of NASA, 1958-1963*, with Foreword by James E. Webb, NASA SP-4101, 1966; for sale by Supt. of Documents ($4.00).

Loyd S. Swenson, James M. Grimwood, and Charles C. Alexander, *This New Ocean: A History of Project Mercury*, NASA SP-4201, 1966; for sale by Supt. of Documents ($5.50).

CHRONOLOGIES AND SPECIAL STUDIES:

Aeronautics and Astronautics: An American Chronology of Science and Technology in the Exploration of Space, 1915-1960, compiled by Eugene M. Emme with Foreword by Hugh L. Dryden, Washington: NASA, 1961 (out of print).

Aeronautical and Astronautical Events of 1961, with Foreword by James E. Webb, published by the House Committee on Science and Astronautics, 1962 (out of print).

Astronautical and Aeronautical Events of 1962, with Foreword by George L. Simpson, Jr., published by the House Committee on Science and Astronautics, 1963; for sale by Supt. of Documents ($1.00).

Astronautics and Aeronautics, 1963, with Foreword by Hugh L. Dryden, NASA SP-4004, 1964; for sale by Supt. of Documents ($1.75).

Astronautics and Aeronautics, 1964, with Foreword by Robert C. Seamans, Jr., SP-4005, 1965; for sale by Supt. of Documents ($1.75).

Astronautics and Aeronautics, 1965, with Foreword by James E. Webb, NASA SP-4006, 1966; for sale by Supt. of Documents ($2.25).

Project Mercury: A Chronology, by James M. Grimwood with Foreword by Hugh L. Dryden, NASA SP-4001, 1963; for sale by Supt. of Documents ($1.50).

Mae Mills Link, *Space Medicine in Project Mercury*, with Foreword by Hugh L. Dryden and Introduction by W. Randolph Lovelace II, NASA SP-4003, 1965; for sale by Supt. of Documents ($1.00).

Historical Sketch of NASA, NASA EP-29, 1965; for sale by Supt. of Documents ($.25).

www.ingramcontent.com/pod-product-compliance
Lightning Source LLC
Chambersburg PA
CBHW081715170526
45167CB00009B/3580